Introduction to Product Design and Development for Engineers and Scientists

Introduction to Product Design and Development for Engineers and Scientists, Second Edition, offers a systematic, best-practices approach to mastering the engineering processes that drive successful product design and development across mechanical, electrical, and software domains. This updated edition expands its scope to include disposable and consumable products and introduces a new chapter on measurement uncertainty, exploring its critical impact on test results and decision-making.

This enhanced edition integrates Agile methodologies with real-world case studies, providing practical frameworks to help engineers and scientists navigate ambiguity, avoid scope creep, and prevent product launch delays. Readers will gain tools to implement contemporary approaches while maintaining engineering rigor and quality standards. Comprehensive case studies and hands-on exercises reflect the challenges of modern product development, equipping readers with a practical skill set applicable to the evolving landscape of modern product development.

Tailored for academic audiences and engineering and science students on product design courses, this textbook delivers industry-proven best practices and actionable guidelines for product design and development.

Online PowerPoint presentations and figure slides further support learning at www.routledge.com/9781032895215, offering essential tools to drive world-class innovation and sustainment of cutting-edge products.

Introduction to Product Design and Development for Engineers and Scientists

Second Edition

Ali Jamnia

CRC Press
Taylor & Francis Group
Boca Raton London New York

CRC Press is an imprint of the
Taylor & Francis Group, an **informa** business

Designed cover image: Shutterstock

Second edition published 2027
by CRC Press
2385 NW Executive Center Drive, Suite 320, Boca Raton FL 33431

and by CRC Press
4 Park Square, Milton Park, Abingdon, Oxon, OX14 4RN

CRC Press is an imprint of Taylor & Francis Group, LLC

First edition published by CRC Press 2018

ISBN: 978-1-032-89521-5 (hbk)
ISBN: 978-1-032-91701-6 (pbk)
ISBN: 978-1-003-56450-8 (ebk)

DOI: 10.1201/9781003564508

Typeset in Palatino
by Apex CoVantage, LLC
Access the Instructor and Student Resources: www.routledge.com/9781032895215

Contents

About the Author

Ali Jamnia is an accomplished professional and dynamic consultant with a proven track record in the medical device, automotive and commercial industries in the United States. He is an innovative and dynamic leader with hands-on experience solving design issues through his expertise in mechanical engineering. His experience is complemented by a PhD and MS in Engineering Mechanics from Clemson University, twelve patents, and four books published by CRC Press.

Preface

Design, development, and life cycle management of any modern product is a complex task that requires a cross-functional team spanning across multiple organizations beginning with commercial, design, manufacturing, and service, among others.

In order to effectively design, develop, and sustain devices that a manufacturer produces, hiring qualified individuals is only the first step. In addition, an organizational structure is needed to drive a common and consistent set of outputs via best practices, templates, and checklists. This structure serves in two ways. First, it guides the work of less experienced team members, and second, it provides a mechanism to capture the contributions of the senior members of the design team.

The purpose of *Introduction to Product Design and Development for Engineers and Scientists* is to provide a practical set of guidelines and best practices on driving world-class design, development, and sustaining of modern products. The information provided within this text is applicable across the entire product life cycle management, from initial concept work that can be undertaken within technology pipelines to detailed design and development stages through product support. Additionally, this book provides a hands-on approach for design and development teams to follow the principles of systems engineering.

This book starts with a description of the *fuzzy front end* of product development and the differences between concepts such as invention, innovation, design, and development (Chapter 1). The rest of the text is divided into six parts. This first part explores the soft skills that every team member needs in order to navigate through a successful product development journey. These skills are topics such as team and team development, program and project management, and communication skills, which include giving effective presentations, including technical reviews, and the need for checklists. Chapters 2 through 5 belong to this section.

The second segment (expressed in Chapter 6) explores the product life cycle management models and describes their similarities. Furthermore, it provides application of two popular designs for Six Sigma tools.

The third segment starts by defining systems and requirements. Voice of customer is explained, and its cascade eventually leads to the creation of design documents. The process of developing product requirements from the voice of customer is shown. Finally in this section, quality function deployment is described along with the initial concept development. Chapters 7 through 9 belong to this part.

The fourth segment focuses on the "nuts and bolts of the design" and is the subject of Chapters 10 through 14. Once high-level designs are completed and a concept is selected, a high-level design failure mode and effects analysis should be conducted to identify potential pitfalls of the selected concepts. Based on this analysis, should the need exist, either alternative selections are made or mitigations are incorporated into the detailed design. Next, an understanding of engineering transfer functions and their role in developing a robust design is needed. Finally, the design team should focus on developing the three- and two- dimensional models and drawings for the electromechanical aspect of the product and, in the case of combination products, for chemical and/or biological elements. In developing the detailed design, various elements from requirements decomposition to development of theory of operations down to parts usage and numbering are discussed. For developing the 2D engineering drawings, a review of applicable standards is presented along with a rationale for the importance of GD&T (geometric dimensioning and tolerancing) as a common engineering language. Elements of *Design for X* (DfX) are discussed in Chapter 14. Engineers and scientists should keep in mind that their designs should be manufacturable, serviceable, and both robust and reliable.

The fifth part of this book contains Chapters 15 through 17. In this segment of the product development cycle, the final preparation for market launch takes place. The concept of cost requirements (Chapter 15) is introduced here; its proper place within the V-model is clarified. Separately, a second, albeit more detailed, design failure modes and effects analysis is completed. The results may be used as inputs into a process failure modes and effects analysis (Chapter 16). The outcome of this activity is to develop control plans for manufacturing as well as verification and validation plans to transfer the design from research and development to the manufacturing and operations phase (Chapter 17).

Parts 6 and 7 of this text are centered around responding to changing customer needs and sustaining a marketed product. Once a product is launched into the market, it should be sustained, and eventually, it needs to be retired from the market. However, *sustaining* a product not only means responding to field failures and customer complaints but also changing customer needs. Traditional product development models do not consider these changing needs; however, within the V-model and Agile mindset, there are ways to respond to the changing needs. Chapter 18 tackles this topic. Associated activities with product performance, as well as engineering and supplier notices of change, are discussed in Chapter 19, along with considerations for when decisions about retiring a product must be made. Product retirement is covered in Chapter 20.

The final segment of this book (Part 8) provides a set of guidelines that – though needed in the toolkit – does not readily belong to the other sections. Notice that the book starts with a section on soft skills and particularly a chapter on communication skills. An important and sometimes ignored factor is that engineers or scientists do not document their work. Many a time, elements of design and its intents remain as *tribal* knowledge. Should members of this technical tribe move to other projects – or worse, to other corporations – this knowledge is lost.

Chapter 21 provides a gateway into developing a structure for document collection called a design history file (DHF) and its associated device master record (DMR). DHF is a compilation of "objective evidence" of the thought process used in the development of a product as well as any changes that might have occurred post-launch. Similarly, DMR contains the needed procedures and specifications to manufacture and service a finished product. DHF and DMR are tools that are inherent components of configuration management. They enable the team of design engineers and scientists to properly communicate the design elements of the product as well as the product's manufacturing and service procedures with various cross-functional teams. Furthermore, a DHF and DMR properly enable engineering support teams to effectively manage engineering change orders and supplier notices of change.

Additionally, Chapter 21 introduces configuration and change management strategies. Earlier, I mentioned that verification and validation are required prior to product launch. This activity often requires measurements and tests. Chapter 22 focuses on data collection and test methods. In connection with this topic, Chapter 23 – contributed by Dr. Torabi – provides a brief primer for metrology. Metrology is an essential but often neglected topic in maintaining measurement equipment.

Chapter 24 delivers insights on tolerance design and its placement within the V-model. Finally, the basis of a failure analysis and reporting as a tool for product improvements is covered in Chapter 25.

In closing, I would like to add that the examples used in this work are all products that I have worked on or have developed over nearly 30 years. In instances when needed, I have removed the design elements that might be considered proprietary.

1

Fuzzy Front End

Introduction

In the distant past – and particularly for technology companies – there was not a clear distinction between the time when teams invented new technologies and when those technologies were incorporated into new products.[1] In that mindset, program and project managers would set schedules and communicate them to marketing people, who would then set customers' (and, to a large extent, investors) expectations of when a product would come to market. In those days, a clear definition or process for bringing innovation to market did not exist. Product launches were often delayed as a result of engineering and technical setbacks. Needless to say, delays of this nature eroded investors' confidence in the company, which negatively impacted the level and degree of the organization's investment in innovation. As time went by, the innovation process and product development were scientifically studied. This effort led to a number of theories being proposed and put into practice in order to bring efficiency and optimization to this practice.

Many of these theories had identified three broad spectrums of the innovation process, beginning with the fuzzy front end of the process, followed by the new product development (NPD) step, and finally commercialization.

Within a corporate environment, when the decision to develop a product is made – that is, at the start of the NPD stage – windows of marketing opportunities are identified, timelines are set, schedules are developed, and the product development activities begin in earnest. Due to both market and investors' expectations, there is a rush to meet deadlines and timelines. There is little time to explore technology options or to examine "what-if" scenarios.

Explorations and what-if options belong to the fuzzy front end when there are no set expectations and schedules. Should a corporation ignore the need for this exploration period, the rush to market may cause oversights and mistakes that may lead to program delays. The expression, *there is no time to do it right, but there is time to do it over,* might have its roots in this reality.

There is much to be said about the fuzzy front end and the innovation process, which is beyond the scope of this text. For further reading on this topic, the works of Anthony (2012), Adner (2012) and Furr and Dyer (2014) are recommended.

Innovation Process and Front-End Funnel

> Innovation process has three primary stages: Fuzzy Front End, New Product Development, and Commercialization

Figure 1.1 depicts a typical innovation process funnel. In this model, a corporation invites its employees and stakeholders to provide ideas and suggestions for possible new products. At the same time, small technical teams are formed to either initiate development of new technologies or to search for other

DOI: 10.1201/9781003564508-1

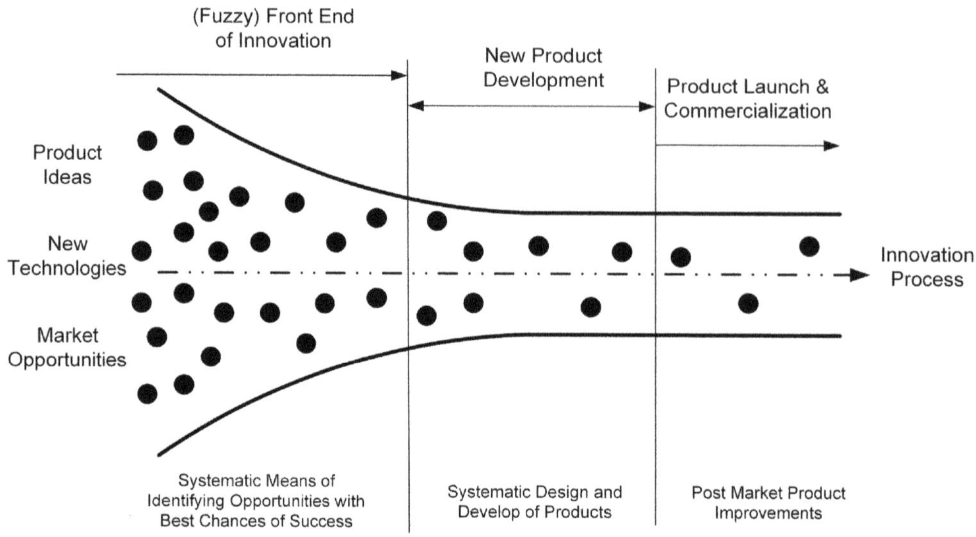

FIGURE 1.1
A typical innovation process funnel

technology start-up companies with promising concepts and even possibly working prototypes. In parallel, the corporation's marketing team may be assessing and forecasting short-, mid-, and long-term market needs and trend.

At this front end, many proposed ideas may be tempting; however, a filtering system must be in place in order to align the proposed concepts with the organization's technology roadmap and long-term strategies (Wheelwright and Clark 1992). These filters, or "idea management processes," help identify both required technological and financial resources and estimate potential returns on investments. From these, a few concepts are selected for further development to reach a proof-of-concept phase. Needless to say, of the many proposals, only a few successfully complete the last stage and are placed in the NPD platform. Up until this point, there are no real deadlines for any given concepts – as any one concept may be eliminated from the race at any time; only, that process of elimination must actively filter the incoming and provide a few "nuggets" for further development. Once the surviving proposals are approved for development and commercialization, the organization's mindset shifts into developing timelines and meeting milestones.

I have used "innovation" within the concept of the fuzzy front end. Often, innovation is used synonymously with "invention." The term represents a slightly different concept, and I would like to explore them next.

Innovation versus Invention

Even though innovation and invention seem to be interrelated, they are quite distinct from one another. The basis of an invention – and therefore applying for its patent – is formed on the three legs of novelty, not being obvious (to someone skilled in the field), and finally being *useful*. Ironically, useful does not imply market success or financial rewards. The following expired patent entitled "Apparatus for Simulating a High-Five" – as shown in Figure 1.2 – was granted in 1994. The inventors provided the following argument for the usability of this patent.

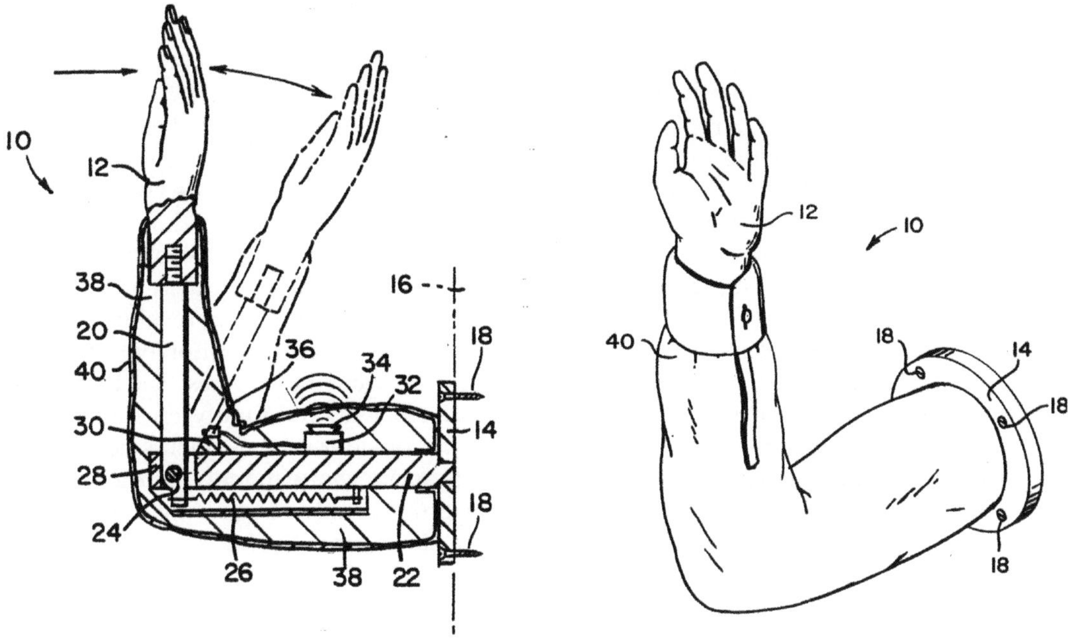

FIGURE 1.2
A depiction of the patent "apparatus for simulating a high-five" – United States patent number: 5,356,330, issued on October 18, 1994

> During a televised sporting event, a "high five" is commonly shared between fans to express the joy and excitement of a touchdown, home run, game-winning basket, birdie or other positive occurrence. Unfortunately, as known in the art, a "high five" requires the mutual hand slapping of two participants, wherein a first participant slaps an upraised hand against the elevated hand of a second participant. As such, a solitary fan is unable to perform a "high five" to express excitement during a televised sporting event.
>
> (Cohen 1994)

Even though one may argue in favor of the novelty and not-obvious aspect of this invention, it would be difficult to imagine that there is wide use for such a product.

In contrast, innovation is "[a] novel combination of art, science, or craft employed to create the goods or services used by society" (Quinn and Zien 1997 as quoted by Evans 2003). Kahn et al. (2005) define innovations as "[a] new idea, method or device. The act of creating a new product or process. The act includes invention as well as the work required to bring an idea or concept into final form." One may also define innovation as developing fresh views of products, services, and/or processes with the idea of increasing efficiency and reducing waste.

> Inventions do not necessarily make for innovations. Innovations need to be providing a useful product or service for the intended market

Hence, one may conclude that invention is only an aspect of innovation. Although an invention is an innovation, an innovation may not need to involve any inventions at all.

Innovation Classifications

If inventions are a subset of innovation, one may ask what the other aspects of innovation are. Before answering this question, I would like to suggest that there are two aspects that one needs to keep in mind. First is the "what" of innovation, and the second is the "how."[2] In this context, "how" is the organizational processes that would put filters in place to select and approve only the ideas that have the best chance of success. The "what" refers to different types of innovation. In this regard, Anthony (2012) suggests that innovation be viewed in one of two ways: the first being the strategic intent and the second being the type of innovation. Recalling that innovation is ultimately about usefulness and market acceptance, strategic intent focuses on more efficient means of increasing revenue by first, making existing processes more efficient. Next, through better collaboration between different internal organizations to improve existing products and/or services. The third step is to evaluate existing products and make them more efficient, and finally, create new categories of products. A second view of innovation is to consider improvements not only to the product offering but also to distribution channels and revenue models (Anthony 2012).

This view of innovations has been expressed in Figure 1.1 at the beginning of the fuzzy front-end funnel. They are *Product Ideas, New Technologies,* and *Market Opportunities.* My challenge with these terms is they are borrowed from the field of management and may not readily resonate with engineering and/or scientific terminology. From a technical point of view, Koen (2005) defines the levels of innovation as *Incremental, Platform,* and *Breakthrough.* If I were to compare the three classifications of Koen to Figure 1.1, I would equate *incremental* innovations to market opportunities, *Platform* to product ideas, and *Breakthrough* to new technologies. So, what does this mean? Let me give a few examples to be more clear.

Many years ago, I was a design engineer at a leading dental instrument manufacturer. The marketing organization had identified that children were afraid of the "industrial" look of the instruments. The marketing opportunity was to offer instruments that did not look "frightening" to children. The *incremental* innovation that we offered was to coat existing instruments with a thin layer of titanium dioxide using a supplier's proprietary process. The outcome was that the instruments seemed to have a rainbow-colored coating that somehow made the bands of color move as the angle of light shifted. This was enough to distract children and reduce their fear. With the exception of shipping the instruments to the coating supplier, there were no other changes to the manufacturing process. There were, of course, different labels and packaging to align with the new product line. As expected, this incremental innovation had minimal impact on existing processes, yet it fulfilled a market need or opportunity.

> There are three categories of innovation: Incremental, Platform, and Breakthrough. The highest form of Breakthrough is *Disruptive.*

The next class of innovation may be called *platform* innovation. As we will read later about concept development and, in particular, morphological analysis in Chapter 9, platform innovation refers to a *natural evolution* of a product line. For instance, in the automobile industry, this evolution has led to platform development of sedans to station wagons, followed by minivans, then to sport utility vehicles (SUVs), and now a platform vehicle called a crossover. All these vehicles do the same basic functions of an automobile, but each satisfies a slightly different market need. Compared to incremental innovations, platform products are substantially larger and more resource and capital intensive.

Finally, breakthrough products – or technologies – are the ones that are new either to the company or to the world. Koen (2005) suggests that with breakthrough products, the level of performance may increase as much as ten times and cost savings of up to 50%.

An example of a breakthrough product in the automotive industry is a contactless fuel level sensor. In the majority of automobiles, the fuel gauge is a combination of a float, a float arm and a simple potentiometer. As the fuel level in the tank changes, the float moves up or down, and along with it, the float arm pivots around a fixed point. The movement of the float arm would describe an arc. By connecting a circular potentiometer to the center of the rotation, it is possible to measure the change in the resistance as the float arm rotates up and down with changes in the fuel level. By correlating the two, the fuel levels in the tank may be measured. The failure mode of these fuel sensors is either corrosion of the potentiometer or collection of debris and gunk on the resistive track, either preventing the arm movement or creating electrical noises causing inaccurate readings. One solution to remove these failure modes is to remove the need for a resistive circuit as a means of measuring the fuel level. This task was achieved by using magnetics and measuring the strength of the magnetic field using a Hall-effect sensor (Jamnia and Chen 2007).

If the product or technology is new to the world, it may be called *disruptive*. Disruptive technologies suggest that existing procedures and paradigms need to be changed or updated in order to incorporate the new product into the market. However, the success of such products depends to a large extent on market acceptance of the paradigm shifts. Figure 1.3 exhibits the patent of a disruptive product and its associated technology. In this patent, Von Seggern (1964) proposed an alternative method of drilling teeth for removing tooth decay and preparing the cavity for filling. The established methods had been to use a rotary motion generated in (or transferred to) a handpiece. The handpiece was then equipped with an abrasive burr. The axial rotation of the burr and its contact with the tooth material would cause material removal. At the time of Von Seggen's invention, the handpieces did not have high speed, and it was accepted that the patient may be uncomfortable or even experience pain. His approach was to take advantage of sonic vibrations of thin metal, as shown in Figure 1.3, to replace the rotary motion of existing handpieces. He summarized his breakthrough as follows:

> it is desirable to have a component of motion of the tool perpendicular to the surface of the progressive cut, even when unbalanced or angled tools are used. In this connection said patent discloses that if desired to vary the pattern of the paths of contact of the surface finishing elements with the surface being treated, the surface may be disposed at any desired angle with the direction of propagation of the sound waves adjacent thereto. This invention is concerned with my discovery of the importance of a particular characteristic pattern for such component motion, and with my further invention of a method and apparatus for accomplishing this characteristic motion.
>
> (Von Seggern 1964)

Briefly, the sonic technology is based on magnetostrictive properties of certain metal alloys. Similar to thermal expansion properties, materials with strong magnetostrictive properties expand or contract in the presence of a magnetic field. Thus, it is conceivable to induce vibration in a magnetostrictive object by alternating the direction of a magnetic field. A sonic or supersonic[3] vibration may be generated by the corresponding frequency of the induced magnetic field. The engineering challenge is to design the vibratory tips in such a way that the vibrations are both predictable and desirable.

It is interesting that for a variety of reasons, the market did not accept this product and this technology as a means of drilling holes in teeth. However, it was widely accepted and in use even today as the preferred method and procedure for cleaning teeth and for calculus[4] removal.

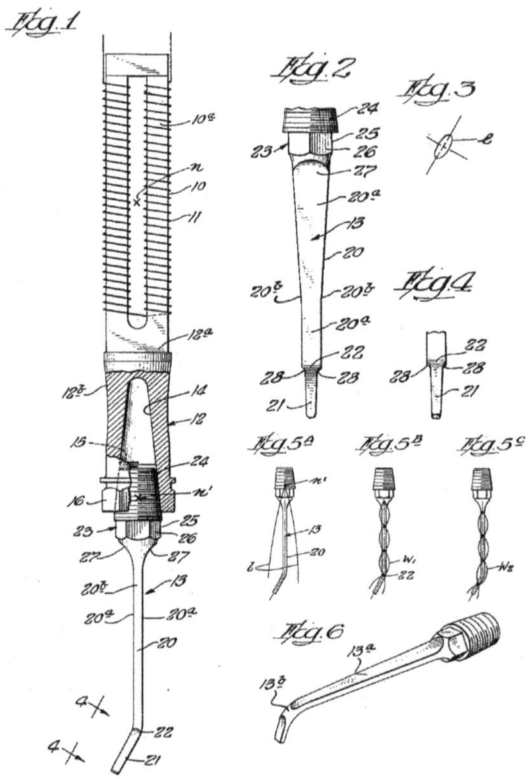

FIGURE 1.3
A depiction of the patent "method and apparatus for sonic dental drilling" – United States patent number: 3,133,351, issued on May 19, 1964

A Systematic Approach to the Fuzzy Front End

As suggested before, the fuzzy from end – particularly for breakthrough innovations – is often chaotic, and the results are unpredictable. This is in contrast with the NPD process, which is quite disciplined and orderly with known budgets, resources, and timelines. However, the financial health of any organization depends on having a healthy stream of concepts into the NPD pipelines.

> A systematic approach to the front end provides a more efficient means of sorting and ranking new ideas. More importantly, it enables the organization to reject concepts that may not have a high chance for success.

The question is whether a systematic approach to the fuzzy front end may be developed. Though most organizations define their own process for innovation, typical stages of the fuzzy front end are depicted in Figure 1.4 and are described as follows:

1) Encourage project or product idea submissions from any level of the organization without judging the quality of the idea.

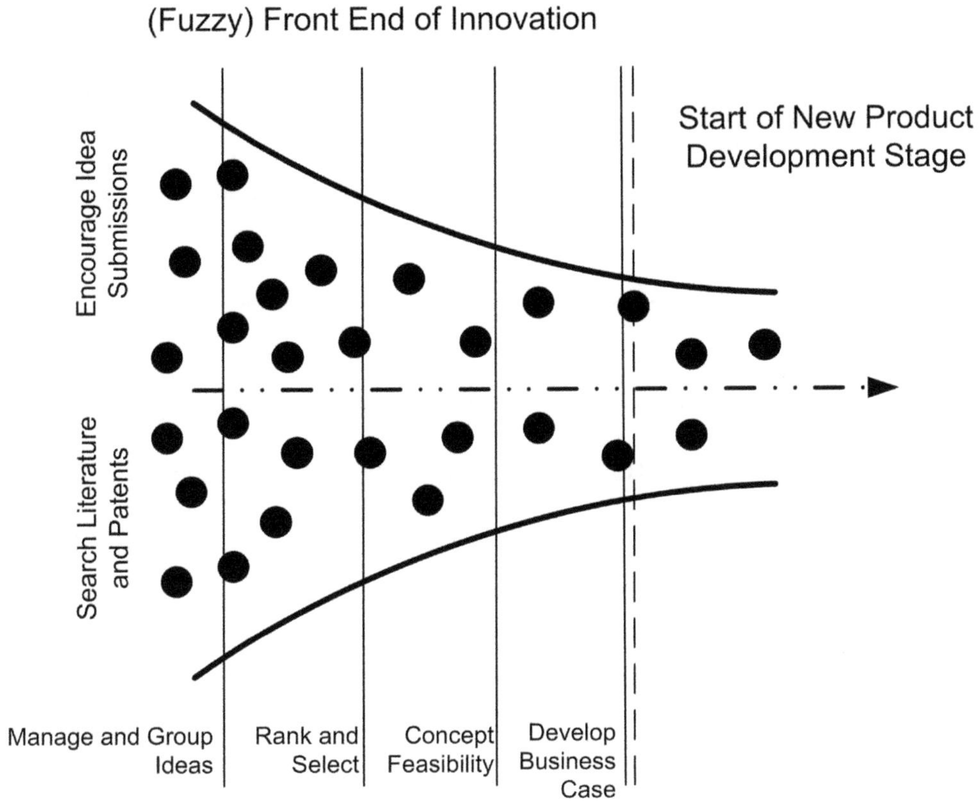

FIGURE 1.4
Typical stages of a systematic approach to front-end innovation

   a. This step is strengthened when the organization provides areas of need or exploration.
   b. An idea submission typically includes three basic parts: the problem being considered, the current state of technology or approach, and finally the proposed solution.

2) Search literature and patents for products and technologies that are either directly applicable or may be modified to the organization's needs.
   a. Similar to step 1, on the basis of found literature or patents, ideas may be formed on how the newly found information may be used innovatively within the organization.
3) Manage ideas to identify groupings and classifications such as existing technologies and new-to-organization technologies. Alternatively, some ideas may point to technologies that need to be invented. An appropriate tool may be the affinity diagram introduced in Chapter 8.
4) Rank various classifications and identify the projects that would need to be studied further.
   a. This step implies that only a select few will proceed to the next steps.

5) Create the initial business case(s) and obtain approvals for concept feasibility studies (if needed).
   a. Concept feasibility stage is valuable not because it would prove certain ideas would work but because it will weed out seemingly promising concepts that would prove difficult to develop.
6) Develop the business cases for the proven concepts and initiate the NPD process.

Closing Remarks

In this first chapter, I provided an overview of the inception of new products. From here on and for the rest of this text, I will focus not only on the NPD but also on the greater product life cycle management (PLCM) process. PLCM starts from the initial NPD steps of writing user requirements to completing the design and manufacturing a product. It continues to product launch and sustain it in the field and ends with retiring the product from its market.

Earlier, I suggested that the fuzzy front end is rather unstructured – particularly, in either engineering or scientific trials and experimentations – whereas the NPD process is quite structured. This structure requires that a larger group of people need to work together in cohesive teams. The pressure of meeting goals and deadlines as well as different personalities requires that various team members have an awareness of certain *soft* skills. By soft skills, I mean a set of know-how beyond technical competencies. These skills enable team members to understand each other's potentials and adjust to one another in order to achieve the desired end results.

Prior to delving into the PLCM process, I would like to share a few areas to develop soft skills in the next four chapters.

Summary

- Any NPD activity has a period at its beginning wherein decisions need to be made about the viability and applicability of the end product to the available markets.
- Inventions are only an aspect of innovations. Innovations need to be providing a *useful* product or service for the intended market.
- Even though the front end is fuzzy by its nature, it is possible to manage the chaos systematically and to ensure that the resulting developed product has a high chance for success.

Exercise

Form a team of four or five people, and identify several industries that you may have familiarity with. This could be either manufacturing or service. Then select one or two products or services and study them for their strengths and weaknesses. Then meet for two consecutive sessions:

1. In the first session, generate at least 20 ideas on improving the product or service. Do not judge them as good or bad.
2. In the second session, generate an additional 20 or 30 ideas. Again, do not judge them.
3. Combine all ideas in one place and classify them into similar groups. These groups may be incremental, platform, or breakthrough.
   a. Create bar chart or dot-plot and compare the results.
   b. For breakthrough product, conduct a patent search using either USPTO.gov and identify any existing patents similar to the one on your list exists.

Notes

1 A product may be either a physical item, software, or a service.
2 I will use the words "what" and "how" throughout this book, particularly in the first few chapters on requirements and concept development. "What" often refers to a desired outcome, and "how" is the means of getting this outcome.
3 In terms of frequency of vibration being in the hearing range or beyond it – not to be confused with speed of motion.
4 Calculus is the solidification of saliva and gingival fluid minerals on teeth.

Part I

Soft Skills Needed in the Workplace

2

Team and Team Dynamics

Introduction

In Chapter 1, I provided an overview of the start of the innovation process called the fuzzy front end. I showed that once a product[1] idea has been successfully vetted and has reached the concept feasibility step, it is assigned to an innovation research and development (R&D) team. This team has the responsibility to show that the selected concept is viable. They do so by developing early prototypes as well as conducting feasibility studies. Once the feasibility of a concept is shown, the idea is passed on to the NPD platform. A design and development team takes the responsibility of creating the required design details to enable manufacturing and launch of the product. Post-market launch, a product-sustaining team accepts the responsibility to ensure that the product is properly maintained in the field and is performing as expected. Different corporations may have different names for these teams or groups. These teams may be large or small. The individuals within these teams may be partially dedicated to one product (or project) and have to manage their time appropriately in between various projects.

In short, there are numerous interactions between various people with a variety of motivations, interests, and personal agendas. And each team member has his or her unique personality, unique ways of approaching solving problems, and view toward one's responsibilities and expectations from the organization. When people with different expectations and outlooks come together, conflicts are bound to happen sooner or later. How one reacts to and manages these situations may make or break one's credibility and ultimately career.

For most engineers and scientists, people management skills are considered to be "soft" skills. These are not technical issues that a textbook formula may provide a ready solution for. In this chapter, I would like to examine the basic elements of these management skills. Fundamentally, good people management starts with an awareness of our own internal thoughts and feelings and how intelligently we manage them. The term for this skill is *emotional intelligence*, and I like to begin this chapter with this topic.

Emotional Intelligence

Most college entrance tests, such as the Scholastic Assessment Test (SAT), have been devised to measure a student's aptitude for learning and hence the chance for success not only in the first years of college but also in life. These are typically variations on a more general intelligence quotient, or IQ test, which is intended to measure an individual's intellectual – and, by association cognitive – abilities. Intelligence is defined as "adaptive problem-solving behavior, where problem solving is defined with respect to practical

DOI: 10.1201/9781003564508-3

goal attainment and some sort of positive development outcome" (Emmons 2000, pp. 5–6). The assertion is that someone who has a higher IQ than his or her peers is smarter and, therefore, can solve problems both at home and at work more easly. Logically, this person would have a much higher chance in succeeding in life. However, as Goleman (2005) points out, studies have shown that of all the factors that contribute to success in life, IQ contributes about 20% – at best. This seems to be counterintuitive and begs the question: "What could be other contributing factors?"

> Studies have shown that of all the factors that contribute to success in life, IQ contributes about 20% – at best.

In 1983, Howard Gardner published his theory of multiple intelligences (MI), theory in which he challenged the notion of IQ as a single metric to indicate one's ability to problem-solve. Gardner considered IQ tests as language- and logic-mathematics based, which did not examine other facets of human cognition. Initially, he proposed intelligence modalities as musical, visual-spatial, verbal-linguistic, logical-mathematical, kinesthetic, interpersonal, and intrapersonal. Later, he added naturalistic and existential intelligences (Gardner 1995, 2000). Each proposed modality meets the criteria of providing the individual with adaptive problem-solving abilities and getting the individual closer to his or her personal goals.

Gardner was among the first psychologists who opened the gateway for expressing alternative theories and models about intelligence. In his popular and seminal work, *Emotional Intelligence: Why It Can Matter More Than IQ*, Goleman (2005) expounded on a model developed primarily by psychologists Peter Salovey and John Mayer called *Emotional Intelligence* (EI). It was defined as "ability to monitor one's own and others' feelings and emotions, to discriminate among them, and to use this information to guide one's thinking and actions" (Salovey and Mayer 1990, p. 189).

The basis of EI (i.e., knowing and managing both one's emotions and others) is very similar to Gardner's interpersonal and intrapersonal intelligences whereby one knows and manages either how one feels or understands the feelings of others. So, to answer the earlier question, if IQ is only a 20% contributor to success in life, then Goleman (2005) asserts that EI is as large of a contributor to being successful, if not larger. The reason is that abilities such as emotional self-regulation and empathy may be stronger factors in success than pure cognitive faculties. Just consider that there are very many logical individuals who may be good at what they do, but because they cannot relate to others, their career growth is somewhat limited. In today's corporate environment, emotions and modes of expression either encourage colleagues and coworkers to align to given organizational goals or, adversely, disperse them in different directions. The dispersion that is aligned with differing individual desires and emotional needs only creates obstacles.

Additionally, emotional intelligence gives us the ability to deal with life's frustrations. In moments of anger, it enables the individual to control his or her impulses. In short, it provides the type of control on the emotions so that one may think clearly. Evidence has shown that those who have an understanding of their own emotions and feelings and can control them tend to excel over those who do not have this ability. They are not only more successful at their careers but also lead happier lives (Goleman 2005).

The conceptualization of emotional intelligence includes the following elements (Grewal and Salovey 2005; Salovey and Grewal 2005)

1) Accurate Perception of Emotions: This is the ability to recognize emotions within ourselves and to classify and express them. Furthermore, it is the ability to recognize various emotions in other human beings through their body language,

the tone of their voices, or even their pictures. Or, as we classify certain music or melodies as happy or sad, we are expressing our abilities to associate emotions with external stimuli.

2) Integrating Emotions with Reason: Typically, emotions are considered to be irrational. Studies have shown that people with happy moods are more creative. Interestingly, people who experience slight sadness are better at doing detail-oriented, meticulous work. Furthermore, other studies have provided that decisions made without the role of the individual's feelings may not be in their best interest.
3) Understanding Language of Emotions: Emotions may be categorized within a spectrum. For example, the emotions of happiness, elation, and ecstasy may belong to the same spectrum. Understanding the language of emotions not only gives us the skill to express our simple emotions, such as being happy, sad, or irritated but also the aptitude to realize that one emotion, if not checked, may lead to another. For instance, anxiety may lead to depression if not controlled. This aspect of EI is particularly important in the corporate environment, as I will discuss later in this chapter. When team members clash over differing opinions – which may appear to be personal, it would be advantageous to be able to express one's *emotion* in response to someone else's *action* in an effort to resolve conflicts. For example, *I feel offended when you call me inexperienced.*
4) Managing Emotions in Self and Others: This aspect of EI is simply the aptitude to manage our own emotions or those of others. The simplest form of this skill is to control our anger in a disappointing situation so that it does not turn into rage. More sophisticated aspects of managing emotions are to maintain a level of, say, sadness or rage in order to instill the same feeling in an audience to call them to action. On a personal level, as we develop a deeper appreciation of and sensitivity to emotions, our capacity to express empathy increases, and we begin to develop personal connections with people around us. In the business world, to influence others, this skill needs to be well developed.

If I were to summarize the lessons of Emotional Intelligence, I would say that through knowing our emotions better, we can make more informed and better grounded personal and career-oriented decisions. Once we develop cognizance of our feelings, we can better control potentially runaway feelings and not do or say something that we would regret at a later time. Additionally, we can put our negative emotions due to life's setbacks and frustrations into an appropriate prospective and, hence, easily bypass life's setbacks. As we develop a deeper understanding of our own emotions, we develop an appreciation of emotions of people around us and are able to read their nonverbal signs of how and what they feel. Finally, we come to influence others around us, either in our personal or professional lives. Being popular or having a knack for others to follow requires our empathy and an understanding of managing emotions in ourselves and others. Leadership abilities such as organization and negotiation skills require high levels of emotional intelligence to develop what people feel and properly respond to those feelings and possible apprehensions.

Through knowing our emotions better, we can make more informed and better grounded personal and career-oriented decisions.

As we speak of EI and developing the ability to interact appropriately with the people around us, we need to understand that in addition to varying degrees of EI, individuals have different modes of communication. Some want details, while others prefer only summaries; some are talkative, while others are quiet. In short, as the ancients have classified it, there are different personalities that people exhibit.

In the next section of this chapter, I will review personalities[2] as they pertain to the workplace and the need for clear communications. This information will help us better understand team dynamics from its formation to its release.

Personality Types and Traits

The Four Temperaments

> A branch of psychology called personality psychology focuses on the study of human personality, which is the expected behavior of individuals that others have observed over a period of time.

For nearly two millennia – if not more – human beings have recognized different behaviors in the members of their societies and have attempted to classify and understand them. Early Greeks such as Hippocrates (460–370 BCE) and Galen (130–210 CE) associated human behavior with human nature and physiology. Later, Avicenna (980–1037 CE) – the Persian philosopher – extended the influence of the four temperaments to one's emotional and mental states. In those days, the general belief was that the body is ruled by the balance of black and yellow bile (fluids excreted by the liver and spleen, respectively) along with phlegm (a fluid that is produced by the respiratory system, particularly during sickness) and blood. These four fluids were also called the four humors. They kept the balance between the mind and the body. An imbalance leads to sickness.

In their mindset, black bile was associated with a *melancholic* temperament, yellow bile with *choleric*, phlegm with *phlegmatic*, and finally, blood with *sanguine*. Hippocrates associated melancholic temperament with being sad but at the same time thoughtful; choleric with being confident but domineering; phlegmatic with being easygoing but sluggish; and finally, sanguine with being happy, hopeful but irresponsible (Littauer and Sweet 2011).

Admittedly, the ancients did not believe that individual behaviors are solely governed by one temperament only. More properly, I should say that the general belief was that individuals are primarily under the influence of one temperament. However, other secondary temperaments do exert their influence from time to time. In other words, while one disposition may be dominant, others are still at play, and their dominance may shift and change from time to time.

Now, let us assume that the ancients were right. Imagine what a team may look like with a mix of people exhibiting these various temperaments as either their leader or members. Imagine that the team leader had a phlegmatic disposition – easy but slow going – and the rest of the team were choleric and sanguine. How would the team function when all individual activities and deliverables were running smoothly? What would the team interaction be in case of a conflict or crisis?

The outcome is not hard to imagine. Should the team leader be of a phlegmatic disposition, he is easy but slow-going. In periods when all is well, one would imagine that the team lead leaves the people under his care alone to do their jobs. He would be happy with periodic reports, which he would provide to his management. Team members would be happy because they are not being micromanaged; in other words, they are being left alone.

Now, imagine that there is a crisis requiring quick decision-making and actions. Remember that some of the team members have a sanguine (happy, hopeful but irresponsible) disposition, and some are choleric (confident but domineering). Chances are the team will be in total dysfunction. The leader – by his disposition – needs time to evaluate and understand the situation, while those with a choleric mindset are trying to solve the problem, possibly through bypassing their leader. They may even try to remove him from his post. All along, the sanguine folks may wonder what the fuss is all about because eventually all will be well.

Clearly, this is a simplistic example. But its point is as follows. In team dynamics, we need to be aware that individuals exhibit varying behaviors under different stressful conditions. This knowledge is important from two points of view. The first point of view points to ourselves. We need to understand our own reactions and responses to different stresses. We can only change and improve if we have an understanding of what we need to work on. The second point of view points to others. By accepting that others do not have to show a prescribed reaction to external stresses, we will be better equipped to exhibit emotional intelligence in response to others' unprofessional behaviors.

I like to emphasize that regardless of what the ancients might or might not have believed, by developing self-awareness as well as EI, it is possible to change our negative responses and improve our relationships with the people around us. Self-awareness and EI are extremely important tools to understand the team's dynamic at any given point in time and to properly navigate the path forward.

Modern Psychology

A branch of psychology called "Personality Psychology" focuses on the study of human personality which is the expected behavior of individuals that others have observed over a period of time.[3] This includes exhibition of one's thoughts, behaviors, and social responses and interactions. There are two branches of this field, namely *trait* and *type* theories.[4] Examples of human traits are *openness to experience*, *self-esteem* (high or low), and *rigidity*. Personality type theory may find its roots in the four temperaments of the ancients. In the 1950s, Type A and Type B personality theory was introduced, with Type A people being "achievers" and Type B being "easy-going." This theory has since fallen out of favor.

A psychologist whose theory continues to thrive and be expanded is Carl Jung. He felt that human beings orient themselves with the external stimuli and adapt to them using their psyche, which may exhibit a combination of the following four basic functions:

1. Perception of the world and the working of its elements. This was called *sensation*.
2. Perception of the feelings, thoughts, and motivation of others. This was called *intuition*.
3. Formation of conclusions and the functioning of the intellect. This was called *thinking*.
4. Formation of values based on emotional responses. This was called *feeling*.

The *thinking* and *feeling* pair, Jung named "rational" or "judging" functions, and the *sensation* and *intuition* pair, he called "irrational" or "perceiving" functions. The rational functions are used in decision-making and perceiving functions are the basis of knowledge and information gathering. Additionally, Jung used the concept of extraverted or introverted attitudes and submitted that there are two pairs of the cognitive functions that are

> Once we develop an understanding of our colleagues' personalities, we understand their modes of communication, patterns of thinking, and decision-making.

expressed in one of the two attitudes; for instance, extraverted thinking or extraverted feeling. In all, he defined eight different psychological types. Further, he suggested that typically, people have a dominant pair, which he postulated is the basis of their conscious behavior. Additionally, their unconscious behaviors are influenced by an opposite set (Monte and Sollod 2003; Hall et al. 1985).

I like to note here that there were other psychologists who focused on human personality and developed instruments (tests) to categorize various personality types. These instruments, in a way, enable one to understand and ultimately predict how people with certain behavior types may respond in given situations or external stimuli. A discussion of personality theories put forth by modern psychologists is beyond the scope of this text.

In short, "What people wear, how they walk, talk, sit and stand can give clues to their personality . . . Physical postures and tones of voice speak the language of personalities without words" (Littauer and Sweet 2011, p. 83). Hence, once we develop an understanding of our own and our colleagues' personalities, we understand the required modes of communication, patterns of thinking, and decision-making. This knowledge enables us to provide and receive information and effectively communicate in our work environment. The question is how we can develop this understanding.

Personality Type Tests

In the 1920s, the US Armed Forces was among the first organizations that expressed an interest in and developed tests and means of understanding individuals. These tests were administered in order to assign appropriate tasks and jobs to the applicants. Since these early times, personality tests (also known as instruments) have come a long way. Today, even though exact figures are not known, it is estimated that personality testing is a $2 billion to $4 billion industry (The Economist Editorial 2013). Another astonishing fact is that over 2.5 million people take the Myers-Briggs Type Indicator (MBTI) each year (Baer 2014). This test is based on Jung's theories reviewed in the previous section of this chapter. Nearly 90% of the Fortune 100 companies administer MBTI to their potential employees (Baer 2014). Often, employers use personality tests to make hiring decisions. These decisions are not always made for the purpose of keeping low performing candidates out or to identify candidates whose loyalties last until a better career opportunity comes along. Hiring decisions are also made to ensure that the new employee would fit the organization's culture and mindset or be assigned appropriate tasks and responsibilities. At other times, these tests are used to develop employees, improve teamwork, and even, at times, in succession planning.

Admittedly, some of these instruments have been error-prone. As a result, there are a number of opinions both for and against the use of personality tests to make hiring decisions (Baer 2014; Clapperton 2015; Stabile 2002). On a personal note, in two corporations that I have worked for, personality tests were administered so that upper management understood their employees' types and could judge their fitness for particular assignments. They were also used as a means of letting the employees know what the preferred communication style of the upper manager may be. I will come to this topic later in this chapter. For now, I will briefly review three available tests that I have taken in the past and share my own results as a way of introducing these types of tests and the information that they provide.

Myers-Briggs Type Indicator

Among various personality tests, the Myers-Briggs Type Indicator (MBTI) is the most widely used.

This test is based on Jung's personality types and is probably the most widely used instrument. This test examines an individual's inclination to either introversion or extroversion, sensing or intuition, thinking or feeling, and judging or perceiving. Both commercial and free versions of this test are available. I have taken this test several times using both the commercial and the free online versions. While there are several websites that administer this test for free, I have taken mine at https://www.truity.com/test/type-finder-research-edition and at https://www.16personalities.com.[5]

The test taken at www.truity.com indicated that I am borderline between being an *extravert* and *introvert*. It also identified me as borderline *Intuitive* and *Sensor*. These two aspects of personality describe how I manage my energy (extravert vs. introvert) and process information (sensing vs. intuitive). The extravert enjoys being and interacting with people and engages with the outside world. In contrast, an introvert enjoys being in calm surroundings, solitude, and being quiet. As for processing information, *sensors* rely on facts, figures, and their details. They are realistic and practical and count on information from their past to guide their future activities. *Intuitives* focus on patterns and connections and are typically creative and innovative. I have to admit I do see both sides in me, and I am not surprised that the test could not determine my tendencies in these two areas.

As far as my personal values were concerned, this test indicated that I am a *feeler*, meaning that I consider the emotions of others and value authenticity. The last aspect of my personality was to measure my self-management style. I was determined to have a *judging* (as opposed to *perceiving*) personality. This suggests that I value order and organization; that I like to plan and see them to their conclusion. This is certainly true in my work environment; however, in my leisure, I would much rather have spontaneity and do things as they come about.

In contrast, the test that I took at www.16personalities.com placed me at being *ESFJ* (extravert, sensor, feeling, and judging). This test had an additional factor called *identity*: Assertive versus Turbulent. I was judged as *assertive*. To be clear, the set and style of questions were different from those used in the first website.

This instrument measures the individual for ten different aspects of their personality. One of these ten aspects includes *strengths and weaknesses*. For instance, one of my strengths is having very practical skills, which makes me a good manager. Another (according to this test) is being warm and sensitive – which enables me to be a strong team player as I look for win-win situations. One of my weaknesses is a lack of flexibility, particularly when I am pushed into a corner.

Another element of personality provided by this test is called the *workplace*. Considering that the focus of this chapter is on teams and team development within a work environment, this section of the personality profile is of particular interest. My type, as depicted by this test, enjoys social order and harmony. A person with this type ensures that each member of the team knows their responsibilities and gets done what needs to be done. The workplace for people of my type should be safe and provide structure as observed through workplace processes and guidelines.

By comparing the slight differences between the two instruments that I took, it may become obvious why some feel strongly that MBTI is not appropriate for making hiring decisions and job placements (Baer 2014; Clapperton 2015; Stabile 2002). Having said this, it is quite clear that this information may be used effectively to create very productive teams. For instance, an aspect of my personality is that I tend to think in terms of strategy and generalities. Under high stress situations, I show a tendency to not "see" details. In important projects, I try to pair myself with individuals whose type is to pay attention to details. This pairing brings a certain degree of balance to the team, and projects are often completed within time and budget.

Insights Discovery Profile

Similar to the MBTI, this profile is based on Jung's typology. However, this test is only available commercially. Unlike MBTI but similar to Hippocrates's, Insights Discovery offers four classes of personalities associated with four colors. Each class is associated with a specific mindset and communication mode, as shown below (Lotian 2000):

Color	Mindset	Communication Mode	Preferences
Fiery Red	Let's do it NOW.	Be brief, be bright, be gone	Extraverted/thinking
Sun Shine Yellow	Let's do it together	Involve me	Extraverted/feeling
Earth Green	Let's do it in a caring way	Show me you care	Introverted/feeling
Cool Blue	Let's do it right	Give me the details	Introverted/thinking

Briefly, this instrument provides guidelines for communicating to and with people who are primarily influenced by character traits as depicted by one of the four colors. For instance, people who belong to the Fiery Red category are typically competitive and strong-willed. In communicating with them, one should be direct and to the point. The wrong approach would be to hesitate on the point to be made or to dive into too many details. In contrast, people who belong to the Cool Blue category need to see the details. In communicating with them, one should be well prepared and thorough (Lotian 2000).

Additionally, Insight Discovery has introduced the idea of a *Personality Wheel*, whereby eight different traits are introduced. These are director, motivator, inspirer, helper, supporter, coordinator, observer, and reformer. These traits are mapped to a color spectrum that gradually shifts from red to yellow, then to green and blue and finally back to red. For instance, director is associated with a solid red, whereas, reformer is purple – a mix of blue and red.

Relatively early in my career, I took this test along with another 20 colleagues from engineering (2), accounting (2), and marketing (16). Though we were all working at the same corporation, the selection of people to take the test was quite random. Figure 2.1 shows the first and second color distribution for this group. It should be noted that these results are only accurate for this team and may not be generalized.

It is interesting to note that blue has the highest count as the first color, and green was the highest as the second color. Recall that the first color is indicative of the conscious mind, and the second color is indicative of the less (or sub) conscious mind.

Figure 2.2 depicts the distribution of the wheel traits. Note that the highest trait is observer, which is consistent with the blue personality.

As mentioned earlier, the importance of this information is in sharing it with the team so that better collaborations may be made. For instance, my conscious wheel position is scored at 53, which indicates an observing coordinator (accommodating), and my subconscious is scored at 12, which is a supporting coordinator (focused). This places my colors at blue (conscious) and green (subconscious). The implication of this classification is that

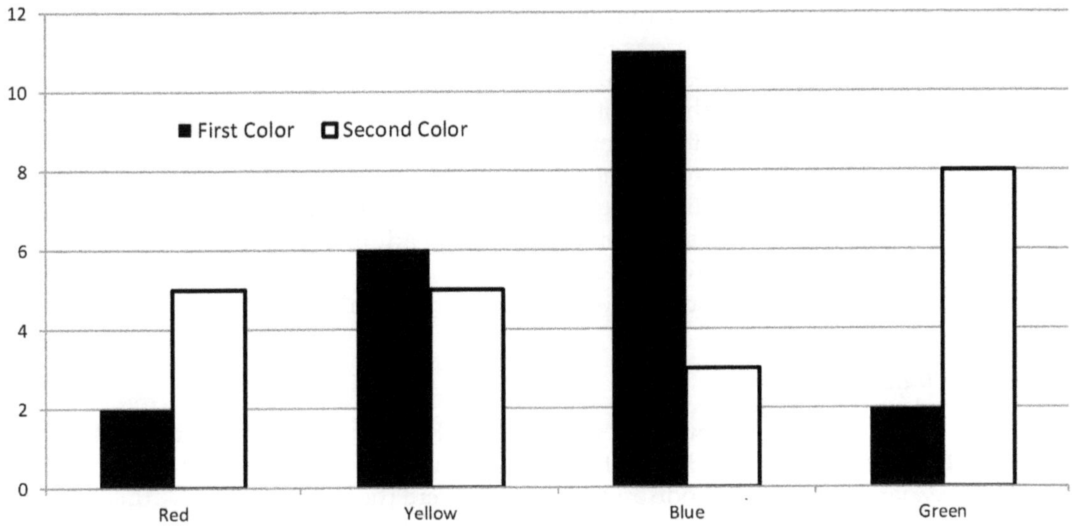

FIGURE 2.1
Distribution of personality characteristics in a group of 21 colleagues

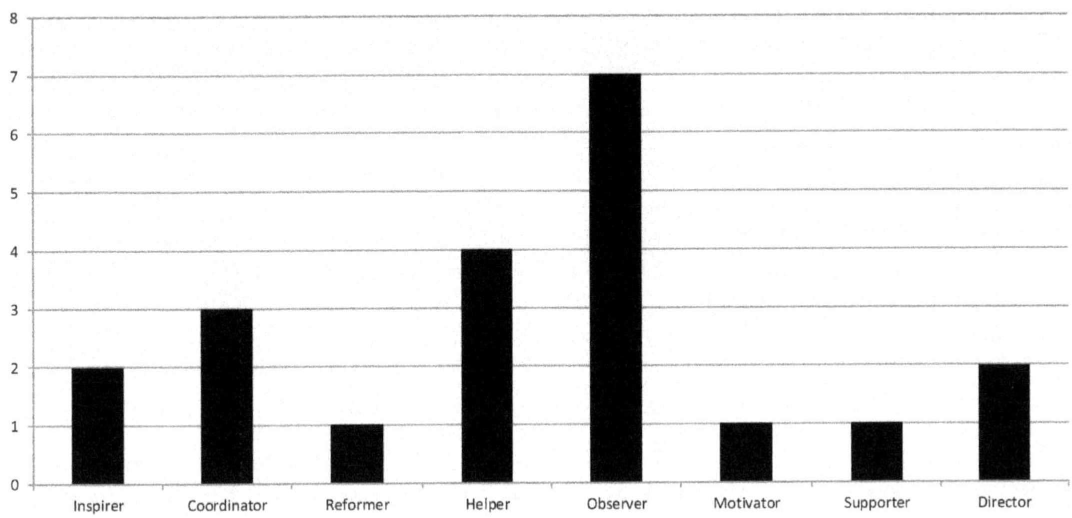

FIGURE 2.2
Distribution of wheel traits in a group of 21 colleagues

> The Insights Discovery profile seeks to help team members understand their communication styles better by understanding their personalities

I find my ideal work environment when I have good access to relevant sources of information and have enough time to prepare for meetings and presentations. Additionally, I thrive in environments where systematic processes and structure are prevalent. For me to be motivated,

I like to be given the opportunity to show my deep technical knowledge (hence writing this book!) and have access to various technical journals and sources.

My management style – based on this evaluation – is to be systematic and procedure-oriented. Additionally, I enjoy the opportunity to take on complex tasks and projects. This test indicated that I take things too seriously, am unwilling to bend rules, and may become stubborn if pressured. There are two additional weaknesses that this test has identified. These are (1) I may have difficulty in quick or unprepared articulation, and (2) I can be overpowered by the assertiveness of others. I have to admit that once I look back at my career, I have seen both the strengths and weaknesses. Over time, I have learned to be mindful of my weaknesses in order to contribute to the team's overall goals.

Strengths Finder

The advantage of this test is that it is neither free on the internet nor does one need to be a part of a corporation to take it. By purchasing a copy of the *Strengths Finder 2.0* book (Rath 2007), the reader is provided a code and a website where the reader may take this test and obtain the results. The book provides the narration needed for the test outcome and ideas for action.

Unlike the other two instruments, the foundation of *Strengths Finder* is based on identifying one's talents and developing one's strengths as opposed to learning more about one's weaknesses and trying to overcome them. This concept is so fundamental to this model that the author quotes Peter Drucker[6] as follows: "Most people think they know what are good at. They are usually wrong. . . . And yet, a person can perform only from strength" (Rath 2007 p. 15).

In order to achieve this goal, the *Strengths Finder* identifies 5 inherent talents from a pool of 34 themes. When I took this test, my talents were identified as *empathy, adaptability, intellection, positivity,* and *relator* from the highest ranked to the lowest, respectively. With each talent, Rath (2007) identifies actions that may be taken in order to mature the given talent into a strength. For instance, as *empathy* is my number one talent, one idea for action – that Roth proposes – is to mentor others or be available for my team members so that they may vent on issues that may bother them at work; in other words, be their confidant. Conversely, *adaptability* may be matured by "cultivating your reputation as a calm and reassuring person when others become upset by daily events" (Rath 2007, p. 46). Intellection is the desire to think. It is the desire to see the logical connection between events; to reflect, to analyze, and to solve problems. To develop this talent into a strength, Rath (2007) suggests that people (like myself) with this talent develop relationships with people who are considered to be "thinkers" who could then help us advance either in personal or professional lives.

I will not expound on the last two talents because I believe it is clear how the Strength Finder is different from the other two personality type indicators. Furthermore, it is obvious how each team member may contribute to the strength and effectiveness of any team by recognizing the talents that they may bring to the team.

Team Formation for Product Development

The foregoing discussion was meant as an introduction to an individual's personality and modes of behavior. This is to help each individual function best with the team that the individual is deployed to. Since the focus is on teams now, it is suitable to explore team formation and team dynamics in this segment.

In a sense, a corporation is a team that consists of the president at the top, followed by a layer or layers of managers, eventually down to individual contributors. This is a static team. It is static because the basic members remain the same. The management structure does not dynamically shift and change to work on different projects or tackle existing crises. New management models have encouraged that product development (and its post-launch sustaining) teams be dynamic and only be formed for a limited duration to complete specific tasks. Let me elaborate a bit more.

Traditionally, functional teams would form around specific expertise: mechanical engineers would be members of the mechanical engineering team, electrical engineers belonged to the electrical engineering team; chemists and biologists to their respective teams, and so on. Needless to say, other functions such as marketing or finance also existed with their own prospective members and managers.

In the old days, once a business would make a decision to develop a product, it was the responsibility of functional managers, who would assign various elements of the project to different subject matter experts (SMEs), and it was up to these SMEs to work with one another as well as other cross-functional team members to bring the product to life. This approach was rather inefficient, and often product launches were delayed due to miscommunications between individuals and teams.

With the advent of systems engineering, the concept of *matrix* organizations began to gain currency. In this model, product development teams are no longer function-based. Rather, teams are formed around specific programs (or projects). A lead systems engineer (LSD),[7] along with a lead systems integrator (LSI) and a program manager (PM), has the responsibility for the timely completion of the program. The LSD is responsible for technical aspects, the LSI for the schedule, and the PM for the financial aspects of the program. Typically, the LSD leads a cross-functional team consisting of the functions required for the completion of the project. These functions include not only various R&D disciplines – possibly human factors and compliance – but also program management, marketing, and quality. Should the industry be a regulated industry (e.g., medical, avionics, or nuclear), representatives of the regulatory departments may have a voice as well. In this model, functional managers are responsible for developing both the technical skills of their prospective team members as well as their interpersonal skills.

The matrix structure enables quick assembly and disassembly of teams depending on the needs of a given project. It is also possible for a team member to belong to different project teams or join a team for a short period of time to accomplish specific tasks.

Successful teams get results within the determined schedule and budget.

Elements of a Successful Team

Successful teams get results within the determined schedule and budget. However, it takes effort for a team to be successful. In the traditional functional teams, the members have the opportunity to develop rapport with each other over time and learn how to work effectively with one another. The functional manager develops an understanding of the makings of each individual in his (or her) team and would normally assign projects that – in a way – would fit the team member's personality. This is no longer possible in a matrix organization. As these cross-functional teams are formed, it is possible for members to have known and worked with each other on previous projects; or they may be totally unfamiliar with one another. Additionally, previous working knowledge of each other does not suggest a good and amiable relationship either; there might have been misunderstandings or clashes of egos.

Regardless of past history, "If you won't change for the team, the team may change you" (Maxwell 2010, p. 1). Various authors have identified the qualities of ideal team players. Lencioni (2016) has identified three virtues, and Maxwell (2010) enumerated seventeen. Whether three or seventeen, the premise of these qualities is to put the interest of the team and its goals ahead of one's own ambitions and objectives. However, this does not mean that individual aspirations should be ignored. On the contrary, in an ideal matrix organization, it is the responsibility of the functional manager to learn and identify the goals and ambitions of individual contributors and to help align them with various ongoing programs and projects. Hence, there will not be any hidden agendas for the team members to worry about. This type of work environment fosters qualities such as humility, adaptability, thirst for knowledge, courage, and the smarts to tackle difficult problems (Lencioni 2016; Maxwell 2010).

To have a world-class team, a necessary (but not sufficient) condition is for the team members to possess the ideal team player virtues and for each team member to have an understanding of personality types and traits. Teams may break down into dysfunction and achieve very little if certain team dynamic are not maintained. Lencioni (2002, 2005) has identified five factors to ensure healthy team dynamics. These are developing trust, conflict resolution, making commitments, having accountability, and getting results. Transparency and honesty lead to developing trust among team members. This is the first element of having a dynamic team.

Developing Trust

We all make mistakes for a variety of reasons. Making mistakes may be embarrassing; others may take advantage of this mistake to poke fun at us or project an unfavorable image to our superiors. It takes emotional intelligence to accept our mistakes as well as our weaknesses and be transparent to our team.

Conflict Resolution

Once trust is established within a team, differences of opinion may be debated frankly and passionately without the fear of repercussion or backbiting. Team members speak their thoughts, and each would present rational arguments in favor of or against an approach. I used the term "frankly and passionately." I like to note that this term does not imply violence that may often be associated with passionately.

> Qualities such as humility, adaptability, thirst for knowledge, courage and the smarts to tackle difficult problems are needed to be fostered in team members to ensure effectiveness.

There is another aspect of conflict resolution. Misunderstandings occur quite frequently. By establishing trust, we enable each other to both give and receive productive feedback.

As a part of constructive feedback, often misunderstandings may be cleared and resolved. I will revisit Conflict Resolution again at the end of this chapter.

Making Commitments

Because everyone trusts one another, everyone recognizes that all arguments are made for the sake of the project and the team's eventual success. Hence, once a decision is made, everyone – including those who had differing opinions – would support the decision and stand behind it. Since everyone was involved in the decision-making process, everyone feels the transparency of the process and makes the commitment to the decisions made, including the details of plans and deliverables. Often, plans assign action items to various team members with clear due dates and timelines. Commitment means agreeing to deliver one's action items on time.

> Elements of a dynamic team are the ability to develop trust, resolve conflicts, commit to team decisions, keep each other accountable, and finally get results.

Having Accountability

Since decisions have been made with every team member's input and feedback and everyone has committed to it, it would be easy to hold each other accountable for delivering on promised action items within the specified timelines. I would like to clarify a point. In making commitments to given action items, we often take a leap of faith in that we assume we have control over all that we need as inputs. At times, there are unforeseen circumstances and situations beyond our control; we may not be able to meet our commitments. Should this happen – and aligned with the developed trust – we need to let the team know the moment that a commitment may not be made. This is how we would keep ourselves accountable to the team.

Getting Results

It goes without saying that once a team operates on the basis of trust, resolves conflicts and disagreements openly and transparently, makes commitments to deliver on assigned action items, and keeps each other accountable, getting the results becomes unavoidable. It is vital that, in addition to everything else, everyone has an eye on the final goal and the ultimate objective of what the team tries to accomplish.

What has happened during this process is that individual team members have placed the good of the team ahead of their own individual needs, such as recognition or their individual department or division. As a result, the greater organization wins, and once that happens, everyone wins.

Feedback: Praise and Criticism

The last topic that I would like to review in this chapter is feedback. There is an expression that *perception is reality*. In other words, other people's opinions of us have a huge influence on our professional future and career growth. Hence, it is important to be knowledgeable of what others think of us and, at the same time, manage those perceptions. At the same time, to ensure the effectiveness of our teams, it is – in a way – our responsibility to provide feedback to others.

> There is an expression that *perception is reality*. In other words, other people's opinions of us have a huge influence on our professional future and career growth.

Giving proper feedback is not easy. It is not easy in two distinct ways. First, giving positive feedback – as easy as it seems – is difficult. Why? Because by just saying *good job* or *great presentation*, what have we really said? Has the work been so perfect that it may not be improved at all?

Second, criticizing someone is hard. Can we just say to our colleague (friend, relative, etc.) that "*you always. . .* (fill in the blank) *and that annoys / bothers/ disgusts me*"? How do we make sure that we do not hurt their feelings and turn them against us? After all, we may need their help and support sometimes in future. The issue at the heart of praise or criticism is its personal nature. Often, when a criticism or praise is given, we tend to speak to the person: "good job *man*!" or "didn't anyone teach *you* not to slam the door?"

There is another aspect that may be equally as hard, and that is receiving criticism.[8] For those of us who have not trained ourselves, we may get angry, annoyed, and dismayed. Many of us may even show a tendency to give a quick retort to put the offender in their place.

Earlier in this chapter, I introduced the concepts of developing trust, personality types, and emotional intelligence. The question is how these concepts may be used to enable us to give and receive *effective feedback*. Notice that I did not use either praise or criticism because in effective feedback, the focus is on the behavior and not the person. Furthermore, in effective feedback, the attention is on the future and not the past.

Let me provide examples of positive as well as constructive feedback, and then I will provide some general rules of both giving and receiving feedback.

CASE STUDY 1

Jane has just given a presentation on a new manufacturing process for one of Corporation XYZ's product lines to reduce scrap. Kathy, the line manager of a different product line, approached Jane after her talk.

Kathy: Thank you for your presentation. I particularly liked the part of your work when you provided the cost savings connected to the scrap reduction. Do you think this approach may be employed on other lines?

Jane: I appreciate your feedback, thank you. Yes, I have studied several of our product lines, and I believe the approach that I am proposing is applicable. Though, I have to admit, I am not familiar with your product line.

Kathy: That is not a problem. Let me set up a meeting with you and discuss the details of what you have done and see if it may work for me as well.

CASE STUDY 2

Same scenario as before, but this time, her boss, Michelle approaches her – smiling.

Michelle: How does it feel to present to the regional president?

Jane: Quite honestly, I was a bit nervous. Did it show?

Michelle: No, you had everything under control. The part where you tied the cost savings to scrap reduction really caught his attention. I believe we may need to take your show on the road and enlighten a few other top-ranking people as well. I am happy for you; this was a successful presentation. It drove the point home.

Jane: Thanks, many hours of hard work went into it.

Michelle: And it showed. As I said, it was a solid piece of work, but if I may make a suggestion, your analysis will become even stronger if we can somehow show how scrap reduction ties into the overall company's bottom line. Would you consider this extra analysis?

Jane: I should have thought about that. Thanks for suggesting it.

CASE STUDY 3

Same scenario as before, but a day later, Jane's colleague, Mark, approaches her.

Mark: Congratulations (sounding a bit sarcastic) I saw your presentation yesterday. I noticed that you had everyone's attention. Good job. I see that you have stepped on your career growth ladder; just be careful not to step on too many people on your way up!

Jane: Thanks, Mark. I appreciate your feedback. But let me ask. Am I mistaken or do I pick up a note of disappointment?

Mark: Actually, I am a bit disappointed. Jerry and I worked very hard so that you have the data that you need for your analysis, and . . ., you did not even bother to recognize our efforts at all.

Jane: Oh! I am sorry, you are right! I am very appreciative of both your and Jerry's contributions. It was a total oversite because I was so focused on making sure that the presentation covered all areas of interest to our regional president. Thank you for giving me this specific feedback. It is not too late yet. I have to email a copy of the presentation to the attendee. I will make sure to acknowledge your contributions.

Mark: Thank you, Jane. This makes me feel a lot better.

How to Give and/or Receive Feedback

In giving or receiving feedback, there are a few common rules. The most important of these are the following (Harvard Business Review 2016; Harley 2013; Cabane 2012; Stone and Heen 2015):

1) Feedback should not be personal; it should focus on either behavior or a deliverable (such as, say, a report).
2) Feedback should be delivered in a relatively short period of time – say a couple of minutes.
3) Feedback should focus on future activities or work. Reminiscing on the past is hardly fruitful.

Receiving Constructive Feedback

In addition to the general rules of feedback that I enumerated above, there are other pointers that we would benefit from keeping in mind when we receive either constructive feedback or criticism. Either criticism or constructive feedback does not generally give us a good feeling. This is one area where we should particularly exercise the rules of emotional intelligence to ensure that our emotions are in check. Additionally, by receiving feedback graciously, we allow others to provide a "mirror" to us and to help us see ourselves as others see us. It shows people care enough to let us know what they think. Feedback provides us great opportunities for personal and professional growth.

1) If we find ourselves to be angry or upset, it is best not to respond immediately. Typically, a deep breath and a long exhale help us check our own emotions.
2) Instead of narrowing our attention on the negative, we need to try to understand what it is that the person is trying to say.
3) It would benefit us to ask clarifying questions to get to the root of the problem. Then, we need to indicate whether we agree with the root problem or if there is a misunderstanding.
4) Then, a common agreement and understanding should be reached on how to proceed in future incidents.

Giving Constructive Feedback

Giving feedback is not any easier than receiving it. First, if not properly handled, it can backfire and turn a relatively neutral situation into a bad and potentially toxic one. Harley (2013) suggests that as we enter new relationships – such as new teams – we ask our colleagues about their preferred methods of communication, particularly feedback. At the same time, we can allow our new colleagues to know more about us and how we like to communicate.[9]

> When delivering positive or constructive feedback, it is essential to focus on behavior and not the person

It is paramount that we think through the feedback that we are about to give beforehand. Giving impromptu feedback is never a good idea. To the extent possible, the location should be as neutral as possible and the timing convenient to both parties. Once this is thought through, Cabane (2012) suggests that we be as empathic as possible and always start on some positive points to assure our recipient that we do value their worth and validate them. Then, we should only state facts and make suggestions for improvements. Then, we need to ask for the other's opinion and whether they agree with the points that we have made. We need to engage our colleague and work through any defensiveness that may present itself. Again, Cabane (2012) suggests that should we observe any defensiveness, we can show how we care about our colleague by showing a degree of warmth. Above all, we need to make sure that the conversation does not become personal. And, finally, we should close our conversation first by agreeing on the course of action to be taken in the near future and once again an assurance of the colleague's worth to the team and the organization.

More on Conflict Resolution

The focus of this chapter has been on teams, team dynamics, and the different personalities that make up the team. As I mentioned earlier, an element of a successful team is its ability to resolve conflicts. Here, I would like to expand on this notion.

As product ideas move from the fuzzy front end into the development stage, management begins to form R&D teams of scientists, engineers (both design and manufacturing), and human factors specialists. There are also marketing and quality teams involved. Each group brings its own mindset. R&D is concerned with how the product works and whether it performs the same when mass produced. Human factor designers focus on usability and user experience, while marketing demands speed to market. Meanwhile, the quality representative on the team insists that documentation must be complete and organizational processes be followed. Needless to say, while having diversity of personalities potentially leads to developing better products, inevitably, it is a source of conflict and disagreements.

If handled properly, conflict can become a tool for bringing team members closer together. However, poorly managed or unmanaged conflict can fracture teams, diminish credibility, and lead to toxic work environments. Technical people are often trained in their areas of expertise, problem-solving, and scientific methods – areas that are often called "hard skills." Soft skills that focus on the know-how for conflict resolution are often learned on the job.

In this section, I draw upon three influential works: *The Fearless Organization: Creating Psychological Safety in the Workplace for Learning, Innovation, and Growth* by Edmondson, A.C. (2018); *Crucial Conversations* by Grenny, J. et al. (2022); and *Mind Stuck* by McQueen, M. (2023).

In *The Fearless Organization*, Edmondson (2018) explains how her initial research led to the conclusion that when team members are free to express their opinions without the fear of repercussions, hidden issues that often derail progress are exposed and resolved. She coined the term *psychological safety* to describe environments where these expressions may be made freely. In her book, Edmondson provides a framework for creating psychological safety, the cultural foundation that allows individuals to speak up, learn, and innovate without fear.

In *Crucial Conversations*, Grenny et al. (2022) highlight that when strong emotions, along with high stakes and opposing opinions, are present, the individuals involved are in a crucial state of communication. Thus, if not managed appropriately, individuals fall into rigid patterns of thinking. By understanding the underlying factors and reframing, teams can identify solutions and unlock new possibilities.

In *MindStuck*, McQueen (2023) begins by explaining the neurobiology of fight-or-flight and how, by understanding this aspect, we can develop a better understanding of how emotions play a role in decision-making and ultimately negotiating solutions to our differences.

Emotional Intelligence and Shared Purpose

Conflicts may escalate when participants lose sight of their shared purpose, particularly when there is a lack of communication or miscommunication and bruised egos. Product-related discussions may easily devolve into *my way versus your way; I know what I am talking about;* or, *I have been doing this for a long time,* resulting in a contest of personalities. Emotional intelligence helps to redirect this energy, by giving the other side the benefit of the doubt and assuming good intent and to restate the common goal that has brought the individuals together.

This anchoring effect is particularly important in cross-functional teams where priorities naturally diverge. By framing the conflict in terms of collective responsibility, *our goal is to create the best possible product given the constraints that we have.* Through emotional intelligence and by proper anchoring, team members can once again see the true north and their differences as complementary rather than antagonistic.

A Psychologically Safe Environment

Edmondson (2018) defines psychological safety as the belief that one can speak up with questions, concerns, or dissenting views without fear of humiliation or punishment. For product development teams, this concept is paramount to success. Any member of the team must be able to point out potential flaws even when the project timeline is under executive pressure. Without such openness, critical information is lost, and risks remain hidden until they become failures.

Creating psychological safety requires deliberate leadership and consistent behavior from all team members. Edmondson (2018) refers to three major elements of psychological safety. The first is acknowledging fallibility, where leaders and senior team members admit that they might have missed something and invite others to raise gaps. This mindset signals that raising concerns is not only safe but expected. The second is showing appreciation when a dissenting opinion is raised. When someone surfaces a difficult issue, the first response should be gratitude, not defensiveness. Finally, the major element is to normalize and encourage questions. Teams that treat inquiry as a strength rather than a weakness develop resilience against groupthink.

Psychological safety provides the cultural ground upon which *crucial conversations* can unfold productively. Without it, even the best conversational techniques risk collapsing under fear and silence.

Safe Conversations

Once psychological safety is established, verbal jabs, microaggressions, and bruised egos become elements of the past. By having awareness of when conversations become crucial and employing proper techniques, team members move from judgment to evidence, and from defensiveness to curiosity. One such technique is to avoid confrontational conversations by using data and posing questions. For instance, instead of saying "This design won't work" (confrontational), say, "we have demonstrated 30% variability in performance. Is this acceptable?" Asking questions may also help to transform the interactions from arguments into joint problem-solving by seeking other views. For instance, one may ask: "What risks do you see?" Finally, in safe conversations, we need to strive for clarity of intent and prevent misunderstanding. For instance, we can always clarify by saying: "Please help me understand."

When handled this way, dialogue strengthens rather than weakens trust, even in moments of sharp disagreement.

Breaking Free from Stuck Thinking

Even when people feel safe to speak, teams may remain trapped in narrow thinking. As McQueen (2023) suggests, individuals insist on the first solution coming to them, frame choices in false black and white binaries, or defend old positions instead of generating alternatives.

The antidote to stuck thinking is reframing the point of conflict. Team members may raise the question, "How can we move forward while managing risk?" In doing so, the team begins to widen the problem space and consider other aspects of a solution. By shifting the team's energy from defending assumptions to testing them, it would help the team advance.

Conflict often persists because arguments alone rarely resolve technical arguments. Data, however, changes the dynamic. Analysis, testing, and/or direct customer feedback reduces the emotional stakes and makes the discussion objective. Evidence-based approaches protect relationships while advancing the project. Disagreement becomes less about winning and more about learning.

When the product development team integrates these lessons, conflict ceases to be a drain on energy and becomes a driver of creativity, learning, and progress. In the long run, it is not the absence of disagreement that distinguishes high-performing teams. It is the ability to transform disagreements into actions to move forward.

Summary

- IQ is no longer considered to be a predictor of success in life.
- Emotional intelligence (also known as EQ) is considered to be just as important, if not more important.
- Philosophers of ancient times and psychologists of modern times have been trying to understand human personalities and their types.
- The most widely known and used instruments to identify human personality type are based on Carl Jung's 16 personality types.
- Emotional intelligence and knowledge of personality types are necessary conditions for having an effective team. The sufficient condition is a combination of developing trust, resolving conflicts, making commitments, keeping team members accountable, and finally keeping an eye on the complete picture and getting results.
- When giving constructive or positive feedback, the focus should be on the behavior and not the person.

- When giving constructive or positive feedback, the focus should be on the future and not what happened in the past.
- When receiving constructive feedback, keep in mind that the feedback is an expression of how others perceive us.
- In conflict resolution, we need to ensure psychological safety and understand when conversations turn crucial and why individuals get stuck in their mindset

Exercises

1. Take a moment for self-reflection. Where do you stand on the emotional intelligence scale? Do emotions often get the better of you? Or are you in control of your emotional reactions?
2. Do an internet search and find a couple of free websites to conduct your own personality type. Take at least two different tests on two different days. How close are the results? Share the information with your colleagues.
3. Speak to a few (two or three) colleagues and ask them to provide feedback for you on a recent project that you may have worked on together. What can you learn from their feedback? Ask them if they are willing to hear your feedback for them. If so, provide feedback, but first, contemplate how they would be able to grow as a result of your feedback.
4. In light of the points made in the two sections on *Receiving and Giving Constructive Feedback*, examine each case study and discuss with your classmates.
5. Review some of your conversations from the past week. Identify the times when the conversation became emotionally charged. Recall the reasons. How could it have been managed differently?

Notes

1 The approaches and models that I mentioned in Chapter 1 and will discuss in this and future chapters are applicable to both tangible and intangible products (i.e., devices and services), but admittedly the bias of the text is on electromechanical/combination products.
2 It should be noted that the popular notion of personality may be different from a psychological definition. My purpose in introducing the concept of personality here is to discuss its influence on teamwork within a work/social context.
3 For a reference, see Hall et al. 1985.
4 The distinction between these two branches is rather technical and beyond the scope of this book.
5 I have no connections to these sites and have chosen them randomly.
6 Peter Drucker has written a number of books on business and management.
7 Different organizations may use different terms such as *technical lead* or *chief engineer/scientist* for essentially the same role.
8 Hearing we have done great is always easy and elating!
9 Recall the section on Personality Types and the conclusions drawn there.

3

Program and Project Management

Introduction

It is not hard to argue that corporations and businesses – be they large or small – can only continue to exist if they are profitable. A common understanding of profit is what is left of sales revenues when the product costs and other organizational expenses (such as employee salaries, overhead, etc.) are taken away. The marketing and sales organizations of any corporation strive to increase the sales volume and contribute to the profit growth of the organization. They are outward looking. In other words, their focus is on the outside of the organization and its customers and potential markets that may exist.

Another method of increasing profit levels is to either reduce the overall cost of the product or increase productivity. By "increasing productivity," I mean doing more with fewer resources and continually striving to reduce waste. Commonly, the organization that oversees productivity improvements is called the Program Management Office (PMO). A *program manager* is responsible for implementing executive management's strategies for bringing about increased performance and overseeing activities and/or projects that would enable and realize the desired improvements leading to increased profits.

Executive management's strategies (or programs) are implemented through and by means of a number of tangible activities called projects. Project managers are responsible for leading the projects to successful completion and are involved with day-to-day activities. Hence, one may consider *program management* to be the discipline of developing and implementing an organization's strategies through tactical implementation and *management of projects.*

As engineering and science students graduate and enter the workforce, they will interact with both program managers and project managers, and many will eventually become project leads (or technical leads), which requires a deep level of collaboration and cooperation with the PMO. Some may eventually join the ranks of program and project managers. For this reason, it is important to understand the responsibilities of program and project managers and the tools they use to manage their respective programs and projects. Interestingly, many of these tools are quite versatile and may be used within cross-functional or technical teams.

Program Management and Its Tools

As program management is involved in the strategic planning and future revenue development and growth of an organization, its main key concerns are to align proposed programs with this strategy and to allocate both financial and human resources appropriately.

DOI: 10.1201/9781003564508-4

Among the program management tools that are beneficial to scientists and engineers, the following may be enumerated:

1) Strengths, Weaknesses, Opportunities, Threats (SWOT)
2) Stakeholder Analysis Matrix
3) Risk Analysis and Profile
4) Cause and Effect Diagram

These tools are described next.

Strengths, Weaknesses, Opportunities, Threats

SWOT analysis is an activity to understand business and/or market opportunities for an organization along with the organization's enablers as well as its potential obstacles and risks. In this activity, business leaders first identify a business opportunity and then assess the strengths of their organization in regard to committing the required resources and capital to harvest the rewards of entering new markets or developing new programs. Along with the evaluation of the opportunity and the strength of the enterprise, business leaders evaluate the organization's weaknesses as well as external threats such as competition (Cooper et al. 2001). Although SWOT is mainly a marketing and program management tool, it may be easily used in project management as well as product development to access related concerns that leaders in these areas may have.

Opportunities

In this section, the identified or proposed opportunity for the enterprise is presented with as much detail as possible. The opportunity has to be related to either increasing revenues or reducing costs and waste. Alternatively, other types of enablers, such as developing footholds for future expansions and growth, may also be presented as opportunities. As an example of a future opportunity for growth, consider the relatively well-known concept of driverless cars. Many well-known organizations, such as Uber or Google, are investing heavily in this future technology.

An example of an opportunity for cost savings is to incorporate lean manufacturing practices in existing production. It is important that as an opportunity is described, an estimate for its impact be provided. For example, an opportunity statement may read as follows:

> A Strengths, Weaknesses, Opportunities and Threats (SWOT) analysis is an activity to understand business and/or market opportunities for an organization along with the organization's enablers as well as its potential obstacles and risks.

> By incorporating lean manufacturing practices, the business could realize 39% reduction in scrap leading to $50,000,000 in savings. A portion of the savings may be passed on to the customers enabling a 5% market expansion.

Strengths

Clearly, once an opportunity has been identified, the ability of the organization to take advantage and reap its benefits should be properly assessed. For instance, many recognize the opportunities that driverless cars may offer, but there are very few companies with the required expertise to even consider wagering in this arena. Fortunately, for the majority of businesses, growth opportunities are within reach, and the *strengths* evaluation helps business leaders focus on the available resources that may be dedicated and assigned to the activity. For example, a strength statement may read as follows:

> Five (5) lean manufacturing subject matter experts with ample experience are in the employment of the organization. Company is in a strong financial position to absorb associated expenses.

Weakness

In this segment, any perceived organizational weaknesses are outlined and addressed. The purpose of identifying these weaknesses is to evaluate any remediation or risk mitigation measures that may need to be implemented. Later in this chapter, I will review a common technique to evaluate program (or project) risk and a means of ranking each risk item. For example, a weakness statement may read as follows:

> The operators are rather seasoned and set in their ways. They may have difficulties in adjusting to the lean manufacturing methods and inadvertently block the success of the program.

Threats

Finally, in this segment, business leaders identify factors that may be threatening the success of the program. Some may allude to threats as the external forces and or stressors that may adversely impact the organizations should the proposed activity not be undertaken. For example, a threat statement may read as follows:

> Our main competitor is launching a product in six months. Their product will be positioned to cannibalize our product and its market share.

Typically, the SWOT is presented in a four-blocker format as shown in Figure 3.1. I will provide an example of how to utilize SWOT in product development in Chapter 8.

Stakeholder Analysis Matrix

Once a program has been identified and its merits evaluated using a SWOT analysis, the next steps of evaluation must be undertaken. One of these steps is an assessment of the people who are either impacted by the outcome of the program or may exert their influence. Called *stakeholders*, they may be members of upper management, or heads of various departments; even external to the organization, major suppliers or customers may be included in this analysis.

The impact or influence of these stakeholders may be either positive or negative; and hence, they may be in a position to support the program, be neutral, or even, in some cases, oppose the proposed initiative. Thus, at the onset of the program, it is important to understand the stakeholders and their mindsets.

Elements of Stakeholder Analysis Matrix

Since stakeholders backing or opposition may have a detrimental impact, their attitude toward the program as well as their relative power to support or derail the proposal needs

Opportunity

By incorporating lean manufacturing practices, the business could realize a 39% reduction in scrap, leading to $50,000,000 in savings. A portion of the savings may be passed on to the customers, enabling a 5% market expansion.

Strength

Five (5) lean manufacturing subject matter experts with ample experience are in the employment of the organization. The company is in a strong financial position to absorb associated expenses.

Weakness

The operators are rather seasoned and set in their ways. They may have difficulties in adjusting to the lean manufacturing methods and inadvertently block the success of the program.

Threats

Our main competitor is launching a product in six months. Their product will be positioned to cannibalize our product and its market share.

FIGURE 3.1
An example of a SWOT in a four-blocker format

TABLE 3.1
An Example of a Stakeholder's Analysis Matrix

Stakeholder	Symbol	Influence/Power	Support	Communication/Action Plan
Stakeholder 1	♦	4.0	1.0	
Stakeholder 2	Δ	2.0	9.0	
Stakeholder 3	▲	8.0	7.0	
Stakeholder 4	●	4.0	2.0	
Stakeholder 5	O	3.0	7.0	
Stakeholder 6	□	6.0	6.0	
Stakeholder 7	■	7.0	3.0	

to be understood. First, their relative influence (or power) as well as their level of support may be summarized. Next, the program manager should take the time and effort to understand if there are any conflicting concerns between various stakeholders and make any efforts to resolve any disagreements that may exist or overcome any objections and resistance. Table 3.1 is an example of such a matrix. Additionally, this information may be presented in a graphical format – again, in a four-blocker format. An example of this presentation is shown in Figure 3.2.

In this four-blocker format, four quadrants are identified. There are stakeholders with low support and low power, low support and high power, high support and low power, and finally, high support and high power.

It goes without saying that the key stakeholders are those with high power and high support who play a major role in the success of the program. These individuals should be

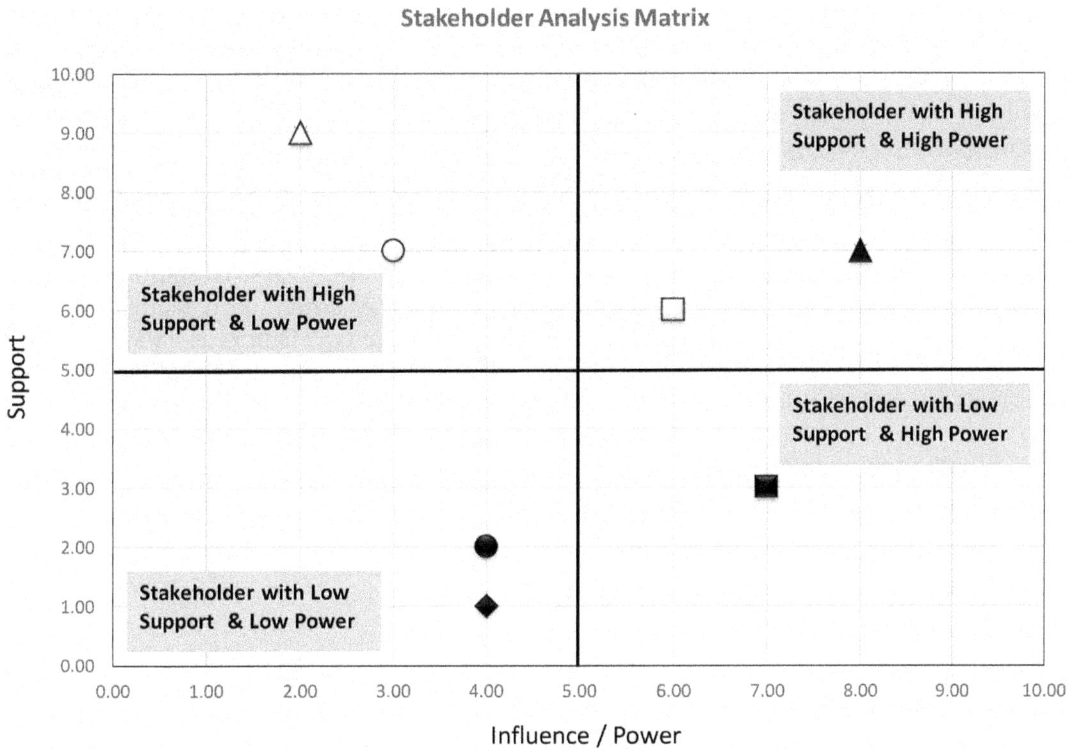

FIGURE 3.2
A graphical view of the stakeholder analysis matrix

kept in the communication loop, and their input should be sought. The opposite quadrant belongs to a group of individuals who have both relatively low influence and a low degree of support or interest in the proposed program. Even though this group should not be ignore; they require a relatively low degree of communication and involvement. The individuals in the quadrant of high support and low power only require to be informed of the progress of the program. By being in the know, they may be able to provide their expertise in crucial stages of the program and help solve problems and remove roadblocks. Finally, the opposite and last quadrant includes the individuals who have a low degree of interest and support in the program but with a high degree of power and influence. These stakeholders should be managed appropriately, and their concerns properly vetted and acted upon. Additionally, any progress and gains should be properly communicated to them. It is possible to influence and improve the opinion of these stakeholders once they learn of the progress made.

> Risk analysis and risk management refer to a set of activities that would anticipate unexpected events and estimate their relative ability to derail the program along with their probability of occurrence.

Risk Analysis, Profile, and Management

Undertaking any new program involves facing and dealing with elements and factors that are unknown at the start. Risk analysis and risk management refer to a set of activities that would anticipate unexpected events and estimate their relative ability to derail the program along with their probability of occurrence. Risk is the product of the probability of occurrence of an event multiplied by the severity of the event's effect. It is possible that an element within a program may have a low impact (severity), but it may occur so frequently that the program may be adversely affected. For instance, consider the common cold or flu. If one or two employees are sick and stay at home, an organization's productivity may not be impacted; however, should half of the organization stay home due to sickness, production schedules or customer service may be greatly impaired.

Thus, in evaluating program risks, two factors need to be considered simultaneously; namely, severity (or impact) and likelihood of occurrence. Often, each factor is scored between 1 (for low) and 5 (for high). The product of these two factors is called the *Risk Priority Number* (RPN); the higher the RPN, the higher the risk. In calculating the RPN, five questions need to be entertained:

1. What is an adverse event?
2. What is its likelihood of occurrence?
3. What is its impact on the program?
4. What is the associated score?, and finally,
5. How should the RPN be used?

Risk Response Planning

Figure 3.3 provides a guideline on how to determine the likelihood of occurrence as well as impact. In general, likelihood of occurrence is categorized into one of *Remote, Not Likely,*

Risk Response Planning Guidelines

Likelihood of Occurance	Nearly Certain (> 90%)	5	10	15	20	25		
	Likely (50-90%)	4	8	12	16	20		
	Probable (30-50%)	3	6	9	12	15		
	Not likely (< 30%)	2	4	6	8	10		
	Remote (< 10%)	1	2	3	4	5		
		< 2 weeks	2-4 weeks	1-2 months	2-3 months	> 3 months	Time / Schedule	Risk Category
		< 2%	2-3%	3-5%	5-10%	> 10%	Budget	
		Acceptable impact	Minor impact	Partially limited	Seriously limited	Compromised	Scope / Quality	
		Negligible	**Minor**	**Serious**	**Critical**	**Catastrophic**		
		Impact						

FIGURE 3.3
Risk response planning guidelines, depicting three different demarcations (red, yellow, and green [in the figure, red, yellow, and green are shown as black, gray, and white])[1] associated with three different risk levels

Probable, Likely, and *Nearly Certain*. Typically, impact is categorized into one of *Negligible, Minor, Serious, Critical,* and *Catastrophic*. However, to provide some degree of clarity, Figure 3.3 defines three different risk categories for a program. These are impact to time (or schedule), impact to the budget, and impact to either program scope (or quality).

Most often the level for each of the categories is agreed upon a priori. For instance, it is agreed that a *Probable* likelihood of an occurrence means that the event may occur 30% to 50% of the time. Or, that a *Remote* likelihood of an occurrence means that the event may occur less than 10% of the time. In this case, Remote is given a score of 1, whereas Probable is given a score of 3. Similarly, depending on the risk category, impact is scored from Negligible (1) to Catastrophic (5).

The next step is for the team to decide what the RPN values mean. Typically, RPN values for three levels of green, yellow, and red are defined. In Figure 3.3, red (shown as black) is identified with RPN values equal to or greater than 12. Values between 4 and 10 are marked as Yellow and finally, RPN values equal to or less than 3 are deemed as green. The red, yellow, and green demarcation is used to identify the risk level and the type of actions needed to deal with the identified risk.

A *red* risk level indicates that the identified risk is high and the program may be derailed as a result. Should a risk be scored as "Red," its action plan should be to either mitigate, avoid, or transfer the risk. I will explain these terms shortly.

A risk scored as *yellow* indicates that enough information may not be known about either its impact or occurrence that would place it in either a red or green categories. The implication is that the risk owner (e.g., program manager) may need to conduct additional investigations to assign a risk action. As a result, the risk actions (mitigate, avoid, transfer, or accept) may apply.

A risk scored as *green* implies that either no other actions are needed (i.e., accept) or in rare circumstances, it may be "transferred" to an external party.

Now to define various risk action terminologies. To *Mitigate* a risk means the team should take action to either eliminate the risk, to reduce its probability of occurrence significantly (proactive mitigation), or reduce its impact (reactive mitigation).

To *Avoid* a risk requires that the risk be removed altogether. Depending on whether risk analysis is conducted at program, project, or task levels, implications of "Avoid" may vary. For instance, a proposed project within a program may be considered risky and is eliminated. At a smaller scale – say a product design – failure of a design feature may be deemed as risky; if risks cannot be mitigated, the team may decide to eliminate the feature and hence avoid the risk. In a global manufacturing environment, different regions may be subject to natural disasters or political unrest. In these cases, risk avoidance would mean placing manufacturing in more stable regions.

To *Transfer* risk implies that another entity will be held responsible in case the identified risk materializes. Typically, it involves an arrangement with another organization to make them responsible (or liable) for any adverse outcome. In program development, adverse outcomes may be financial losses (such as fines) associated with delays or expense/cost overruns, etc. Through transferring the risk to a different organization, the responsible party manages the risk without any impacts to the program's health.

To *Accept* a risk implies that the impact of the associated risks, should they occur, is low enough that no specific actions will be needed.

Risk Tracking

Now that the elements of program risk management have been described, the next step is to develop a table and define various program risks. The columns of this table include

TABLE 3.2

An Example of a Risk Tracking Matrix

Risk#	Risk Description	Risk Category	Likelihood 1 (low)–5 (high)	Impact 1 (low)–5 (high)	RPN	Overall Risk	Descriptions of Potential Causes	Tasks to Accept, Avoid, Transfer or Mitigate Risk
1	Identified Risk 1	Time	3	1	3	Green	Causes leading to identified risk	Accept - no other actions needed
2	Identified Risk 2	Budget	4	2	8	Yellow	Causes leading to identified risk	Tasks to avoid, transfer, or mitigate Risk 2
3	Identified Risk 3	Scope	3	3	9	Yellow	Causes leading to identified risk	Tasks to avoid, transfer, or mitigate Risk 3
4	Identified Risk 4	Scope	1	3	3	Green	Causes leading to identified risk	Accept – no other actions needed
5	Identified Risk 5	Time	4	3	12J	Red	Causes leading to identified risk	Tasks to avoid, transfer, or mitigate Risk 5

a description of the risk, its category, followed by the likelihood of occurrence along with the risk impact. Then, a column for the calculated RPN value followed by the associated overall risk. The last column typically contains potential causes along with the recommended risk actions.

Often, other program-specific columns are added to address specific needs. For instance, some add a "Status" column in order to track any risk action activities. Once a risk is retired, it may be marked as "Retired" in the status column, or the line may be removed altogether. An example of risk tracking is shown in Table 3.2.

Cause and Effect Analysis

The two right columns in Table 3.2 speak to causes leading to identified risks and the decision to accept, avoid, transfer, or mitigate them. With experience, one may be able to identify the causes and actions needed to retire them; however, this approach is good for the "tried and true" programs. It is not a methodic. A systematic approach would be to use a Six-Sigma tool called Cause and Effect (otherwise known as "Fishbone" or "Ishikawa") analysis. This tool may be applied not only in program and project management but also in design, product development, manufacturing, even in sales and marketing.

The premise of this tool is that first, the effect or the problem (here identified as risk) should be defined and placed at the head of the diagram. Next, relevant cause categories are selected. For instance, should the problem be within technical design or manufacturing, cause categories are typically identified as manpower (or people), machine, method, material, measurement, and mother nature (or environment).[2] In a marketing environment,

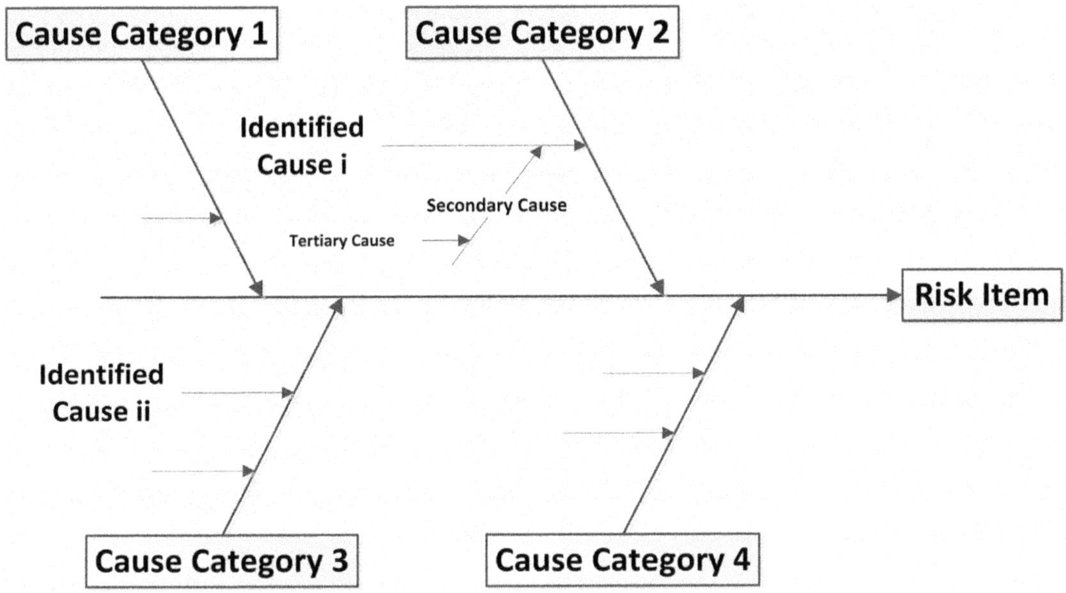

FIGURE 3.4
A typical fishbone or cause and effect diagram

the categories may be people (or personnel), product (or service), place, promotion, process, and price.[3] Relative to this chapter and topic, in program management, cause categories may be external stakeholders, internal stakeholders, project scope, corporate structure, management process, and management team.

Once this basic structure is defined, brainstorming sessions need to be held to identify potential causes and place them in each cause category. It may even be possible to identify deeper-level causes as well. Armed with this information, the Cause column in Table 3.2 may be filled out with the outcome and data produced through a detailed analysis. Similarly, associated actions may be identified without any "guess" work. Figure 3.4 is an example of a generic Fishbone diagram.

Project Management

In the previous segment, I reviewed the role of program management and explained how this discipline seeks to set strategies for moving an enterprise into the future and enabling its growth and success. I also explained that an organization's strategy is realized through a set of tactical activities called projects; or, as it was suggested earlier, programs are a collection of related projects (PMI[4] 2017).

Considering that projects are the means by which an organization drives its strategic objectives forward, they should have defined start and end dates along with a clear set of goals and objectives as well as an agreed cost. Often, the relationship between these variables is expressed as the triangular relationship between *time*, *quality* (or *performance*), and *cost*. A lack of clarity in any of them may lead to delays and cost overruns at best and failures and disappointments at worst. In the language of project management, quality is defined and measured by what are called project *requirements* – or *specifications*; time is defined by the project *schedule*; and finally, cost is defined by the project *budget*.

These three factors are interdependent and may not be set independently. As an example, consider making a homemade apple pie from scratch. The steps are: first, to purchase the ingredients such as apples, butter, flour, etc.; second, to peel and slice the apples; third, to make the pie crust dough using flour and butter; fourth, to preheat the oven; fifth, to mix sugar, cinnamon, flour, and butter and coat the apple slices with the mixture; sixth, to roll out the crust dough to fit the pie pan bottom; seventh, to pour the apple mixture into the dough-lined pie pan; eighth, to roll out another crust dough to cover the top; ninth, to seal the edges and cut steam vents in the top crust; and finally the tenth step is to bake the pie. These are the steps needed to bake a homemade pie from scratch. Each step takes a particular length of time to complete, leading to the total time to bake a pie from scratch. The cost of the pie includes not only the cost of the ingredients but also the overhead associated with paying the cook as well as maintaining a kitchen.

> Program management is the discipline of setting strategies for moving an enterprise into the future and enabling its growth and success. These strategies are realized through a set of tactical activities called projects. In other words, programs are a collection of related projects

Now, suppose that it is desired to reduce the time that it takes to bake a pie. There are two ways to accomplish this. One way is to hire another cook so some tasks are done in parallel. For instance, while one is peeling and slicing the apples, the other would make and roll the dough. Hiring another cook will reduce the time but will increase the cost. Another approach is to purchase prepared ingredients. For example, canned apples and prepared dough may be purchased, so the cook would be required to fit the dough into the pan, pour the ingredients, and bake it. This solution may save time and keep the cost the same but it will impact the quality. The pie may no longer be called *homemade from scratch*. This is an example of how the three variables of quality, time, and cost are interrelated.

As Rosenau and Githens (2005) suggested, it is the role of the project manager to point out the trade-offs between these three variables. To either understand the trade-offs better or to explain them sufficiently, project risks must be understood and their impact on the program be evaluated.

Project Life Cycle

As I mentioned earlier, a project should have clear start and end dates. In the language of project management, these are two stages of the project life cycle – though expressed by different terminologies. The basic stages of a project life cycle – as identified by Project Management Institute (PMI) (2017) are

1) Initiation
2) Planning
3) Execution, and
4) Closure or Completion

During each of these stages, the project manager is expected to balance conflicting requirements and constraints. These constraints may be placed in broad categories of scope, schedule, cost, quality, human and capital resources, communications, risk, etc. (Badiru 2012). As I explain each of these four stages briefly, I will speak to each of these constraints as appropriate.

Project Initiation

At the start of any project, the role of the PM is to develop a proposal that would outline what the project aims to achieve and how, who and what the needed resources are, and the associated cost. And, finally, how long the project will take. Additionally, the approach as well as project risks are identified in the proposal. The project proposal should also determine which aspect of the triple constraints is significant and drives the project. Would the project be time- or cost-driven? Or, are specifications and quality more important than the other two?

While a comprehensive review of project initiation is beyond the scope of this present volume, I would like to focus our attention on two facets of project initiation: first, project roles, and then, project approach. The first deals with human interaction and team dynamics and is related to the topics that were discussed in Chapter 2. The project approach for product development is introduced in Chapter 6 and beyond.

Project Roles

There are five critical roles within the project management structure. The obvious one is the role of the *project manager*; the others are *sponsor, steering committee, team leader,* and finally *team members* (HBR Staff 2016).

> The basic stages of a project life cycle are Initiation, Planning, Execution, and Closure or Completion.

To a large extent, this chapter is about the role of a project manager, so in this section, I will only speak about the other roles.

A *sponsor* is typically a member of an organization's executive management who advocates for the project and has the authority to not only provide leadership and direction to the project but also the ability to remove any potential obstacles and roadblocks. Obstacles generally manifest themselves in terms of either a lack of human or financial resources.

The *steering committee*'s role within project management is to provide overview and leadership for the project. At times, it may be called a *core team*. The responsibility of this team is to provide oversight for the project and to deal with conflicts and risks that often arise as the project progresses. The project sponsor, project manager, and team leader form the basis of the core team; however, as the need arises, it may include members of other teams, such as marketing.

The position and role of a *team leader* is fundamental to the making or breaking of a project. As a general rule of thumb, the person who fills this role should be very seasoned in the category of the projects that have been undertaken and have served as a team member on several successful projects in the past. Furthermore, this role requires a high degree of emotional intelligence in order to understand people's mindsets. A team leader generally sets the technical direction and is responsible for technical requirements. In this regard, while he (or she) is a working member of the team, he or she coaches other team members to maximize their potentials.

Finally, without having *team members*, nothing gets done. Generally speaking, there are three criteria for a strong team member. These are technical, interpersonal, and organizational skills. Quite often, people are selected into a team on the basis of their technical skills; however, shortcomings in the other two areas often lead to either delays or personality conflicts. Just imagine the damage that may be done if a team member does not have the acumen to communicate problems and issues at all, particularly issues that may impact the output of other team members. The outcome is often chaos, disarray, and conflicts.

Project Approach

Another aspect of project management is to determine the approach to be taken to complete the project. While an implicit approach may be to develop the work needed in-house, it may also be possible to have an external organization carry the tasks forward. For instance, suppose that a company decides to develop and market a new product line. One approach is to design and manufacture the product line using internal resources. Another approach is to hire an external company to design, develop, and manufacture the product line. Yet, a third approach may be to buy an existing company with a product line similar to the desired one.

Once the decision is made to run the project in-house, the project manager attempts to optimize the triple constraints of time, cost, and quality by selecting a fitting (or appropriate) project life cycle model to follow in the course of the project. Collectively, these models may be placed in three broad categories; namely, *single-version, incremental,* and finally, *iterative.* In the single-version category, the end-product is released to the customer in its entirety; whereas the incremental and iterative categories require that the end-product be released to the customer gradually – each release provides either an updated or redesigned version of the product. A well-known example of the incremental and iterative models is the cell phone industry. New upgrades (incremental changes) are released nearly every twelve to eighteen months, and every three to five years a new design (a new iteration) is introduced to the market.

Considering that the focus of this text is on product design and development, I would like to discuss the single-version category further.[5] Within this category, three models have been defined, namely, the *big-bang* model, the *waterfall* model, and the *V-model.* The big-bang model is based on the assumption that the development team receives the instructions from management to develop a particular product. The team would then engage in the development activities and deliver the product without engaging external resources or seeking the customer's feedback during the development process. The end results may or may not be what the customer has asked for. On a personal note, I have never seen this model at work in my professional life.

Once one stage is complete, the results "pour" into the next segment as in a waterfall.

In the waterfall model, as shown in Figure 3.5, there are distinct stages of the project starting with *requirements* – specifics of what the project is aiming to achieve in measurable terms. The team advances to the *concept* development only if requirements are fully defined.[6] This progression is made until the project is completed and delivered. While this approach may be used in well developed (and understood) projects – such as, say construction – it is hardly used in product development.

Testing and verification are conducted to ensure that the end product meets customer expectations.

A more refined approach is called the V-model as shown in Figure 3.6. Here is the project requirements are defined, first, at the system or the highest level; then high-level concepts are developed to achieve them. Next, the high-level concepts are segmented into smaller and more manageable sections. Then, the requirements for the smaller sections are developed. This process continues to the smallest components that would enable designing the

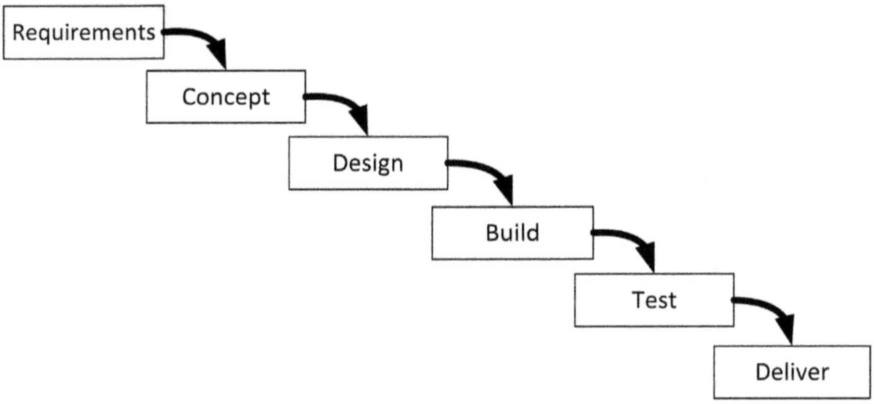

FIGURE 3.5
Waterfall model

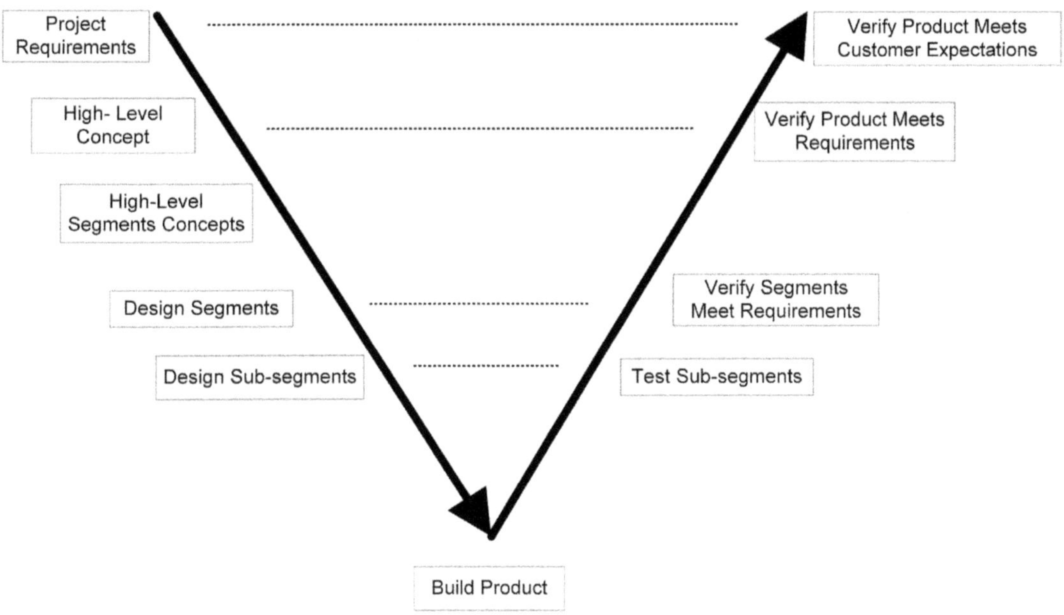

FIGURE 3.6
The V-model

conceptual building blocks. The feature of the V-model is that as the building blocks are developed and built, they are tested to make sure that they conform to their requirements. At this stage adjustments may be made. Then smaller building blocks are assembled into larger blocks – and again tested. This iterative process continues until the entire end result is completed and tested as per the customer's needs. The V-block is not only a project model but also a product development model and the focus of this text.

A methodology that, in a way, is representative of the incremental or iterative project management approaches is called *Agile*. Agile is gaining attention for responding quickly to market needs and for being lean. It is worthwhile to review this method here as well.

> There are three broad categories of project management models: single-version, incremental, and iterative. In the single-version category, the end product is released to the customer in its entirety. The incremental and iterative categories require that the end product be released to the customer gradually – each release provides either an updated or redesigned version of the product.

Similar to the V-model, Agile was first developed within the software development industry in the 1970s (Alexander 2017). In 2001, a group of seventeen individuals issued a manifesto for Agile software development along with the twelve principles of Agile Software (http://agilemanifesto.org/). The basis of this manifesto is to give priority to people, their collaboration, and their communication over project plans and business processes. In a way, the authority and decision making authority are given to the teams who are actually responsible for deliverables.

In order to appreciate the advantages of both the V-model and the Agile methodologies, we need to keep in mind that most established product development models follow a "Waterfall" project management model with clearly defined levels or gates. At the end of each gate, extensive project reviews are conducted that may take a span of several days. Once a stage gate is crossed and its deliverable completed, it is difficult, if not impossible, to go back and make any changes. However, in an Agile (and V-model) environment, focus on the stage gates is removed. Instead, by providing the decision-making authority to smaller development teams, flexibility is infused into the development cycle in order to deliver a better product. These smaller teams have the ability to meet more frequently with less "procedural" overhead to discuss issues, review, and solve problems. Should the need be, they can go back up the "V" and modify requirements.

In literature, Agile is often contrasted with the Waterfall model to show its advantages. As I have alluded, the Agile methodology is much more aligned with the V-model. As we will see in the rest of this volume, the V-model allows for iteration between consecutive steps. Furthermore, it emphasizes testing of components and assemblies to ensure that requirements have been met and to make changes to requirements should errors or oversights be discovered.

Project Planning

Earlier, I mentioned that the role of a project manager is to balance conflicting requirements and constraints during each stage of the project's life cycle. Once the project is approved and initiated, the project manager develops an *integrated project plan*, which, in a way, is a proposal of how these requirements and constraints will be managed. As shown in Figure 3.7, an integrated project plan includes a summary of the activities associated with each of its elements, such as scope, schedule, cost, quality, human and capital resources, communications, risk, change control, etc. Needless to say, each element of the Integrated Project Plan, will have its own plan of activities.

Of these elements, the only plan that impacts design and development teams directly – and, hence, within the scope of this text – is developing the project schedule. Other elements are in the core discipline of project management and will not be discussed here. Arguably, change control management may also impact design and development teams as well. I will discuss this topic in Chapter 21 as it is related to managing design changes.

FIGURE 3.7
An illustration of the elements of an integrated project plan; other element (not shown) may be stakeholder management plan

Scheduling

> Work breakdown structure is the identification of sequential and parallel tasks that are need to be completed and their relative priority so that the project may come to a successful conclusion.

Developing the schedule for any project begins with understanding the *work breakdown structure;* that is, identifying various tasks that need to be completed and come together so that the project may come to a successful conclusion. The work breakdown structure is best visualized graphically. By developing a graphical view of the work to be done, the interdependencies of the tasks can be clearly identified. Additionally, the duration of each task may be assigned, and finally, tasks that may be conducted in parallel, as opposed to sequentially, are identified.

Once tasks and durations are identified, human resources are assigned to each task. At this stage of project planning, the responsibility of the project manager is to ensure that resources are not over-allocated. For complex projects, commercial software tools are used for this purpose. A common tool for managing project schedules is Microsoft Project software.

Figure 3.8 is an example of a work breakdown structure for the Bake Apple Pie project mentioned earlier in this chapter. The first activity in this project is to buy all the ingredients. Once this task is completed, then the four tasks of peeling apples, mixing sugar and ingredients, making crust dough, and preheating oven may follow. Note that in this

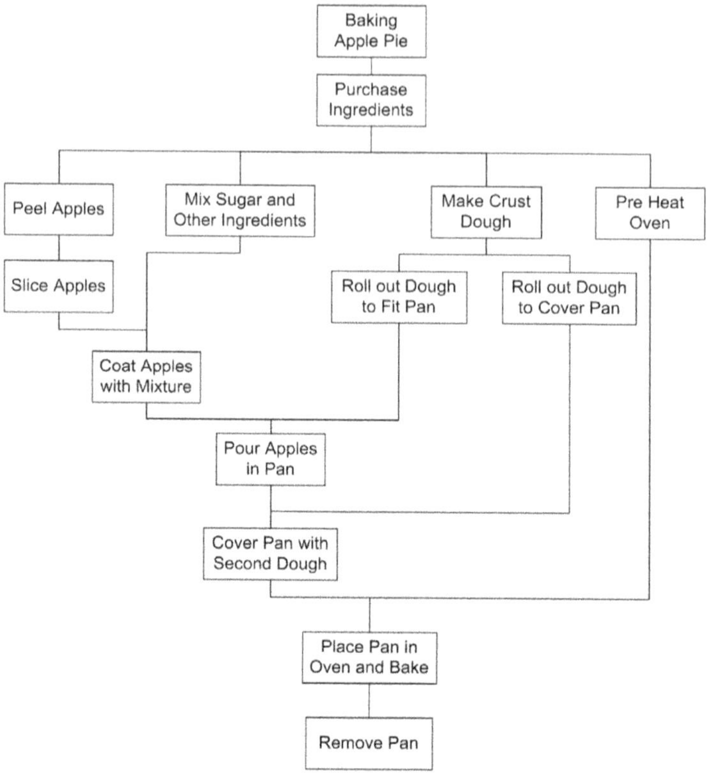

FIGURE 3.8
Work break down structure example

breakdown, it is assumed that these tasks are parallel. The implication is that four different resources will be assigned to each. Should only one cook be available, then the tasks will be done sequentially,[7] though some tasks still need to take precedence over others.

Work breakdown structure shown in Figure 3.8 also provides a view of task dependencies. For example, it is clear that apples may not be coated with sugar and other spices if the either mixture is not ready or the apples are not sliced. Once task interdependencies are clearly identified, the final activity is to assign duration to each task. This should be done in collaboration with subject matter experts who have prior experience with similar tasks. Once task durations are agreed upon, a Gantt chart may be developed. An example of the Gantt chart developed for the Bake Apple Pie project example is shown in Figure 3.9.

Project Execution

Once the plan is developed, resources are identified, and tasks are assigned, the project may begin in earnest. However, there are unforeseen issues and concerns that arise that may impact project outcomes. As much as possible, these issues are anticipated and accounted for in a risk plan. However, it is prudent to review and report the progress of various tasks as measured against the expected to ensure objectives of the project are met as defined in the plan. This activity is called monitoring and control.

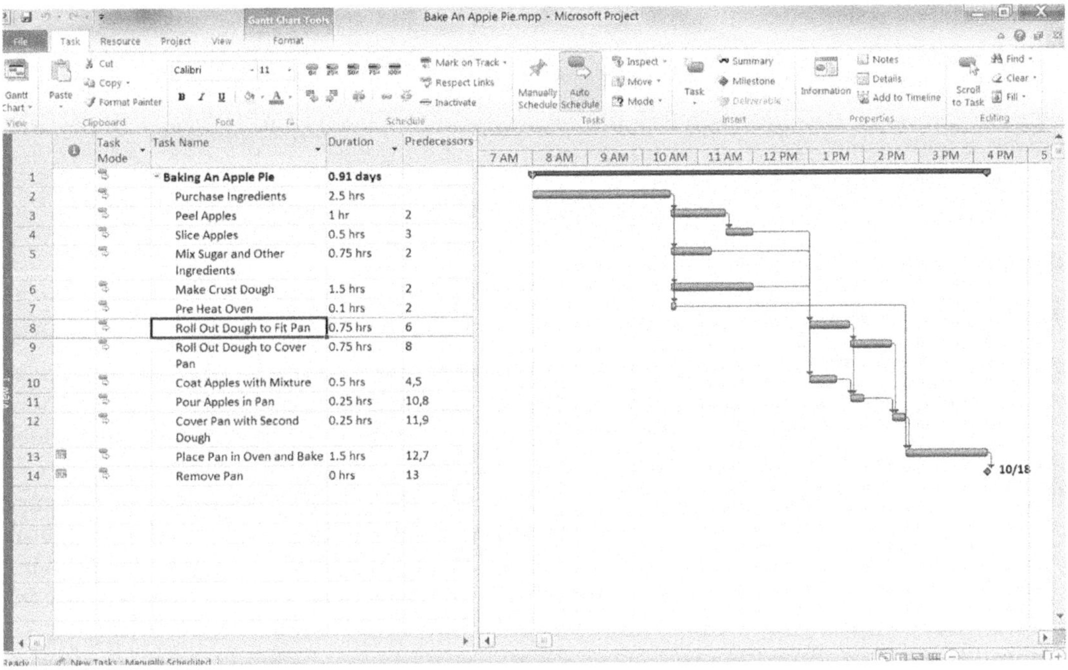

FIGURE 3.9
Project schedule example (using Microsoft project software) showing the Gantt chart for the Apple Pie Bake project

Monitoring and Control

I recall having an interview early in my career when the interviewer asked me the following question: "Suppose you were managing a group of engineers. What would you say if one of your engineers would report an issue to you first thing on a Monday morning?"

I had no idea! Seeing my ignorance, he responded, "The question should be: *do we meet the schedule*?" Then, he continued by saying, "If your engineer suggested that the schedule is not impacted, what would be the follow-up question?" Again, I was dumbfounded. He answered his own question: "Are we on budget?" He was a very gracious interviewer and finished interviewing me with many other technical and nontechnical question – many of which I answered confidently and comfortably, but needless to say, I did demonstrate to him that I knew nothing of project management at the time.

One may recall that budget and time form two sides of the triple constraint that was mentioned earlier; thus, it is natural for a project manager to also inquire about the quality of the work being performed. There are other questions that a project manager should constantly ask. Among these, a project manager should be diligent about project risks, unauthorized changes, and team conflicts and/or issues.

Project Completion

Once assigned tasks are completed and the objectives of the project are met, there are still lingering activities that need to be finished before the project may be formally closed. The first and foremost activity is transferring the project team's knowledge to other stakeholders. I will discuss this transfer of knowledge in Chapter 17. This is done by providing documentation and training,[8] obtaining final approvals, and closing accounting activities. This is followed by an evaluation of lessons learned. Finally, there is a celebration and recognition of the work that the team has done and those who have been particularly star team members.

Traditional Project Management versus Agile

What I have described so far may be referred to as traditional or classical project management. Often, Agile is thought of as the approach that will shorten product development cycle time. Yet in practice, many organizations are puzzled when, in practice, Agile does nothing to reduce timelines. I will discuss this issue later in Chapter 6 and in Appendix A. For now, I would like say that the principles of Agile are based on understanding customers and the ability to respond to their changing needs. Additionally, it was based on teams consisting of individuals who were cross-trained and – in a way – could do each other's work as well as being self-directing. Within software development – where Agile was conceived – this structure may be easily constructed. However, it will be more challenging – if not impossible – to apply the same approach and mindset to developing complex electromechanical combination products. In my experience, the traditional project management tool sets – such as work breakdown structure or task scheduling – will be needed even though an Agile framework may be employed.

The root of any Agile failure may be found in an incomplete understanding of the Agile mindset as well as the corporate culture needed for it to flourish. Agile thrives in environments where focus is placed on people and team members and not merely plans and processes. I have discussed this topic in Appendix A.

Summary

- Program Management is involved in the strategic planning and future revenue development and growth of the organization.
- A SWOT analysis is an activity to understand business and/or market opportunities for an organization.
- The people who are either impacted by the outcome of the program or may exert their influence are called *stakeholders.*
- Risk analysis refers to a set of activities that would anticipate the unknowns and estimate their relative ability to detail a program and its probability of occurrence.
- A program is typically a grouping of several projects.
- A project's triple constraint is the balance between *quality* (or *performance*), *time* and *cost.*
- The basic stages of a project life cycle are initiation, planning, execution, and closure or completion.

Exercises

1. Consider organizing a class trip to a local manufacturing site. This is a project. How would you apply the principles of project management to this trip.
2. Develop a proposal for this trip. Who are your stakeholders? Whose approvals do you need?
3. Develop a work breakdown structure to make this trip happen. Who do you need to help you? Would you be able to estimate durations for each segment? Would you be able to develop a Gantt chart?
4. How would you estimate the cost of this class trip? Who would sponsor the trip?
5. Try to make the trip happen. How would you monitor the progress?
6. Once the trip happens, consider having a lessons learned session with your teammates. What would you do differently?

Notes

1 In Figure 3.3, red, yellow, and green are shown as black, grey, and while.
2 To simplify, this is known as 6Ms.
3 This is known as 6Ps.
4 The Project Management Institute (PMI) maintains and publishes the body of knowledge for project management.
5 If I were to rename the single version to initial version; then, one could argue that the life cycle model may be considered as a continuum and not truly separate from one another.
6 In a so-called waterfall with backflow model, if discrepancies or errors are found in one stage, it is possible to go back one stage to make the needed corrections.
7 Not shown here, and not very exciting either!
8 In the product development terminology, this is called design transfer.

4

Technical Writing and Communication Skills

Introduction

I recall a couple of standard jokes among engineering students from my college days:

> *Who sayz enginers cant spel or right.*
> or
> *I am an engineer, I can't spell.*

I am not sure if these so-called jokes are still popular on college campuses, but, I do know that effective communication is in fact a problem that many team members experience at work, particularly those who are recent graduates. In this chapter, I would like to share some thoughts on effective communication skills in the workplace based on my own experience – though I am sure marketplace is quite rich with literature on this subject.

In today's workplace, with ambiguous corporate procedures followed by an abundant number of meetings and ever-changing assignments, it is not too difficult for various team members to forget details of decisions that have been made. As a result, I am of the opinion that if a thought, idea, discussion, or decision has not been documented, it did not happen.

Communication, by which I mean clear and concise, is not only fundamental to any documents that we write but also to any verbal presentations that we may hold, be it in presentation halls, conference rooms, or hallway conversations with a colleague or with a supervisor.

Features of Technical Communications

Language

> At the workplace, the language for communication should be kept at a simple level and free of technical jargon and flowery word.

There are certain features that distinguish technical communications at work from other types of communications. The first of these characteristics is language. For those who have a good command of language and literature, there is a tendency to develop a literary document with a flowery use of words. For those who have advanced technical backgrounds, there is a tendency to be overly technical with the text peppered with technical jargons – as if a presentation is being given to a group of like-minded colleagues. It is important to keep the audience in mind when writing technical communications in the workplace. Particularly when the audience is upper management. Typically, documents developed for upper management should be kept to a sixth-grade level. This is no insult to their intelligence but an indication that they are more interested in the facts and not particularly details and explanations; however, the details should be available in such a fashion that they may be easily found, read, and understood.

DOI: 10.1201/9781003564508-5

Purpose

The target audience needs to know the purpose of what is being communicated. Is the document or conversation about a new issue or problem, providing a status update, or reporting a solution and/or its implementation? Furthermore, by the time the reader (audience) comes to the end of the communication, he or she should have a clear understanding of the conclusions drawn by the author. For instance:

> *The root cause of low production yield on Line 3 is contributed to an intermittent failure of a spray nozzle. This line needs to be brought down for a period of three hours on Saturday (given date). Line 1, 2 and 3 managers should make appropriate arrangements so that the work completed by line 3 during this time period is properly diverted to lines 1 and 2.*

Or, we are communicating that a project has been completed:

> *It is my pleasure to let the team know that Project xyz was successfully completed today and all approvals are in place. I like to express my appreciation to the entire team for their dedication but particularly to John Doe who developed the mechanical design, to Jane Frank who lead the electrical problem resolution and to Randy and his team – Jack and Jill – for their IT support.*

Structure

There are two basic formats for communication, namely, written and oral. Typically, oral communicators use a form of visual aid called a slide deck, though slide decks are frequently shared electronically as if they are a form of written communication. Written communications typically take the following forms:

- Memos – These are typically written to provide status updates and are done via email. They are usually communicated internally within the team or externally with suppliers or customers.
- Letters – These are more official documents that represent a corporation's view on a special topic or issue. It is important these documents are handled with care and often cleared by the legal counsel or an officer of the corporation.
- Technical Reports – These documents are generally the output of scientific and engineering work, whether solving a problem or designing a new product. The most important aspect of a technical report is that it is readable and tells a story. It defines who the players are and their roles. It uses the principle that less is more. It provides enough details to relate the point but does not adore them to the point that the storyline is lost. A balanced combination of text and graphics is used to communicate the pertinent aspects of the story. In place of having tables of large amounts of data, it is more appropriate to use graphs, which represent data and its variations better than groups of numbers, leading to easy communication of conclusions. Should the point of the report focus on *instructions* and *process flow,* consider that a large proportion of people communicate through visual cues. Instructions and process descriptions are better suited to a combination of text and graphics. This also brings the old adage that "A picture is worth a thousand words."

Often, prior to publishing a technical report, it is customary to provide a presentation and an overview of the problem solved and its results to various team members and stakeholders. This gives an opportunity to the author to solicit feedback and answer questions or to clarify any concerns that reviewers and approvers may have. At other times, in the course of completing a project, an issue arises that requires attention and possibly some quick actions. In both cases, ideas, findings, and concerns need to be properly and clearly communicated to various stakeholders. There are two tools for this purpose that I would like to introduce and elaborate on briefly. The first is called a four-blocker and the second is called the "A3 Method."

Communication is either written or oral. Written documents often take the form of a memo, a letter, or a technical report. Oral communication often makes use of slide decks as a tool.

Four-Blockers

In Chapter 3, I introduced SWOT, and the format for its communication was a four-blocker (Figure 3.1). There are two more common uses of the four-blocker format. The first is to state or escalate a problem clearly, offer a solution or a path forward, ask for the needed help or support, and finally provide a risk assessment should the problem be ignored. The use of this tool helps its author not only to think through the issue and provide a clear statement of the problem but also to propose an appropriate solution. Additionally, the author is forced to consider the help needed to bring the proposed solution to fruition and what possible risks may exist should the problem not be tackled. This is a powerful tool to enable one to "sift" through a myriad of concerns and put them in a logical order so that others may develop an understanding of the problem. Figure 4.1 is an example of such a four-blocker format.

Visual communication tools such as the four-blockers and A3 reports provide a simple yet effective means of reporting or escalating issues to others.

Problem Statement During testing of the new 1.2-liter engines, excessive engine heating has been observed	Proposed Solution It is proposed that the path of the engine coolant be redesigned in order to reduce the wall thickness between the coolant and the critical areas. This will enable a greater degree of heat transfer and lowering of critical temperature
Help Needed Help is needed to conduct a computational fluid dynamics analysis in order to examine the efficacy of the proposed method prior to design change	Risk This issue has the potential to derail the project

FIGURE 4.1
Example of a four-blocker to state a problem and propose a solution

In Figure 4.2, one may notice that the blocks have different headings. The format of a four-blocker is not unique, and it may be changed to reflect what needs to be communicated. In the format used in Figure 4.2, concerns about field complaints and their associated risk are being communicated. In Chapter 10, I will provide some details on product risk and its classifications; however, for now, consider that the issue that needs to be raised is placed in the block under "Description of Issue." Next, its "Impact" is discussed, followed by "Decisions or Actions" that need to be made or taken. The last block provides a "Status" of actions that are taken. Again, this format enables a clear pathway for communicating pertinent information and facts.[1]

Description of Issue 5 complaints exceeded the group risk threshold of 7 for leaky fuel detection on 1.2-liter engines	Impact No driver involvement or injury was reported in any cases over the previous two years. Car fire is possible, but the possibility of occurrence is negligible (>1:1,000,000 but <1:10,000,000)
Decisions or Actions This problem code group was triggered in the monthly product review report on 5/26 for 1.2-liter engines. These were included in the initial scope for investigation, as this is considered to be an engine failure. A decision was made not to initiate a Field Corrective Action	Status Two engines have been returned to the design center's Failure Analysis Lab. Additional units are expected to arrive within the next week.

FIGURE 4.2
Example of a four-blocker to escalate an issue

The A3 Method

The world owes a great deal to the Japanese for a number of innovations in *Lean Thinking*. One of these innovations that was pioneered by Toyota is now commonly referred to as the *A3 Method, A3 Reporting,* or *A3 Problem Solving*. The term A3 refers to the European paper size[2] that was initially used in Toyota to either present proposals, status reports, plan deployments, or solve problems. Figure 4.3 provides an A3 template that may be used in problem-solving. This template may be modified slightly to be better suited for other purposes.

The A3 Method provides a harmonized approach to telling a technical story. It starts by introducing and defining issues and continues through the journey to actions taken and needed resolutions. Its purpose is to capture and communicate facts along with the implications of what has been learned through the studies of facts and observations. The A3 blocks shown in Figure 4.3 are described below.

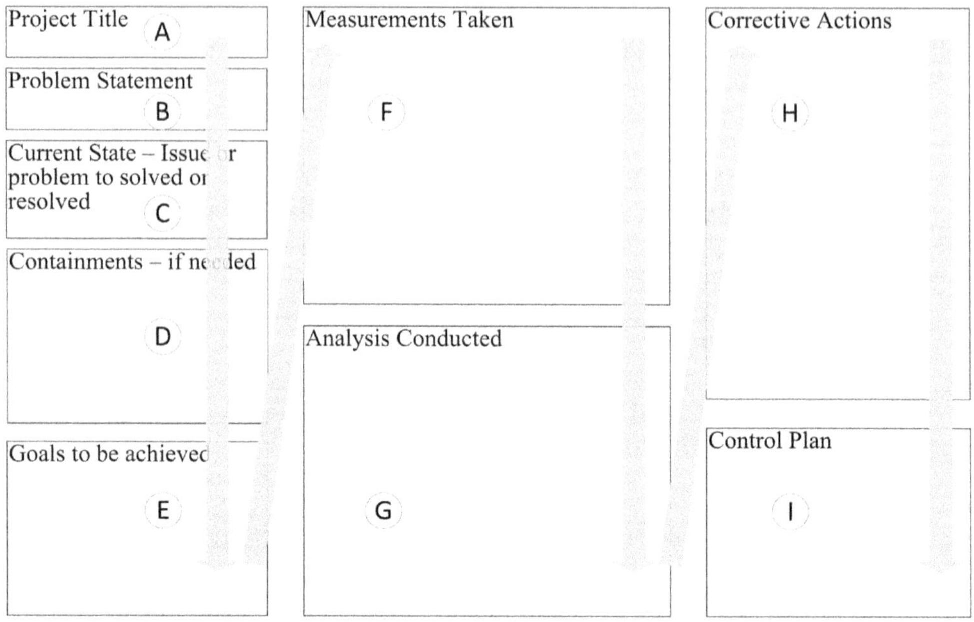

FIGURE 4.3
An example of the format used for the A3 method

A) In developing the A3 Report, it is important to indicate who the leader or the owner of the activities is and the time frame of interest along with the title of the project. This will enable others to quickly identify who they should to contact if they have questions or need additional information.

B) A clear and concise problem statement is provided based on available data.

C) An explanation of the current state is provided and compared with what the current state should be. In other words, what should be achieved once the problem is solved.

D) Depending on the problem, it may need to be contained. For instance, if a fault in a marketed product has been identified that may cause a danger to its users, it may need to be pulled from the market to prevent harm to its users. This needs to be done before the root causes are fully discovered.

E) Goals to be achieved is the block after containment gives an overview of the activities that need to be undertaken. It is typically characterized as *Do What, To What,* and *By How Much,* and *By When.*

F) Collect data and capture results, and their summary is provided in the blocks named measurements taken.

G) In the analysis conducted block, provide the analysis of the measure data. The results may raise other questions requiring either additional data to be collected or updating the goals as expressed in Block E. The three blocks (E, F, and G) are generally intertwined because as data is collected and analyzed, it often leads to further questions and the need for additional data to be collected.

H) In a problem-solving activity, eventually, the root causes are discovered and captured. Once root causes are identified, short- and long-term corrective actions need to be devised and put in place.

I) Finally, control plans are developed and measures are put in place to ensure that the solved problems do not appear again.

At every stage of the A3, the use of graphs, figures, and pictures is greatly encouraged in order to make the story more clear. For additional reading on this subject, *Understanding A3 Thinking: A Critical Component of Toyota's PDCA Management System* (Sobek and Smalley 2008) or *The A3 Workbook: Unlock Your Problem-Solving Mind* (Matthews 2010) are recommended.

Data, Information, and Decision-Making

For the remainder of this chapter, I will be focusing on technical reports and the elements required for reporting excellence; however, before doing so, I would like to discuss the data, information, and knowledge paradigm. The reason is that a fundamental – and often unspoken – element of writing (or presenting) technical reports is to convey information along with its associated knowledge for the purpose of either justifying a decision made or encouraging a decision to be made.

I used the words *information* and *knowledge*. I believe these terms may be elusive and require further exploration. For instance, if a pen costs $5.00 to purchase, does this constitute information or knowledge? Or is this a data point? I suppose I could bring in examples that might argue for this to be either a datum, information, and/or knowledge. Here is what I mean: it is a data point if I am looking at setting up an office and I need to consider the overall cost of running a business. Therefore, the cost of a pen is a data point in the overall cost structure. This cost may be considered information if I want to buy a present for someone and I am trying to decide what to buy. Finally, it is knowledge if I need to buy a pen for personal use.

The point is that as authors of any technical report or presentation, we need to be aware of this distinction and provide the level of detail that our audience would expect. Interestingly, scholars in the field of information systems and technology have debated this point since the late 1980s. See, for example, Ackoff (1989) or Rowley (2007).

From an engineering and/or scientific point of view, let me define terms and their interrelationship relative to the topic of data-information-knowledge.

- Signals: singular observations made that attract our attention but have no meaning or relevance in and by themselves. For example, an LED light may turn on (or off).
- Data: a larger collection of signals – that is not yet cohesive – collected about a subject within a given duration of time or span of space. For example, an LED light turns on in a car along with a sluggish engine performance.
- Information: an outcome of creating a cohesion between various data points and the ability to answer questions that begin with who, what, when, and how (Rowley 2007; Ackoff 1989).
- Knowledge: application of critical thinking to information and developing an understanding of the cause-and-effect relationships that have led to the generation of the signals to the collected data.
- Decision-Making: acting on knowledge and the cause-and-effect relationships to set a course of future actions and pathways.

Example

To illustrate the point made above, let us consider the ACME ProudGlue company that has been making a specialty glue for the last 25 years. Recently, they have noticed an uptick in their complaint rates. Based on this signal, one of the project SMEs responsible for this product family – let's call them Kim – collected the complaint data from the last few years and provided it in Table 4.1.

Note that this table does not provide any specific information other than the complaints are primarily from the United States and a few from Canada. To develop further information based on this data, we need to conduct further investigation. Next, Kim extracts the manufacture date of the glue and compares it to the use date in an attempt to understand how long it has taken for the glue to become thick and unusable. This information is provided in Table 4.2.

While the project team may be able to extract information from Table 4.2, it is difficult to develop a sense of patterns from this set of numbers. As a result, Kim decided to look at this data[3] graphically. The result of the graphical analysis is shown in Figure 4.4, which reveals new information when combined with another data point: the glue has a nine-month shelf life and a four-week open life. Based on the dot-plot in Figure 4.4, there are three patterns associated with this complaint. These are complaint groupings that are significantly below the shelf life (at 180 days and below), the second category that clusters around the 9-month shelf-life (between 220 and 290 days), and beyond the one year.

TABLE 4.1

Specialty glue complaint data for the last four years

Use Date	Country	Ref. #	Lot/Batch #	Description	Complaint	Category
3/16/2022	Canada	Glue A	060921	Thin Glue	Glue is too thick to use	Quality
7/7/2020	USA	Glue B	112219	Medium Glue	Glue is too thick to use	Quality
7/18/2022	USA	Glue A	120221	Thin Glue	Glue is too thick to use	Quality
7/13/2022	USA	Glue A	120221	Thin Glue	Glue is too thick to use	Quality
7/14/2022	USA	Glue A	022322	Thin Glue	Glue is too thick to use	Quality
7/13/2022	USA	Glue A	060921	Thin Glue	Glue is too thick to use	Quality
9/30/2021	USA	Glue A	031020	Thin Glue	Glue is too thick to use	Quality
3/14/2022	USA	Glue A	031020	Thin Glue	Glue is too thick to use	Quality
9/17/2021	Canada	Glue A	031921	Thin Glue	Glue is too thick to use	Quality
7/14/2020	USA	Glue B	112219	Medium Glue	Glue is too thick to use	Quality
6/26/2020	USA	Glue B	112219	Medium Glue	Glue is too thick to use	Quality
6/26/2020	USA	Glue B	112219	Medium Glue	Glue is too thick to use	Quality
2/17/2022	USA	Glue A	060921	Thin Glue	Glue is too thick to use	Quality
4/15/2022	USA	Glue A	033021	Thin Glue	Glue is too thick to use	Quality

(Continued)

TABLE 4.1 (*Continued*)

Specialty glue complaint data for the last four years

Use Date	Country	Ref. #	Lot/Batch #	Description	Complaint	Category
11/3/2021	USA	Glue A	031020	Thin Glue	Glue is too thick to use	Quality
12/23/2021	USA	Glue A	031921	Thin Glue	Glue is too thick to use	Quality
2/10/2022	USA	Glue A	081321	Thin Glue	Glue is too thick to use	Quality
4/30/2021	Spain	Glue B	112219	Medium Glue	Glue is too thick to use	Quality
3/25/2022	Canada	Glue A	120221	Thin Glue	Glue is too thick to use	Quality
3/25/2022	USA	Glue A	101521	Thin Glue	Glue is too thick to use	Quality
7/25/2022	USA	Glue A	022322	Thin Glue	Glue is too thick to use	Quality
3/24/2022	USA	Glue A	031020	Thin Glue	Glue is too thick to use	Quality
9/1/2022	USA	Glue A	120221	Thin Glue	Glue is too thick to use	Quality

TABLE 4.2

Specialty glue complaint: Time of manufacture to getting too thick

Opened Date	Manufacture Date	No. Days
3/16/2022	6/9/2021	280
7/7/2020	11/22/2019	228
7/18/2022	12/2/2021	228
7/13/2022	12/2/2021	223
7/14/2022	2/23/2022	142
7/13/2022	6/9/2021	400
9/30/2021	3/10/2020	570
3/14/2022	3/10/2020	734
9/17/2021	3/19/2021	183
7/14/2020	11/22/2019	236
6/26/2020	11/22/2019	218
6/26/2020	11/22/2019	218
2/17/2022	6/9/2021	254
4/15/2022	3/30/2021	382
11/3/2021	3/10/2020	604
12/23/2021	3/19/2021	280
2/10/2022	8/13/2021	182
4/30/2021	11/22/2019	526
3/25/2022	12/2/2021	114
3/25/2022	10/15/2021	162
7/25/2022	2/23/2022	152
3/24/2022	3/10/2020	745
9/1/2022	12/2/2021	274

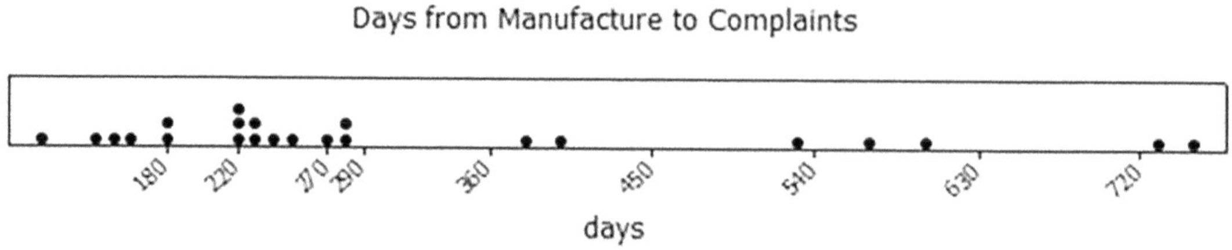

FIGURE 4.4
This dot-plot enables the project SME to discover a pattern in the complaint data

Based on this knowledge, Kim (and their management) may decide to ignore complaints older than 1 year and focus their investigations on two areas: root causes of complaints that are below 180 days and, separately, those that are just below the 270-day shelf life.

Now that we have a view of the differences between data, information, and knowledge, we can focus on developing technical reports that either justify a decision made or encourage a team to set a course of action forward.

Technical Reports

The most important aspect of a technical report is that it is readable and tells a story. It defines who the players are and their roles. It uses the principle that *less is more.* It provides enough detail to relate the point but does not adorn them to the point that the storyline is lost. A balanced combination of text and graphics should be used to communicate. Consider the following two cases:

- *Large amount of data.* When presenting large amounts of data, use graphs. Plots and graphs provide a clearer representation of data and its variations than groups of numbers. Most people do not process large amounts of data in a short period of time. Graphical presentations lead to the easy communication of drawn conclusions.
- *Instructions and process flow.* When presenting instructions and process flows, consider that a large proportion of people communicate through visual cues. Instructions and process descriptions are better suited to a combination of text and graphics.

A Few More Words on Writing

> Writing a technical report is like telling a story. The audience should remain engaged and be fully aware of how the story begins, evolves, and ends.

As I mentioned, writing a technical report is very much like telling a story. We may focus on the plot at first, but it may not be a good starting point. It is the intended- or even the unintended-audience that is important. It is relatively easy to determine who the intended audience may be if we are presenting a paper at a technical conference. However, clearly communicating a status report in an open cross-functional team may reach more people than initially intended.

Having said this, we need to be mindful of the following point: what do we need the audience to do with the information that we are presenting to them? Is it merely for them to know? If so, we need to be very clear on what we want our audience to know and, most importantly, remember. Here, I like to use the adage: *tell them, repeat it, and finally tell them again.*

Or is there a problem that should be escalated and solved? I call this the *call to action* story. Once the audience hears/reads our report, they are charged to do something with the provided information. For this call to action to be effective, we, as the authors, need to consider three factors:

1. What is the audience's interest in the topics being discussed? Who will likely benefit from this information? In other words, what is in it for them? This is an important point because if the recipient of the information has no interest in the topic, the call to action will be ignored.
2. Who is likely to be opposing the report, its conclusions, etc., and why? Suppose that an improvement may save the corporation a large sum on an annual basis but would require extensive tooling changes in a different department with no apparent benefit to them. It should be expected that the members of that team may consider the activity as frivolous.
3. Are there any potential cultural differences? This may be readily applicable to larger global corporations where words and their connotations may have certain levels of associated sensitivity across various world cultures.
4. Are there other company-specific sensitivities? This may be the case in heavily regulated industries such as medical or avionics. There are terminologies that may cause scrutiny by auditors. I recall a case when my team was conducting an aging study of a plastic housing. One morning as I walked into the lab, I heard a nervous conversation that the housing had *failed*. When I investigated further, I learned that the housing had not, in fact, *failed*. Meaning, it had not cracked or lost its structural integrity, nor was any crazing observed. The perceived failure was that its color had shifted. It is important to appreciate that words carry certain connotations and to use them with those connotations in mind.

After considering the audience, we can develop the story. But, to develop the story, we need to have a plot. The plot needs to demonstrate the beginning, the middle, and how the story ends. The beginning is developed based on researching and knowing the background information and the extent needed for the report. It should clearly point out what is the purpose of the report, what is in scope, and what is out of scope. Any cited reference (external or internal) should be properly documented. This is particularly important if the report is based on previously made decisions.[4] Additionally, we need to provide an executive summary in which we will tell our audience what to expect in the body of the report.

Our plot continues with developing the middle. Here, we bring all the pertinent data at an appropriate level of data reduction or analysis at an appropriate language level (no greater than sixth grade). The secret sauce of reports is to develop an appreciation for the right balance between details and overall conclusions that need to be drawn. This competence is typically developed with experience along with receiving good mentoring. As needed here, we need to highlight what the audience needs to know gradually as we come across each topic.

Finally, our story comes to an end by providing a summary of what they were told (called a conclusion) and a list of actions that are expected of them.

As we are planning the plot of our story, we need to keep in mind that there are different ways of structuring it. Depending on the topic, the plot may be developed in a chronological order, in a hierarchical order, as a description of a physical object in spatial order, or in a possible combination of these. To clearly convey the storyline, the author may experiment with these methods and choose the best one.

A Few Words on Presentations

While we are on the topic of writing, I would like to share a few words on the format of presentations. It is natural to prepare for what we are about to say and to ensure that our delivery is smooth. However, we need to keep in mind that giving a presentation is as much a story as a written report is. That it, too, requires a well-developed plot. In fact, not only should the entire presentation have a plot, but also, each slide within the presentation should have its own subplot with a beginning, middle, and end. Examples of these subplots are provided in Figures 4.5 and 4.6.

In Figure 4.5, the beginning of the story questions the reader on the importance of a systems approach to NPD and design for X (DFX). The body of the story (middle) provides data for the final conclusion (end of the story), which answers the question initially asked: it provides 30% savings.

Figure 4.6 brings the readers' focus to the cost of making changes within a product's life cycle in the beginning of the plot, provides cost data at various stages (the middle), and ends the plot with the cost of making changes after the product is launched. The author of this page allows the readers to draw the final conclusion, that is, catch any defects prior to launch.

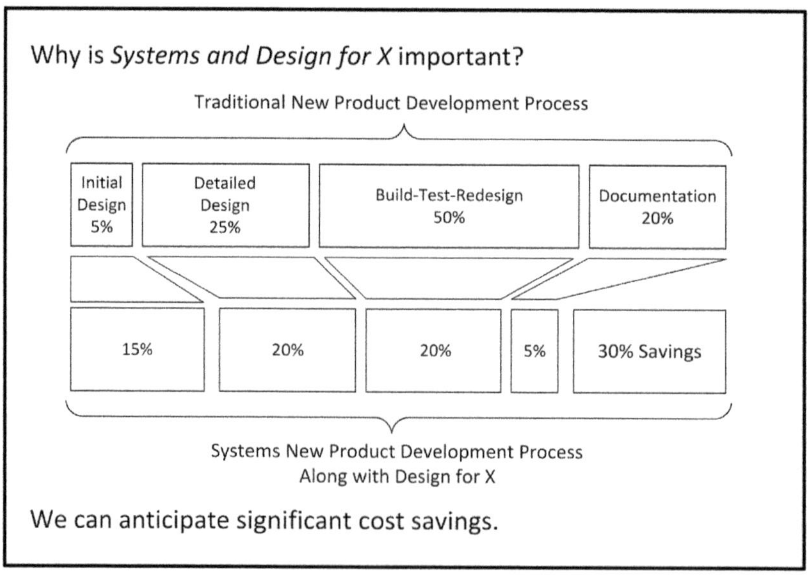

FIGURE 4.5
First example of a presentation slide having a subplot with a beginning, middle, and end

Reliability and The Cost of Making Changes

According to its president, ACME Company decided against design for reliability during their NPD process. As a result, a defect was not recognized until units were shipped to customers.

Had the problem been corrected before the start of production, the estimated cost of correcting the problem at various stages would have been:

- At the design phase: $50
- Before procuring parts: $200
- Before start of production: $1,000

Instead, the cost of correcting this problem was: $600,000

FIGURE 4.6
Second example of a presentation slide having a subplot with a beginning, middle, and end

Working with the Team

It may sound strange that I speak of a writing team. After all, aren't we reporting on the results of our own work? The hard fact of the matter is that at a bare minimum, even though we may be the only ones working on a project, the outcome of our work needs to be approved by at least one other person – either officially or unofficially. In most companies, our reports should be approved by a subject matter expert along with a member of the quality assurance team to ensure that the document is compliant with corporate procedures.

Draft Document

So, at the onset of writing a document or preparing a presentation, we need to identify who the stakeholders are and what requirements we are trying to fulfill. In other words, who will approve the document and what purpose will it fill. Once that is determined, use the following as a guideline:

- As mentioned earlier, use simple language, free of technical or local jargon. This should also include the use of any abbreviations. If abbreviations are necessary, a definition table should be included.
- Avoid long sentences. Technical team members write or speak to an audience who may already have difficulty following the technical content. Trade literary work for a simple sixth-grade-level writing. Furthermore, allow different paragraphs for the expression of different ideas or points.
- Use headings and subheadings liberally to separate the story into its logical settings.
- Use graphs and illustrations to drive your point. However, they should be merged with the text. I have seen too many report drafts where the author has just dropped

figures in, leaving it to the reader to make sense of them. If a figure is used, draw the reader's attention to it and tell the reader what you want him or her to see – both in the text and the captions.

- Include an executive summary at the beginning and a conclusion at the very end. The executive summary and the conclusion should have the "Call to Action." The point that you – as the author – want the reader to take away.

Feedback and Revision

Once a draft document is developed, it is routed for review and feedback. However, depending on the style of the document, feedback may take different forms. For instance, for a simple memo, a verbal go-ahead is all that is needed. For a technical report, there may be an additional document (typically Excel) that accompanies the original report to capture various reviews, corrections, and comments. The expectation is that the author will identify the common threads and update the document as needed.

In the case of a letter being sent to outside organizations or regulatory agencies, it is quite commonplace that a cross-functional team, including the legal counsel, will review the draft in a series of meetings to ensure that everyone's input is properly debated and captured.

Finalization and Approval

Once a document has been reviewed sufficient times and all stakeholders' concerns have been addressed, it is routed for approval. Approval does not necessarily mean that someone would place their signature on the document. It may be a simple acknowledgement by the team that the work is done. And that, everyone stands behind it.

Closing Remarks

The reader may notice that I did not speak about making sure that the grammar or spelling has been checked and that there are no errors of this nature. The ideas discussed in this chapter are universal and language independent. Today, we live in a multilingual, multiculture environment. With globalization, we may find ourselves relying on others with English as a second language to develop reports; hence, there may be superficial flaws in grammer.

The ideal of communication, though, is an international concept with its regional and cultural nuances. Perfection comes only with practice and patience.

Summary

- Clear and concise communication is not only fundamental to any documents that we write but also to any verbal presentations that we may have.

- Technical communications in the work place is different from communication elsewhere. The primary difference is that simple language should be used at work place.
- The target audience needs to know the purpose of what is being communicated within the first paragraph of any document.
- Two powerful tools for visual communication are the four-blocker and A3 method.
- Technical reports or presentations are similar to stories. It should have a clear beginning, a logical journey, and a proper closure.
- Today, we live in a multilingual, multiculture environment. With globalization, we need to be sensitive to cultural differences and sensitivities.

Exercise

1. Apply and use the A3 Method to a research project. Following the completion of the project, seek feedback from your colleagues on the effectiveness of what was communicated.

Notes

1 Particularly when dealing with complaints associated with fielded products (Chapter 19).
2 This is roughly equivalent to the 11 x 17 tabloid size used in the United States.
3 Notice that I switched from calling the table as, "information" to calling it "data" for the next layer of analysis. This should indicate the fluid movement between data and information as we attempt to move towards knowledge.
4 An undocumented decision is the same as a decision not made!

5

Technical Reviews and Checklists

Introduction

The purpose of a *technical review* meeting is to evaluate a particular design element by a group of technical experts in order to identify pitfalls and/or areas for improvement. Design teams need to ensure that pertinent technical areas have been properly vetted in the course of design activities by considering the impact on various disciplines. Technical reviews are formal means to document that in-depth reviews by various pertinent stakeholders have taken place. In a gated product development process, technical reviews of design elements are needed prior to moving to the next stage of development.

Guidelines

> A checklist is a tool to ensure that no pertinent fact or information has fallen through the cracks or is missing.

It should be noted that not all technical reviews carry the same weight. Some reviews are considered to be gate reviews. These are typically milestone reviews and may *make or break* a program. Needless to say, there is a great deal of visibility associated with these meetings. The other end of the spectrum is a gathering of two or more colleagues to provide an update on a specific subject. Regardless of the type of meeting being held, it is prudent that notes be captured, along with any reviewed documents and decisions made.

If I were to classify technical reviews into high and low-level reviews, then a *design evaluation checklist* should be used with the high-level technical reviews to demonstrate full coverage of evaluation. Another feature of a high-level review is that, in addition to stakeholders, independent reviewer(s) should participate. In a way, an independent review will provide an unbiased view on whether what is being reviewed meets the standards of good product development practices. Furthermore, a designated scribe should take notes and record any action items. Technical reviews (both high and low levels) should be successfully completed prior to the release of any design element(s). In other words, any action items must be clearly and completely closed prior to moving on to the next phase of development.

Table 5.1 provides a suggestion on the typical technical reviews that would occur during the course of a program. Additional reviews will likely be added depending on the project plans, integration plans, and release plans of the individual programs. A single technical review may cover a number of different areas.

DOI: 10.1201/9781003564508-6

TABLE 5.1

A Suggested Guideline for Choosing a Level for Technical Reviews

	Components or Subassembly	Integrated Subsystem or System	Other Product Specifications	Fixtures/ Aids
Release for Prototype	L	H	N/A	L
Release for Production Quotation	L or H	L or H	N/A	N/A
Release for Tooling	H	H	N/A	N/A
Release for Production	H	H	H	H
Design Transfer	H	H	H	H

H = High-level technical review
L = Low-level technical review

Design Evaluation Checklists

The purpose of design evaluation checklists is to provide a list of areas to consider when evaluating various design elements from a technical perspective.

There are a wide range of areas to consider within the technical discipline(s) with respect to the evaluation of a design element. The list of pertinent areas for each particular evaluation varies widely depending on the design element and phase of development. Thus, there is a need for a tool to manage and to simplify the evaluation process. A checklist is intended to capture the typical areas of consideration with respect to design, but it is otherwise nonspecific. Each design team is encouraged to develop checklists that are relevant to their industry and market. Checklists may be used at any point during a program for any design element: at the system level, subsystem level, or component level.

Here, I would like to share a thought. I have seen corporations that consider checklists as a divine gift – that every element of it must be considered and documented; that it is a mortal sin should a portion be ignored or dropped. I have also been around corporations whose cultures have been so against checklists that the personnel stay away as if checklists are the plague incarnate.

Throughout my career, I have found checklists to be useful. They have helped to keep me honest, particularly as an independent reviewer. I say this in the sense that due to the number of projects, I do not often remember all the points that need to be reviewed in a particular session. Having my own personal checklist has helped both the team and myself to turn every possible rock to identify potential design flaws.

Elements of a Checklist

There are typically three columns in a checklist. The first column speaks to a general area of concern. For instance, should the topic be a printed circuit board assembly (PCBA), areas of concern may be electrostatic discharge (ESD), and thermal and reliability considerations. The second column asks specific questions within the areas of concern mentioned in the first column. For instance, in the area of ESD, specific questions may be the minimum distance of any copper trace to external housings or minimum air gaps. Specific questions on reliability may be the percentage of derating or any specific life tests that have been required. The third column provides a response to the questions asked in the second column. Often, there are additional columns for comments or even for action items and action owners as well.

Example 1 – Developing a Checklist for PCBA Layouts

While I am on the subject of PCBAs, I would like to complete developing a sample checklist. This checklist (Table 5.2) is by no means exhaustive.

Example 2 – A Checklist for Plastic Component Designs

Table 5.3 is an example of a checklist that may be used when reviewing the design of a plastic component. Again, this list is by no means exhaustive.

Example 3 – A Checklist for Developing Engineering Drawings

Table 5.4 is an example of a checklist that may be used when reviewing engineering drawings. Again, this list is by no means exhaustive.

Example 4 – A Checklist for Design Review of a Chemical Solution

Table 5.5 is an example of a checklist that may be used during a design review. Again, this list is by no means exhaustive.

TABLE 5.2

A Possible Checklist for the Layout Design of PCBAs

Area	Concerns	Response	Comments
Electrostatic Discharge (ESD) Protection	Is there an ESD shield in place?	Yes No	
	If no, is the air gap sufficiently long?	Yes No	
Reliability	Has a derating analysis been done?	Yes No	
	Have theoretical failure rates been calculated?	Yes No	
	Has component life testing been done?	Yes No	
Thermal Analysis	Has a board-level temperature mapping been done?	Yes No	
	Are maximum temperatures below critical thresholds?	Yes No	
	Are critical components in the wake of high temperatures?	Yes No	

TABLE 5.3

A Possible Checklist for Design of Plastic Parts

Area	Design Consideration	Response	Comments
Material Selection	Does the material meet chemical resistance requirements?	Yes No	
	Does the material meet load bearing requirements?	Yes No	
	Does the material meet fire resistance requirements?	Yes No	
	Does the material have any regrinds?	Yes No	
Wall Thickness	Is the wall thickness at its optimum level?	Yes No	
Draft Angles	Are draft angles specified?	Yes No	
	Are at least three degrees added to all shut-out faces?	Yes No	
	Are the draft angles sufficiently large for textured or cosmetic walls?	Yes No	
Ribs	Is the ratio of rib thickness to wall thickness at about 0.5?	Yes No	
Undercuts	Are there internal undercuts?	Yes No	
	If yes, can the design be modified to remove them?	Yes No	
Ejection Pins	Has the location of ejection pins and their shape been considered?	Yes No	
Gate Location	Has the location of injection gate(s) and their shape been considered?	Yes No	

TABLE 5.4

A Possible Checklist for Reviewing 2D Engineering Drawings

Area	Concerns	Response	Remarks
Title Block	Is the change order number specified?	Yes No	
	Is the angle projection mentioned?	Yes No	
	Is the proper revision level indicated in the title block?	Yes No	
	Is Scaling Ratio Indicated properly?	Yes No	
	Is material specified?	Yes No	
	Is surface finish mentioned or required?	Yes No	
Tables	Has the revision table been filled and placed correctly?	Yes No	
	For assembly levels and higher: Has the BOM table been filled correctly?	Yes No	
	For assembly levels and higher: Is the quantity of items in the BOM correctly specified?	Yes No	
	For assembly levels and higher, do assembly drawing designations (such as numbered balloons) match the BOM?	Yes No	

(Continued)

TABLE 5.4 (*Continued*)

A Possible Checklist for Reviewing 2D Engineering Drawings

Area	Concerns	Response	Remarks
Dimensions	Are there any dangling dimensions?	Yes No	
	Are there any redundant dimensions?	Yes No	
	Are dimensioning features consistent, i.e., not a mix of fractions with decimal dimensions?	Yes No	
	Are all hole dimensions specified?	Yes No	
	Is hole diameter tolerance callout needed?	Yes No	
Symbols	Are all necessary axis lines for the holes in all the views shown?	Yes No	
	Are there any repeated label letters for section views, detail views, and datums?	Yes No	
	Do the symbol and text in notes have the same font size?	Yes No	
Notes	Are all notes complete and clear?	Yes No	
	Should the tolerance table be mentioned in the notes section?	Yes No	
	For plastic parts, should notes cover topics such as flash height, sink marks, flame-retardant grade, etc.?	Yes No	
	For sheet metal or machined parts, should notes cover topics such as burr removal, breaking sharp edges/corners, etc.?	Yes No	

TABLE 5.5

A Possible Checklist to Be Used at Design Review for a Chemical Solution

Area	Design Consideration	Response	Comments
Product Definition	Product objectives clearly stated	Yes No	
	Technical requirements defined	Yes No	
	End-use application validated	Yes No	
Chemistry and Formulation	Formulation/composition verified	Yes No	
	Stability data reviewed	Yes No	
	Critical to quality (CTQ) attributes identified	Yes No	
	Degradation pathways assessed	Yes No	
Environment, Health, and Safety (EHS)	Hazard identification completed	Yes No	
	Process safety assessment	Yes No	
	Regulatory compliance review	Yes No	
Process Engineering	Scale-up feasibility checked	Yes No	
	Equipment compatibility verified	Yes No	
	Process control strategy defined	Yes No	
Supply Chain	Supplier qualification status	Yes No	
	Raw material impurity profile reviewed	Yes No	
Quality	Analytical methods validated	Yes No	
	Product specifications drafted	Yes No	

Example 5 – A Checklist for Program Gate Review

Table 5.6 is an example of a checklist that may be used during a program review. Again, this list is by no means exhaustive.

TABLE 5.6

A Possible Checklist to Be Used at Program Review

Focus	Area to Consider	Summary of Finding	Support Documents/ Comments
CTQs	Identify top items critical to quality		
ToO	Theory of Operation		
	Other Design Documents		
Technical Transfer Function(s)	Engineering/Scientific Analysis		
	Kinematics and Dynamics Studies		
	Computational Fluid Dynamics		
	Finite Element Analysis		
	Thermal Analysis and Management		
	Tolerance Worst-Case Analysis		
	Sensitivity Studies		
	Vibration Analysis		
	Shock Analysis		
DFMEA	Report		
	Critical/Safety/Significant Characteristics		
Design for Reliability	Reliability Plan/Report		
	Fatigue Evaluation		
	Wear Evaluation		
	New Failure Modes Identified?		
Design for Manufacturability and Assembly	Mold Flow Analysis		
	Manufacturing and Fabrication Optimization		
	Timing Studies		
	Tear Down Studies		
	Motion/Time Studies		
	Kaizen		
Design for Serviceability	Serviceability Review		
	Timing Studies		
	Tear Down Studies		
	Kaizen		
PFMEA	Report		
	Key Process Characteristics		
	Control Plan(s)		
	Supplier Quality Review: Right suppliers for the job?		

(Continued)

TABLE 5.6 (***Continued***)

A Possible Checklist to Be Used at Program Review

Focus	Area to Consider	Summary of Finding	Support Documents/ Comments
Design Transfer	Component drawings/Specifications		
	Assembly drawings/Schematics		
	Bill of Material		
	Assembly Instructions		
	Inspection Specifications		
	Assembly and Inspection Fixture Designs		
	Tooling Approvals		
	Process Validation		
Verification and Validation	Assembly Verification		
	Integration Testing		
	Subsystem Verification		
	System Verification		
	HF and Product Validation		
	Safety and Regulatory Testing and Certification		
	Traceability Matrix		
Production Concerns	Final Bill of Material Cost		
	Final Labor Cost		
	Final Total Cost including overhead		
	Total Tooling Cost		
	Production Throughput		
	Production Yield Efficiency		
Technical Design Reviews	Product Requirements Document Review		
	Architecture Design Review		
	Assembly Design Review		
	Subsystem Design Review		
	System Design Review		

Summary

- The purpose of a technical review meeting is to evaluate a particular design element by a group of technical experts in order to identify pitfalls and/or areas for improvement.
- Technical reviews do not carry the same weight. Some reviews are milestone reviews and may make or break a program.

- Each design team is encouraged to develop checklists that are relevant to their industry and market.
- There are typically three columns in a checklist. The first column speaks to a general area of concern. The second column asks specific questions within the areas of concern mentioned in the first column. The third column provides a response to the questions asked in the second column.

Part II

Product Development Life Cycle Management and Roadmap

6

Product Life Cycle Management Models

Introduction

A product begins with an idea, and through technical efforts, this mental concept turns into a physical reality. Should this embodiment of the concept fulfill a market need – be it real or imagined – it may become a financial success. Throughout the last six or seven decades, a number of approaches have emerged in order to organize the way products are developed; on the one hand, to bring a certain degree of efficiency, and on the other hand, to ensure that the finished product is what the customer had asked for and wanted. The majority of these methodologies start with an effort to understand the market and/or customer requirements. And the last step in them ends with a litany of best practices to launch the product. Some authors place a greater emphasis on the front end, i.e., the marketing and business needs, while others concentrate on the back end, i.e., product launch. Typically, sources that focus on design or technical practices tend to neglect non-technical aspects of product development.

Product Development Models

Stages of product development and its life cycle are defined differently by different people. For instance, Crawford and Di Bennedetto (2003) define the life cycle of a product to have five stages; namely, *prelaunch, introduction, growth, maturity* and finally, *decline.* This view reflects someone who has a bird's eye view. Cooper (2001, 2005) provides a five-stage, five-gate Stage-Gate®[1] model for developing new products. The five stages include *scoping, build business case, development, testing and validation,* and finally, *launch.* In his model, the product development activity would not proceed unless it passes through management review gates. Cooper focuses on management activities leading up to and through launching a product. De Feo (2010) proposes a six-step process defined as follows: *project and design goals, customer identification, customer needs, product* (or *service*) *features, process features,* and finally, *controls and transfer to operations.*

If we were to examine these authors' point of view, their emphasis is on developing a new product and launching it into the market. They do not consider any post-launch activities such as support and/or service, nor do they talk about engineering and/or scientific undertakings.

> Proposed product life cycle models place their emphasis on various subjects, but they all start with an assessment of market opportunities.

Hooks and Farry (2001) consider that product life ends with its "disposal" as the last stage in their life cycle model: *manufacturing, verification, storage, shipping, installation, upgrading,* and *disposal.* Similarly, Pancake (2005) extends the steps of product life cycle

DOI: 10.1201/9781003564508-8

beyond launch by including "support" as the last step of the seven steps of product development: *discovery process and product strategy, product requirements, design and development, verification and validation, manufacturing, quality assurance and regulatory,* and finally, *product launch and support.* Stockhoff (2010) cites several different models: in some, the first step is *idea generation,* and in others, *establishing market concepts,* or *project initiation.* The last step is either *commercialization, launch,* or *post-implementation.*

Clearly, there is no standard definition or a "one-size-fits-all" process. However, in a simplistic way, product life cycle may follow these generalized steps to a greater or lesser degree: As shown in Figure 6.1, first, the idea for a product is conceived (*market opportunity*). On the basis of this idea, some analysis is conducted to provide evidence that the idea has commercial merits (*scoping/business case*). When investors are willing to support the idea financially, technical teams are engaged to first develop concepts (*concept generation*) and then provide detailed design (*design*).

The next step is to fabricate and test prototypes (*verification and validation*). When all parties are satisfied with test results, the product is introduced in one or two small markets (*limited launch*) and customer response along with the behavior of the product – particularly any failures – is studied closely. Eventually manufacturing ramps up to full production (*full launch and production*). Shortly after, the service organization begins to service failed units. As time goes on and manufacturing continues, issues associated with product failures and reliability, process and/or design improvements, component changes, and part obsolescence would require design changes (*technical/engineering notices of change*).[2] Eventually, the business recognizes that the product is no longer profitable and decides to first end manufacture (*end of manufacture*) and then stop any service activities (*end of service*). In some instances, the manufacturer may even actively remove the product from the field (*retire product*).

The V-Model for Life Cycle Management

A product development model that has gained some currency in the United States is called the V-model (Forsberg and Mooz 1998). A pictorial view of this model is presented in

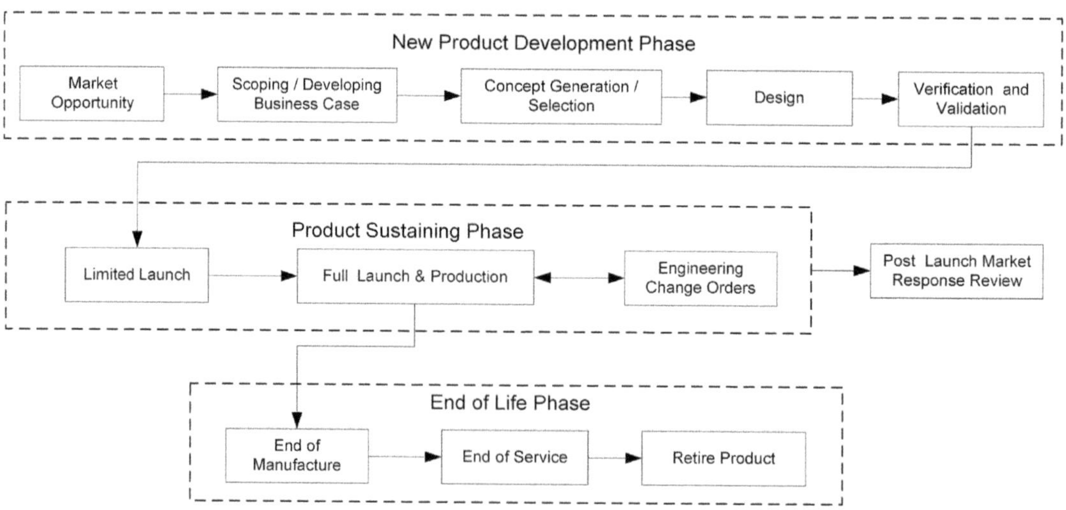

FIGURE 6.1
Typical stages of a product's life cycle

Figure 6.2. This model has been referred to as the V&V model as well – to represent that verification and validation of the outcome is an integral aspect of this model.

Briefly, the left-hand branch of the "V" corresponds to a series of activities to define the tasks to be completed – starting from a very high level overview of what should be done, broken down to smaller tasks, and eventually down to manageable pieces. Once all tasks are properly defined, implementation may begin at or near the apex of the "V." Once the smaller tasks are completed, they are integrated into larger pieces and tested to ensure that they work as expected. This aspect of the work is represented by the right-hand branch of the "V." This integration and testing continues with larger assemblies until the entire system is put together and tested to ensure that it functions as expected. Finally, the end product is validated to ensure that it is the right response to the initial question (or need). Even though Figure 6.2 outlines 12 steps from the beginning to the end, the number of these stages is rather arbitrary. The intention is to show the progression from the big picture to the development of the details and then expand back to a confirmed full picture.

The versatility of this approach is that it is a task-oriented model. It should be readily obvious that the V-model is not specific to developing only physical products. Rather, it has been used extensively for project management as well as software development.

So, what do these 12 stages of the V-model mean to a design team? To answer this question, let me repeat the first five stages below:

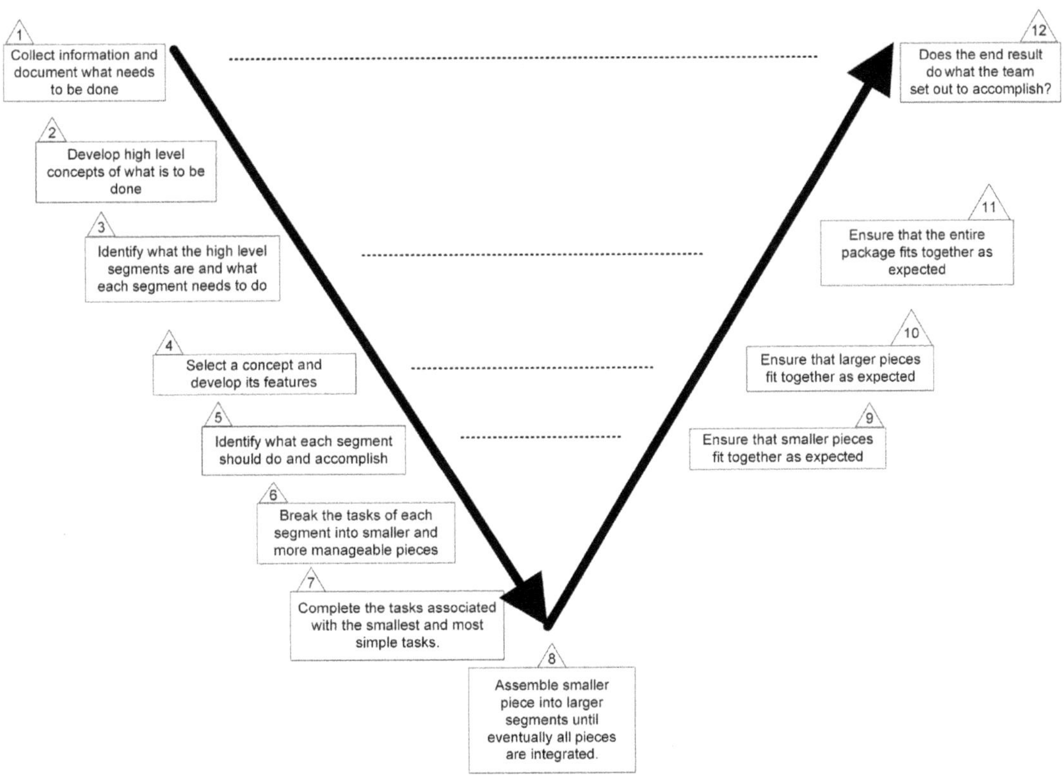

FIGURE 6.2
The steps and stages of the V-model

1. Collect information and document what needs to be done.
2. Develop high-level concepts of what is to be done.
3. Identify what the high-level segments are and what each segment needs to do.
4. Select a concept and develop its features
5. Identify what each segment should do and accomplish

> The V-model for product development begins with an understanding of user needs and continues by cascading them level by level to eventually define specifics of the design.

From an electromechanical product development point of view, the first item is clearly a marketing, business, and ultimately financial question and concern. I often say this is a question and concern for the *decision-makers.*

Stages 2 and 3 involve technical input and work. This process begins with the development of system-level requirements (Stage 3) cascaded from product requirement definitions (Stage 2). From the system-level requirements, further decomposition occurs to produce subsystem and possibly sub-subsystem requirements and the functions (i.e., mechanical, electrical, and software) allocated to them. This is akin to Stages 4 and 5 of Figure 6.2 and is presented in more detail in Figure 6.3.

Incidentally, there is an element that is implied but not explicitly called out. A lower step cannot be attempted unless the upper level is completed and approved. In other words, a subsystem requirement may not be developed if system-level requirements (and architecture) are not fully defined. This mindset ensures that design inputs are properly defined before the R&D team attempts to develop design outputs (i.e., solutions).

This approach and model has given birth to a discipline called systems engineering which is the discipline of understanding what market needs are and translating these needs into technical terms not only at the product level but also translating how various

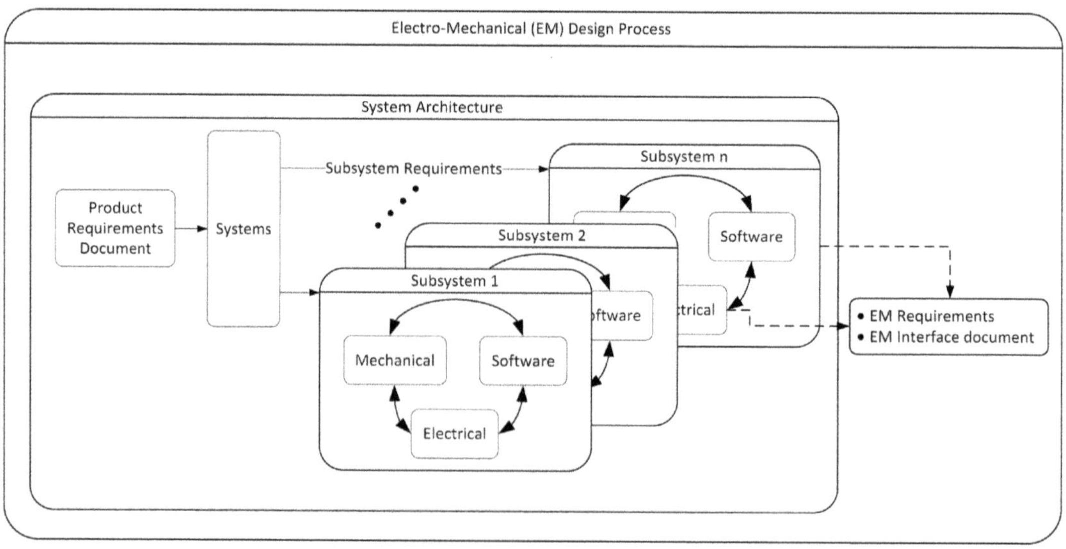

FIGURE 6.3
A requirements-based electro-mechanical (EM) engineering design process flow

lower-level assemblies and components should function and interface with one another. This is done to ensure that at the top level, everything works as intended.

The outcome of Figure 6.3 is *EM Requirements* and *EM Interface Document* – EM being an abbreviation of *electromechanical*. When I was first introduced to this concept, my reaction was: "What does this mean and what does that have anything to do with my design!?!" Often, the response to this question is that "they are the initial steps in product development model." For the uninitiated, this is no more clear. This lack of clarity, as so properly pointed out by Ogrodnik (2013) is because the process has not provided guidelines and details on the required set of activities. In a way, this shortcoming may be due to the fact that systems engineering has not been actively and properly integrated with other functional (i.e., traditional) technical fields.

Engineering and Scientific Activities

One may note that I have not mentioned any related *engineering* processes or activities as an integral part of the proposed product development models. If mentioned, it has been *glossed* over as a given set of workflows. There are authors who attempt to explain product development methodology from an engineering point of view. Among these are Ullman (2010), *The Mechanical Design Process;* Gerhard Pahl et al. (2007), *Engineering Design: A Systematic Approach;* and Stuart Pugh (1991), *Total Design: Integrated Methods for Successful Product Engineering.* Ogrodnik (2013) provides a critical review of the models proposed by Pugh as well as Pahl and Beitz. He eventually concludes that "While these models are worthy, in reality they do not give a full picture of actual activity. They visualize the processes but not the activities. To this extent I shall present a didactic model" (p. 31). And then provides a *divergent–convergent* model of his own. Xiao (2021), in his *Analytical Scientists in Pharmaceutical Product Development: Task Management and Practical Knowledge,* offers a task-based approach to developing analytical chemical formulations. He does provide a brief overview of Agile versus waterfall, but his focus is on developing various classes of chemicals.

I am not really sure if there is a great deal of difference between these various models, per se. As I mentioned earlier, almost all begin with a reference to market need and end with the embodiment of a product. A few have even mentioned the need for sustaining the product and eventually removing it from the market. An often forgotten (or ignored) factor is that even though most product development models present their various steps and stages in a linear fashion, in reality, product development is a highly iterative process, either within each step or in between two or several stages. Another overlooked factor is that product development is never a purely engineering, scientific, sales and marketing, or business activity. It is truly an interdisciplinary and interwoven set of activities whose success depends on various teams working together.

The shortcoming of many models may be due to the fact that the activities required in the product development process are not *visualized*. Or, if they are referenced, they are considered within a specific discipline such as, say, engineering, finance, or marketing.

I agree with Ogrodnik (2013) that any proposed model should present both the process and the activities. Needless to say, this book sets out to do exactly that. The set of activities - required from an engineering/scientific point of view- is outlined in a roadmap and will be presented shortly. I am emphasizing that the roadmap for the purposes of this book is for design engineers and scientists; though, I am fully aware that, as a part of product development activities, various cross-functional disciplines have their own undertakings and associated roadmaps.

Before embarking on the roadmap, I would like to discuss two popular tools from the Design for Six Sigma (DFSS) toolkit and explore whether design activities may be integrated within the stages of these product development models and bridge the gaps that exist.

Quality by Design

Typically, as shown in Figure 6.1, steps from the market opportunity through verification and validation are named *new product development*; steps through launch and production, which include activities such as manufacturing, service, and more generally operations, are called *product sustaining*. Needless to say, activities that bring an end to manufacturing and service are called the *end-of-life phase*.

> Quality by design means doing things right the first time. That implies that design, development, manufacturing, and service must follow processes and activities that do not generate waste.

The question is this: What is the most effective way to conduct these life cycle activities? The answer may be *doing things right the first time*. That means the design, development, manufacturing, and service must follow processes and activities that do not generate waste. In this regard, De Feo (2010) employs the *quality by design* model as a tool for doing things right the first time. It is noteworthy to consider that he associates the word *quality* with customer satisfaction and lack of product defects, that is, very few failures.

If we were to consider developing a design from a *quality* point of view, we would first consider what makes our customers happy and satisfied and, at the same time, strive to develop processes and procedures that make product failures, on the one hand, rare and, on the other hand, predictable. In today's world, this mindset is referred to as either Lean or DFSS. In short, DFSS is a data-driven methodology that strives to proactively identify sources of variation and attempts to reduce them. There are a number of tools within the DFSS toolkit with different steps having been proposed. In this book, two relevant approaches are introduced that may be integrated easily with the V-model. These two are DMADV and DMAIC. DMADV, which stands for *define, measure, analyze, design,* and *verify,* and DMAIC, which is an acronym for *define, measure, analyze, improve,* and *control.* For more information on the DFSS toolkit, Munro et al. (2008), Juran and De Feo (2010), Kubiak and Benbow (2009), and Creveling et al. (2003) are recommended.

> DMADV and DMAIC are two acronyms and tools used in Six Sigma methodology. DMADV is a design tool that stands for Define, Measure, Analyze, Design, and Verify. DMAIC is an abbreviation for Define, Measure, Analyze, Improve, and Control. It is often used in failure analysis.

Within the NPD process, DMADV is a prominent tool. It helps to drive focus on the development of customer needs and their fulfillment through design and ultimately variation reduction. DMAIC is better suited for use in the product-sustaining phase. It is designed to place attention on driving to root causes of failures, problem solving, and/or continuous improvements. While in the following chapters I will provide working details on these two tools, the following is a brief overview of their steps.

DMADV

Figure 6.4 provides a map that links the steps in DMADV to the steps of the NPD model as well as its related activities.

1. *Define.* In general, this stage forces the team to think about the market reasons for the product to be developed. In the NPD map, this step is associated with market opportunity and how to respond to a market need. Hence, once this step is completed, the market opportunity is fully defined and project goals are set. By project goals, I mean the team identifies what the market desires versus what it already has. Then, a decision on the scope of the project is made. Finally, an assessment of the project risks is made. In other words, it is determined what the business risk would be if no actions are taken.
2. *Measure.* Once the project and its risks are identified, the marketing team creates the business case, whereas the technical team sets out to develop an understanding of customer and use processes and to capture and prioritize both customer and business needs. The next steps are to develop product requirements and identify elements that are critical to quality (CTQ). Finally, based on the CTQs, verification and validation plans are developed.
3. *Analyze.* Here, marketing and technical teams come together to generate concepts through cross-functional brainstorming, select certain designs, and conduct an initial cost analysis. Systems Engineering conducts a further allocation of requirements and tracks how CTQs are maintained. Finally, at the close of this step, transfer functions have been developed and tolerance analysis has been conducted.
4. *Design.* Some may refer to the previous step as *design* and to this step as *optimize.* Regardless of what this step is called, herein, the team has developed a concept of what will be launched in the market. Now is the time to understand the gaps between what was desired, and what is designed, identify potential trade-offs, and ask whether any design changes need to be made. Product variability is understood, and design optimizations are completed. Other activities within this step are poke yoke, or mistake-proofing, and reliability testing. Before this step is closed, a final cost analysis is completed.
5. *Verify* (and *validate*). The final step – in NPD prior to launch – is to verify that all product requirements are met. Furthermore, an end-user study may be needed to validate that all user needs have been met. At the same time, DFSS requires that all manufacturing and service processes be written, tested, and validated, and process control plan(s) put in place before product begins.

DMAIC

As mentioned earlier, DMAIC is better suited to be used in the product-sustaining phase, particularly in systematically driving to root cause analysis of failures and improvements. As shown in Figure 6.5, it has five steps or stages as follows:

1. *Define.* In this step, the problem statement is defined clearly and concisely. This is crucial to the success of DMAIC activities. In fact, the team should dwell in this stage as long as possible and come back to it in order to ensure that the problem

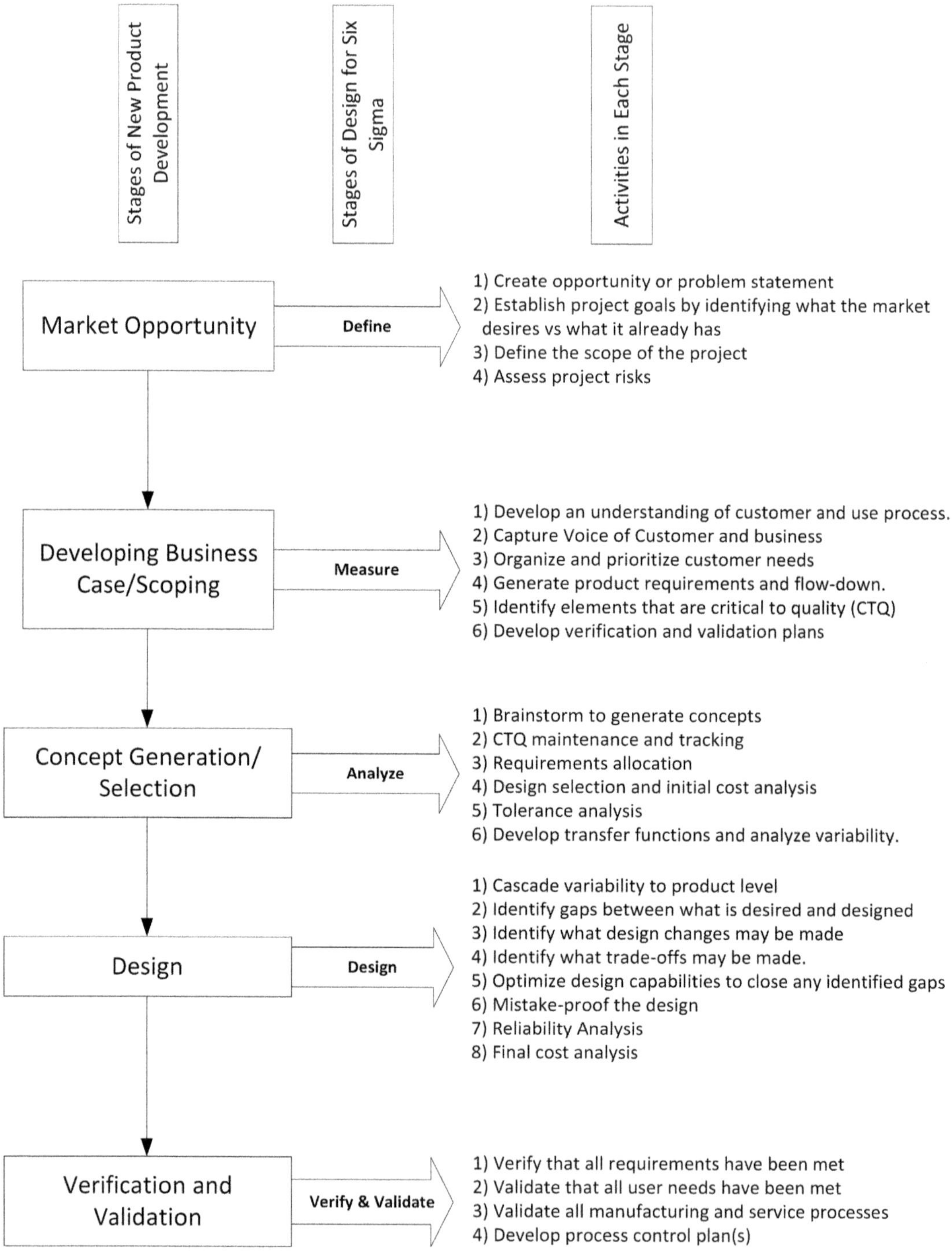

FIGURE 6.4
DMADV steps used in product development

statement is complete. It should focus on what has failed (or needs to be improved), who has observed the issue, and its extent. Here, the question is, *specifically, what is important to the customer?* is answered. In DFSS language, "what is important..." is called *critical to quality* or CTQ. Finally, a strategy to collect data and a list of questions or inquiries are developed.

2. *Measure.* As is determined by its name, in this stage, relevant data are collected; CTQs are measured and evaluated. If additional facts are identified that affect the problem statement, *define* is revisited and the problem statement is updated. The more comprehensive the problem statement, the richer the collected data.
3. *Analyze.* Once sufficient levels of data are collected, analysis begins. The effort will concentrate on identifying factors that have a significant impact on the CTQs. By studying these factors and their impact, a root cause of failure or a potential process bottleneck is generally identified in this stage.
4. *Improve.* On the basis of what has been learned, a design change (and/or process improvement) may be proposed. Prototypes (and, in the case of process improvements, pilot runs) are created and evaluated for expected outcomes.
5. *Control.* The DMAIC process identifies the root cause(s) of the failure. However, if the problem has happened once, it is likely to happen again. To close the loop, DFSS requires that proper control mechanisms must be identified to prevent a relapse and a return to the state under which failure took place.

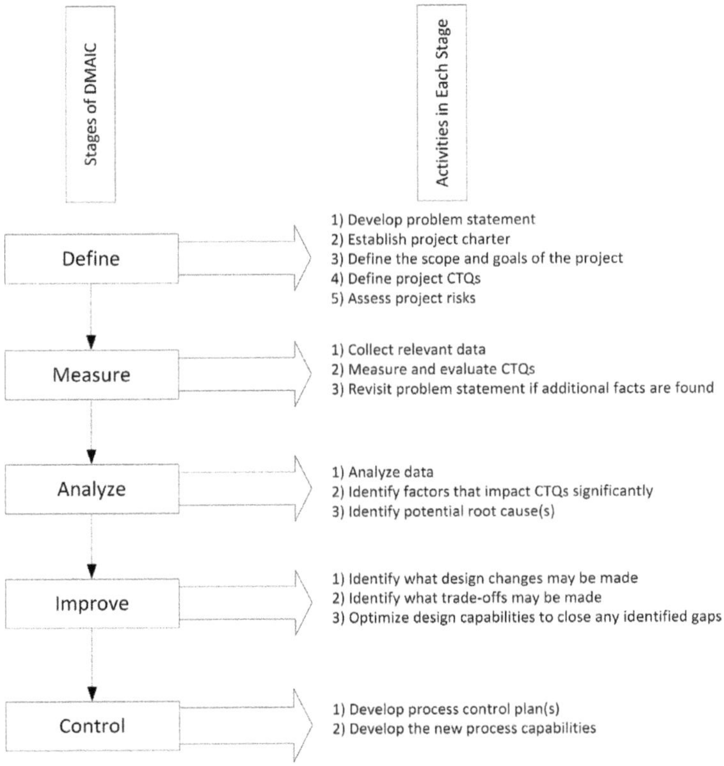

FIGURE 6.5
DMAIC steps used in sustaining products

V-Model, Agile, and Product Development

The terms "Agile" and "Agile product development methods" have gained significant currency in the last two decades since the inception of the Agile Alliance as the approach to shorten product development timelines. Many product developers find themselves knowing terms such as "epics" and "stories" along with "scrum" and "sprints"; yet, in practice, many in the hardware development world may not be aware of how to apply this mindset to their work in order to shorten their product development cycles. At the same time, there are those who have indeed applied this methodology but have not experienced the expected results.

In Appendix A, I have provided a brief overview of Agile, its history, and the *plan-do-review* process. In Appendix B, I have provided an example where Agile was successfully applied. In the rest of this chapter, I will discuss how the Agile process of plan-do-review may be integrated with the V-model and the DFSS. Often times, V-model and DFSS are considered to lack flexibility; once the one step of the model is complete and the gate approved, the tool will not allow a return to a previous set of activities. However, it has been my experience that inflexibility is rather an element of corporate culture or its quality management system. As we will see, flexibility is inherent in the V-model, DFSS, or the roadmap that I will introduce shortly.

Co-relationship between V-Model and Agile

In the last two decades, Agile has been celebrated as a methodology that is responsive to changing customer and market needs. In general, this is true, but as the market and the number of customers grow, so does the diversity in what they demand. Back in the 1980s and 1990s, a number of software development companies had – in a way – single customers, particularly as those customers were trying to convert their data processing software to be Year-2000 ready. Responding to changing requirements over simply following a plan or focusing on working software over comprehensive documentation[3] is much simpler for a single customer as opposed to addressing the needs of a multitude of customers in a market.

Consider Figure 6.6 depicting the V-model in light of an Agile mindset. As we will see in the next few chapters on cascading user needs to product and system requirements, a single user need may fan out into several requirements at lower levels. Additionally, a single requirement or lower-level specification may support multiple upstream requirements and user needs. The V-model gives us the ability to return to previous stages in order to review and/or modify any upstream requirements with the understanding that the change may have ripple effects. Impacts that need to be considered and managed.

It is true that the Agile method allows flexibility to respond to market changes; so does the V-model. However, while in the software world, this response may be trivial, in the hardware world, this has the potential of becoming more impactful. Imagine, for instance, that the market may require a much lighter device. Weight reduction and its ripple effect on structural integrity or cost are not trivial when product development is well underway. A change in requirements either in Agile or V-model should not be incorporated blindly without understanding the full impact.

An element of change within the V-model is the ability to make partial changes. By way of an example, and as shown in Figure 6.7, elements at Stage 5 fan out into a number of segments. Development of each segment may continue or be revised as needed somewhat

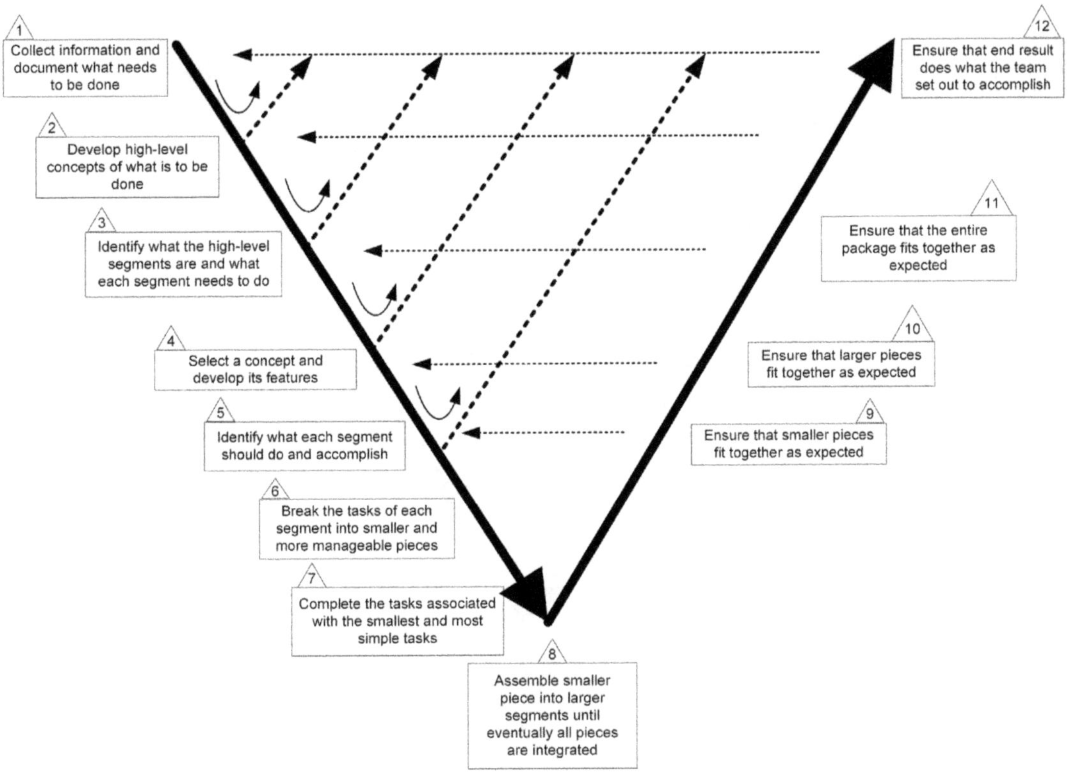

FIGURE 6.6
The V-model depicting return to a previous state as needed to address change

independent of each other until such time that module or subsystem integration needs to happen.

We should not be under any illusion that change does not impact timelines. It does! The difference between an Agile mindset and a traditional mindset is that Agile gives the ability to revise what is to be delivered by reducing or altering a product's feature set, whereas a traditional mindset insists that what was originally planned has to be delivered.

Traditional versus Agile Development Cycle Times

In Chapter 3, I mentioned that Agile is often thought of as the approach that will shorten product development cycle time. Yet in practice, many organizations are puzzled when, in practice, Agile does nothing to reduce timelines. The root of this failure may be found in an incomplete understanding of the Agile mindset as well as the corporate culture needed for it to flourish.

Sahota (2012) concluded that Agile culture is about collaboration and cultivation. In other words, Agile thrives in environments where focus is placed on people and team members and not merely plans and processes. A thoughtful review of Agile values and principles as presented in Appendix A will affirm this conclusion. The importance of

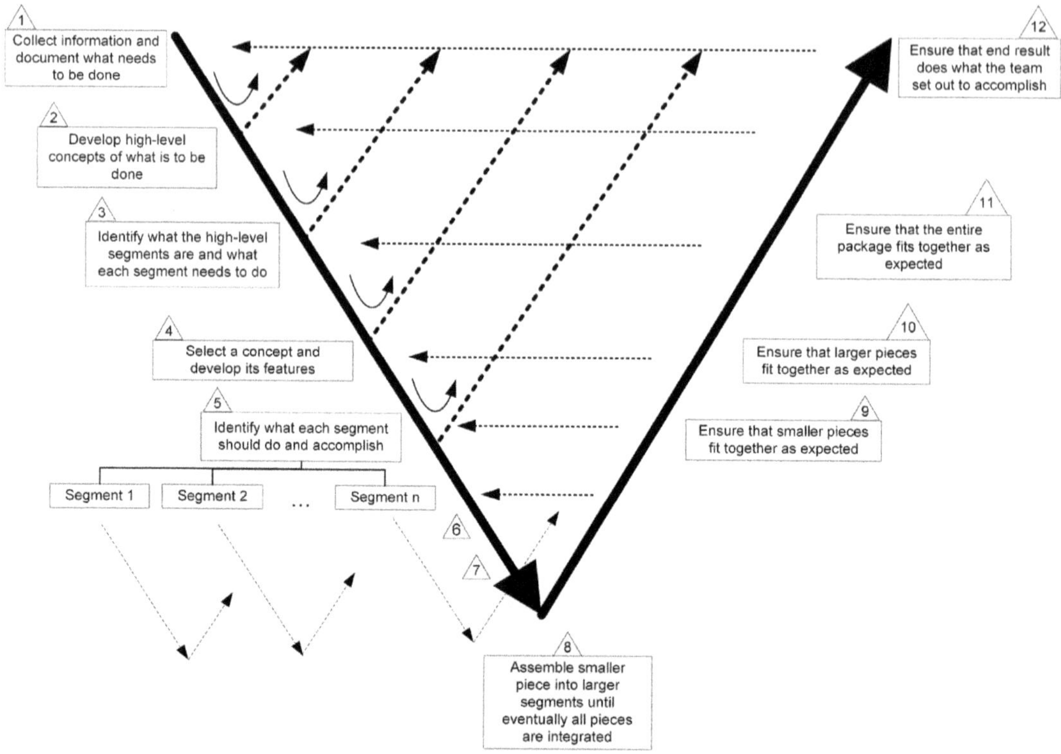

FIGURE 6.7
An example of the ability to make partial updates within the V-model. In here, elements at Stage 5 fan out into a number of segments. Development of each segment may continue or be revised as needed at different stages

corporate culture and its role in making or breaking an Agile transformation cannot be overemphasized. Such a transformation may need to happen in stages, possibly by redefining the agile values and principles as needed or adjusted to their own organization.

In the next section, I will provide a design and development roadmap. Should an organization be using an Agile methodology, this roadmap may serve as a backlog for planning either the Kanban or the Scrum. Otherwise, the roadmap just will define the milestones that need to be met in the product development journey to product launch.

Engineering/Scientific Roadmap

Figure 6.8 provides a roadmap and a set of guidelines for the entire electromechanical or combination product design and development process. In fact, the rest of this book is an explanation of this roadmap with the needed details and examples. Chapters and sections will provide specific details of the steps needed for each portion of the roadmap. In addition, ancillary information needed to accomplish these tasks will also be provided. Briefly, the roadmap is comprised of three sections. In the first part, the design team participates in understanding the overall requirements of what is to be developed and cascaded. At the end of this segment, concepts are developed, and not only are their risks understood, but also the impact of variations on expected outcomes. In the second part of the roadmap, the concepts are fully developed as functional designs. The impact of cost and buy versus make decisions on the business is examined, and design output documents such as formulation details, engineering drawings, or component specifications are developed. Finally, in the third part of the roadmap, steps are taken to prepare for product verification, validation, launch, and market activities. In a nutshell, this book attempts to fill a gap for doing design *right* by understanding the elements of systems and concurrent engineering. Furthermore, by developing a fundamental understanding of the working principles of the product, it is made more immune to changes and variations.

Combination Products

At the risk of being redundant, I like to provide a brief description of a combination product. Consider, for example, an inkjet printer. The system is comprised of an electronic printing subsystem along with an ink cartridge. In turn, the ink cartridge is made up of a quantity of ink along with the housing. The ink is a consumable product, whereas the ink housing is typically disposable.

Another combination product is an automobile. It is made of the electromechanical system plus the fuel, which is a consumable product. The third and the final example that I would like to share is a medical infusion pump. This system is made up of the pump along with the disposable IV sets and the bags, as well as the medicament contained within the bags.

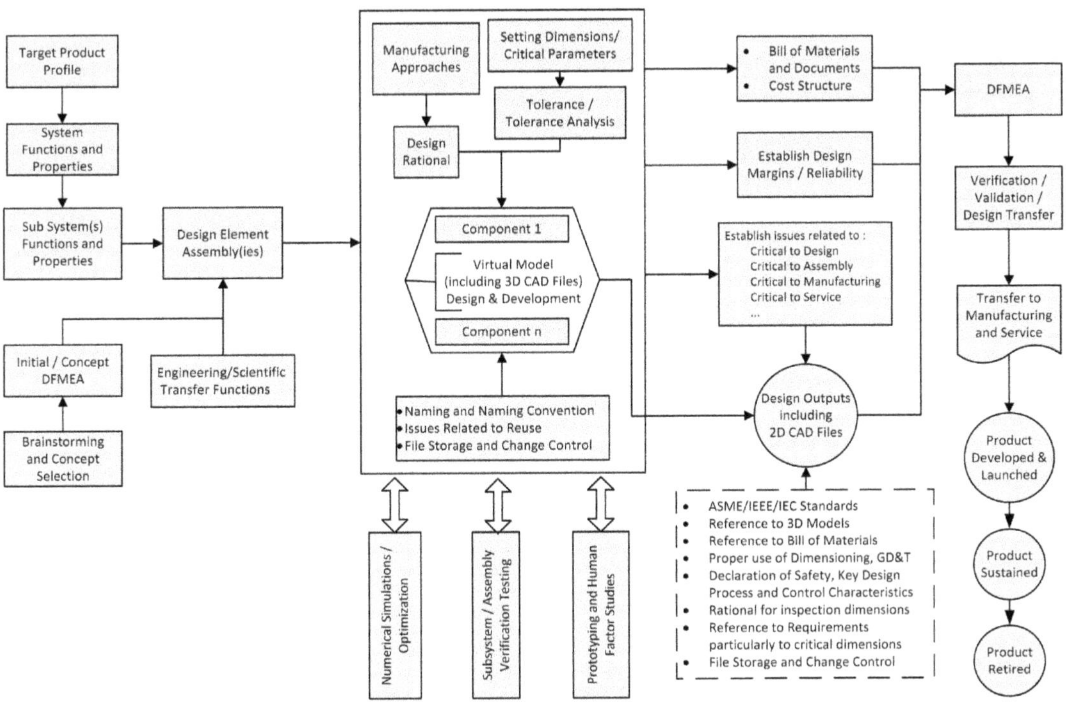

FIGURE 6.8
Technical product design and development roadmap

From a design point of view, it matters very little whether a system under development is purely electromechanical, chemical, or disposable. The principles of product development are the same. What matters is the level of attention needed to ensure that the requirements of each area are properly addressed and their interactions are well understood.

Summary

- In the last six or seven decades, a number of product development models have been developed to bring efficiency and to ensure that the finished product is what the customer wanted.
- A product development model that has gained some currency in the United States is called the V-model.
- This approach and model has given birth to a discipline called systems engineering.
- Quality by design means that the design, development, manufacturing and service must follow processes and activities that do not generate waste.
- Agile has been celebrated as a methodology that is responsive to changing customer and market needs.

Notes

1 ®Stage Gate is a registered trademark.
2 Different organizations may use different technologies.
3 This is a reference to Agile values and principles as explained in Appendix A.

Part III

Requirements and Their Cascade

7

Systems and Requirements

Introduction

In the last chapter, I made a number of references to *systems engineering* and *requirements.* Additionally, the first steps in the product development roadmap (Figure 6.6) begin with product requirements document as well as the systems requirements. The goal of this chapter is to provide a high-level review of requirements and requirements management documents. Associated with these artifacts, there are architecture and design documents. I will define these and provide a definition of their role and place in designing a product.

Systems Definition

Requirements are considered to be one of the seven key elements of systems engineering; the others include the systems engineering management plan, master schedule, detailed schedule, work breakdown structure, technical performance measurement, and technical reviews and audits (Martin 1997).

This sounds impressive, but what does a "system" mean? And how does systems engineering differ from other functional technical disciplines such as chemicals, material science, mechanical, electrical, or software engineering.

Let's explore the meaning of a *system*: In response to a market need, corporations – large or small – work on providing a "product." This product may be a service, a family of devices, or a combination of the two. The novice, often, thinks of the device (or the service) as the system because it is the "product" that the end user interacts with. Strictly speaking, this is not a correct definition of a system.

For example, consider a blood sugar monitor (the "product"). This product is a corporation's response to a customer's need to monitor his or her blood sugar daily at a time and place that is convenient to the user. However, this product did not just happen: it was designed, manufactured, tested, marketed, sold, and, if damaged, serviced; and finally, when the product is no longer used or useful, disposed. One may consider these various organizations and their interactions as the "system." Now consider this: in addition, suppose that this unit uses special strips that capture the end user's drop of blood for analysis in the unit. These strips must be available to the user on a daily basis. Wouldn't the customer support organization and the delivery company (e.g., FedEx, UPS, and/or U.S. Postal Service) be considered as a part of the system as well? As Hatley et al. (2000) point out, "Every system below the level of the whole universe is a component of one or more larger systems." However, to be practical, the "system" and its boundaries must be defined by the design team.

Industry standards define systems in different ways. For instance, IEEE 1220–1998 defines systems as: "A set or arrangement of elements and processes that are related and

DOI: 10.1201/9781003564508-10

whose behavior satisfies customer/operational needs and provides for life cycle sustainment of the products." Whereas ISO/IEC 15288:2008 defines "system" as: "A combination of interacting elements organized to achieve one or more stated purposes." At the same time *NASA Systems Engineering Handbook* (2007, p. 275) gives a more comprehensive definition:

1. "The combination of elements that function together to produce the capability to meet a need. The elements include all hardware, software, equipment, facilities, personnel, processes, and procedures needed for this purpose."
2. "The end product (which performs operational functions) and enabling products (which provide life-cycle support services to the operational end products) that make up a system."

In general, a system is a combination of the "end" product(s) as well as product "enablers." Product enablers are all the ancillary organizations that make it possible for the end user and/or customer to make use of the device or service.

From a more practical point of view, for relatively simple applications, a system is generally defined as the end product or a combination of end products. In more complicated situations, a "system of systems" is defined in an effort to partition complexity into simpler segments. Thus, each system may be considered independently. And the interactions of various systems are considered and accounted for at a higher system of systems level.

In general, a system is a combination of the "end" product(s) as well as product "enablers." Product enablers are all the ancillary organizations that make it possible for the end user and/or customer to make use of the device or service.

In the previous blood monitor example, the *system of systems* (SoS) may be defined as the business of providing blood sugar monitoring to the market. Depending on the decisions made at the SoS level, one of the constituent systems may be the electromechanical monitor. Another may be the disposable items (such as the strips and the sterile scalpel used to draw blood). Figure 7.1 provides a pictorial example of the blood sugar measurement business and its various segments, which may be considered as systems in and by themselves. In this example, the end products are the handheld monitor and the disposables (i.e., strips and lancets); however, these end products can only be realized if they are supported by the enabling systems such as manufacturing, service, or other organizations.

Earlier, I mentioned that industry standards indicated that a *system* is a combination of elements. Let's call these elements as building blocks. For example, the elements or building blocks of an automobile consist of an engine, chassis, wheels, and its body, among others. By way of another example, we could envision that a display, along with a keyboard,

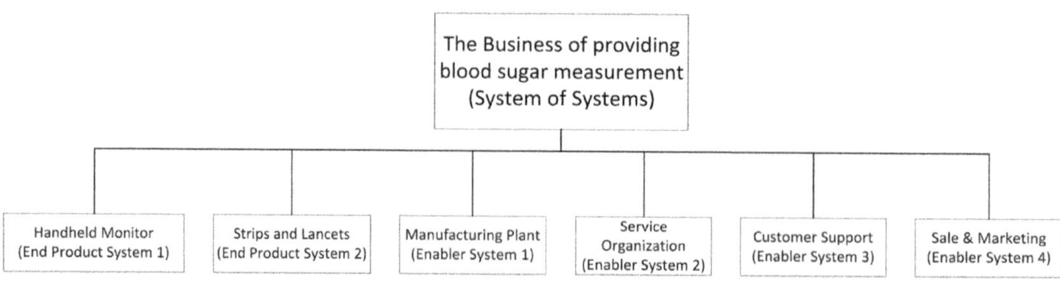

FIGURE 7.1
A pictorial representation of system of systems

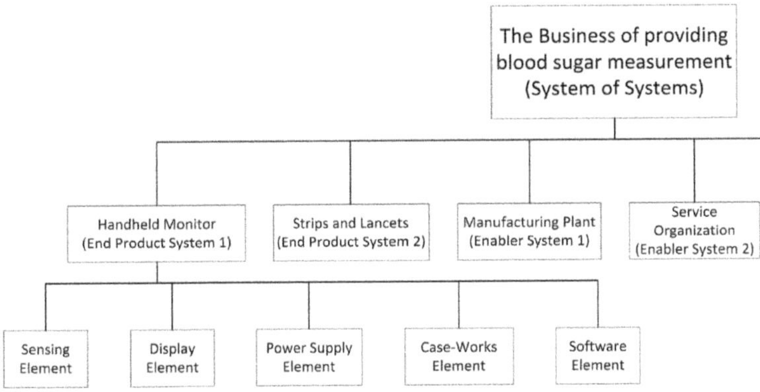

FIGURE 7.2
A pictorial representation of system of systems and its lower-level elements

hard drive, and a motherboard are elements or building blocks of a computer. For the blood sugar monitor example, these elements may be sensing, display, power supply, casework (enclosure), and software. Hence, the systems of systems diagram in Figure 7.1 may be updated as shown in Figure 7.2 to reflect these various building blocks. Later, I will suggest that these building blocks be called subsystems.

Within this context, the role of the systems engineer is to define various building blocks of a system and to ensure that these elements are seamlessly integrated and that they all function with coherence.

Voice of Customer and Voice of Stakeholders

Typically functional (e.g., mechanical, electrical, and software) teams are trained to solve problems. Often, the desire to provide a solution is so strong that it overshadows the problem statement itself. Systems engineers balance this tendency by first focusing on the problem statement and dwelling on it so that it is properly defined. They call this first description of the problem statement the voice of the customer (VOC).

The VOC is a reflection of a market demand and, hence, an opportunity for financial gain. By suggesting that the VOC is a reflection of market needs and demands, we accept the assumption that customers know what they want. While there is debate among entrepreneurs and product innovators on the veracity of this statement,[1] a product that is launched into a market has to meet the recognized or unrecognized/unmet needs of customers to be financially viable. As such, the VOC is considered to be the highest-level requirement.

In reality, VOC is not the only voice that is heard by systems engineers. Business has a voice of its own, and so does manufacturing and service. For regulated industries such as avionics and medical, compliance with regulatory agencies brings unique voices to the table as well. Collectively, systems engineering brings the voice of stakeholders (VOS) together and develops the highest-level design document, typically called a Product Requirements Document (PRD). This document reflects the integration of various voices that go into shaping the initial requirements of a product and its final outcome(s). It is the first product development document.

> Voice of the customer is a reflection of a market demand and, hence, an opportunity for financial gain.

Agile Personas

A systems designer does speak of the voice of the customer, but often there is not a personal interaction between the designer and the customer. That being the case, how can one claim that they have an understanding of the customers' wants and needs beyond a document that is called a VOC?

In his seminal work, *The Inmates Are Running the Asylum*, Cooper (2004) suggests: "If you want to create a product that satisfies a broad audience of users, logic will tell you to make it as broad in its functionality as possible to accommodate the most people. Logic is wrong. You will have far greater success by designing for a single person" (Cooper 2004, p. 124). Thus, he introduced the concept of *personas.*

A persona is an archetypical user of a product – someone who could truly be called a "typical customer" or a "typical user" of the product. It has to be developed based on real knowledge of the end users. Needless to say, as there are a variety of customers, there may need to be a number of personas to reflect a fairly complete picture of customer needs. Having said this, should an excessive number of personas be identified (typically more than three or four), it may be necessary to develop a product family for each family of personas as needed.

Cooper (2004) gives the following suggestions for developing effective personas:

1) Discover personas as a part of market research into user needs; do not imagine them from your office desk.
2) Write specific personas. Personas should have names, likes and dislikes, and strengths and weaknesses. Personas use products in specific ways. Do not adjust personas to your product; rather, adjust the product to your personas
3) Be hypothetical. Understand their skill levels in using the product as well as their pains in coming up to speed in learning how to use it. Understand their goals.
4) Be precise but do not try to be exact; after all, a persona is a collage of a group of individuals having an interest in the product that you are developing.

Cooper (2004) suggests that in his team, they do not speak of a *user*; rather, the persona is often referred to by a given name. Thus, a "primary" persona needs to be identified. One whose needs are not satisfied if the product is designed with another persona in mind. Conversely, it is also important to identify "negative" personas. In other words, the individual who will be satisfied with the product being developed.

I will discuss this issue later in chapter three as we discuss product requirements in the light of Agile and how personas and user stories may be connected.

Developing a Persona

To develop a persona, the design team needs customer data that is not based on assumptions. To collect real-life data, tools such as market research, customer interviews, and feedback may be utilized. Customer surveys and feedback should focus on creating an empathy map (see Bland 2020; Brown 2018) and on the following areas:

1) What do customers hear from their friends, colleagues, and thought leaders?
2) What do customers worry about? What do they aspire to?
3) What do customers see in the marketplace? In their environment?
4) What do customers advocate for?

For more information on developing an empathy map, I do recommend both Brown (2018) and Bland (2020) articles.

Common Pitfalls

It is too easy to assume that we know who the *typical* customer is and write a persona based on our own expectations. This mindset needs to be avoided as it can easily lead to the wrong product feature sets. Additionally, personas should not be confused with *actors* as used in use-case modeling. One way of reconciling the difference between an actor and a person is to think of it as personas are archetypical instances of actors. The design team should be able to relate to personas as if they were humans, whereas actors are often impersonal. Furthermore, personas should not be confused with market segments and demographics. For instance, thinking about 15- to 19-year-olds – or late teenagers – is very different from thinking about Frank, a 17-year-old who is thinking about studying histology.

Finally, keep in mind that in Agile, we develop everything in iterations. Developing personas is no exception. As the team learns more about the archetypical user, its persona needs to be updated and kept "alive."

Example of Personas

The following are examples of two personas. Parvin's persona is that of a special education teacher, whereas Michael's persona is that of an individual with severe autism. I will make further use of these personas in Chapter 8 to set the foundation for developing an electronic system to help educate Michael. What is important to note here is that these two personas – one of a special education teacher and the other of the special needs student – will interact differently with this system. Personas will bring these varying interactions home for the design team. Some would also add a picture to the persona for more impact.

PARVIN JACKSON

Thirty-four-year-old Parvin is married and a mother of two children. She has a master's degree in special education and graduated ten years ago. She works at a center for children with mental health issues and cognitive delays. Parvin aspires to manage the center in the next five years.

She lives close to her elderly mom, cooks a few dishes on Saturdays so that she has something to take for her mom on Sundays and have leftovers for the rest of the week. In her spare time, she likes to knit and do needlework. At times, she donates them to the local charity for the annual sale to raise money or as Christmas presents for disadvantaged children.

At her work, every morning when she comes in, she visits every classroom and greets every student. She wants to make sure that she is aware of how each child feels and comforts children who may be experiencing discomfort. To her, the most important element is that each student feels they are cared for. She prepares lesson plans on a weekly basis.

Parvin tries to keep up with the latest developments in her field. She reads two quarterly scientific journals and shares any worthwhile developments with her team and/or psychiatrists. She tries to influence her superiors to purchase any equipment that she believes would help her students.

She fears that budget cuts will impact students' education and advocates for her team and her center to get what they need.

MICHAEL RAFFI

Michael is a five-year-old nonverbal boy with severe autism. He was diagnosed with the disorder when he was two years old. By the age of three, he was enrolled in special education at his local school.

Michael's routine is as follows. His mother wakes him up around 6:00 AM, helps him wash, brush his teeth, and put his clothes on. While his mother puts his clothes on, Michael likes to play with his mom's ears and nose. For breakfast, he eats peanut butter and grape jelly and a glass of milk. Any changes from this routine will cause distress for Michael, which may result in him throwing a temper tantrum. After breakfast and while waiting to go to school, Michael runs back and forth in the living room in a straight line while making various sounds.

At school, he has his routine as well. When he enters his classroom, he places his backpack on his shelf. He then runs to the crayon box and selects his favorite color of the day. He holds on to it all day and leaves it in the box just before going home in the afternoon. He would throw a violent tantrum if anyone tried to take it away from him.

Michael is non-verbal and functions on a lower cognitive level as compared to his peers. For example, while he recognizes that there are different shapes and colors, he does not identify them when asked.

Document Collection and Control

VOC and VOS are among the first documents that form a collection of files (or documents) that profiles the development of a product from conception to retirement. Some organizations do not have a specific name for this collection of files, while others may refer to it as the design evolution collection. For a systematic approach to product development, it is important that this collection be maintained and updated to keep a record of how a design has evolved from its conception. As various members come into the design team or leave it, it can be tapped as a source of truth for the decisions that have been made and the rationale for them.

In the medical field, this collection is called the design history file (DHF) that is retained under revision control. Because of the simplicity of this terminology and its meaning, in this book, I will adopt this term. Often, if there is a legal dispute, it is the documents in the DHF that are provided as evidence. In addition to protecting the business in the case of litigation, there are very practical reasons why these collections should be properly maintained. On the one hand, in regulated industries such as the medical industry or avionics, these files are subject to regulatory body inspections. On the other hand, engineering or technical change notices can be done much more efficiently if the history file is properly maintained.

Design Documents

Design documents for a product contain the requirements and specifications for that product, that is, they identify what function the product performs – and depending on the hierarchy of the document – how the function is performed. There are typically four levels of documents in a hierarchy: system, subsystem, assembly, and component-level documents. The number of levels could change depending on the complexity of the product (in simple designs, remove a layer; in more complicated designs, add additional layers).

> Design documents for a product contain the requirements and specifications for that product, that is, they identify what function the product performs – and depending on the hierarchy of the document – how the function is performed.

At the highest-level document, the "what" of the product function is described. In a way, the voice of stakeholders in a semi-technical language is expressed at this level. Each lower level provides more detail in describing the "how" of the design. Eventually, the component-level documents provide detailed design descriptions. In the blood sugar monitor example mentioned at the beginning of this chapter, a high-level requirement may be stated as follows:

Product design shall be rugged.

The next-level document may translate the ruggedness requirement as such:

Product design shall survive a drop from a height of 1 meter onto a wooden surface.

This requirement at the lowest level may be satisfied by the following specifications:

Enclosure wall thickness shall be 0.100 inches.
Enclosure material shall be polycarbonate plastic.

Typically requirements flow down from "what" to "how" by "which" (subsystem, assembly, or component); where, *how* are the requirements for the lower level, and *which* becomes the description of the structure. The *what-s* to *how-s* flow down forms the requirements documents, and the *what-s* to *which-s* develop the systems architecture.

Figure 7.3 provides a pictorial view of the requirements and their cascade. The top-level requirement document contains the PRD, which reflects the voice of the stakeholders as well as performance characteristics of the product or product family.

The architecture documented at the second and third levels describes how various requirements are distributed to multiple subsystems or assemblies. In traditional approaches, there may be a tendency to create the architecture along each technical discipline (i.e., mechanical, electrical, or software). In a systems approach, a cross-functional product development team determines the number of subsystems or assemblies, typically along product functionality, or physical or operational boundaries, and develops the architecture documents.

The system requirements document constitutes the top-level technical definition of the product. This document translates marketing needs into a technical language that sets the basis of the product to be developed. At this stage, the voice of the stakeholders is expressed in concrete concepts that are quantifiable and verifiable. This document is usually developed by the technical leads collaborating with the lead systems engineer.

> The system requirements document translates marketing needs into a technical language that sets the basis of the product to be developed.

Along with system requirements, the system architecture is also developed, dictating various subsystems and their relationship to one another. Subsystems, generally, are defined in such a way as to segregate various specific operations or functions of a product or along physical boundaries. For instance, in an automobile, the powertrain belongs to one subsystem and the chassis to another; in a medical insulin pump, the pumping mechanism would belong

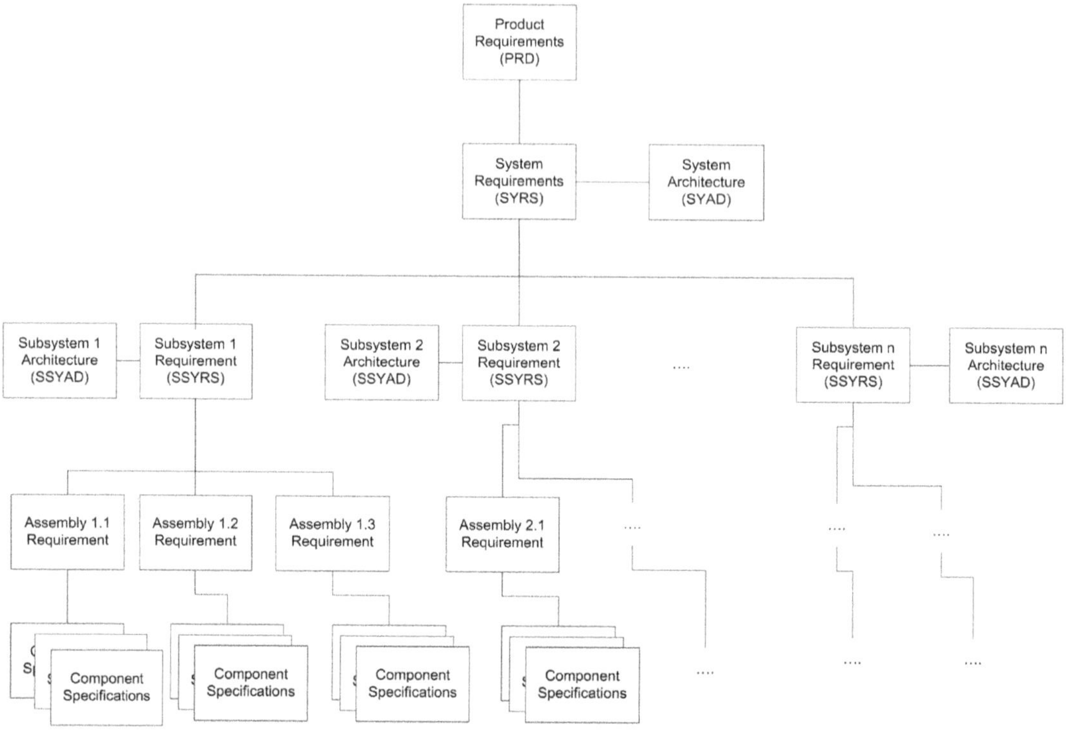

FIGURE 7.3
Requirements and architecture hierarchy

to one subsystem and the power supply to another. Subsystem design documents contain the requirements and functions allocated to them by the system architecture.

The next layer of the requirements cascade is the assembly design document. This level of the requirements tree identifies various functions within a subsystem. For example, within an infusion pumping mechanism subsystem, there is a tube-load assembly, a pumping assembly, an occlusion sensing assembly, and so on. The term "subassembly" may be used to describe lower levels of an assembly. It should be noted that there may be any number of assemblies under each subsystem, and similar to subsystems, assemblies should be created in operational, functional, or physical blocks. Subsystem architecture dictates the distribution of subassemblies and their relationships.

Those with a strong manufacturing background may equate the assemblies that I am speaking of to those defined by manufacturing processes. The two are not the same. While, from a design point of view, an assembly document is the lowest-level document containing requirements. From a manufacturing point of view, an assembly may be one of several intermediate steps to physically build a portion of the product. Even though the design team may define the architecture with its particular subsystem and assemblies, this structure may not reflect the steps required by manufacturing procedures. As a result, manufacturing organizations may define assemblies differently than what is presented in the design architecture diagram. This does not imply that the manufacturing process does not align with design requirements. On the contrary, each manufacturing step does require its own documentation with requirements decomposed from design specifications.

In Figure 7.3, component specifications are shown under the assembly documents. Component specifications (by definition) may be any document that describes the component to be used. It may be specifications from vendors or engineering drawings and/or technical models.

A facet of component specifications is that it should identify significant (or key) characteristics and/or critical dimensions; provide justifications and traceability to the requirements tree and ultimately to the PRD. This tracing provides a logical flow of original design intent into product realization. In the medical field, this traceability is a Food and Drug Administration (FDA) requirement. See, for instance, Teixeira (2014).

Typically, high-level documents, that is, system, subsystem, and assembly documents, are reviewed and approved by members of associated cross-functional teams. Conversely, component specifications are reviewed and approved by specific discipline subject matter expert.

Subsystem Design Document

The subsystem design document consists of the requirements allocated by the system architecture to the specific subsystem. So, it is simply a list of the requirements that apply to the same function. The purpose of this document is to group requirements and to provide operating parameters.

A facet of component specifications is that it should identify significant (or key) characteristics and/or critical dimensions and provide justifications and traceability to the requirements tree and ultimately to the PRD.

Subsystem Architecture Document

The subsystem architecture document divides a subsystem into assemblies and assigns requirements to each assembly. This document describes the design intent and reasons behind design decisions made during the design process. It has four major sections:

1. Theory of Operations,
2. A Description of Integrated Assemblies,
3. Requirements Allocation, and
4. Design Rational

"Theory of Operations" provides a high level description of the subsystem's functions. It is a general overview of what the subsystem does. It briefly describes how the subsystem executes its functions.

Another section of the subsystem architecture model describes the assemblies that are integrated within the subsystem. Subassemblies should be grouped by functions, which at this level (the lowest level) may be strongly correlated with the physical grouping. Generally, block diagrams describing subassemblies and their interaction with each other are provided in this document.

In the "Requirement Allocation" section, requirements are assigned to each subsystem. These requirements, which may span over several assemblies, collectively fulfill the subsystem requirements.

"Design rationale" is a document where justifications and explanations for the design approach and intent are provided.

Assembly Design Document

The assembly design document contains the requirements that are assigned to the subassembly. The document also contains a description of how the assembly works. The major sections of the assembly design documents are:

1. Theory of Operations,

2. Design Rational,
3. Requirements Allocation, and
4. Relevant Drawings or Design Specifications

"Theory of Operations" is a section of the assembly design document where an overview of how the subassembly works is provided. This is a higher-level view of how the 'black box' of the subassembly works.

The "Design Rationale" segment provides justifications and explanations for the design approach and intent. Considering that this is the lowest-level design document, rationales for component-level choices as well as critical dimensions should be provided in this segment. As an example, the material choice of a critical component or the dimensioning scheme of a critical interface should be justified in here.

The "Requirements Allocation" section lists the requirements assigned to the subassembly.

In the "Relevant Drawings" section of design documents, drawings or specifications of critical components are referenced and/or discussed.

Summary

- The first steps in the product development roadmap begin with developing product requirements document as well as the systems requirements.
- A system may be more than just a product. It needs to be defined, and its requirements understood.
- The role of the systems engineer is to define various building blocks of a system and to ensure that these elements are seamlessly integrated.
- Voice of customer is a reflection of a market demand, and hence, an opportunity for financial gain.
- Design documents for a product contain the requirements and specifications for that product.
- The subsystem architecture document divides a subsystem into assemblies and assigns requirements to each assembly.

Exercises

1. Consider a simple household item such as a coffee mug. Develop a list of customer needs. Translate those needs into design requirements.
2. Consider a more complicated household item, such as a television. Similarly, develop a list of customer needs and system-level requirements.
   a. Develop the system and subsystem architecture.
   b. Cascade system-level requirements to each subsystem.

Note

1 The famous "if I had asked people what they wanted, they would have said faster horses" quote, generally attributed to Henry Ford, speaks to this fact.

8

Developing a Product's Requirements

Introduction

In Chapter 6, I introduced a model along with a roadmap of activities that help us (the design team) navigate the product development maze. The initial steps in the process (or road) map required that we start with developing the product requirements document. Chapter 7 helped us develop an appreciation for voice of customer, requirements, requirements cascade, and design documents. In this chapter, I will revisit the topics covered in Chapter 6 and provide more detail to lay a foundation of a systems-level understanding for the design and development of electromechanical products.

Since the roadmap (Figure 6.6) begins with developing PRD and that PRD depends on understanding customer needs, the questions to be entertained are as follows. How do we get voice of customer? How do we develop product-level requirements from them? And, how can we flow these requirements to lower levels?

The answer to the first question is in the realm of those with expertise in marketing – and any detailed treatment of the subject is beyond the scope of this work. However, a simplistic answer is that the business needs to first know who its customers are. Are the intended customers other businesses or individuals? Are customers the end users? By this question, I mean whether the customer would personally use the purchased item or the customer buys on behalf of a third party. The tools for gaining this information are user studies, surveys, and other similar mechanisms.

The answers to the second set of questions on developing the PRD and requirements flow-down to subsystems and assemblies will be explored in this chapter.

Defining the Product to Be Developed

In Chapter 6, the first step in DMADV was called *Define*. Typically, this means that the problem statement is created in this step.[1] This requires a knowledge and understanding of the market and potential financial gains – either through sales and profits or through savings to the bottom line. Writing a clear and concise problem statement may require some practice; however, a complete problem statement is often created through a repetitive process. As we learn more, this statement may be updated and refined further with additional relevant information.

> In the product development process, a problem statement should identify intended customers, added value or benefit to the customer, and other relevant known information.

A problem statement should identify intended customers, added value or benefit to the customer, and other relevant known information. Building on the persona developed in

DOI: 10.1201/9781003564508-11

Chapter 7, the following is an example of a statement addressing problems faced by children with severe autism. Note that the personas for this market were developed in Chapter 7.

> Studies have shown that children with severe autism[2] respond more effectively to instruction by computers that other children or teachers. Furthermore, children with severe autism exhibit behaviors that may lead to computer damage. Providing a robust computer systems that is not easily damaged allows us <the business> to provide learning opportunities to this segment of population and bring significant advantage over our competitors.

As a part of defining the problem statement, the project scope as well as current versus desired states are also developed. An example of current versus desired states for the previous problem statement may be as follow.

Current	Desired
Child with severe autism: • Easily damages computer • Is easily distracted by various computer features • Does not easily operate mouse	Child with severe autism: • Easily interacts with computer • Is task-focused • Does not depend on a mouse
Business: • No market presence • Market presence is needed for competitive advantage	Business: • Strong market presence in three years • Become a market leader

Once the cross-functional team has agreed on the current versus desired states, the project scope is properly defined. Project scope – as the title implies – describes the boundaries of the project, where it begins and ends. Similarly, it is just as important that the out-of-scope activities are also defined. For the previous example, the following in-scope/out-of-scope conditions may be written:

In-Scope	Out-of-Scope
• The new product will use existing technologies to provide a solution • Initial market will be English-speaking countries	• Any custom-made electronic hardware or software • Any CE certification requirements

Often, products are launched into the market, and they fail not because they are bad products, but because they were not scoped properly. Time and resources were dedicated to inconsequential issues. Another factor that potentially could lead to either costly product launches and/or failures is a lack of understanding of what the strengths and weaknesses of the business are. To develop an understanding of a business position and standing, a SWOT analysis is done.

> A SWOT analysis is an honest discussion of an organization's ability to enter a desired market.

Cooper et al. (2001) consider the bulk of SWOT analysis to be conducted as a part of the marketing activities; though, an aspect of this analysis is to make a realistic self-assessment of the business as well. This implies that for the opportunity at hand, marketing and sales channels, along with R&D, manufacturing, and operations, need to be evaluated. The results of this activity are generally reported as a summary in the Define stage. For the hypothetical problem statement mentioned earlier, the following SWOT may be given:

Opportunities	**Strengths**
There is an opportunity to grow the business by 25%	The company is a leading educational electronics manufacturer

Weakness	**Threats**
Expertise on autism is nonexistent within the company. Furthermore, sales and marketing channels are developed primarily in the K through 6 market	Lack of expertise on autism could lead to product launch delays. This, combined with unexpected new products launched by competitors, could jeopardize any market share realization

The final step in the Define phase of DMADV (Chapter 6) is to assess project risks. This could take the shape of technical risks, manufacturing and operations risks, or commercial and financial risks. As is typical with any risk assessment, the cross-functional team identifies risk mitigation factors for each identified risk. For the previous example, the following risk table may be presented as an illustration.

Risk	Impact	Probability of Occurrence	Mitigation
High Manufacturing Cost	Medium	Low	Source components in bulk and outsource manufacturing
Product Launch Delay	High	Medium	Anticipate risk factors early
Inadequate Response to Market Need	High	Low	Hire autism experts early in the project

Before I leave the topic of *define*, I would like to make a point here. This phase is arguably the most important phase of any project and should not be taken lightly. In here, it took only a few pages to provide a brief description and give an example – which incidentally was based on a product called the Learning Station which was developed at Airtronic Services but was manufactured and distributed by an electronics manufacturing company. It is highly recommended to stay in this Define phase as long as possible to ensure that the project is as fully explored as possible.

Measuring What Customers Need

Once the project is properly defined,[3] we move to the *measure* phase. During this phase, the team will collect VOC and voice of business. This is the realm of systems engineering. In my experience, functional teams are not typically very involved in this stage.

To collect VOC, systems engineers first identify their customer with the help of their marketing colleagues and then develop an understanding of them and how they operate a given task or function. For instance, in developing the Learning Station, a systems engineer should first develop an understanding of how an educator would interact with a child with severe autism. How would one teach this individual, and what challenges do they face?

Griffin (2005) provides three techniques for obtaining customer needs. These are: be a user, critically observe users, and finally, interview users for their needs. Each approach has its own benefits and drawbacks; however, this is a task that requires a high degree of diligence. Cooper (2001) provides a checklist of tips and hints. These tips and hints ensure that the customer needs study considers what its objectives are, uses a structured questionnaire, the right information is solicited, a representative sample of the population is interviewed, the technical people are involved, and finally, the use environment is considered.

Rarely would any VOC survey return pure and true customer and/or user needs. Often, this information reflects what the customer may see as a solution to the problem or a different design based on an existing solution.

Another noteworthy factor is that customers are not always the same as end users. For instance, for the Learning Station mentioned earlier, educators and ultimately children with autism are the end users, but the customers are those who actually purchase the product. In the case of a medical device, while the end users are clinicians, the customers may be hospital administrators. The needs and wants of both groups should be considered and studied.

Once sufficient data is collected, it should be organized and analyzed. Rarely would any VOC survey return pure and true customer and/or user needs. Often, this information reflects what the customer may see as a solution to the problem or a different design based on an existing solution. The following is an example of a list of VOCs.[4]

- Once we had a client who threw a tantrum and threw the monitor out the window. Computers and their monitors should weigh about 60 pounds (nearly 30 kg) so that clients won't be able to lift them.
- A couple of years ago, we had this cute little boy who became enamored with the computer speakers. One day, he decided to cut the cables and take them home with him.
- Sometimes our students are more interested in the CD[5] ROM tray moving in and out or the computer knobs and buttons than what is going on on the screen.
- So, what is the use of a computer? Children need to learn to interact with others, particularly those with autism. They need to learn social skills and communication.
- Internet is the way to go. Any computer that does not get on the information highway is a thing of the past. It has to run smoothly and start up just like a Mac Computer.
- I have only a limited time to engage and keep a child's attention. Often times, if the student does not like the program that I have in mind, I cannot stop and change the CD fast enough. I should be able to change the CD in less than 30 seconds or so.
- No one seems to be able to make something that lasts. It seems that the glass on the tablets breaks every other day.
- Many of my younger kids do not get how the mouse works.

In analyzing the VOC list, the first step is to develop "bite-size" statements. Thus, for the previous example set, the following may be developed:

- Thrown and damaged monitor.
- Product weighing 60 pounds (30 kg).
- Damaged and removed speakers.
- CD-ROM tray distracts.
- Develop social skills.
- Develop communication skills.
- Internet is the way to go.
- Similarity to Mac Computers.
- Limited attention span.
- Change CDs in less than 30 seconds.
- Product should last.
- Glass on tablets break.
- Working with the mouse is not obvious.

De Feo (2010) recommends the use of spreadsheets in order to analyze and organize customer statements. The first step is to identify categories for this organization. Typically, *needs* and *requirements* are among this classification. Others may include "preconceived notions" of what the new product should do, look like, etc. The second step is to develop the true needs. With this in mind, we can develop the spreadsheet as shown in Table 8.1.

As the statements from customers are collected and analyzed, there may be a realization that there may be gaps or issues that have not been communicated clearly. In the example of the Learning Station, I needed to go back and learn more about "Internet is the way to go, Similarity to Mac Computers," and "Working with the mouse is not obvious." Here is what I learned:

With this additional information at hand, we can complete our customer statement to customer need spreadsheet. For the example that we have been following, refer to Table 8.2.

- *Internet is the way to go*: I should be able to find information, games, and lesson plans that I can download directly to this computer. It will make my job so much easier than having to download stuff into my own computer and then bring it here.
- *Similarity to Mac Computer*: I like the look and feel of Apple Macintosh computers. They look slick. Besides, everyone knows that they are not susceptible to computer bugs.
- *Working with the mouse is not obvious* When many of my younger kids hold and move the mouse, they cannot split their attention between the mouse and the computer screen. So, they do not realize that when they click on the mouse button, something actually happens on the computer monitor.

TABLE 8.1

Customer Statement to Customer Need Cascade Table

Customer Statements	Need	Requirement	Preconceived Notion	Statement of True Need
Thrown and damaged monitor	X			I need a monitor that cannot be thrown or damaged.
Product weighing 60 pounds (30 kg)		X	X	I need a very heavy unit.
Damaged and removed speakers	X			I need computer speakers that cannot be damaged or removed
CD-ROM tray distracts	X			I need tray-less CD ROMS
Develop social skills	X			I need to work on social skills development
Develop communication skills	X			I need to work on communication skills development
Internet is the way to go			X	*
Similarity to Mac Computers			X	*
Limited attention span	X			I need to manage a limited attention span
Change CDs in less than 30 seconds		X		I need to change CDs quickly
Product should last.	X	X		I need a reliable product
Glass on the tablets breaks	X			I need shatterproof glass
Working with the mouse is not obvious	X			*

TABLE 8.2

Updated and Revised Customer Statement to Customer Need Table

Customer Statements	Need	Requirement	Preconceived Notion	Statement of True Need
Thrown and damaged monitor	X			I need a monitor that cannot be thrown or damaged
Product weighing 60 pounds (30 kg)		X	X	I need a very heavy unit
Damaged and removed speakers	X			I need computer speakers that cannot be damaged or removed
CD-ROM tray distracts	X			I need tray-less CD ROMS
Develop social skills	X			I need to help develop social skills
Develop communication skills	X			I need to help develop communication skills
Internet is the way to go			X	I need to have ready access to Information
Similarity to Mac Computers			X	I need protection against computer bugs.
Similarity to Mac Computers			X	I need to have units with slick looks
Limited attention span	X			I need to manage a limited attention span
Change CDs in less than 30 seconds		X		I need to change CDs quickly
Product should last	X	X		I need a reliable product
Glass on the tablets breaks	X			I need shatterproof glass.
Working with the mouse is not obvious	X			I need an intuitive mouse.

The third step in analyzing and organizing customer needs is to place them into categories. This activity is not trivial and, in general, will require participation of the development team.

This participation is often in the form of brainstorming, and an effective tool for brainstorming is called the affinity diagram. However, before I discuss this tool, first, I would like to say a few words about some related Agile terminologies, and then I would like to review some general brainstorming guidelines.

Stories, Epics, and Personas

A common theme found within Agile methodology used for software development is the concept of stories, epics, and the role of personas in developing them. Typically, epics are considered to be a collection of stories with a common goal or focus. Another view of an epic is a story that is too large to be completed within a single sprint. Epics and stories have the following format:

As a <persona, user, customer>,
I want to <perform what action>
so that I <can expect, complete a task>

As Ullman (2019) has pointed out, *perform what action* refers to a desired function of the product, whereas a *can expect* or *complete a task* refers to the requirement that the product needs to satisfy. In the case of the product being developed for children with autism, the following epic may be written.

> *As a special education teacher, I want a robust computer system with touch-screen so that I can teach cause and effect to my students.*

I have to admit that I do not see much benefit in using epics and stories in place of requirements in non-software product development. What is important in my mind is the accuracy and correctness of understanding of the true needs of the individuals who will be benefiting from the products that we develop. Furthermore, epics and stories are used in software development to help create the backlog. For hardware, this backlog is dependent heavily on the system architecture and its subsystems.

Having said this, there is an element where utilizing epics may prevent scope creep and enable the design team to release products on a more frequent basis.

Minimal Viable Product

The idea of minimum viable product (MVP) is to understand the needs of the primary persona using the product. This primary persona may be representative of 65% to 75% of the users; yet, focusing on their needs alone may reduce complexity significantly. This is in contrast with the traditional mindsets, where no compromises are made and the design team works to develop a complete set of requirements for as large of a user base as possible – a product that may take several years to be developed and launched.

In an incremental product release, MVP determines the minimal set of features along with their associated requirements (epics and stories) that would allow releasing a product sooner while at the same time enabling the design team to remain aligned with changes to market needs and requirements.

There is a caveat that I like to mention here. Incremental hardware release requires that the base architecture have the required foundation for growth. For example, an infusion pump designed to be used at a patient's bedside may not be readily upgraded to be used as an ambulatory pump in a backpack. As the minimal viable product is being developed, the architecture of its future expansion needs to be considered as well. In the language of TRIZ (see Chapter 9), this may present the design team with conflicting requirements, a factor that, in the past, design teams have almost always opted for choosing complexity over simplicity, leading to scope creeps.

Brainstorming Guidelines

Brainstorming has been defined as a group activity to solicit and obtain input from members of a team in an attempt to solve a problem. In other words, the purpose of brainstorming is to stimulate individuals' and groups contributions in cultivating and generating new ideas with the goal of discovering a solution to a problem.

Brainstorming provides a structured approach to generate a myriad of solutions to a given challenge. It is designed to enable free thinking and exploration of ideas. It should be used any time that the collective knowledge of a team is needed to solve a problem.

The mechanics of running a brainstorming meeting are rather simple: in addition to the *participants*, there is an *owner*, a *facilitator/moderator*, and a *scribe*. The person who presents the problem statement and seeks a solution is the *owner*. He or she will evaluate ideas after the meeting is over.

The *facilitator/moderator* organizes the meeting, obtains and/or provides any material needed by attendees or the *scribe*, and keeps order during the brainstorming by enforcing rules. The *scribe* is responsible for distributing material for sketches, notes, or other models/renderings of ideas and makes sure all ideas are recorded and provided to the *owner* after the meeting. The *scribe* should have enough background in the area being discussed to allow him or her to capture information with a minimum of explanation being necessary. The roles may be combined if deemed feasible, with the exception that the *owner should not* be the *scribe*.

> Ideas and concepts expressed during brainstorming sessions should not be judged; rather, they should be heard and accepted. One thought may spark a series of others. Eventually, all presented ideas and concepts need to be ranked and scored.

It is important that the *participants* are invited from a diverse group of people, be they, technical, management, or other unrelated business functions. The "looking in" perspective is valuable. There should also be members from the design or development team that have possible previous knowledge of relevant designs and implementations.

Meetings may vary in size and structure, but it is recommended that they should not last longer than two (2) hours. In each brainstorming meeting, the following eight simple rules should be observed:

1. Clearly define and focus on the problem.
2. Stimulate lateral thinking by encouraging wild ideas.
3. Visualize the ideas using sketches, models, etc.
4. Build on each other's ideas.
5. Generate as many solutions as possible from everyone involved.
6. Defer judgment until after the brainstorming is over.

7. Minimize interruptions to maximize positive collaboration.
8. Record all ideas and concepts.

When the brainstorming meeting is completed, the *owner* should

1. Rank concepts and/or solutions using a list reduction technique (or multi-voting technique) such as an affinity diagram and identify three (3) to five (5) concepts for further exploration.
2. Rational decision-making process may be used to select an ideal approach if the need be. I will describe this method in some detail later.
3. Document the results and provide feedback to participants and management on the progress made.

It should be noted that the problem statements addressed in brainstorming meetings change as the project moves forward, and as the problems encountered become more defined and smaller. For instance, at the start of a project, the problem statement may be very broad, such as "what is the theme coming from a set of customer needs?" At a later meeting, the problem statement may have a narrower scope, one such as "what features will satisfy a customer need? And, eventually, "how can we eliminate the 'X' button?"

Affinity Diagram

This approach of brainstorming is used to capture the input and thoughts of a large group of people. It typically begins by the gathering in a room of sufficient size with an empty wall or a whiteboard.[6] The owner provides the "Problem Statement." In the first round, everyone is given the opportunity to develop their thoughts and express them on a Post-it paper. One by one, participants give a brief description of their concept/idea and stick the paper on the wall or the whiteboard. In the consequent rounds, participants read the ideas and concepts presented by others and build on them – presenting yet newer ideas or different versions of existing ones. Once the team feels that nothing new may be generated, collectively, they begin to group the concepts and ideas in similar categories. Once this work is complete, a hierarchy of objectives may be developed.

Ranking High-Level Needs

Using the affinity diagram or similar techniques, customer-specific needs may be translated into higher-level needs. There are a number of ways to present this information: via, graphs, Pareto charts, or just simple tables. In here for the assistive technology example, Triptych, a commercial software tool, was used in the management and arrangement of the Affinity ideas and their groupings. The result is shown in Figure 8.1. In this figure, light grey boxes are the expressed VOCs, whereas the darker color boxes are higher-level needs.

Note that by arranging the concepts in this manner, a "sanity" check may be done to ensure any misclassifications and/or gaps are identified. For projects where the number of specific needs is much larger, each grouping may need to be reviewed separately. Furthermore, there may be instances that each group may need to be combined so that a higher level may be achieved.

At this point of product development, the elements of the product and what would appeal to the customers and users begin to take shape. The issue facing the development team is to decide which of these "Needs" are more important for the team's focus and

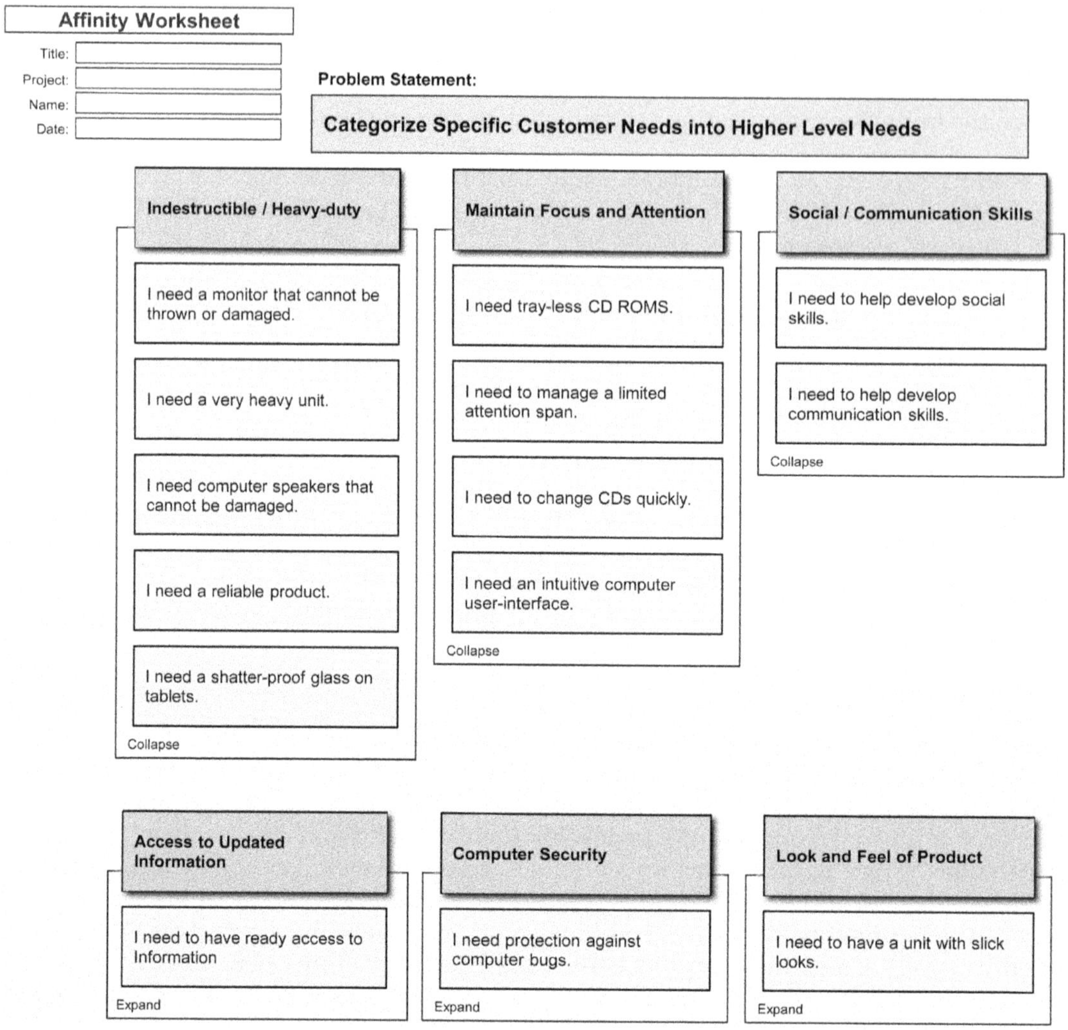

FIGURE 8.1
From specific customer needs to higher-level needs

which ones require a lesser degree of attention. In fact, it may be rather difficult to align the expressed needs in a ranking from the lowest to the highest and measure their relative importance to each other.

A solution for this problem was proposed in the late 1920s. It is called the paired comparison; see Mannan (2012). Later, Saaty (1980) developed the Analytical Hierarchy Process, or AHP, that is widely used today. This technique is utilized not only in product development and design but almost anywhere that the human decision-making process is used, from human factor studies to risk analysis to project management [see Lamb (2003–2004), Badiru (2012), Gawron (2008), and Sen and Yang (1998)]. This technique is discussed in the next section.

Analytical Hierarchy Process

In the context of ranking the relative importance of the high-level needs, Saaty's Analytical Hierarchy Process (AHP) enables the development team not only to put these in order from high to low but also to provide a mathematical weighting for each placement.

This problem statement is defined as follows. Suppose that there are n number of high-level needs denoted by $N_k, k = 1,2$ to n. Using AHP, the relative importance of N_i compared to N_j should be determined.

Without engaging in the details of derivations, the solution to this ranking is the eigenvectors associated with the following matrix equation Pillay and Wang (2003).

$$P = \alpha_{ij}$$

where α_{ij} is the preference of N_i against N_j for $i, j = 1,2$ to n. There are two rules associated with this equation. First, if the preference is that N_i is equal or equivalent to N_j, then, $\alpha_{ij} = \alpha_{ji} = 1$. This implies that $\alpha_{ii} = \alpha_{jj} = 1$. Second, if N_i is preferred to N_j, then, $\alpha_{ij} = p$, where p is the strength of preference. This also concludes that $\alpha_{ji} = \frac{1}{p}$. See Table 8.3.

In order to develop the solution, key customers are asked to rank each pair, and the matrix $P = \alpha_{ij}$ is formed. The eigenvectors (w_k) associated with this matrix provide the desired ranking of the high-level needs. For larger matrices, the exact calculations may prove tedious. Pillay and Wang (2003) provide a good estimate for calculating the preference vector as shown below.

$$w_k = \frac{1}{n}\sum_{j=1}^{n}\left[\frac{\alpha_{kj}}{\sum_{i=1}^{n}\alpha_{ij}}\right] \text{ for } k = 1,2 \text{ to } n \tag{8.1}$$

Any study in human factors by necessity does not produce absolute numbers. These studies typically provide the development teams with various probabilities of what (or how) the end users may do something – or may not do it.

In collecting customer data, care must be exercised to ensure that noise or errors are not inadvertently introduced. Gawron (2008) suggests that the customers must be given clear written as well as verbal instructions. Additionally, evaluation sheets must be developed to collect data, and finally, the data may need to be normalized before Equation 8.1 is used to calculate the ranking.

I'd like to take a moment and acknowledge one fact here. Any study in human factors by necessity does not produce absolute numbers. These studies typically provide the

TABLE 8.3

A Suggested Ranking Value for p

Preference	p	Definition
Most	10	Choice N_i is definitely more important than choice N_j
More	5	Choice N_i is more important than choice N_j
Equal	1	Choice N_i is equally important than choice N_j
Less	1/5	Choice N_i is less important than choice N_j
Least	1/10	Choice N_i is definitely less important than choice N_j

development teams with various probabilities of what (or how) the end users may do something – or may not do it. Often enough, questionnaires and surveys that explore customers' needs and wants are developed by professionals who truly understand this field. I have witnessed enough studies where an inexperienced marketing staff has influenced the interviewee to get the answers that she wanted to hear. Thus, to get realistic data that is reflective of the market's needs, it is best to allow consultants with deep experience to provide us with the results of their studies.

> Each newly developed product should have features that excite its customers. At the same time, it should fully provide features that the users have become accustomed to and expect.

One such study was conducted for the Learning Station, and the results are presented in Figure 8.2. One would notice that the purchase price and its relative importance have also been evaluated in this survey.

The last step is to identify the *exciters*, the *expected* and the *normal* features as defined in the Kano Model analysis. For a description of the Kano model, see ReVelle (2004). Briefly, the *exciters* (also known as *delighters*) are the aspects of the new product that the customers do not realize that they needed or that they did not realize they could get or could afford. Leaving these needs as unmet would not cause customer dissatisfaction; however, meeting them would bring excitement and delight. An example of the exciters is when an airline traveler is given a free upgrade from economy to first class on an international flight.

The opposite of exciters is the *expected*. While meeting these needs may not bring any degree of satisfaction to the customers, their absence will cause grief and complaints. An example of this need is having a spare tire in a car. Regardless of the price, a new car has to have a spare tire. Because a spare tire is needed, no car manufacturer considers removing the spare to save money.

The middle ground between the two is called the *normal* (or performance) need. Customers are used to this class of features, and the more they have, the happier they will be. An example of this class of needs is food portions in restaurants. For a while, many fast food chain restaurants were taking advantage of this normal need to increase profits by offering to "super-size" drinks for a very small fee.

For the Learning Station example, the ranked needs of Figure 8.2 are shown once again in Table 8.4. The list in this table is rearranged on the basis of importance from the highest to the lowest. Furthermore, a *status* and a *critical needs* column have been added. The *status* column is used to note if any of the proposed needs may have a degree of difficulty associated with them or that it may be indicative of new and/or previously unacknowledged market needs. The lines denoted as *critical need* will be used to develop a new product's requirements and features that would differentiate it from its competitors.

To be sure, I am not implying that expected or normal needs will be ignored; on the contrary, the design team should ensure to meet these needs. For instance, much of regulatory needs (such as safety) is classified as expected; they have to be met, otherwise the product cannot be marketed. Another factor worth mentioning is that developing a truly disruptive product, that is, a product that would cause a paradigm shift or a change in the way people do things, is somewhat rare. Frequently, new products to be developed are based on already existing legacy products but with more features or more conveniences. As such, often, we build on what already exists.

AHP / Importance

Title:
Project:
Name:
Date:

	Maximum	Minimum
	5	1

Items		Items: 1 Indestructible / Heavy-duty	2 Maintain Focus and Attention	3 Social / Communication Skills	4 Access to Updated Information	5 Computer Security	6 Look and Feel of Product	7 Purchase Price	Row Total	Relative Importance	Scaled Importance
1	Indestructible / Heavy-duty	1	0.1	0.1	0.2	1	10	5	17.4	13.1%	2.477
2	Maintain Focus and Attention	10	1	1	10	5	10	5	42.0	31.6%	5.000
3	Social / Communication Skills	10	1	1	1	5	10	5	33.0	24.8%	4.077
4	Access to Updated Information	5	1/10	1	1	5	5	1	18.1	13.6%	2.549
5	Computer Security	1	1/5	1/5	1/5	1	5	5	12.6	9.5%	1.985
6	Look and Feel of Product	1/10	1/10	1/10	1/5	1/5	1	5	6.7	5.0%	1.379
7	Purchase Price	1/5	1/5	1/5	1	1/5	1/5	1	3.0	2.3%	1.000

FIGURE 8.2
From specific-customer needs to higher-level needs

TABLE 8.4

A Ranking of True Needs and Critical Demarcations

Need	Expected	Normal	Exciter	Status	Importance	Critical Needs
Maintain Focus and Attention			X	Difficult	5.000	X
Social/Communication Skills			X	New	4.077	X
Access to Updated Information		X			2.549	X
Indestructible/ Heavy-duty		X		New	2.477	X
Computer Security	X				1.985	
Look and Feel of Product	X				1.379	
Purchase Price		X			1.000	

Product Requirements Development

I have to admit that as a young engineer, I did not have much patience with all this work. Why can't we get to the design already?! It took me years to come to appreciate spending time on developing good requirements. Having said this, I would like to share a comment and a personal experience. In my career, I have come across two typical groups of technical people: one group considers design to be the work that leads to the completion of requirements. Any work beyond this level is often not valued and possibly considered as "details" that can be farmed out. The second group – which I belonged to – considered "design activity" what my colleagues and I did. I considered design to be the actual CAD modeling and the creation of the physical embodiment of the product or the details of the formulations that chemists and material scientists develop. After all, it is because of technical people like myself who know the details of what features on a product may or may not be realized. A personal experience as a design engineer led me to change my mind and value having a clear and inconspicuous set of instructions on what needed to be designed. At the time, I was not familiar with the term *requirements*. After this experience, I coined the term *statement of product* and influenced the marketing team to work on developing such a document for the products that they proposed to develop.

I had recently joined a small but international medical instrument manufacturer as a senior design engineer. My first project was to develop a wrench to tighten a small tip onto a handpiece once it was screwed in place by hand. The handpiece would cause the tip to vibrate, which was then used in clinical procedures. The challenge was obvious: on the one hand, do not tighten enough and the tip may loosen and fall off; on the other hand, tighten too much and the threads may be damaged prematurely. Some competitors had ignored this issue and would just provide a flat – un-ergonomic – disc that could be used to tighten the tip, albeit at the customer's discretion. Others had provided a torque wrench. The initial in-house design of a torque wrench had produced a very heavy and bulky unit. When I came on board, the marketing team had remained undecided on whether they wanted a "boring" disc or an "exciting" torque wrench. It is important to note that, in the new product development process, this inability to decide is detrimental.

In my first conversations with the product owner,[7] I was directed to design a simple tool that would only tighten the tip onto the handpiece. By choosing engineering plastics, I could provide two distinguishing factors: first, an ergonomic design, and second, different colors. Different colors could lead to having subclassifications of various tips by their clinical use. This concept was developed, and a production-ready tool was designed, leading to the initial prototypes. Now, with the use of different-colored wrenches, the idea began to flourish that the tips should be shipped with the wrenches in the same package. Before long, it became evident that what marketing had wanted all along was a torque-limiting wrench. It so happened that a novel design was developed and, in fact, patented. Units were developed and tested, and all test results were satisfactory. The initial ramp-up began, and products were manufactured and placed on shelves in anticipation of shipment.

Within a week of ramp-up, there were reports from inventory that some of the tips had been observed to lie in the container as opposed to being held in place by the wrench/carrier. Needless to say, all manufacturing was put on hold in order to remedy the problem. Although the project was relaunched a few months later, there were financial impact on the company as well as negative impact on the reputation of the design engineering team in general and on me in particular. Did the company and the team (including me) eventually recover? Certainly! Once these issues were ironed out, the business venture was quite successful. A short while later, no one seemed to be cognizant of the series of ad hoc decisions made leading to the failure of the initial launch. I came to learn that this was business as usual.

Unfortunately, my story is not unique. We read too many stories about products that fail despite extensive testing. Even some very large corporations miss the mark. Hooks and Farry (2001) retell the story of how IBM missed the mark on developing a product to provide sports news feeds to newspapers during the Atlanta Olympics Games. IBM's computer system was supposed to channel information about the games and their outcomes to a second computer that would then feed the data to various national and international newspapers. At the time that the games began, the system did not work!

"Many of the [World News Press Agency] problems were a case of programming a computer to format a certain sport's information one way, while newspapers were expecting it another way." Hooks and Farry (2001, pp. 83–84)

In reality, there is not enough time or resources to test every aspect of a product in the hope of ensuring that the design team has developed the *right* product. The right product is and should be identified through its requirement.

At the time, IBM personnel felt that had they had a chance to test their system more, they would have been able to figure things out and deliver what the customer wanted. On the contrary, they missed the mark because they did not understand what their customers wanted, expected, and eventually required (same reference).

On a more personal note, would more testing of my product have identified the issues with my design? The sad answer is "probably not!" In reality, there is not enough time or resources to test every aspect of a product in the hope of ensuring that the design team has developed the right product. The *right* product is and should be identified through its requirement. This is why requirements need to be developed, written, and agreed upon. On the basis of these requirements, the design team knows what to develop, and the test personnel know what to test. Hence, through a limited number of tests, the design team can identify whether the right product has been developed. Therefore, the question is how to use customer needs to develop product requirements.

Translation from Needs to Requirements

One method to transition from customer needs to product requirements is for the development team to have brainstorming sessions to consider each need separately and agree on what product feature(s) is (are) a response(s) to a specific need. For simple products and line extensions, this is rather easy; just design the extra feature into the device. For instance, I recall an instance when the marketing team developed a campaign to support breast cancer research and prevention efforts. My task was to find a way to apply a pink logo onto a series of our products. The technical challenge was that the product being used by clinicians was subjected to some severe cleaning procedures after each use. We needed to identify a paint that would endure these cleaning cycles. Once the paint was identified, the verification activity was to complete an established set of test regimens that were common and customary to the product line.

The first step in translating these what-s into how-s, is to identify the performance characteristics associated with these what-s.

Testing to establish a set of test procedures neither worked for my torque wrench nor for IBM's newsfeed product. Why? Because a new set of tests needed to be developed and no one had thought about it. It was not that the participants did not have the experience or the expertise to develop the test. There were other so-called fires burning. The challenge was that the expectations for the product features were evolving as the launch date was drawing near, and there was not clear communication among the development team. As the expectation for the product function evolves during the detailed design phase of development, the possibility that the finished good meets its intended expectations may be a hit or a miss. Why? Because this approach does not follow a robust process. A robust process, if properly followed, tends to flush out all the factors that need to be considered.

Earlier, I asked how one transitions customer needs to product requirements. In other words, my question is *how* to deliver *what* the customer needs. I like to note that the terms *how* and *what* are a part of the systems engineer's vocabulary, and I am deliberately using them here.

The first step in translating these *what*-s into *how*-s, is to identify the performance characteristics associated with these *what*-s. For instance, a customer need in developing dental hand instruments such as dental picks is that they have to be lightweight. Also, for ergonomic reasons, the instrument's body should not be too narrow (i.e., small diameter). By conducting user studies or based on prior experience, the development team learns that if a dental instrument weighs between 9 and 12 grams and has a diameter approximately between 8 and 11 millimeters, it has an ideal size. Anything else is either too light, too heavy, too small, or too bulky. This is what I mean by translating *what*-s into *how*-s. In this instance, lightweight translates into 9 to 12 grams, and ergonomic means 8 to 11 millimeters.

Another example may be that customers want bright colors on their instruments in order to easily differentiate their categories. This need can be translated into a combination of light frequency (associated with a color), such as, say, 680 nm for a particular shade of red or 460 nm for a shade of blue. Note that merely specifying red or blue is not sufficient because the frequency range of red is about 622 to 780 nm and for blue is 455 to 492 nm.

TABLE 8.5
Customer Needs versus Performance Characteristics

	Performance Characteristics						
	Lack of Any Distractive Features	Ruggedized Unit	Integrated User Input Devices	Integrated Features for Reward	Integrated Features for Turn Taking	Hidden Communication Ports	Industrial Look
Maintain Focus and Attention	X		X			X	X
Social Skills				X	X		
Communication Skills			X		X		
Access to Updated Information						X	
Indestructible/Heavy-Duty	X	X	X				X

Or, in a more simplified and practical sense, the associated Pantone color may be specified; say, Pantone Red 032M. Similar to the previous example, *bright color* translates into Pantone Red 032M.

Recall the Learning Station that I have mentioned previously. As shown in Table 8.4, I identified its customers' needs and the order of their priorities. A cross-functional team met to review the list, brainstorm, and identify associated performance characteristics. Table 8.5 provides the first pass and outcome as well as a set of needs versus performance characteristics. It suggested that through seven (7) *how-s*, the five (5) *what-s* may be satisfied.

Effectively, in Table 8.5, the "Needs" have been translated to "Requirements" and *what*-s have been correlated to *how*-s. Furthermore, note that through the cross-functional brainstorming sessions, the "Social/Communication Skill" need was further expanded to two separate needs, namely, "Social Skill" and "Communication Skill" needs.

Let's dwell on this example a while longer. Are these *how*-s as clear and measurable as possible? The concepts of *ruggedized unit, integrated user input devices, hidden communication ports,* are relatively clear, even though there may be several different means of achieving them. The terms *integrated features for reward, integrated features for turn taking,* and *industrial look* may be ambiguous for those who do not have the required familiarity with terms in the field of education such as turn-taking or rewards. However, the term *lack of any distractive features* may be interpreted in a number of different ways. Upon further review with the team, this *how* was modified to read: Minimize the number of user-facing buttons and colors – as shown in Table 8.6.

It is easy to infer that if I were to treat the how-s in Table 8.4 as the what-s in a new table, the development team could brainstorm more specific aspects of the intended product. As we repeat this cycle – that is, placing the how-s of one matrix as the what-s of a follow-up matrix – more details of the final product emerge. Typically, by the second or third cycle, the final details emerge.

A PRD captures the outcome of the product-related decisions, including the team's understanding of the performance characteristics.

TABLE 8.6

Customer Needs versus Updated Performance Characteristics

	Performance Characteristics						
	Minimize the Number User-facing buttons and colors	**Ruggedized Unit**	**Integrated User Input Devices**	**Integrated Features for Reward**	**Integrated Features for Turn Taking**	**Hidden Communication Ports**	**Industrial Look**
Maintain Focus and Attention	X		X			X	X
Social Skills				X	X		
Communication Skills			X		X		
Access to Updated Information						X	
Indestructible/Heavy-Duty	X	X	X				X

Product Requirements Document

At this point, it is important to initiate what is called a "Product Requirements Document" and capture the outcome of the product-related decisions, including the team's understanding of the performance characteristics. By the time this document is completed, the following are captured:

- Intended use of the product
- Customer and/or end-user performance needs
- Business needs including voice of manufacturing, service, etc.
- Specific markets for the product and any special needs of these markets, such as language variations.

As I had alluded to earlier, it is important to resist the urge to develop a solution at an early stage of product development. It is important that the product requirements are developed and defined in a "solution-free" context. When this task is developed, various concepts need to be explored. Again, it is important to remain at the concept level and not drive down to details prematurely. Once the system is completely defined, then further details of subsystems and assemblies may be developed.

In the next chapter, I will review the means by which performance characteristics may be further developed into concept generation and system and subsystem-level requirements development and design.

Summary

- A problem statement or product definition should identify intended customers, added value or benefit to the customer, and other relevant known information.
- Project scope – as the title implies – describes the boundaries of the project, where it begins and ends.
- The define phase is the most important phase of any project and should not be taken lightly.
- Customer or user needs studies should ensure that the study objectives are clearly known; a structured questionnaire is used; the right information is solicited; a representative sample of the population is interviewed; the technical people are involved; and finally, the use environment is considered.
- Any study in human factors by necessity does not produce absolute numbers. These studies typically provide development teams with various probabilities of customer behaviors.
- One method to transition from customer needs to product requirements is for the development team to have brainstorming sessions to consider each need separately and agree on what product feature(s) is (are) a response(s) to a specific need.
- A "Product Requirements Document" captures the outcome of the product-related decisions, including the team's understanding of the performance characteristics.

Exercises

1. Create a project to improve a household product such as, say, a coffee maker or toaster oven.
   a. First, conduct a survey of what people think of the product that they use. What do they like? What bothers them?
   b. Collect and collate responses.
   c. Conduct brainstorming sessions on what may be improved and rank them.

Notes

1 I use the term "problem statement" but in reality, it may also be called "opportunity statement" if a new market is to be created through this effort. For consistency's sake, I will uniformly use "problem statement."

2 Autism is on a spectrum from high-functioning to severe. Although the term autism is used as an umbrella, working with individuals at various locations on the spectrum presents unique challenges for the educators.

3 Notice that I do not say "fully" defined. We may come back and add to Define in just about any phase as we learn more about the project.

4 This VOC was collected in the early 2000s. I am using it here as an example. I presume if we were to do this again, a different list may be generated.

5 I do recognize that CDs are no longer in use. But at the time this system was developed, it was an essential part of any new computer system.

6 Admittedly, with the advent of new software such as Miro®, participants may collaborate virtually.

7 I am not sure if this is a universal term. A product owner is someone who takes ownership of a product family, generally from a marketing point of view. The proverbial "buck" stops at this person's desk.

9

Concept Development and Selection

Introduction

In the previous chapter, I examined a method by which customer needs may be identified and ranked. Then, I developed a matrix to relate the *what* of a customer need to the *how* that need is satisfied in a product. Developing the matrix of what-s against how-s is the basis of developing and using a tool that in the English language is called *quality function deployment* (QFD). This technique has gained a wide degree of acceptance as a robust tool in the initial stages of requirements and concept development. See, for instance, Juran and De Feo (2010), Weiss (2013), or Ullman (2010).

Quality Function Deployment

Hinshitsu kinoe tenkai, translated as *quality function deployment* was first developed at the Tokyo Institute of Technology by Shigeru Mizano. The word *hinshitsu,* in addition to "quality," also means "natural disposition"; the word *kinoe* means "first in rank"; and, finally, *tenkai* means "expansion." Hence, the intention behind *hinshitsu kinoe tenkai* may be said to be "to first expand on natural disposition" of what the customer needs.

While different authors may develop the QFD matrix differently, the construct essentially contains a set of primary elements and a set of secondary elements. As shown in Figure 9.1, the primary elements are the what-s and the how-s, just as shown in Tables 8.5 and 8.6 in Chapter 8. In addition, there is a list of importance associated with each *what*, along with the completeness list and the technical weights. These three metrics measure the relative importance of *what*-s, the technical weight of each *how* as it relates to all the *what*-s that may be impacted by it, and finally the completeness of the design response to each *what* through the *how*-s. This may be confusing at the moment but will become more clear as I provide an example later.

The secondary elements are shown in Figure 9.2 and include the interaction matrix, which appears as the roof of a house; target values, target direction, and technical benchmark rows that give the impression of the lower floors and the basement of a house; and finally, columns associated with requirement planning and competitive benchmark. Because of this look, this construct has been called a House of Quality (HOQ) as well (ReVelle 2004; Ogrodnik 2013).

QFD is the action of developing a matrix of the "what" of customer needs against "how" the product will deliver it.

The competitive benchmark column and technical benchmark row provide a means of measuring the proposed how-s against the competition and their degree of success in

DOI: 10.1201/9781003564508-12

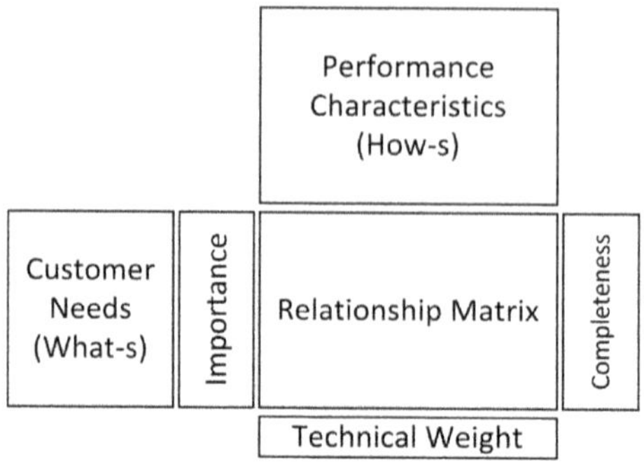

FIGURE 9.1
The primary elements of the QFD

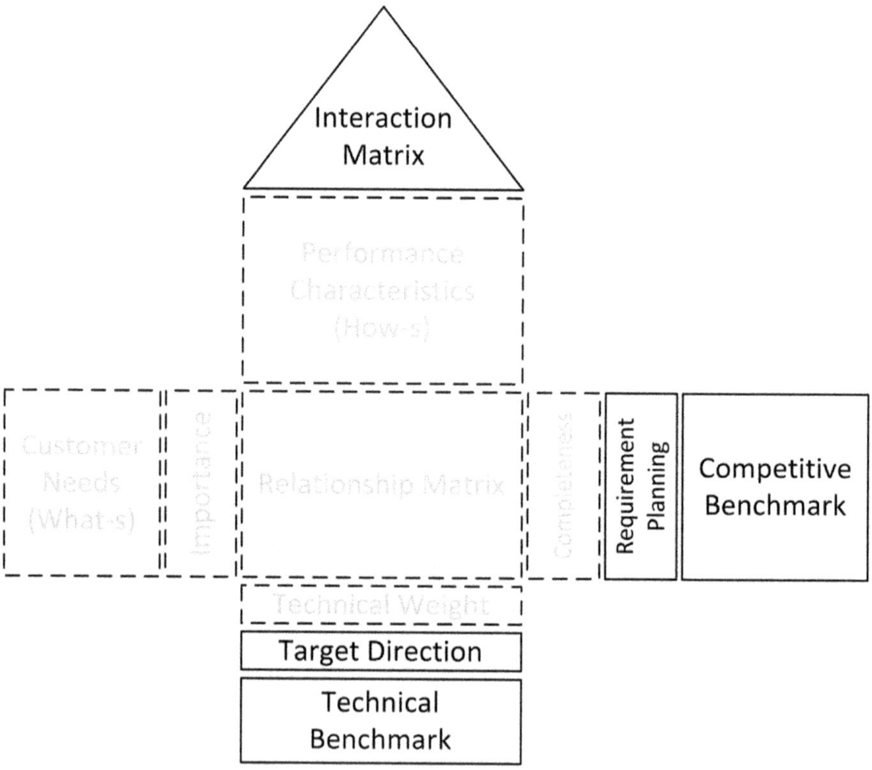

FIGURE 9.2
The secondary elements of the QFD

responding to customer needs. The target value and direction rows encourage the product development team to focus their attention on quantifying the measures of how to achieve the *what* and whether a higher or lower value is better. For instance, the need for a dental probe to measure pocket depth around teeth may be stated as "cause less pain." The *how* to achieve this may be to have a small tip diameter and a smooth surface finish. The target value and direction fields encourage the development team to assign a value, say 0.1 mm, to the value field and "↓" (lower is better) to the direction field. Similarly, smooth may get a target value of 0.1 μm and again "↓." This mindset enables the team to think in concrete and objective terms, as opposed to in more subjective terms. The interaction matrix shown as the roof of the house explores if two different *how*-s complement one another, or they are contradictory and compromises may be needed.

Using QFD in Product Development

In Chapter 8, we began to take the initial steps to build the primary elements of a House of Quality for the product that was called the Learning Station. These initial steps were:

1) Develop *what*-s using voice of customer surveys and other tools.
2) Using tools such as the AHP, identify the rank and importance of each *what*.
3) Identify the *how*-s through brainstorming session(s).
4) Identify the *what* to *how* relationship (e.g., Tables 8.5 or 8.6 in Chapter 8)

Now, we are ready to take the steps required to finish the first HOQ. These are:

5) Develop the *what* to how relationship matrix (as shown in Figure 9.1)
6) Calculate the technical weight row and the completeness column.

Once these steps are done, optionally the following two steps may also be taken:

7) Develop and include the secondary elements,
8) Flow down the HOQ to its second generation.

The easiest way to demonstrate this technique is to continue with the Learning Station example. To use the QFD tool, we form a matrix as shown in Figure 9.1 and use the *what-s* and *how-s* that were developed in Tables 8.5 and 8.6 in Chapter 8. Next, their relative rank (importance) from AHP calculations is entered into the matrix. For the sake of simplicity, a tool such as Triptych software may be used.

Then, the cross-functional development team populates the relationship matrix by assigning an importance level between each what and how. Typically, three values of high (H), medium (M), and low (L) with numerical values of 9, 3, and 1, respectively, are assigned. The completed QFD matrix is shown in Figure 9.3.

In this case study, the team recognizes that minimizing the user-facing colors and buttons (Column 1) strongly impacts maintaining focus and attention; hence, a high value (H) is assigned to this cell. Similarly, the team indicates that there is a low impact from a "ruggedized unit" on "maintain focus and attention." Thus, a low value (L) is assigned to that particular cell. The third cell worth mentioning is the one to indicate the relationship between "maintain focus and attention" and "integrated user-input devices." The team has assigned a medium (M) to this cell because how user-input devices are integrated into the product may cause distractions, but a well-integrated device may not necessarily add to increased attention. Cells that indicate that no relationship between a *what* and a *how* exists are left blank.

Once the relationship matrix is populated, the technical weights row and the completeness column may be calculated. Assuming that the relationship matrix is represented as

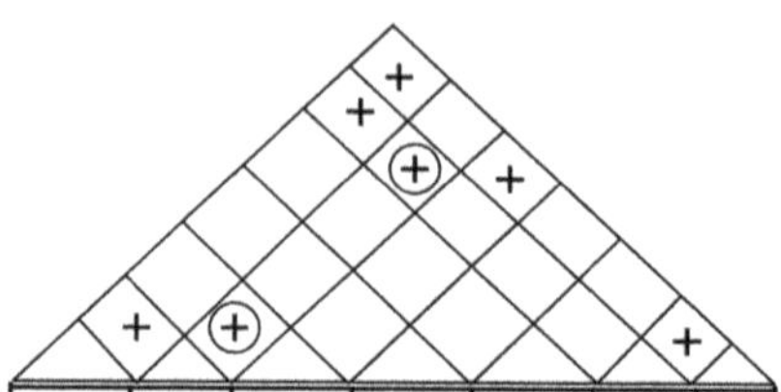

			Importance	1 Minimize the number of User-facing buttons and colors	2 Ruggedized unit	3 Integrated user input devices	4 Integrated feature(s) for reward	5 Integrated features for turn taking	6 Hidden ports	7 Industrial look	Completeness
What-s	1	Maintain Focus and Attention	5	H	L	M	L	-	H	M	130
	2	Social Skills	4	-	-	M	L	H	L	-	56
	3	Communication Skills	4	-	-	H	M	L	-	-	52
	4	Access to Updated Information	3	-	L	-	-	-	H	-	30
	5	Indestructible / Heavy-duty	3	M	H	M	-	L	M	M	66
	6	Computer Security	2	-	H	-	-	-	M	L	26
	7	Look and Feel of Product	1	H	L	M	L	M	M	H	20
		Technical Weights	Absolute	63	54	75	22	46	94	35	
			Relative	16%	14%	19%	6%	12%	24%	9%	
			Importance	3	3	4	1	2	5	2	

FIGURE 9.3
The first QFD for the Learning Station

R_{ij} where i and j are the rows and columns, the technical weights (T_j) are calculated as follows:

$$T_j = \sum_{k=1}^{n} R_{kj} A_k$$

where n is the number of *what*-s and A_k, $k = 1$ to n is the importance column. Similarly, the Completeness (C_i) column is calculated as shown below:

$$C_j = A_j \sum_{l=1}^{m} R_{lj}$$

where m is the number of *how*-s. These calculations are typically done using commercial software such as Triptych.

Recall that the completeness column is indicative of how well the what-s are satisfied. In this example, the completeness cell (C_1) is 130. This means that the selected design approach will provide a high degree of customer satisfaction. Similarly, other user needs that were considered as *exciters* or *normal* are getting high completeness scores.

> The interaction matrix of the HOQ provides a tool for comparing and contrasting two solutions for satisfying a need (a what) and for exploring whether the proposed solutions are complementary or contradictory.

The next topic to discuss is the interaction matrix or the HOQ roof. As mentioned earlier, this matrix provides a tool for comparing and contrasting two solutions for satisfying a need (a *what*) and for exploring whether the proposed solutions are complementary or contradictory. In this example, it may be seen that the "Ruggedize unit" (column # 2) has a strong complementary relationship (as indicated by a circle and plus sign) with both "Integrated user-input devices" (column # 3) and "Hidden Ports" (column # 6), but it has just a complementary relationship (as indicated by a plus sign) with "Minimizing the number of user-facing buttons and colors" (column # 1). In this example, there are no contradictory relationships.

It is difficult to speak about "how" to respond to a "what" without forming a concept of a product. As the *how-s* and *what-s* are discussed and contemplated, it is natural that concepts should begin to be formulated. It is, however, important that the team remain disciplined and maintain a "full solution-neutrality" mindset.

Now that the first pass of the QFD matrix is completed, this information may be flowed down to the second layer of QFD. This is done by treating the *how-s* of the first QFD as the *what-s* in the second QFD. Furthermore, the *technical weights* of the first QFD become the *importance* column of the second QFD. Now, armed with some formed concepts, the *how-s* of the tier 2 QFD may be populated by the team as shown in Figure 9.4. Note that as the first QFD cascades down to its second tier, product requirements begin to take shape.

In Chapter 6, the concept of CTQ was discussed. Here, the team can readily identify the CTQs for the learning station; maintain focus and attention, social and communication skills, and indestructible/heavy duty. One may ask: "[I]s not compliance to regulatory agencies a CTQ?" The answer to this question is both negative and affirmative. On the one hand, the response is negative because being compliant with regulatory agencies ensures product safety, but it is the "wow" factor – the differentiating elements – that sells the product. On the other hand, the answer is affirmative in the sense that if the product is not compliant with safety and/or regulatory rules, it may not be marketed and sold. In the new product development stage, CTQs are generally associated with the "exciters," the "wow" factors.

In the analysis of Figure 9.3, the completeness column was discussed. It was mentioned that the higher the value, the higher the degree to which the "how" satisfies the "what." Here, I like to draw attention to the two low values of 14 and 31 (in Figure 9.4) belonging to "integrated feature(s) for reward" and "industrial look" (lines 4 and 7). The low scores are indicative that enough "how-s" are not available that would satisfy the corresponding "what-s." The design team may make note of this situation and reconsider whether to add additional columns for additional solutions to address them in more detail or to separate the "what" and develop them separately.

For simplicity, the interaction matrix is not discussed here and is shown as blank. In reality, the "roof" will be populated as required and considered by the design team.

Developing System Architecture and Function Diagram

Once the QFD has flowed down once or twice, possible product concepts begin to emerge. At this stage of design, the team should begin to create an architecture of the product. It is not advisable to create a high degree of detail; only what the major branches of the product may look like. This is decomposing the product into its major building blocks (or

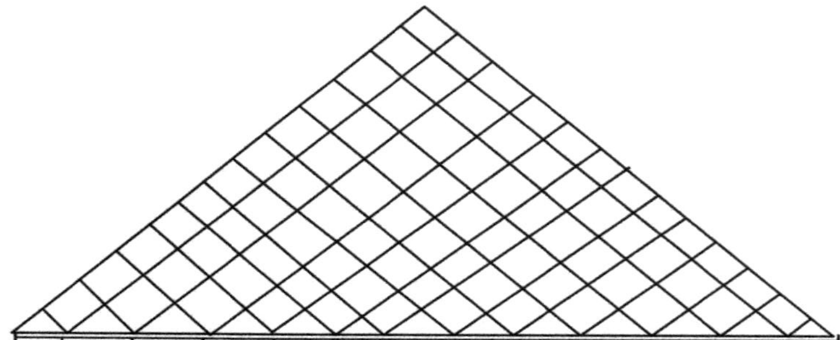

			Importance	1 Use only one exterior color	2 Two-button mouse operation	3 Enclosure with with heavy wall thickness	4 Integrated inductrial touch screen	5 Large integrated trackball	6 Token operated unit	7 Base enclosure with turn-table	8 Multi disk optical drive	9 Enclosure with cover for ports	10 Shock mounted sensative components	11 Industrial single color buttons	12 Enclosure with tab to physically secure product	13 A "brick" Shape	Completeness
What-s	1	Minimize the number of User-facing buttons and colors	3	H	H	-	-	L	-	-	-	-	-	H	-	L	87
	2	Ruggedized unit	3	-	M	H	M	M	-	-	-	M	H	-	M	M	108
	3	Integrated user input devices	4	-	H	-	H	H	-	-	-	-	-	M	-	L	124
	4	Integrated feature(s) for reward	1	-	-	-	L	-	H	L	M	-	-	-	-	-	14
	5	Integrated feature(s) for turn taking	2	-	M	-	M	L	-	H	-	-	-	-	-	L	34
	6	Hidden ports	5	-	-	-	M	M	-	L	H	H	-	-	-	-	125
	7	Industrial look	1	H	L	M	L	L	-	-	-	M	-	M	L	H	31
		Technical Weights	Absolute	45	80	33	69	67	9	24	48	60	27	45	11	36	
			Relative	8%	14%	6%	12%	12%	2%	4%	9%	11%	5%	8%	2%	6%	
			Importance	3	5	2	4	4	1	2	3	4	2	3	1	3	

FIGURE 9.4
The second QFD – the *how-s* become requirements

subsystems). In Chapter 7, three different categories of decomposition were mentioned. There was decomposition along operational, product functionality, or physical boundaries. Let's explore these three approaches briefly.

Operational boundaries are formed around how the product is expected to operate. For instance, for a motorcycle, engine sound may need to have a "signature" quality and be in a particular sound frequency domain. The same may also be said about the "ride" experience of a luxury cars, which would require the chassis to have a certain vibrational frequency. Other operational boundaries include cost or certain levels of quality. Often, component suppliers ask either procurement or the design team for the cost target of the assembly being developed. In systems language, the conversation is about the operational (cost) boundary of the product.

Functional boundaries or decomposition is on the basis of the functions of the product to create the desired user experience. In the case of a motorcycle, fuel and its delivery to the engine may be one function, and starting the engine another function.

From a physical point of view, it is relatively easy to draw the boundaries. Following the previous motorcycle example, physical decompositions may take place around the chassis, engine, brake system, etc. This approach is easily suited to products that already exist on the market. For instance, an electromechanical device always needs to have a power supply, or a dental instrument almost always requires a handpiece.

After the QFD flow-down of the Learning Station (Figures 9.3 and 9.4), the following subsystems emerged:

1. Enclosure – for the Learning Station, enclosure is primary means of developing a "rugged" unit, one of the main user needs.
2. Input Devices – for the Learning Station the input devices and their design are one of the product "exciters."
3. Main Electronics – electronics needed to operate various components of the product.
4. Power Supply (including wiring harnesses) – any electromechanical system needs this subsystem.
5. Software – majority of sophisticated electromechanical systems need to be operated with a variety of commands. This includes both the operating system and a customer-driven special-purpose software.

In the next step of design, the *how*-s developed in the last QFD may be assigned to each of the subsystems. At this stage, the "expected" features of the product (e.g., complaince with various regulatory agency rules) may also be added. For instance, it is expected that the main electornics subsystem will have a motherboard and memory cards. The choice of the motherboard and the amount of memory may be specified here if they are rooted in a voice of business (i.e., if the manufacturer of the Learning Station may happen to manufacture a certain brand of motherboards or memory chips). This linkage of *how*-s to subsystems for the Leaning Station is shown in Figure 9.5. For completeness' sake, the "expected" and "normal" needs are also added to the requirements. It should be noted that the requirements described in Figure 9.5 are copied from the top row of Figure 9.4.

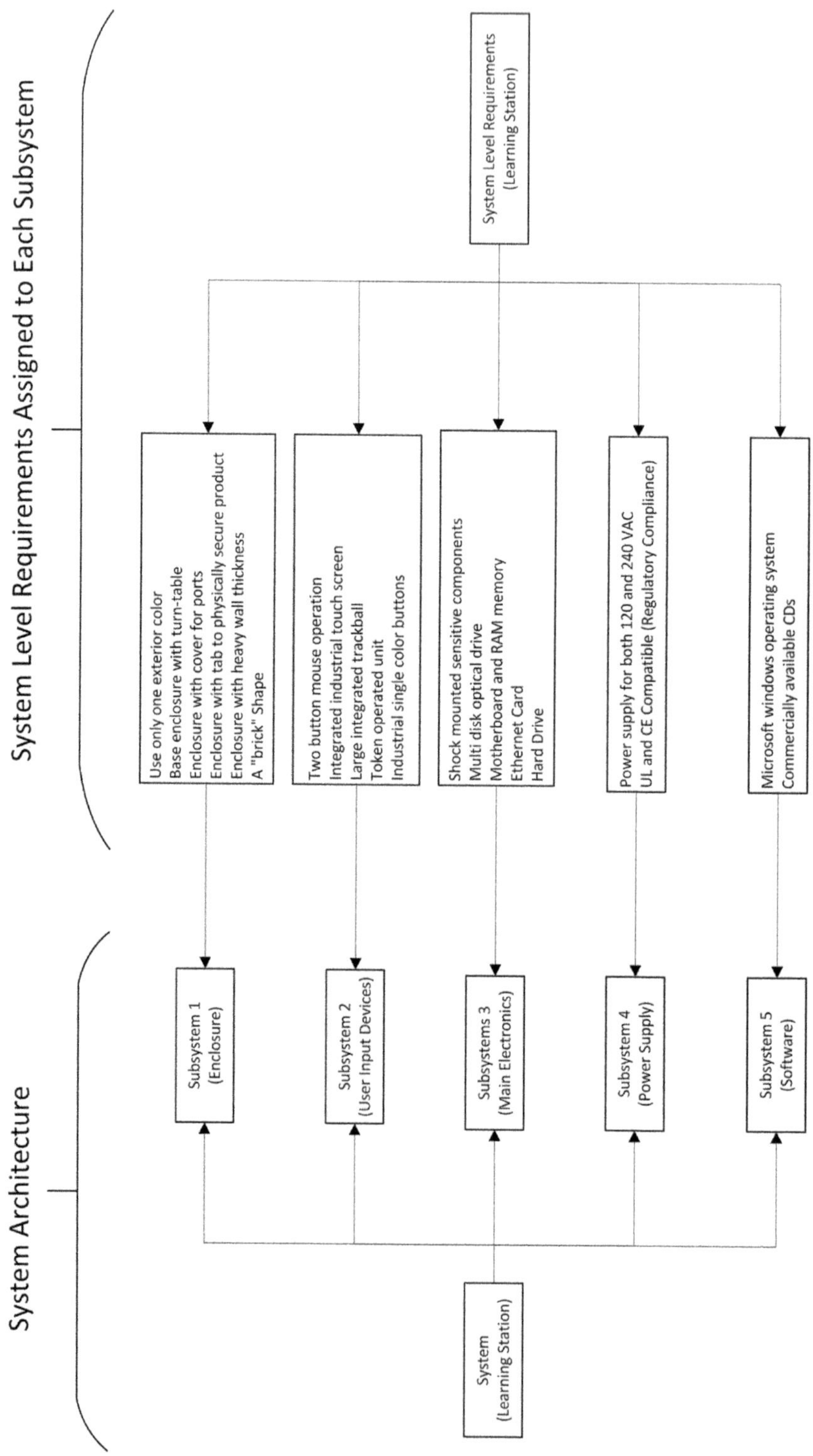

FIGURE 9.5
The system architecture and requirements linkage

For clarity's sake, it should be noted that the requirements assigned to each subsystem are not subsystem requirements. They are the *what-s* for each subsystem that would require *how-s* to be realized. These newly developed *how-s* become each subsystem's requirements. Furthermore, the system-level CTQs may now be associated with each subsystem.

> In function diagrams, three types of signals should be considered. These are energy, information, and material, denoted by a solid-lined arrow, a dashed-line arrow, and a heavy-lined arrow, respectively. These are the desired interactions that are transferred from one body to another.

Boundary and Function Diagram

As the system architecture is being developed, the system function diagram should also be considered. To begin with, the system boundaries need to be defined – or drawn. As shown in Figure 9.6, a system boundary is that imaginary line that separates the system from either its end user or environment. While often the system boundary is located at the physical boundary of the product, it does not always need to be so, particularly if the product is software and/or a service. For this reason, virtual boundaries are shown with dashed lines, while physical boundaries are shown as solid lines.

A system communicates with either the environment or the end user through its boundary. Typically, this communication is either in the form of a desired interaction (called a signal, which I will define shortly) or in an undesired interaction called noise.

Figure 9.6 defines three types of signals, namely energy, information, and material, denoted by a solid-lined arrow, a dashed-line arrow, and a heavy-lined arrow, respectively. These are desired interactions that are transferred from one body to another. Symbolically,

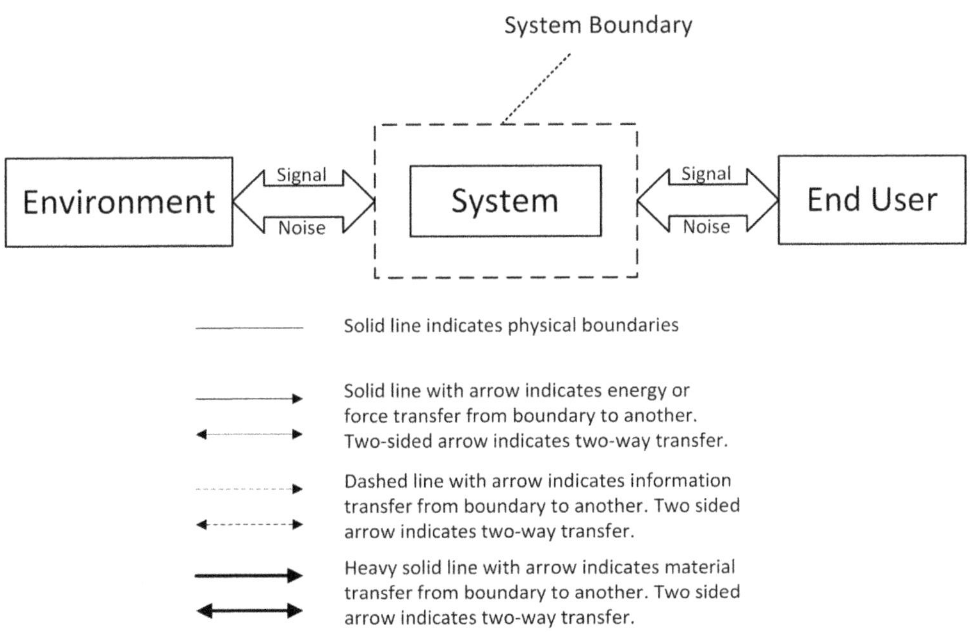

FIGURE 9.6
System boundary diagram

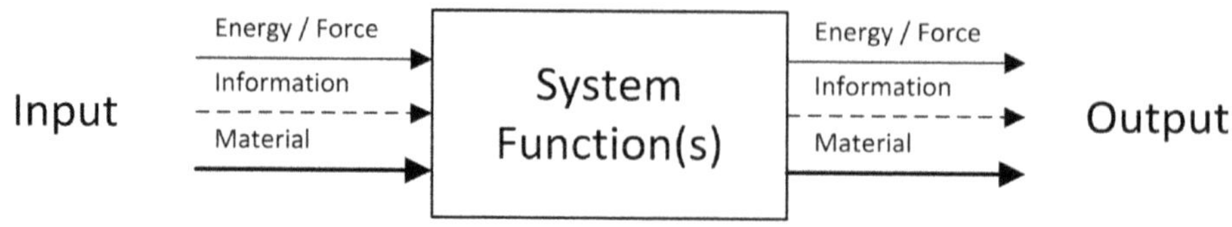

FIGURE 9.7
System function diagram

signals (or interactions) are shown as arrows. Needless to say, a double-ended arrow indicates a two-way transfer of a signal.

With this knowledge of system boundary and its diagram, a function diagram may be developed as shown in Figure 9.7. As input, a system may take in any or all of the three forms of the signals and then produce any or all of the three forms of the signals. For instance, a gas-powered engine takes in gasoline as a form of material but outputs mechanical energy and a different material in the form of exhaust gases. Developing a solid understanding of the functions of the product becomes the basis of concept development. Once this is understood, concepts begin to be formed and formulated as a response to fulfilling the function. In the words of Professor David Ullman, "form follows function" (Ullman 2010).

For this reason, let's dwell on the forms of these signals and the impact a system (and subsequently subsystems) may have on each signal. I am using the term "signal" rather loosely, and herein, by using it, I mean to generically denote either, energy (or force), information, and/or material. One may conceive that in a generic sense, as signals move from one point to the next, they may merge with similar signals and combine; that they may flow through undisturbed; or that they may be distracted and diverge from their initial path; and/or they may dissipate and disappear altogether. Hence, the more specific the design team can be about how the system should manipulate, transform, transfer,. . ., hold, rotate, position,. . ., respond, amplify, compare,. . . energy/force/information/material, the easier it would be to develop the detailed design later. Ullman (2010) suggests that the words associated with flow of energy are action words such as "store," "supply," etc.; words associated with material flow are "position," "lift," "mix," "separate," etc.

Figure 9.8 depicts the initial function diagram for the Learning Station. It depicts its five subsystems. Furthermore, it shows that the enclosure (subsystem 1) has physical connections to the rest of the subsystems, but it is linked to the environment by material (heavy solid line) and energy (light solid line). The reason the enclosure has a physical connection to the rest of the subsystems is because the enclosure is the physical body of the product. It is the structure to which the rest of the subsystems are attached. This is shown by the use of the light solid lines that connect this subsystem to the rest. Additionally, the enclosure should protect the rest of the subsystems from the environment, whether it is liquid spills (material) or intentional misuse of the product due to behavioral issues (energy). The system – whose boundaries are shown in dashed lines – receives electrical energy from the environment via the power supply (subsystem 4), which in turn supplies the user interface and main electronics subsystems. The end user and the system exchange information via the user interface which in turn communicates information to and from main electronics and software. It is also shown that main electronics and software have a two-way information exchange as well. Environment and software are connected via information input. Finally, environment provides material input (in the form of CDs or DVDs) into main electronics. As a summary, Figure 9.8 indicates that system inputs are material, energy, and information. The system output is information.

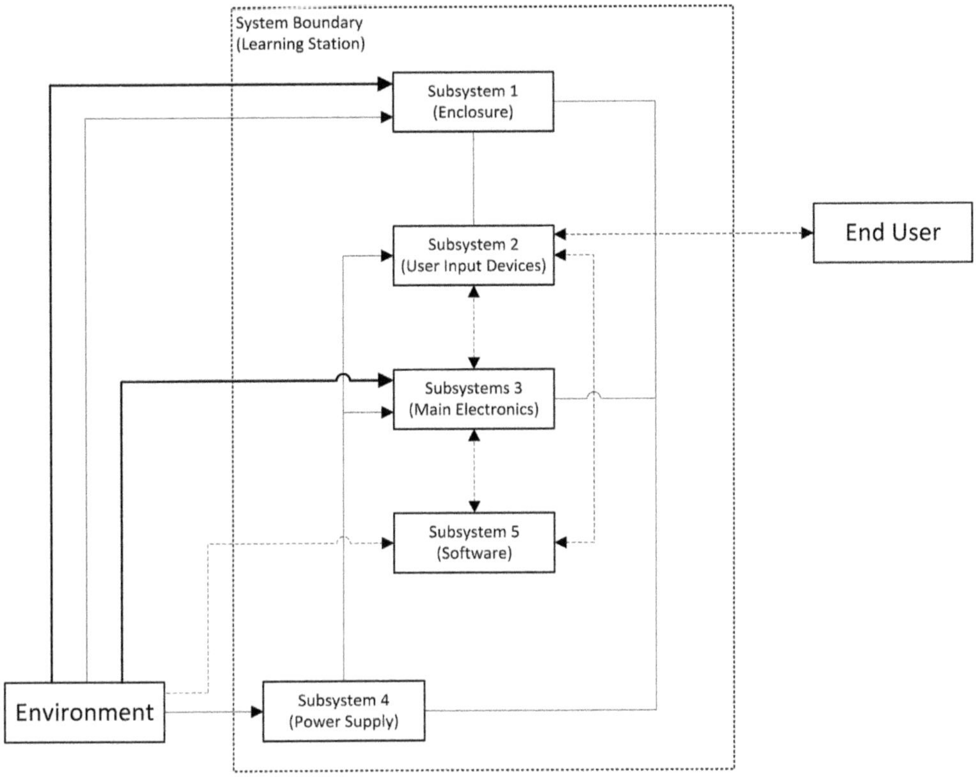

FIGURE 9.8
The Learning Station boundary diagram

Now that the system and subsystems and their boundaries, along with their inputs and outputs, have been identified, the next step is to develop the function diagram for the system and subsystems. It should be no surprise that system functions are related to user needs:

1. Function 1: Run Educational Software (this was based on an expected need; hence, it was not mentioned earlier).
2. Function 2: Enable Maintaining Focus and Attention.
3. Function 3: Enable Developing Social Skills
4. Function 4: Enable Developing Communication Skills
5. Function 5: Protect from Environment
6. Function 6: Maintain Cyber Security

Assign Functions to Subsystems

Once the system functions are defined, they should be assigned to each subsystem or groups of subsystems. For the Learning Station, the following assignment is done – as shown in Figure 9.9.

As the larger group of functions is assigned to subsystems, a simplification begins to take place. For instance, the subsystem-1 lead may consider both the assigned functions

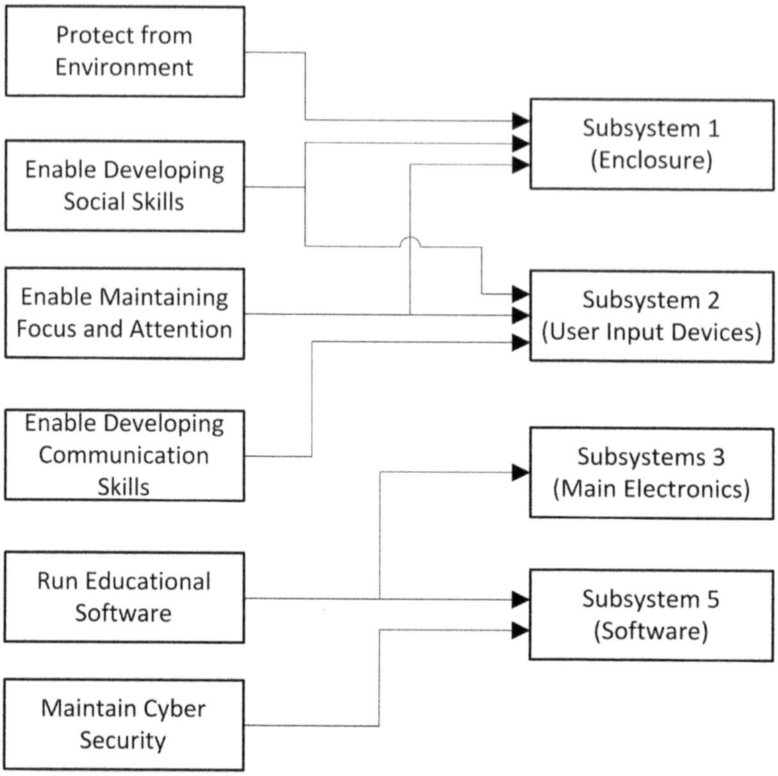

FIGURE 9.9
Assignment of system functions to subsystems

and requirements (Figure 9.5) and begin to develop subsystem requirements and functions. In this regard, if the assigned functions are the "what-s," one has to ask the "how" questions. As the design team digs deeper for answers, the how-s become more grounded in physics and technology. For instance, *protect from environment* and *develop social skills* become:

Protect from Environment	Stop liquids from entering internal compartments
	Shield sensitive internal components against Electrostatic Discharge
	Shield users from reaching (or touching) internal components

and,

Develop Social Skills	Devise a turntable for turn-taking between users
	Devise system operation by tokens to enforce social skills development

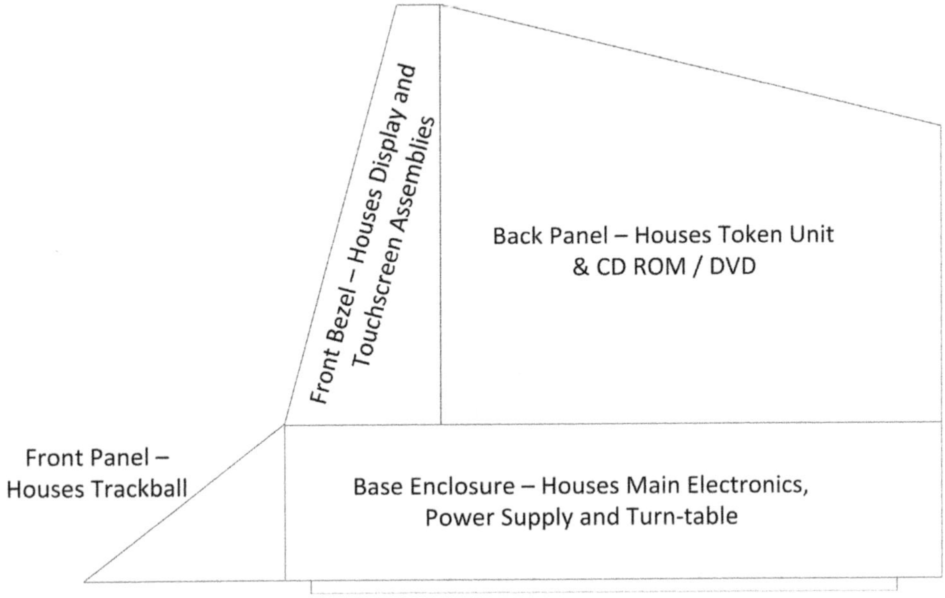

FIGURE 9.10
A proposed high-level design configuration of the Learning Station

One could easily expand the work done so far to develop system-level requirements and functions for subsystem development. It is not a leap of faith to realize that as further decompositions take place, the design team needs to develop concepts for delivering the assigned functions and requirements. As higher-level concepts are developed, a decomposition to a lower level becomes necessary. With this decomposition, the assembly requirements and functions will need to be developed and defined. This activity will continue to the component level. For instance, the Learning Station design team may begin with the *High Level Design* configuration proposed in Figure 9.10.

As requirements are decomposed to lower levels, the design team needs to develop concepts for delivering the assigned functions. As higher-level concepts are developed, a decomposition to a lower level becomes necessary. With this decomposition, the assembly requirements and functions will need to be developed and defined. This activity will continue to the component level.

High-Level Design and Concept Selection

Before entering a discussion of concept development and selection, let me review the V-model that I introduced in the first chapter. Figure 9.11 depicts a typical V-model showing the stages of product development. In this and previous chapters, requirements and their cascade (steps 1 through 3) were discussed. Once step 3 is well understood the design team's work begins in earnest by developing one or several high-level designs. On the

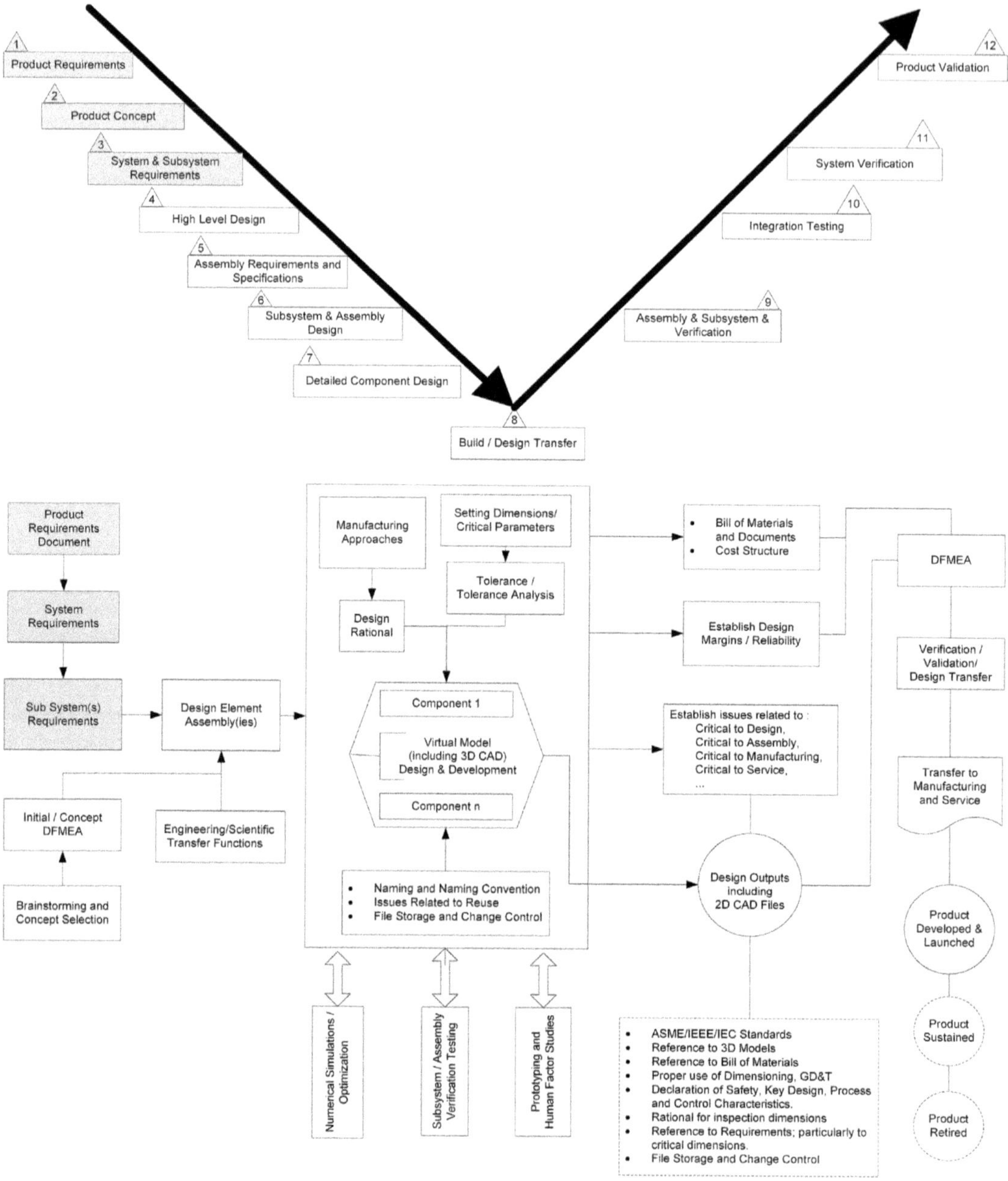

FIGURE 9.11
A typical V-model for product development along with the roadmap. The highlighted area indicates the segments that has already been discussed

basis of a high-level design selection (step 4), focus will be placed on developing assembly requirements and specifications (step 5). With the specifications at hand, the actual three-dimensional (3D) modeling begins (step 6), leading to detailed component design (step 7). While Figure 9.11 gives the impression that these steps are taken sequentially, in practice there is generally a high degree of interaction between steps 6 and 7, with occasional looping back to step 5 to update assembly requirements and specifications. Needless to say, if the assembly requirements are modified, the higher-level requirements need to be checked to assure uniformity and traceability.

For the remainder of this chapter, I will maintain focus on *high-level design* selection tools and provide descriptions of a few that may be utilized in this pursuit. Steps 8 through 12 of the V-model and the rest of the roadmap will be discussed in subsequent chapters.

The *high-level design* stage is the first step in the execution of requirements toward the actualization of a product. Should a robust QFD process be followed, eventually, the how-s define more details of the product.

The *high-level design* stage is the first step in the execution of requirements toward the actualization of a product. Should a robust QFD process be followed, eventually, the how-s define more details of the product. High-level design requires the resolution of factors such as comparing and contrasting existing (or off-the-shelf) technologies to provide the *how* to the asked *what*. For instance, how would the design team rank a series of existing components, such as, say, display units, and choose the best one? How would one compare the merits of one technology, say, resistive touchscreens, to another, such as capacitive screens? A tool for making a decision by logically studying various options against a set of critical criteria is called a *Pugh matrix*. I will explain this technique shortly.

Any realistic product will have conflicting requirements. For instance, a car cannot be both fuel-efficient and have high engine horsepower at the same time, or a computer cannot be lightweight and have a long-lasting battery at the same time. During the QFD decomposition, conflicting requirements are captured and noted in the roof of the house of quality. These existing conflicts need to be resolved at the high-level design. A technique enabling a design team to work through conflicting requirements is called the Theory of Inventive Problem Solving (also known as TRIZ). This technique will be reviewed as well.

At times design team needs to look at the evolution of a design prior to developing the next generation of their product. Morphological analysis is a technique that is based on studying how products evolve. This knowledge may then be extended to similar products. One may consider "the good ol' fashion engineering" as an aspect of morphological analysis. The design team can learn a great deal by studying the works of other engineers and scientists and adopt their approach to the particular design at hand. There are many ways to go about this. For one, there are numerous books that cover various engineering design – either mechanical or electrical/electronics. There is also a review of published patents for the novel works of others. Another approach is to study similar designs available on the market.

Once a design option is selected, a high-level design failure modes and effects analysis (DFMEA) should be conducted to identify potential flaws in the design where failures may take place.

Once a design option is selected, a high-level design failure modes and effects analysis (DFMEA) should be conducted to identify potential flaws in the design where failures may take place. By identifying these potential areas, the design team can properly address these concerns early in the process and provide controls in the detailed design stage. I will defer the discussion of DFMEAs to Chapter 10.

For the rest of this chapter, I will review the Pugh matrix, TRIZ, and morphological analysis.

Pugh Matrix (A Rational Decision-Making Model)

Once several viable models (or solution potentials) emerge from brainstorming, they need to be ranked in order that a subset of them may be pursued further. The approach for making logical choices is called *rational decision-making* and a prominent tool is the Pugh matrix developed by Stuart Pugh (Tague 2004).

Also known as a selection matrix, decision grid, opportunity analysis, criteria rating form, or criteria-based matrix, the Pugh matrix enables individuals to make technical decisions based on a variety of options such as risk, reliability, manufacturability, cost, etc.

Developing this matrix is rather simple. A two-dimensional matrix is developed with design or technical options placed on the top row and the selection criteria placed on the first column of this matrix. Then each box corresponding to a particular combination of technical option and selection criterion is given a score. Once the matrix is populated, the scores are compiled and the solutions are ranked accordingly.

It should be noted that scoring may be done in different ways. Two options are as follows. If various options are compared to a baseline, then "Better, Worse, Same" may be used. It may be that while two options may be "Better" than the baseline, of the two, one is *best*. In these situations, a numerical score may be used where a value of zero (0) is assigned to any option that is the same as the baseline. *Better* is given a positive (+) value, and *worse* is scored negatively (-). If there is no baseline comparison, a straight scoring may be used.

In addition to scores, a selection criteria weight factor may also be used. The use of a weighting factor helps differentiate the importance of various selection criteria. For instance, a cosmetic selection criterion may have a weight factor of one (1), whereas product risk may have a weight factor of 5.

To illustrate using this tool, assume that the design team needs to select an air-in-line sensor technology in the development of an infusion pump. Furthermore, the following technologies have been considered: visual, ultrasonic, laser, and optical. The cross-functional design team has also agreed that the section criteria are as follows: low complexity, reliability, time to market, cost, manufacturability and service. Table 9.1 shows the basic Pugh matrix without a weighting factor.

It has been agreed that a 0 to 5 scoring will be used. Using this scoring, the team may populate Table 9.1. The results are shown in Table 9.2.

Based on this information, the best option for air-in-line sensing is visual inspection (by the clinicians). However, this is not a practical solution because of its low reliability. If a weighting factor were used to emphasize the importance of this selection factor, the results may be different. Table 9.3 shows the same matrix, which includes a weighting factor with

TABLE 9.1

A Simple Pugh Matrix

		Technical Options			
		Visual	Ultrasonic	Laser	Optical
Selection Criteria	Low Complexity				
	Reliability				
	Time to Market				
	Cost				
	Manufacturability				
	Service				
	Score				

TABLE 9.2

A Populated Simple Pugh Matrix

$$Score_j = \sum_i X_{i,j}$$

		Technical Options			
		Visual	**Ultrasonic**	**Laser**	**Optical**
Selection Criteria	Low Complexity	5	3	2	1
	Reliability	1	4	4	5
	Time to Market	5	4	3	1
	Cost	5	3	2	1
	Manufacturability	5	3	3	2
	Service	5	4	4	1
	Score	26	21	18	11

TABLE 9.3

A Populated Pugh Matrix with Weighting Factor

$$Score_j = \sum_i W_i X_{i,j}$$

		Technical Options				
		Weight	**Visual**	**Ultrasonic**	**Laser**	**Optical**
Selection Criteria	Low Complexity	1	5	3	2	1
	Reliability	9	1	4	4	5
	Time to Market	3	5	4	3	1
	Cost	1	5	3	2	1
	Manufacturability	1	5	3	3	2
	Service	1	5	4	4	1
	Score		44	61	56	11

the scores of 1, 3, and 9. In this instance, the clear option to be evaluated further is the ultrasonic approach.

The Pugh matrix is a powerful tool, and its application is in making design decisions. It enables the team to have constructive debates about the importance of each criterion and its weight on the overall design outcome. This approach is more effective when every member of the team participates in the discussion and has an equal voice and influence. There are times, however, when one or two dominant voices may prevent others from participating. In this case, the meeting facilitator should ensure that techniques such as the Delphi method (Linstone and Turoff 1975) are utilized to ensure that everyone participates equally in forming this matrix.

Theory of Inventive Problem-Solving (Also Known as TRIZ)

While the acronym for Theory of Inventive Problem Solving should logically be *TIPS*, TRIZ is the Russian abbreviation (*Teoriya Resheniya Izobretatelskikh Zadach*) used by its inventor Genrikh Altshuller beginning in 1946 (Altshuller 2005; Rivin and Fey 2005).

TRIZ is founded on the theory that technical systems evolve along certain trajectories with objective laws. In other words, inventions are not purely inspired but in reality follow certain evolutionary patterns. Shulyak (2005) summarizes the principles of TRIZ as follows:

1. The first principle of TRIZ is that the purpose of a product is to deliver a set of functions. These functions need to be properly understood and defined. For instance, the function of an electric drill may be enumerated as follows:
    a. Receive 120 VAC electric power.
    b. Receive and securely hold a drilling bit.
    c. Rotate the bit at a given rotational speed when the trigger is squeezed.
2. The second principle of TRIZ is that there are four different levels of innovation. These may be enumerated as follows.
    a. First level of innovation is to solve a technical problem to improve the product line. The technical knowledge required to bring such improvements is typically possessed by subject matter experts of the particular technical field.
    b. The second level of innovation involves knowledge sharing among various experts of the same field. This typically concerns resolving technical contradictions that are found in the roof of the HOQ.
    c. The third level of innovation is to find a solution when conflicts have a physical basis. For instance, in an airplane, wings are required to generate aerodynamic lift. However, wings at the same time produce drag, resisting the forward motion. In some designs, the resolution has been to sweep the wings backward. Other designs have offered moveable wings that may be swept back during flight in order to increase the flight speed.
    d. The fourth and final level of innovation may be termed disruptive innovation in that new technologies are developed with the aid of truly understanding interactions of different disciplines of science.
3. The third principle of TRIZ is that innovative activities always tend to remove complexity away from the mechanisms by which a product delivers its function. In other words, the same product delivers more functions. An example of this increased functionality is today's smartphones compared to the basic cell phones of the late 1990s.
4. The fourth principle of TRIZ is that improvements to one aspect of a system's characteristics cause a decline of another aspect of the same system. An example is the compromise between the fuel efficiency of a car and its available horsepower: the higher the horsepower, the lower the fuel efficiency, and vice versa. Another example is a component that is used for both shock and vibration isolation. As isolation requirements are different for shock and vibration, often the solution becomes somewhat of a compromise.
5. The fifth and final principle is that all products evolve. Shulyak (2005) suggested that the founder of TRIZ had established eight ways by which a product evolves.

> One of these eight is the product evolution from macro to micro. In other words, products are developed without regard to their size at first, and then their developers tend to make them smaller in size and weight. Examples of these may be the evolution of a relatively large and bulky Sony Walkman tape player to the small size of the iPod shuffle, which incidentally has a much larger capacity to store songs.

Altsuller, the founder of TRIZ, postulated that all inventive problems have 39 *design parameters* in common and their solutions similarly exhibit 40 common *inventive principles.* In other words, contradiction – mentioned earlier – are the product of any two of the 39 design parameters. To resolve these contradictions, any combination of the 40 Inventive Principles may be employed.

In short, to be inventive, one has to first understand the inherent contradictions of his/her system and/or subsystems. The path to this understanding is to formulate the functions of systems, and subsystems down to assemblies and components.

An Understanding of Function within TRIZ

Earlier, the concept of specifying functions was introduced. From the point of view of TRIZ, there are the *common* ways of formulating functions, and there are *correct* formulations (Rivin and Fey 2005). Common formulations typically speak to what the end users expect to get done, such as, say, *dishwashers wash dishes* or *pencils draw lines on paper.* From an engineering and/or technical (or functional) point of view, a pencil provides *structural support for a thin graphite rod to be dragged across a surface,* or a *dishwasher provides a regulated dispersion and drainage of water and soap in a cavity.* By formulating functions by their physical (or technical) nature, the design team will enable a fundamental shift in perceiving technical challenges and potential approaches to their solutions. In the context of TRIZ, at the most elemental basis, *function* is the result of the *action* (or *energy*) of a *tool* on an *object*. Hence, actions may be classified as useful/harmful, adequate/inadequate, absent/present. For a more in-depth study of this triad, Rivin and Fey (2005) is recommended.

By formulating functions by their physical (or technical) nature, the design team will enable a fundamental shift in perceiving technical challenges and potential approaches to their solutions

Whether TRIZ methodology is used or not, by expressing functions on the basis of their physical applications, the design team can develop a more grounded concept of the product. This concept will then be used as the basis for the first system-level design failure modes and effects analysis to help derive design details at subsystems. Presently, morphological analysis, as yet a different concept generation tool, is introduced.

Morphological Analysis

"Morphological analysis is simply an ordered way of looking at things." As defined by its founder, Fritz Zwicky (quoted from Ritchey 1998). The term "morphology" has its roots in the Greek terms *morphe* and *-logy* meaning *study of form* and/or *shapes.* To be simplistic about this approach, the design team studies the evolution of a particular design (i.e., how its shape transforms from earlier concepts to more developed ones). A good example of this study may be the progression of the early automobiles, which were nothing more than horseless carriages, to the shapes of the 1960s (looking like airplanes) to some of the futuristic cars that do not have steering wheels. Another example may be made of an instrument clip that is used in either orthopedic surgical toolkits or dental instrument cassettes.

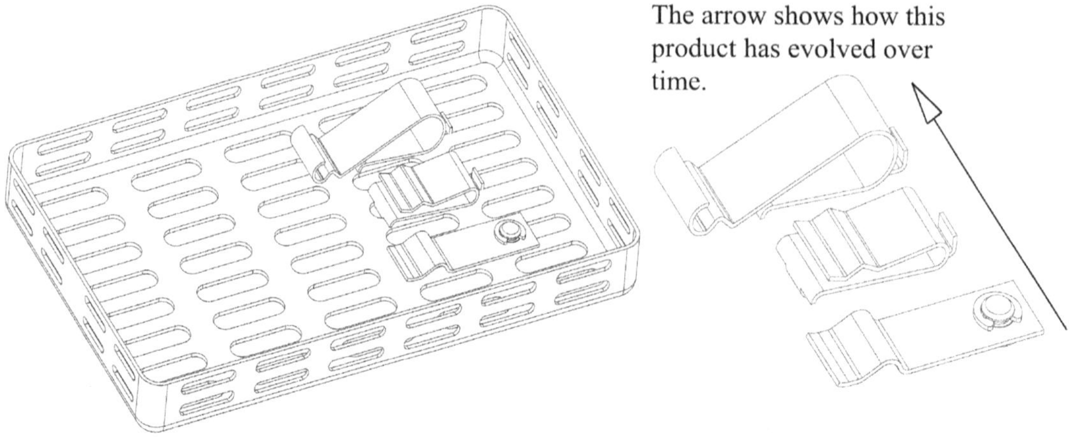

A) Clips are shown in the cassette. B) Clips are shown separately.

FIGURE 9.12
An example of morphological analysis: A medical device cassette and three clip designs

These cassettes are typically used to manage a set of instruments used for a particular clinical procedure. Once the procedure is completed, the instruments are returned to the cassette, which is used to wash and sterilize them. At times, clips are used to hold these instruments in place in order to prevent them from moving or rattling in place and possibly damaging other instruments. Figure 9.12A shows a progression of three different clips in a cassette, and Figure 9.12B shows the clips alone. The disadvantage of the first design is that it is held in place by a pin. As such, a cassette may not be easily configured for different instruments. This problem is solved by the middle design, which enables the clinician to move the clip around. However, the drawbacks of this design are twofold. First, the geometry is such that it can only accommodate a limited range of instrument thicknesses; and second, there is a risk of the clinician cutting his or her finger on the sharp frontal edge of this instrument. Both of these drawbacks have been resolved in the third embodiment, where the longer arm of the clip as well as the larger back diameter allow a larger range of instruments to be used, and a curved front end removes the risk of any injuries.

Methodological analysis is typically used to investigate problems that cannot be easily cast into mathematical models and/or simulations. It decomposes a larger problem into a group of smaller parts. Once each part is analyzed and understood, proposed solutions may be assembled to provide solutions for the larger problem (Ritchey 1991). This approach has been applied not only to solving technical design concerns but also to solving problems in the fields of humanities (Ritchey 2005). The general steps in morphological analysis are as follows:

1) Concisely define and formulate the problem to be solved.
2) Identify and study the totality of factors that play a significant role in the solution.
3) Construct the domain containing all the possible solutions for each part or factor.
4) Examine each solution within this domain and select appropriate options.
5) Reassemble the solution from the sum of the selected parts.

While this methodology has been applied to complex technologies such as jet propulsion, its application to solving design problems may prove cumbersome.

Selected Concept

Earlier, I mentioned that system functions and requirements need to be decomposed into subsystem and ultimately assembly and component functions and specifications. For this purpose and to aid us in this decomposition, techniques such as TRIZ and morphological analysis were introduced. Regardless of the technique or approach used, once the concept is selected, the system architecture as well as the final assembly structure should be defined, though the detailed design is yet to take place.

For the Learning Station, the proposed high-level design as shown in Figure 9.10 may have the architecture as shown in Figure 9.13 and a final assembly structure as shown in Figure 9.14. It should be noted that the structure shown in Figure 9.14 is a reflection of Figure 9.10.

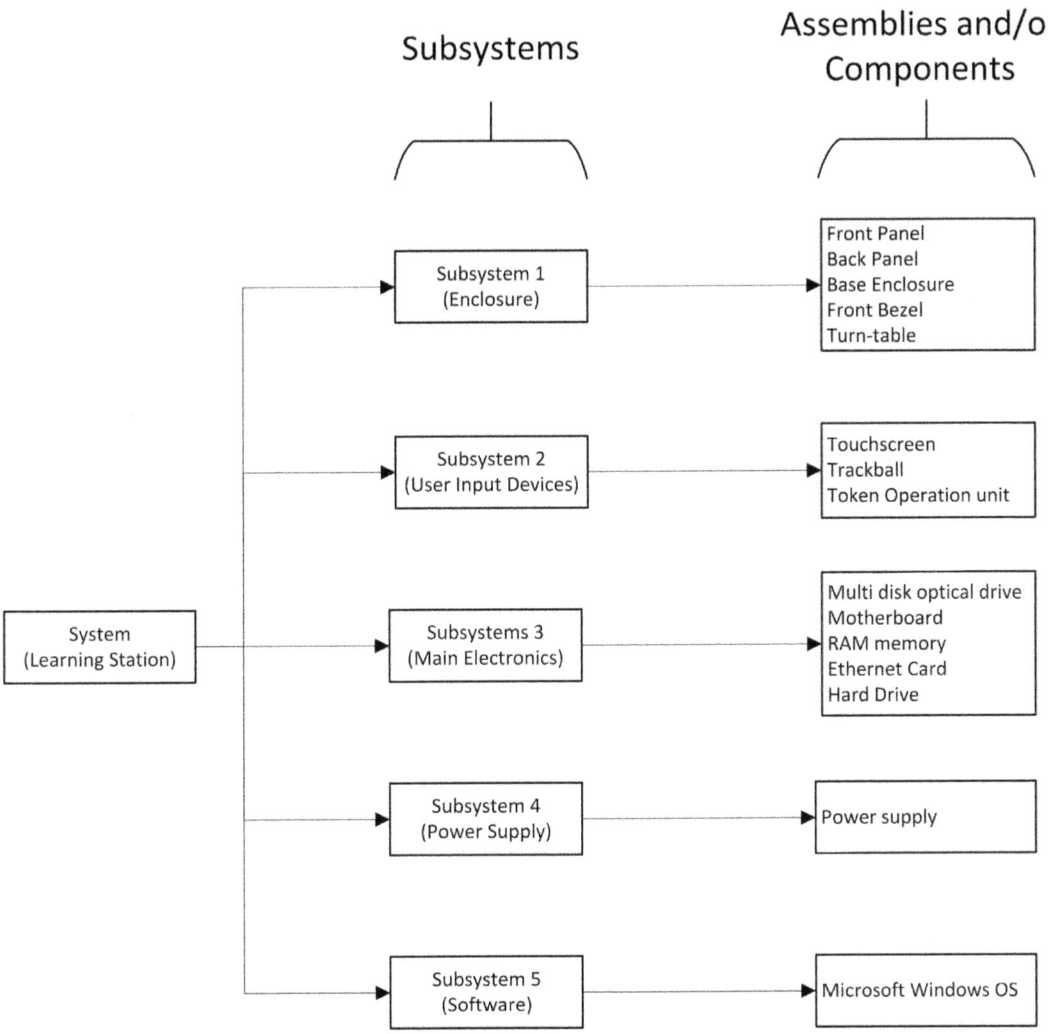

FIGURE 9.13
Learning Station architecture at subsystem and assembly levels

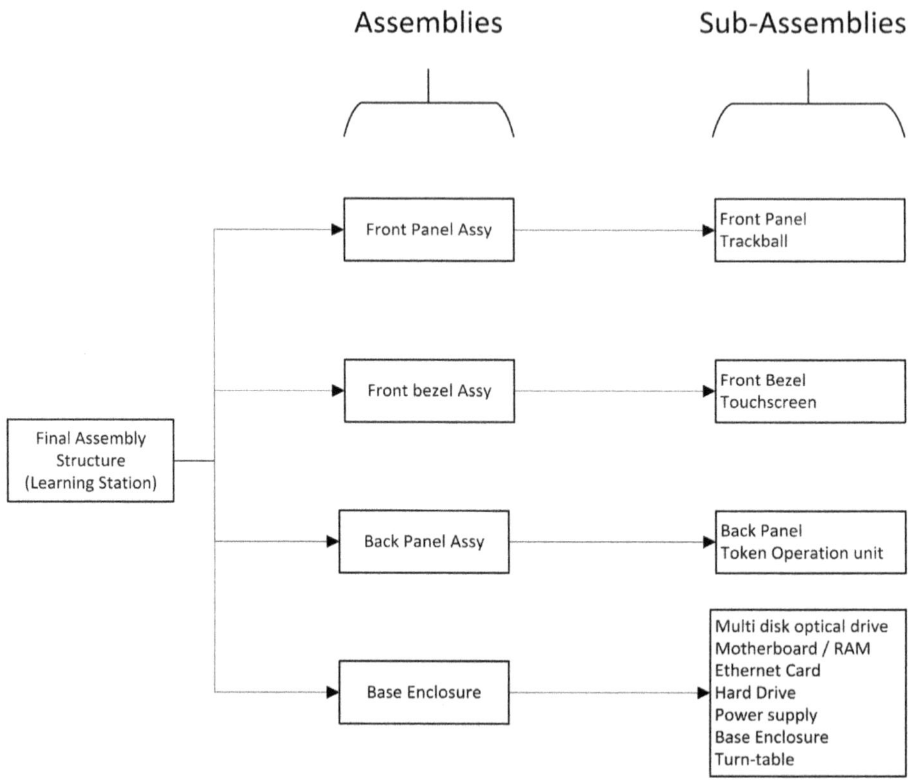

FIGURE 9.14
Learning Station final assembly structure

At this moment, the concept design may reside on the back of a napkin, on a piece of paper, or virtually in a CAD system. It is easy – and, in fact, exciting – to move forward into the detailed design and finalize the product. However, my recommendation is to wait a bit longer. Up to this point, the design team focuses on developing the product and the design. Now, the team should look for what could go wrong and what might be overlooked. In Chapter 10, I will offer a technique on how to watch for potential pitfalls.

Summary

- Developing the matrix of what-s against how-s is the basis of developing and using a tool that in the English language is called quality function deployment.
- Once the QFD has flowed down once or twice, possible product concepts begin to merge.
- As the system architecture is being developed, the system function diagram should also be considered.
- Developing a solid understanding of the functions of the product becomes the basis of concept development.

- Prior to developing the function diagram for the system and subsystems, their boundaries along with their inputs and outputs must be identified.
- Design concepts may be ranked using the Pugh matrix.
- Two popular tools in concept design are TRIZ and Morphological Analysis.

Exercises

1. Continue the design work for the concept selected in the Chapter 8 exercise.
   a. Develop the QFD diagram and cascade it at least two levels.
   b. Use either TRIZ or morphological analysis to generate design concepts
   c. Use Pugh matrix to rank and select one of them.

Part IV

The Nuts and Bolts of the Design

10

Initial DFMEA and Product Risk Assessment

Introduction

In the previous chapter, I examined various steps as well as techniques in developing initial concepts of a new product. These concepts may be captured on the so-called back of the envelope or virtual three-dimensional renditions. Often new products are not necessarily novel concepts. They can be line extensions or short-run promotions. Whatever the circumstance, there is typically good energy and excitement associated with reaching this gate of product development. Enthusiasm (plus scheduling pressures) drives the design team to push forward and complete the detailed design and enter into production. Experienced component design engineers (or scientists) ask the right questions about how components come together, what the tolerances are, etc. If there are ambiguities, they may have brainstorming sessions with colleagues or suppliers to resolve whatever the issue may be. And, relatively quickly, a final production-ready design is created.

Before rushing into finishing the detailed design and at the time when a concept is selected, it is prudent to ask what may go wrong, either in manufacturing, use, or even during service. Failure to ascertain design pitfalls often leads to low production yields, excessive field returns, long service turnaround times and ultimately dissatisfied customers.

Initial Concept Failure Modes and Effects Analysis

Now that both the system-to-subsystems architecture and a physical concept of the new product have been developed, to ensure robustness and provide a systematic approach, an early-stage design/concept failure mode and effects analysis (FMEA) should be conducted. This technique enables the design team to uncover potential shortcomings in the conceptual design by identifying and evaluating

> The first step in creating the DFMEA charts is to ask: What are the requirements and/or functions? The next step is to ask: How will these functions fail?

1. Design functions and requirements
2. Foreseeable sequence of events both in design and production
3. Failure modes, effects, and potential hazards
4. Potential controls to minimize the impact of the end effects

This document may then become the basis of a more formal and rigorous DFMEA and process failure mode and effects analysis (PFMEA). It should be noted that both DFMEA

DOI: 10.1201/9781003564508-14

and PFMEA are living (version-controlled) documents that are initiated before design requirements have been fully established, prior to the completion of the design, and updated throughout the life cycle of the product, including design changes, where appropriate. In Chapter 16, I will provide a more detailed explanation for developing FMEA matrices. For now, let's see how we can take advantage of the DFMEA to help us at the early stages of design.

The first step in creating the DFMEA charts is to ask: What are the requirements and/or functions? The next step is to ask: How will these functions fail? I will cover this area in some detail later in Chapter 16 but for now, let it suffice to say that functions fail in one or a few of the following ways (though rarely more than three ways):

1) No function – for example, consider that the function of a vacuum cleaner is to produce suction to remove dirt. No function means that you turn a vacuum cleaner on and nothing happens!
2) Excessive function – by way of the vacuum cleaner example, the suction is so strong that you cannot move the cleaner around as it is stuck to the surface being cleaned!
3) Weak function – there is hardly any suction present.
4) Intermittent function – vacuum cleaner works fine, but unexpectedly it loses power. and shortly after, its power is restored.
5) Decaying function – vacuum cleaner starts working fine but over time (either short or long term) loses its suction, and it does not recover.

Consider the Learning Station that was developed in the previous chapters. Three of the functions (or requirements) associated with the Learning Station are shown below in Table 10.1. Notice that associated with this table, failure modes associated with each function are also indicated.

Table 10.2 contains the failure modes and effects associated with the Learning Station. In a typical FMEA table, the module to be studied is mentioned. By suggesting that the module is the system, the design team will be concerned with the system-level failures. Also, the specific function under observation is mentioned. In this example, three functions are

TABLE 10.1

Learning Station Functions and Failure Modes

General Function	Specific Function	Failure Mode
Protect from Environment	Stop liquids from entering internal compartments	Liquid leaks inside
	Resist electrostatic discharge from reaching sensitive internal components	Electrostatic discharge reaches sensitive components
	Stop users from reaching (or touching) internal components	Users reach (or touch) internal components
Develop Social Skills	Devise a turntable for turn-taking between users	Turn table fails to turn over time.
	Devise system operation by tokens to enforce social skills development	Tokens fail to work
		Token operation fails to work over time
Run Educational Software	Execute Commercially Available Software	Software does not run Software begins to run but stops

TABLE 10.2

Learning Station Initial Design Failure Modes and Effects

Item	Module	Function	Potential Failure Mode	Potential Effects of Failure	Severity	Potential Causes of Failure
1	Learning Station	Protect from Environment	Liquid leaks inside	Unit stops working temporarily		Lack of gasket and/or drain paths in the design
				Unit stops working permanently		
2			Electrostatic discharge reaches sensitive components	Unit stops working temporarily		Lack of ESD/EMI barriers in the design
				Unit stops working permanently		
3			Users reach (or touch) internal components	Electric Shock		Excessively large openings in the enclosure
				Bodily Injury		
4		Enable Developing Social Skills	Turntable fails to turn over time.	Unit does not rotate on its axis		Turntable deflects under unit weight
						Turntable bearings have corroded
5			Tokens fail to work	Unit stops functioning		Wrong tokens were used
6			Token operation fails to work over time	Unit stops functioning		Component life is too short
7		Run Educational Software	Software does not run	Software does not accept commands		Incompatible software with operating system
			Software begins to run but stops	Software does not accept commands		Systems is overheated

mentioned; namely, protect from the environment, enable developing social skill, and run educational software. associated with each failure mode, one or two effects are suggested. Finally, a potential cause of failure is also provided. At this point, the Severity column is left blank, though it can just as easily be filled out.

I like to make an observation at this moment, though I will elaborate on this point later: I am using the term "Failure Modes and Effects"; however, at the early stages of design, I have more interest in potential causes. Why? Because once I can identify the root causes of failures, I can attempt to make changes to the design to either remove the failure mode(s) or attempt to reduce their frequency or mitigate them in such a way as to reduce their impact. As I will explain later, *effects* impact product risk while *causes* influence product design configuration.

At the early stages of design, the emphasis should be placed on potential causes of failures. The reason is that once the root causes of failures are identified, attempts can be made to change the design to either remove the failure mode(s), reduce their frequency, or mitigate their impact on the end user.

In the context of the Learning Station design, the first item in this initial DFMEA reminds the design team of the possible need for either a gasket or drain paths in the housing (enclosure) to guide liquids away from the inner cavity. The second item ensures that both electromagnetic emission interference (EMI) and electrostatic discharge (ESD) have been properly considered and mitigated. In fact, a clever design may be to integrate both functions in one component.

Another point worth noting is the potential conflict between mitigating overheating (item 7) by having large openings in the unit's housing and mitigating users reaching/touching internal components by having no openings (items # 3) at all. Here is an example of a conflict that may be resolved using TRIZ – as introduced in Chapter 9. Openings are needed to remove heat generated by the electronics, and at the same time, any opening provides an opportunity for the user to reach the electronics inside and hurt themselves. Clearly, this conflict should be resolved satisfactorily.

Finally, this initial DFMEA provides other design insights. First, before specifying the turntable willy-nilly, features such as its weight-bearing capacity as well as its resistance to corrosion should be considered. Another insight is that products do not last forever (item # 6). Product and component life expectancy and their impact on the service organization should not be overlooked. One last point that is brought to light is that certain failures may not be avoided altogether and require mitigations. For instance, an identified cause in item 7 is software incompatibility with the operating system. It is next to impossible to design a software for all computer operating systems. For this reason, a sensible solution and mitigation is through labeling. Typically, the retail packaging of many software applications provides information about the operating systems they are compatible with. To be clear, labeling is not just a sticker that is applied to a product. In general, any printed material either attached to the product or included within the packaging, or even the packaging material, is considered as labeling.

Preparing for the Detailed Design

Suppose that through this exercise, the Learning Station design team decided to design a gasket at the interfaces of the housing. The DFMEA matrix is modified as shown in Table 10.3.[1] One may notice that this DFMEA matrix has an extra column compared to Table 10.2. This column called *Prevention* is a direct response to potential causes and provides a plan of what needs to be done to either eliminate the cause of the failure or to mitigate it.

Generally speaking, once potential causes of failure are recognized, a control plan should be developed to minimize their impact. At the concept level, this control plan would dictate the elements of detailed design. For instance, for the Learning Station as shown in Table 10.3, these elements may be:

A. Once a gasket is designed, a finite element analysis (FEA) should be conducted to ensure that the gasket does not warp under the clamping load of the enclosure in such a way as to create openings for penetration of liquids inside.
B. The size and shape of any openings should be evaluated by human factor engineers to ensure that any potential harm or injury is minimized and mitigated.
C. Proper load and material analysis should be conducted to ensure the proper workings of the turn-table.
D. Through the judicious use of labeling, any product misuse is mitigated and reduced.
E. Finally, both a thermal and reliability analysis of the product (in the design phase) should be done to ensure any thermal or aging failures are either resolved or remediated.

For the sake of providing another example for developing the control plan, consider the infusion pump example of the previous chapter. One of the requirements of any pump infusing a therapeutic solution into the human body is to detect air in the IV tube and to stop any further infusions until the air has been safely removed. In Chapter 9, through the application of the Pugh decision matrix, the design determined that an ultrasonic air-in-line sensor was the technology to be used for a particular product to be developed.

The system requirement relative to air detection is as follows: *The product shall detect air bubbles larger than 10 ml.* There are five potential failure modes associated with this function (or requirement). These failure modes, effects, and causes are shown in Table 10.4. All air-in-line detection failures affect the alarm mechanism embedded in almost all medical device pumps. On the basis of the *Prevention* column, a control plan may be formed. In particular, I like to draw our attention to line item 4. It points to three factors:

A. The gap between transducer and receiver.
B. The lateral alignment of transducer and receiver.
C. Component (mechanical and electronics) tolerance interaction.

In a way, these questions may be related to a bigger issue: How well do we know the behavior of the selected sensor and its response to variations from ideal conditions? This leads to the topic of *transfer function* which I will explore in more detail in Chapter 11. In general, a transfer function speaks of the relationship between input variables into a system (and/or subsystem or component) and its response. For instance, to conduct a finite element analysis of a gasket, the inputs would be applied loads to the gasket, its geometry, and material properties, and the response would be the deflected shape. For the ultrasonic sensor example, the inputs may be the frequency and power of modulation of the transmitter as well as the distance between the transmitter and the receiver. The response may then be the frequency and power sensed at the receiver.

TABLE 10.3

A Populated Initial Design Failure Modes and Effects Matrix for the Learning Station

Item	Module	Function	Potential Failure Mode	Potential Effects of Failure	Severity	Potential Causes of Failure	Occurrence	Prevention
1	Learning Station	Protect from Environment	Liquid leaks inside	Unit stops working temporarily		Excessive warping of gasket		Finite element analysis
				Unit stops working permanently				
2			Electrostatic discharge reaches sensitive components	Unit stops working temporarily		Inadequate ESD/EMI barriers in the design		Design review
				Unit stops working permanently				
3			Users reach (or touch) internal components	Electric shock		Excessively large openings in the enclosure		Human factor study
				Bodily injury				
4		Develop Social Skills	Turntable fails to turn over time	Unit does not rotate on its axis		Turntable deflects under unit weight		Load analysis
						Turntable bearings have corroded		Material compatibility
5			Tokens fail to work	Unit stops functioning		Wrong tokens were used		Labeling
6			Token operation fails to work over time	Unit stops functioning		Component life is too short		Reliability analysis
7		Run Educational Software	Software does not run	Software does not accept commands		Incompatible software with operating system		Labeling
			Software begins to run but stops	Software does not accept commands		Systems is overheated		Thermal analysis

TABLE 10.4
A Populated Initial Design Failure Modes and Effects Matrix

Item	Module	Requirement(s)		Potential Failure Mode	Potential Effects of Failure	Severity	Potential Causes of Failure	Occurrence	Prevention
1	Infusion Pump	Detect air bubbles larger than 10 ml		Bubbles are not detected	No warning alarm		Combined component electronics tolerances may be excessive		Tolerance analysis
2				Bubbles are not detected over time	No warning Alarm		Components have exceeded their life span		Reliability analysis
3				False bubbles are detected	False warning alarm		Improper transducer has been specified		Design review
							Electronics components have been improperly specified		Electronics simulation
							Detection software has not been verified properly		Design review
4				Bubbles are detected intermittently	No warning alarm		Ultrasonic transducers are not mechanically aligned		Tolerance analysis
							Required gap between transducers varies excessively		Tolerance analysis
							Combined component electronics tolerances may be excessive		Tolerance analysis
5				Only bubbles much larger 10 ml are detected	Unreliable warning alarm		Improper transducer has been specified		Design review
							Combined component electronics tolerances may be excessive		Electronics simulation
							Detection software has not been verified properly		Design review

All in all, in my opinion, an outcome of the initial or concept DFMEA is that various aspects of the design and its component behaviors are well understood and controlled. If these relationships are agreed on by the design team, then in the detailed design phase, proper design margins are set to ensure design robustness. The relationship between inputs and outputs is called the transfer function, which will be discussed in Chapter 11. However, before I shift gears, there is one more related topic that needs to be explored at this early stage of product design.

Product Risk Assessment

As the design team progresses in their development efforts, there comes a time when the team should ask some fundamental questions. These are:

1) What can potentially go wrong with the product?
2) How serious will it be when it happens?
3) How often will it happen?
4) And, finally, will the product benefits outweigh its risks?

It is easy to recognize that the first question is addressed by the failure modes and effects columns of the DFMEA matrix. The second and third questions are addressed by the *Severity* and *Occurrence* columns of the DFMEA matrix as well. These two fields are expressed numerically and as such require a more in-depth understanding of their implications within the context of Risk.

Hazard, Hazardous Situation, and Harm

When the question of what can potentially go wrong is asked, the real but unexpressed concern is this: what kind of harm is the end user exposed to?[2] But, just because something can go wrong, on the one hand, it does not always go wrong; and on the other hand, if it goes wrong, it does not necessarily cause harm. By way of example, consider a car developing a flat tire. Depending on whether the flat tire is the front or back tire or if it happens when the car is parked in a driveway or going down the highway at 65 mph, the outcomes could be very different. Another example is having a slippery walkway after a rain shower. It is obvious that not every walkway is slippery when it rains, nor does everyone who walks on a slippery walkway slip and fall. So, how should we look at what can go wrong and the potential for harm?

Harm (or *mishap*) is defined as "an event or series of events resulting in unintentional death, injury, occupational illness, damage to or loss of equipment or property, or damage to the environment."

Let us consider this: *harm* (or *mishap* as used in MIL-STD-882E) is defined as "an event or series of events resulting in unintentional death, injury, occupational illness, damage to or loss of equipment or property, or damage to the environment." However, before harm comes to anyone or anything, there has to be a potential source of harm defined as a *hazard* (MIL-STD-882E). Following the same logic, for harm or mishap to take place, either people or property have to be exposed to one or more hazards. This exposure is called a *hazardous situation.* For instance, lightening is a hazard; being in a storm with lightening is a hazardous situation; harm is getting hit by lightning. Finally, *severity* is the indication of the consequences of harm if it were to befall.

Now that four terms associated with risk have been defined, I would like to note that there are two other factors to be considered in determining risk. The first is the frequency of *occurrence.* This metric provides a measure of the likelihood of occurrence of the specific hazardous situation (or the failure mode). The second, factor called *detection,* is a measure of how easily the hazardous situation (or failure mode) can be detected. In other words, if a shark is present in the area of the beach with people swimming, how easily can it be seen and pointed out.

> *Risk* may be measured as the product of *severity, occurrence,* and *detection.*

Risk may be measured as the product of *severity, occurrence,* and *detection.* On the DFMEA, there is typically a column known as RPN (Risk Priority Number), which is a product of these three metrics. Clearly, the higher the RPN, the more catastrophic the risk may be. The question remains how we assign a numerical value to the individual constituents of RPN.

Some sources – particularly in automotive – choose a 1 to 10 scale (Engineering Materials and Standards 1992). For severity, a numerical value of 1 indicates the lowest level of harm, typically associated with end-user discomfort or minimal damage to property and/or the environment. A numerical value of 10 indicates death or extremely serious injury or extreme damage to environment and/or property.

For occurrence, a numerical value of 1 indicates the lowest level of occurrence, typically identified as *almost impossible.* How should we quantify "almost impossible?" Stamatis (2003) suggests using a numerical value of 1 when reliability has to be 98% (or better). In contrast, Engineering Materials and Standards (1992) suggests a probability of occurrence equal to 1 in 1,500,000 opportunities for a rating of 1. The same source suggests using a rating of 10 where occurrence is almost certain (better than 1 in 3). Again, Stamatis (2003) recommends using the reliability measure, and for a rating of 10, reliability should be less than 1%.

For detection, a rating of 1 means that the defect is easily discernible and that there are proven methods for detection. A rating of 10 refers to defects that are almost always undetectable with no known methods.

On the basis of these numerical values of severity, occurrence and detection, the RPN may be computed as low as 1 and as high as 1000.

Table 10.5 is an example of how these ratings may be put to use for the air detection sensor of the infusion pump of Table 10.4. In this table, there are two columns called "Detection." One is a numerical rating, and the second is a description of the methodology (Anleitner 2011). In addition, there is a classifications column that I will explain shortly.

This initial assessment of Risk provides an early warning to the design team on mechanisms to prevent the causes of failure and to detect potential effects of the failure (Anleitner 2011). In the air-in-line sensor example, the highest RPN belongs to line number 2, "Sensor does not detect bubbles over time." This only means that the design team, as a part of the detailed design, will focus on conducting reliability analysis and life testing to improve the design in an effort to reduce the occurrence rating from 7 to a lower number. As will be shown later, if this reduction is not possible, then, a preventive maintenance of this component should be developed to alleviate the risk.

The *classification* column is used to "mark" which design characteristics may have *critical* or *significant* impact on the end user. In this instance, item number 2 may be

TABLE 10.5

An Initial Design Failure Modes and Effects Matrix With RPN

Item	Module	Requirement(s)	Potential Failure Mode	Potential Effects of Failure	Severity	Classification	Potential Causes of Failure	Occurrence	Prevention	Detection	Detection	RPN
1	Infusion Pump	detect air bubbles larger than 10 ml	Sensor does not detect bubbles	No early warning/ Death or serious Injury	10		Combined component electronics tolerances may be excessive	3	Tolerance analysis	Verification test	1	30
2			Sensor does not detect bubbles over time	No early warning/ Death or serious injury	10		Components have exceeded their life span	7	Reliability analysis	Life test	2	140
3			Sensor detects false bubbles	False early warning/ Delay of therapy	7		Improper transducer has been specified	2	Design review	Verification test	3	42
							Electronics components have been improperly specified	2	Electronics simulation		3	42
							Detection software has not been verified properly	4	Design review		1	28
4			Sensor intermittently detects bubbles	No early warning/ Death or serious injury	9		Ultrasonic transducers are not mechanically aligned	3	Tolerance analysis	Verification test	2	54
							Required gap between transducers varies excessively	3	Tolerance analysis		2	54
							Combined component electronics tolerances may be excessive	3	Tolerance analysis		2	54
5			Sensor only detects bubbles much larger 10 ml	Unreliable early warning/ serious injury	8		Improper transducer has been specified	2	Design review	Verification test	2	32
							Combined component electronics tolerances may be excessive	3	Electronics simulation		2	48
							Detection software has not been verified properly	4	Design review		1	32

classified as critical and item 4 (sensor intermittently detects bubbles) grouping may be considered as significant. The decision where to draw the lines between these classifications is generally made by the design team and some input from management and/or the legal team.

Other sources – in particular MIL-STD-882E – choose different scales for severity and occurrence as shown in Tables 10.6 and 10.7. As may be observed, this approach does not easily integrate with the DFMEA matrix. Nonetheless, the risk associated with each system or subsystem may be evaluated as shown in the Risk Assessment Code Table (RACT) as shown in Table 10.8. Some (especially in the medical field) prefer this approach to risk evaluation because of the clear classification of risk types.

TABLE 10.6

A Table of Severity Levels

Description	Level	Criteria for Harm
Catastrophic	1	Death, extreme bodily injury, permanent environmental damage or extreme financial loss
Critical	2	Substantial bodily injury, significant environmental damage, or extensive financial loss
Marginal	3	Bodily injury, moderate environmental damage, or significant financial loss
Negligible	4	Minor bodily injury, minimal environmental damage, or some financial loss

Source: (Adapted from MIL-STD-882E)

TABLE 10.7

A Table of Probability of Occurrence Levels

Description	Level	Specific Individual Item
Frequent	A	Likely to occur regularly
Probable	B	Will occur several times
Occasional	C	Likely to occur sometime
Remote	D	Unlikely, but possible to occur
Improbable	E	So unlikely, it can be assumed occurrence may not be experienced
Eliminated	F	Incapable of occurrence. This level is used when potential hazards are identified and later eliminated

Source: (Adapted from MIL-STD-882E)

TABLE 10.8

A Risk Assessment Code Table (RACT)

	Severity			
Probability	**Catastrophic**	**Critical**	**Marginal**	**Negligible**
Frequent	High	High	Serious	Medium
Probable	High	High	Serious	Medium
Occasional	High	Serious	Medium	Low
Remote	Serious	Medium	Medium	Low
Improbable	Medium	Medium	Medium	Low
Eliminated	Eliminated			

Source: (Adapted from MIL-STD-882E)

TABLE 10.9

An Initial Design Failure Modes and Effects Matrix with Risk Assessment Code (RAC)

Item	Module	Requirement(s)	Potential Failure Mode	Potential Effects of Failure	Severity	Classification	Potential Causes of Failure	Occurrence	RAC
1	Infusion Pump	Detect air bubbles larger than 10 ml	Sensor does not detect bubbles	No early warning/ Death or serious injury	Catastrophic		Combined component electronics tolerances may be excessive	Remote	Serious
2			Sensor does not detect bubbles over time	No early warning/ Death or serious injury	Catastrophic		Components have exceeded their life span	Probable	High
3			Sensor detects false bubbles	False early warning/ Delay of therapy	Critical		Improper transducer has been specified	Remote	Medium
							Electronics components have been improperly specified	Remote	Medium
							Detection software has not been verified properly	Remote	Medium
4			Sensor intermittently detects bubbles	No early warning/ Death or serious injury	Catastrophic		Ultrasonic transducers are not mechanically aligned	Remote	Serious
							Required gap between transducers varies excessively	Remote	Serious
							Combined component electronics tolerances may be excessive.	Remote	Serious
5			Sensor only detects bubbles much larger 10 ml	Unreliable early warning/ serious injury	Critical		Improper transducer has been specified	Remote	Medium
							Combined component electronics tolerances may be excessive	Remote	Medium
							Detection software has not been verified properly	Occasional	Serious

Briefly, Table 10.8 indicates that if the severity of a particular event is *catastrophic* and its probability is *remote,* then the associated risk is *serious.* Whereas if the severity is marginal and its probability is occasional, then the risk is considered as medium.

The DFMEA matrix may be updated with these ratings and in place of the RPN, RACT may be used. The recommendation is that any *high* or *serious* risk should be reduced and any *medium* risk to be investigated for a possibility of reduction. This is done for the air-in-line sensor example as shown in Table 10.9. It should be noted that in this approach prevention and detection do not play an upfront role. For the same reason a direct comparison of the outcomes of the RACT with RPN may not be appropriate. Once risks are mitigated, the table can be updated to reflect the reduced risk levels.

Before we close this chapter, I like to share that DFMEAs are typically associated with electromechanical products. While they do exist for chemical solutions, the ones that I have seen are based on product *requirement* FMEA. A DFMEA based on requirements does not make sense to me since requirements are cascaded from user needs. A function-based DFMEA should work the same way if the chemists designing the solution are aware of the function of the end results as well as the functions of each component that make up the solution. For example, consider Schiff reagent that is used in histology. Its function is to create a sharp contrast of biological tissue. Hence, each component used in its make up should have known function(s) as well. Admittedly, I am not a chemist and for this reason, I have not provided a DFMEA example for a chemical solution.

Summary

- Before rushing into finishing the detailed design and at the time when a concept is selected, it is prudent to ask what may go wrong; either in manufacturing, use, or even during service.
- An early-stage design/concept FMEA should be conducted when system to subsystems architecture, and a physical concept of the new product have been developed.
- The first step in creating the DFMEA charts is to ask what the requirements and/or functions are.
- Once potential causes of failures in a concept are recognized, a control plan should be developed to minimize their impact. At the concept level, this control plan would dictate the elements of detailed design.
- An outcome of the initial or concept DFMEA is that various aspects of the design and its component behaviors are well understood and controlled.
- In the design process, there comes a time when the team should ask what potentially can go wrong with the product and how serious it will be when it happens.
- When the question of what can potentially go wrong is asked, the real but unexpressed concern is what kind of harm the end user is exposed to.

Exercises

1. Continue the design work for the concept selected in Chapter 9 exercise.
   a. Develop a system level DFMEA for the selected concept.
   b. Identify the potential hazards of the conceptual design. How can these hazards be eliminated or mitigated?
   c. Develop a RAC table.

Notes

1 I have left the two Severity and Occurrence columns blank on purpose. I will pick this topic up once again.

2 Regulated industries such as avionics and medical devices are very focused on risk. Recent events such as toys that are harmful to children or hoverboards that catch fire are reminders that even consumer product developers should be mindful of the impact of their products on the public.

11

Transfer Functions

Introduction

In Chapter 6, I introduced a roadmap and a design for a Six Sigma tool called Define, Measure, Analyze, Design, and Verify (DMADV). Before I explore the concept of *transfer function*[1] further, I would like to review the steps of this tool first and then indicate our position on the product development roadmap and along DMADV. Briefly, the steps of DMADV are as follows.

1. *Define* – Properly define the customer issues or needs that require a set of solutions.
2. *Measure* – Once the project and its risks are identified, the next steps are to develop product requirements and identify elements that are CTQ. Often CTQs are mathematically represented as Y's.
3. *Analyze* – In this step, product concepts are generated and initial cost analysis is conducted. The technical team conducts further allocation of requirements and tracks how CTQs (Y's) are maintained, and the factors impacting them are identified (typically called project X's). Through either experiments and/or simulation, mathematical equations describing the relationship between Y's and X's are established. These equations are called the *transfer function* and mathematically may be shown as

$$\{Y\} = [F](X_1, X_2, \ldots, X_n)$$

   This establishes how variations in {X} impact {Y}. Finally, at the close of this step, transfer functions have been developed. At the design stage, transfer functions can be used to conduct tolerance or sensitivity analysis.
4. *Design* – Based on the developed transfer functions, needed features in the subsystem are designed in this step, and the product variability is understood and design optimizations are completed.
5. *Verify* (and *Validate*) – The final step – in new product development prior to launch – is to verify that all product requirements are met and that all manufacturing and service processes have been written, tested, and validated, and process control plan(s) are put in place.

In Chapters 8, 9, and 10, the three steps of *define, measure,* and *analyze* have been explored. Although I did not specifically call them CTQs in Chapter 10, the elements that were explored and analyzed for failure modes and effects were in fact the elements that were critical to the quality of the intended product. The RACT or RPN classification prioritizes

DOI: 10.1201/9781003564508-15

the list of CTQs. Development of transfer functions is the last step in the *analyze* phase and the starting point of the *design* phase. Figure 11.1 provides a pictorial view of the relative location of transfer functions on the product development roadmap.

Transfer functions are the relationship between a set of variables and system or subsystem response. The set of activities to develop transfer functions may also be known as *characterization.*

A Closer Look at Transfer Function

From a theoretical point of view $\{Y\} = [F](X_1, X_2, \ldots, X_n)$ is the transfer function enabling us to provide needed design details? In Chapter 9, I talked about system (product) functions. It was depicted as a "box" with arrows going in on the input side and arrows leaving on the output side. Let's look at that diagram again in Figure 11.2.

Figure 11.2A speaks to the functions that have been derived on the basis of user needs, whereas Figure 11.2B provides a relationship between how the user and the environment interact with system functions and the output that the user experiences. The assumption is that this relationship and interaction may be quantified in a mathematical equation. Hence, a mathematical relationship may be written to describe the relationship between VOC and product requirements and between product requirements and system requirements, and so on down to the component level.

While in a theoretical sense there is always a transfer function between various steps of product development – that is, let's say between the first QFD and the second – in my experience, those transfer functions have been too elusive and high-level to be expressed in any practical mathematical forms.

However, at the subsystem and assembly levels, with a little effort, various transfer functions can be developed. From a practical point of view, the design team needs this information to identify the variables impacting the design. To do this, a mathematical form of the transfer function is needed. Often, governing equations derived on the basis of chemistry, physics, and engineering constitute these functions. For instance, consider a cantilever beam with an applied load at one end. If the deflection of the free end is critical to quality, then the transfer function between the applied load and deflection is as shown in Figure 11.3.

Note that for a given load *P*, the endpoint deflection may be influenced (and controlled) by four different variables; namely, material selection (*E*) and geometry (*b*, *h*, and *L*). At the same time, any changes or variations in these influencers would cause changes and variations in the output (i.e., Y). Figure 11.4 provides a simple RLC circuit and the development of its transfer function. Similarly, variations in any of the circuit component values may impact the output voltage, particularly if conditions are near resonance. I will examine this in Chapter 24 under tolerance analysis.

A mathematical form defining the relationship between the inputs and outputs may not always be readily available. If this is the case, a series of simulations or experiments needs to be conducted to establish how the desired response changes with variations of input variables. Once this is done, a *design surface* or *response surface* may be identified and mathematically formulated. This surface is the desired transfer function from which an optimum segment may be identified along with the corresponding values of the input variables.

A mathematical form of a transfer function may not always be available. In these cases, a design or response surface may need to be developed using simulations or experiments.

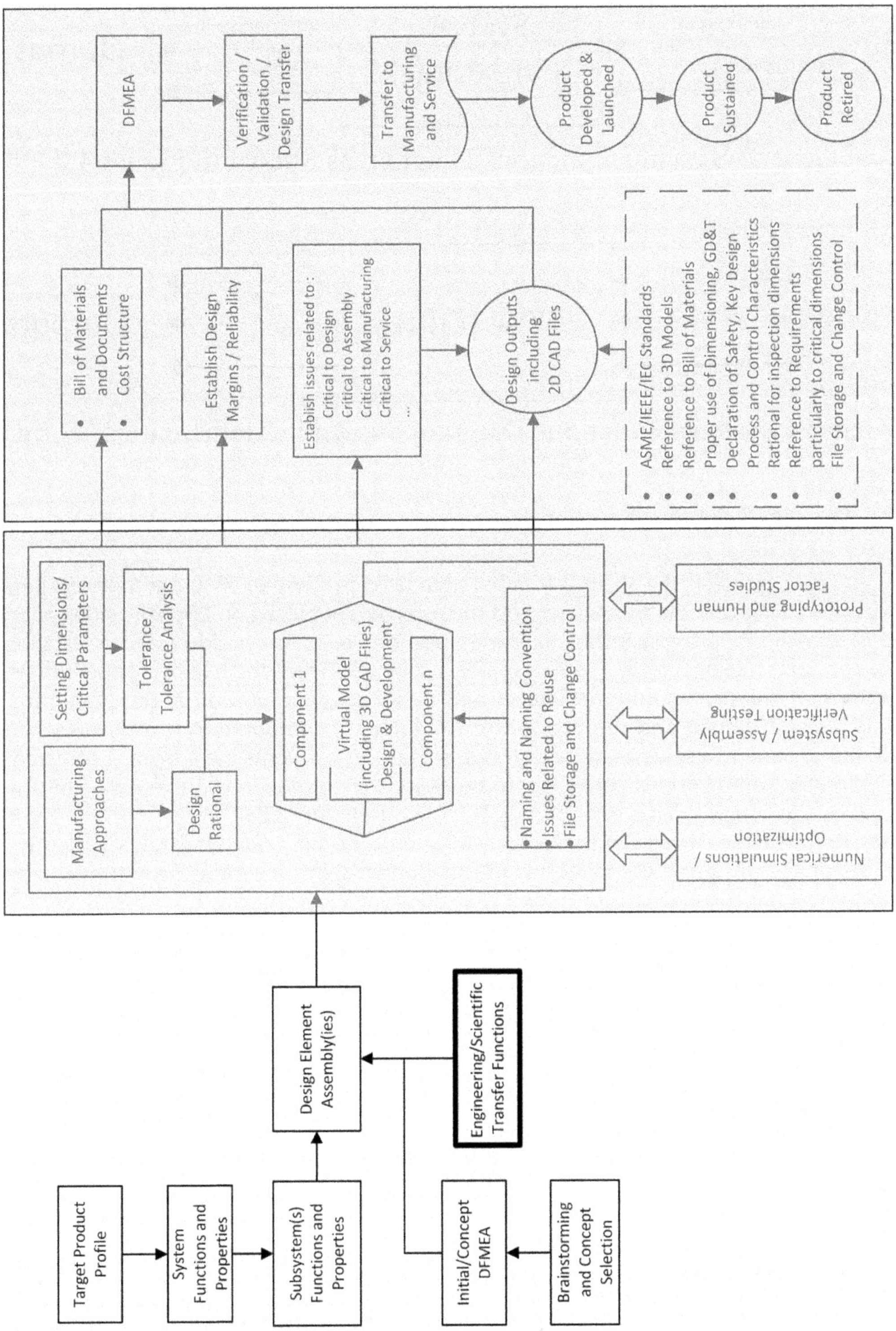

FIGURE 11.1
Relative location of transfer functions on the product development roadmap

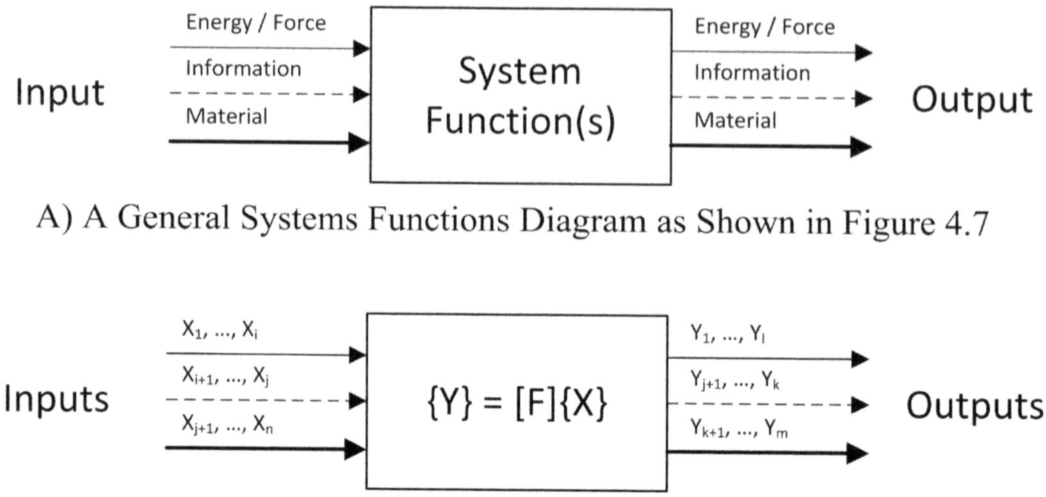

B) A General Systems Functions Diagram Shown in Mathematical Form.

FIGURE 11.2
A generalization of system and transfer functions

As an example, consider the air-in-line subsystem used in Chapter 10. From a design perspective, considering that the transducer and the receiver are placed on opposite sides of an IV tube as shown in Figure 11.5, the following question needs to be addressed: What tube diameter should be used (or what is the nominal distance between the two components be)?

The nominal distance should be determined either by computation if the governing physics is well developed and understood or by testing if computation proves to be too difficult. For argument's sake, assume that two sets of experiments have been conducted, one set in water and a second set in air, to measure the signal strength received by the receiver as a function of *d*. This relationship is shown in Figure 11.6.

Bearing in mind that this is a hypothetical situation, a design team will quickly identify that for his or her design, there is a clear distinction between a signal through water as opposed to a signal through air, provided that the distance between the transducer and the receiver is larger than 1.25 mm. This curve (obtained mathematically or empirically) provides objective evidence that this sensor will in fact identify a pocket of air within the flow of water in a tube because a significant shift in the signal may be detected.

Furthermore, the transfer function enables the design team to set various parameters and helps the team to understand the interaction of one component (or assembly) with its neighboring components. This point may be made more clear by considering Figure 11.6. The signal behavior is highly nonlinear up to a distance of about 1.1 mm. Furthermore, the signal behavior for distances larger than 1.5 mm is almost linear, and at the same time, the same ratio of signal strength may be observed in the ranges of 1.5 mm to 4 mm. Now, on this basis, the minimum tube diameter should be set to 1.5 mm. Should a smaller tube be desired, a different sensor with a different range must be considered. Similarly, a maximum tube diameter may also be set using a similar logic.

Note that the configuration in Figure 11.5 is an ideal configuration with an ideal alignment of its components. As the design team, we need to make sure we do not get trapped in this "ideal configuration" mindset. In reality, misalignment happens as shown in Figure 11.7; the relative position of components may shift laterally, rotate, or even shift and rotate. Fortunately, having an early-stage DFMEA helps us avoid design errors of this

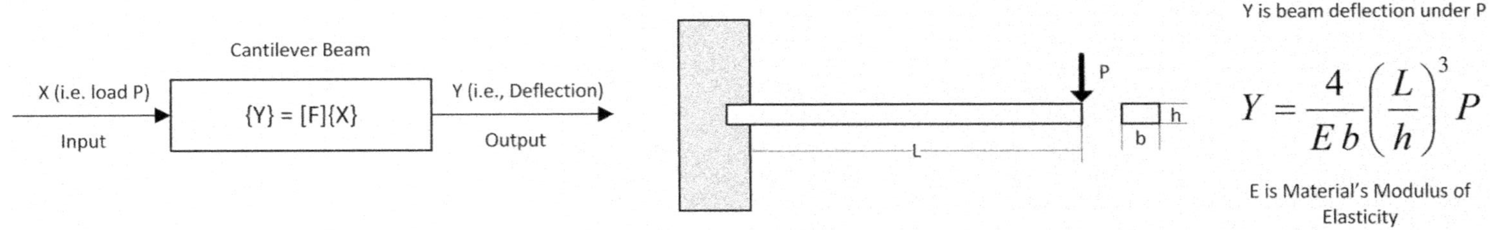

FIGURE 11.3
Transfer function for a cantilever beam

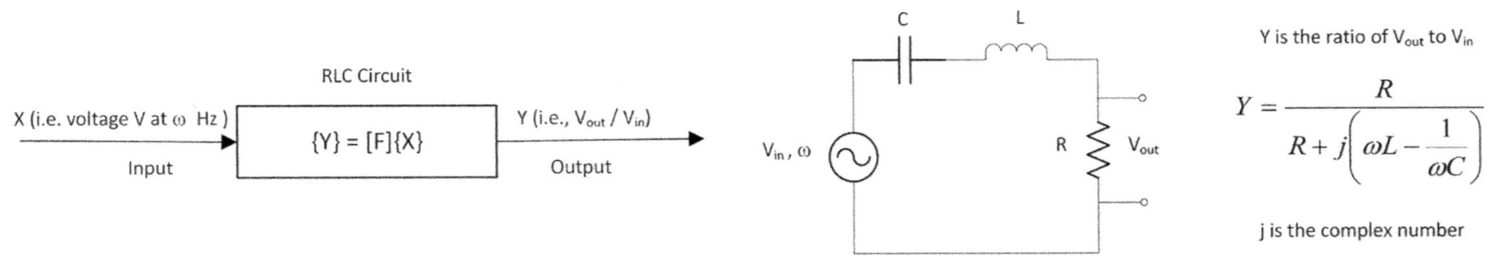

FIGURE 11.4
Transfer function for an RCL circuit

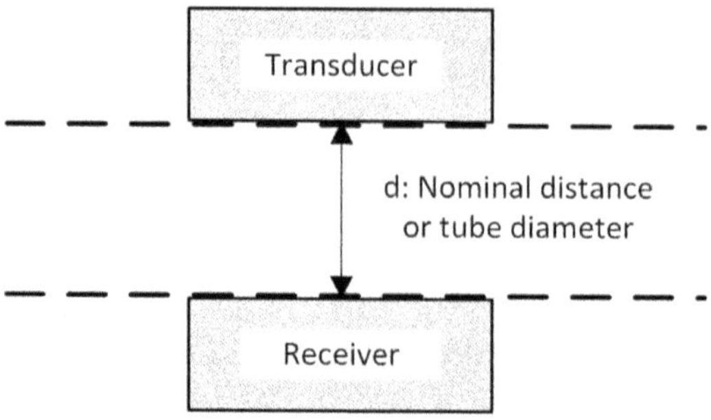

FIGURE 11.5
An ultrasonic air-in-line sensor configuration

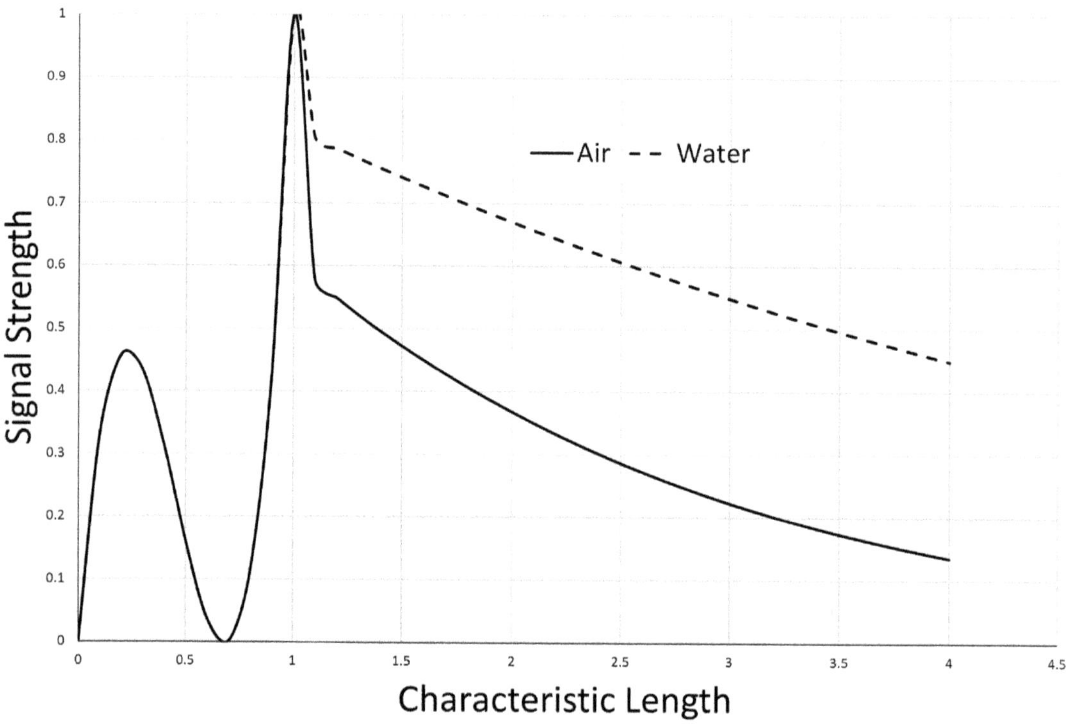

FIGURE 11.6
Hypothetical air-in-line transfer function in both air and water media (signal strength as a function of the distance between the two transducers)

nature. For this particular example, the DFMEA developed in Chapter 10 identified the following three issues:

1. Proper gap between transducers.
2. Proper lateral alignment of transducers.
3. Component (mechanical and electronics) tolerance interaction.

I will consider the tolerance interaction in Chapter 12. For now, let's consider the gap and alignment issues. From a gap and alignment point of view and depending on the design, one of the four configurations as shown in Figure 11.7 may take place.

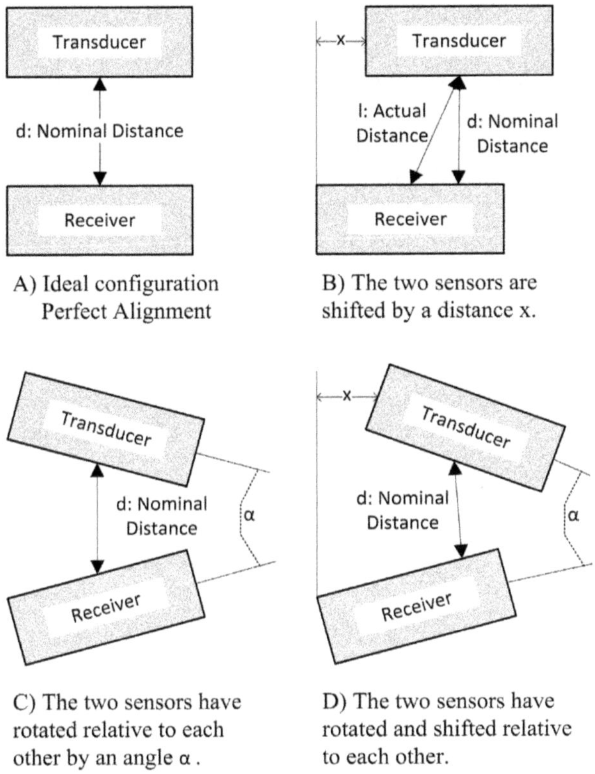

FIGURE 11.7
Realistic configurations of the air-in-line sensor

The transfer function of a general configuration (Figure 11.7D) must be developed in order to help the design team develop a robust design. To establish this transfer function, design of experiments (DOE) is a tool set that may be used to study the relationship between independent variables (i.e., the distance between the two sensors and any lateral and/or pitch misalignment) and the expected outcome (i.e., the response curve).

Design of Experiments

> DOE is a tool set that may be used to establish transfer functions. In other words, DOE may be used to study the relationship between independent variables and the expected outcome.

Design of Experiments is a structured technique to study the impact of a set of factors (inputs or X's) on the response of a system (outputs or Y's). In a nutshell, the basis of DOE is to run a set of experiments (or simulations) on the basis of nominal and/or maximum/minimum values of its inputs. Once this is done and data is collected, regression (i.e., curve fitting) tools are used to develop a mathematical model that relates input variables to the response(s). This technique is well studied and documented. See, for instance, Del Vecchio (1997), Hicks (1973), Morris (2004), and Mathews (2005). A full description of DOE is beyond the scope of this work. As such, in this section, I will only attempt to use the DOE techniques as a part of the air-in-line example to demonstrate how to develop a transfer function from a set of input variables.

The approach in DOE is relatively simple to explain: first, the inputs to DOE should be identified. For the air-in-line sensor, these input variables are *d*, *x*, and α. One may suggest that electrical power and circuitry may also be considered as input variables. While this statement holds true, it may be suggested that the electronics and mechanical design in this particular example do not interact and may be decoupled.

Second, the degree of linearity (or nonlinearity) of the model should be decided. Typically, in DOE, one of the linear, quadratic, or cubic models is used. The simplest model is the linear model. For a threevariable input such as the ones of the air-in-line example, a linear model is expressed as below:

$$Y = A_0 + A_1X_1 + A_2X_2 + A_3X_3 + A_4X_1X_2 + A_5X_1X_3 + A_6X_2X_3 + \epsilon \tag{11.1}$$

The $X_i's$ are independent variables (e.g., d, x and α) and Y is the transfer function. ϵ is the error term that results from casting a general equation into a specific form, as in Equation 11.1.

The third step is to conduct m number of experiments (or numerical simulations, if running a finite element analysis) using different values of X_i and obtain corresponding values of Y. To ensure that there are an appropriate number of equations to solve for the unknowns (i.e., $A_j's$) m should be equal 2^n; where n is the number of independent variables. For the three-variable example $(i = 3, m = 2^3 = 8)$, this results in the following set of equations:

$$Y_1 = A_0 + A_1X_{11} + A_2X_{21} + A_3X_{31} + A_4X_{11}X_{21} + A_5X_{11}X_{31} + A_6X_{21}X_{31} + \epsilon_1$$

$$Y_2 = A_0 + A_1X_{12} + A_2X_{22} + A_3X_{32} + A_4X_{12}X_{22} + A_5X_{12}X_{32} + A_6X_{22}X_{32} + \epsilon_2$$

$$\cdots\cdots$$

$$\cdots\cdots$$

$$Y_8 = A_0 + A_1X_{18} + A_2X_{28} + A_3X_{38} + A_4X_{18}X_{28} + A_5X_{18}X_{38} + A_6X_{28}X_{38} + \epsilon_8$$

In matrix form, this set of equations may be written as

$$\begin{Bmatrix} Y_1 \\ Y_2 \\ \cdots \\ Y_m \end{Bmatrix} = \begin{bmatrix} 1 & X_{11} & X_{21} & X_{31} & X_{11}X_{21} & X_{11}X_{31} & X_{21}X_{31} \\ 1 & X_{12} & X_{22} & X_{32} & X_{12}X_{22} & X_{12}X_{32} & X_{22}X_{32} \\ . & . & . & . & . & . & . \\ 1 & X_{18} & X_{28} & X_{38} & X_{18}X_{28} & X_{18}X_{28} & X_{18}X_{28} \end{bmatrix} \begin{Bmatrix} A_0 \\ A_1 \\ A_2 \\ A_3 \\ A_4 \\ A_5 \\ A_6 \end{Bmatrix} + \begin{Bmatrix} \epsilon_1 \\ \epsilon_2 \\ \cdots \\ \epsilon_8 \end{Bmatrix}$$

The general form of this equation may be expressed in the matrix notation:

$$\{\mathrm{Y}\} = [\mathbb{X}]\{\mathrm{A}\} + \{\epsilon\} \tag{11.2}$$

Finally, this matrix equation should be solved by minimizing the error term $\{\epsilon\}$. The least squares method is typically used to minimize this term and calculate the regression coefficient vector $\{A\}$ (Myers and Montgomery 2001):

$$\{A\} = \left([\mathbb{X}]^t [\mathbb{X}]\right)^{-1} [\mathbb{X}]^t \{Y\} \tag{11.3}$$

The transfer function (or the response) may be calculated as follows:

$$\{\hat{Y}\} = [\mathbb{X}]\{A\} \tag{11.4}$$

Morris (2004) has developed a template that conducts these calculations in Microsoft Excel. In addition, he provided means of calculating the goodness of fit as well as other statistical measures, such as the confidence level associated with each value of the regression coefficient. Needless to say, many analysis software programs, such as Minitab, solve these types of equations.

Coded Space

In selecting the values of the input factors (or independent variables) to run the experiments, it is customary to choose the maximum or minimum values if a linear relationship is assumed. For nonlinear models, additional points are required. Considering the air-in-line example, suppose that the design team anticipates the following parameter ranges:

> The four steps in design of experiments studies are:
>
> 1) Identify inputs,
> 2) Decide the degree of linearity,
> 3) Conduct experiments, and
>
> Solve the matrix equation.

$$1.5 \le d \le 4.0\ ,$$
$$0 \le x \le 1.0\ ,$$
$$0 \le \alpha \le 10.0$$

Since there are three variables, $8(=2^3)$ experiments need to be set up. Table 11.1 depicts these values along with the corresponding signal strength measured from the hypothetical experiments.

Mathews (2005) warns against forming the $[\mathbb{X}]$ matrix on the basis of the actual input values. It is possible that the resulting matrix is ill-conditioned, leading to erroneous

TABLE 11.1
Experiment Set with Input Values and Corresponding Measurements

	Input Factors (Xs)			Transfer Function (Y) Signal Strength
Experiment	d (X_1)	x (X_2)	α (X_3)	
1	1.5	0.0	0.0	0.47237
2	4.0	0.0	0.0	0.13534
3	1.5	1.0	0.0	0.28650
4	4.0	1.0	0.0	0.10688
5	1.5	0.0	10.0	0.41992
6	4.0	0.0	10.0	0.12912
7	1.5	1.0	10.0	0.26608
8	4.0	1.0	10.0	0.10246

results. Instead, he recommended transferring the variables into a *coded* space. In this space, variables range from -1 to +1. The transformation equation is as follows:

$$C_i = \frac{2}{X_{i\,max} - X_{i\,min}} X_i + \left(1 - \frac{2}{X_{i\,max} - X_{i\,min}} X_{i\,max}\right)$$

Hence for the air-in-line example:

$$C_1 = \frac{2}{4-1.5} d + \left(1 - \left(\frac{2}{4-1.5}\right) 4\right)$$

$$\Rightarrow C_1 = 0.8\,d - 2.2$$

$$C_2 = \frac{2}{1-0} x + \left(1 - \left(\frac{2}{1-0}\right) 1\right)$$

$$\Rightarrow C_2 = 2x - 1$$

$$C_3 = \frac{2}{10-0} \alpha + \left(1 - \left(\frac{2}{10-0}\right) 10\right)$$

$$\Rightarrow C_3 = 0.2\,\alpha - 1$$

As a result of this transformation, Table 11.1 turns into Table 11.2 – note that Y's have not changed:

Now the $[\mathbb{X}]$ may be set up in the coded space, and the coefficient vector may be calculated using Equation 11.3. The results are presented in Table 11.3.

There are two additional metrics that need to be calculated and evaluated prior to accepting the coefficient values (or A_j). The first is a metric indicative of goodness of fit or R^2. This metric ranges between 0 and 1; 0 means that the data fit was very poor, and 1 indicates perfect fit. The second metric is the confidence level $(1-p)$. This value is typically chosen at 0.9 or higher (90% confidence or better). Most available statistical software packages provide this information. For developing an algorithm in Excel, see Morris (2004). Table 11.4 provides this information for the previous example.

TABLE 11.2

Experiment Set with Coded Input Values and Corresponding Measurements

Experiment	Input Factors (Cs)			Transfer Function (Y) Signal Strength
	$d\,(C_1)$	$x\,(C_2)$	$\alpha\,(C_3)$	
1	-1	-1	-1	0.47237
2	+1	-1	-1	0.13534
3	-1	+1	-1	0.28650
4	+1	+1	-1	0.10688
5	-1	-1	+1	0.41992
6	+1	-1	+1	0.12912
7	-1	+1	+1	0.26608
8	+1	+1	+1	0.10246

TABLE 11.3

The [𝕏] for the Air-In-Line Sensor Example in the Coded Space Along with the Calculated Coefficients

Intercept	d	x	α	d x	d α	x α		
C_0	C_1	C_2	C_3	C_1C_2	C_1C_3	C_2C_3		
1	-1	-1	-1	1	1	1	A_0	0.2398
1	1	-1	-1	-1	-1	1	A_1	−0.1214
1	-1	1	-1	-1	1	-1	A_2	−0.0494
1	1	1	-1	1	-1	-1	A_3	−0.0104
1	-1	-1	1	1	-1	-1	A_4	0.0356
1	1	-1	1	-1	1	-1	A_5	0.0078
1	-1	1	1	-1	-1	1	A_6	0.0042
1	1	1	1	1	1	1	–	–

TABLE 11.4

The Confidence Level and R^2 Values Associated with the Calculated Coefficients

	p-value	Confidence Level	R^2
A_0	0.010	99.0%	0.9992
A_1	0.020	98.0%	–
A_2	0.049	95.1%	–
A_3	0.221	77.9%	–
A_4	0.067	93.3%	–
A_5	0.288	71.2%	–
A_6	0.464	53.6%	–

The $R^2 = 0.9992$ shows that the calculated curve fits the data very well. However, the confidence levels of A_3, A_5, and A_6 are low – much lower than the typical 90 or 95 percent. These coefficients are associated with α – the angle between the transducer and the receiver. A low confidence level is indicative of a low degree of influence of the variable on the response. For this reason, the contribution of α may be ignored. Finally, the response equation or the transfer function in the coded space may be written:

$$Y = 0.2396 - 0.1214C_1 - 0.0494C_2 + 0.0356C_1C_2$$

In the physical space, the transfer function for the signal in air may be written as

$$Y_{\text{Air}} = 0.635 - 0.126\,d - 0.255\,x + 0.0569\,d\,x \tag{11.5}$$

This equation indicates that the signal strength is a function of not only the distance and the pitch between the transducer and receiver but also the product of the two. Similarly, the transfer function for the signal in water may also be developed.

$$Y_{\text{Water}} = 0.891 - 0.111d - 0.178x + 0.0346dx \tag{11.6}$$

In Chapter 12, I will demonstrate how to use these functions to develop an understanding of variations in the design and their impact on the factors that are critical to quality.

Chemical Reagent Case Study

To demonstrate that DOEs may be applied to consumable products, consider the following case study.

In medical diagnostics and post-biopsy, the removed sample is processed and stained for increased contrast prior to being examined by a pathologist. Staining samples enables the pathologist to view biological cell structures along with the nuclei. Two common stains are hematoxylin and eosin (H&E); a third stain is called Schiff, which overcomes some of the limitations of H&E stains.[2] Schiff reacts with aldehydes in biological samples to produce a bright red-colored specimen. Any additional stain that has come into contact with a biological sample may need to be disposed.

Manufacturing the Schiff reagent is rather simple: Dissolve fuchsine dye in distilled or deionized water and mix well. Add potassium or sodium metabisulfite along with hydrochloric acid and mix well. Finally, use activated charcoal to filter the solution. While fuchsine dye produces a magenta colored solution, the filtered Schiff is colorless. To test its staining strength, formaldehyde is added to a small sample quantity of Schiff solution, which turns into a reddish-colored fluid. Examined in a spectrometer, the light absorptance (α) level at a given wavelength is indicative of the Schiff's dye strength; too high of an absorptance would cause the cell structure to be too dark for evaluation; too low of an absorptance would not create enough of a contrast in the cell structure.

While making Schiff is simple, maintaining consistency and quality in a production environment is not. For instance, fuchsine dye is not pure. Its dye content may depend on its supplier or may degrade over time. Also, it has been suggested that "To obtain perfectly colorless Schiff reagent, it is necessary that fresh activated charcoal be used" (Lillie 1951). How fresh is fresh?

To study the impact of the ingredients on the quality of the Schiff, the following DOE was developed:

1) Fuchsine: two levels of dye content; one at 95%, the other at 75%.
2) Activated Charcoal: two levels of aging; one freshly produced, the other one-year-old.
3) Metabisulfite: two levels of aging; one freshly manufactured, the other one-year-old.
4) Formaldehyde: two levels of concentration; one at 40%, the other at 30%

This DOE was set up in Minitab with the number of replicates set to 3, a resolution of IV (i.e., two-factor effects confounded with other two-factor effects) for a total of 24 randomized experiments. Table 11.5 contains the order of the runs as well as the measured absorptance. Figure 11.8 provides the main effects factorial plots for the absorptance (α). Clearly, as expected, fuchsine dye has the highest impact – the higher the dye content, the stronger the stain. Activated charcoal age has little to no effect on absorptance, which seems to be

TABLE 11.5

The Order of the Experiments as Well as the Measured Absorptance (A) A Stands for Fuchsine Dye, B for Activated Charcoal, C for Metabisulfite, and D for Formaldehyde

A	B	C	D	α
1	1	1	1	1.06
1	1	1	1	3.33
1	1	1	1	2.21
1	1	-1	-1	4.51
1	1	-1	-1	5.27
1	1	-1	-1	4.44
1	-1	1	-1	4.69
1	-1	1	-1	3.19
1	-1	1	-1	4.54
1	-1	-1	1	4.45
1	-1	-1	1	1.43
1	-1	-1	1	1.81
-1	1	1	-1	1.06
-1	1	1	-1	0.56
-1	1	1	-1	0.73
-1	1	-1	1	0.69
-1	1	-1	1	0.69
-1	1	-1	1	0.93
-1	-1	1	1	0.67
-1	-1	1	1	1.27
-1	-1	1	1	0.65
-1	-1	-1	-1	1.06
-1	-1	-1	-1	1.20
-1	-1	-1	-1	1.00

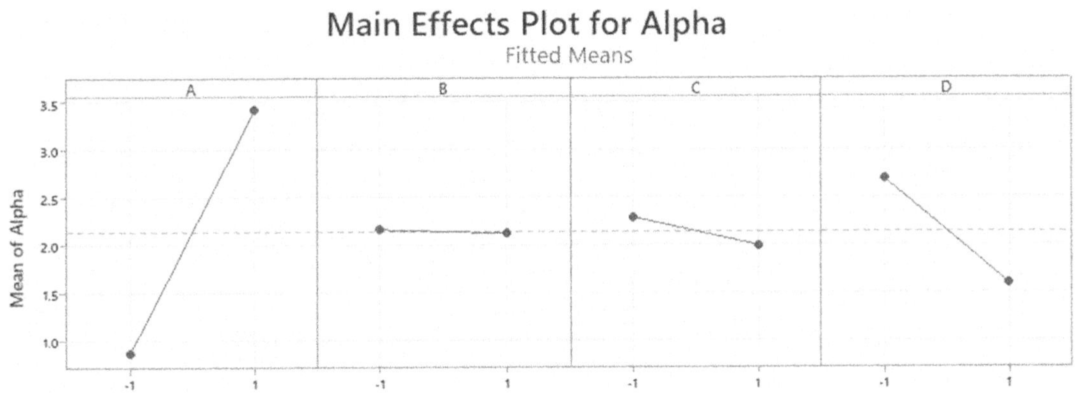

FIGURE 11.8
Main effects factorial plots for the absorptance (α) A stands for fuchsine dye, B for activated charcoal, C for metabisulfite, and D for formaldehyde

TABLE 11.6

Analysis of Variance A Stands for Fuchsine dye, B for Activated Charcoal, C for Metabisulfite and D for Formaldehyde

Source	DF	Adj SS	Adj MS	F-Value	P-Value
Model	7	52.22	7.46	11.68	0.000
Linear	4	46.21	11.556	18.09	0.000
A	1	38.59	38.56	60.42	0.000
B	1	0.01	0.01	0.02	0.902
C	1	0.51	0.51	0.80	0.384
D	1	7.10	7.10	11.12	0.004
2-Way Interactions	3	6.01	2.00	3.14	0.055
AB	1	0.15	0.15	0.23	0.636
AC	1	0.21	0.21	0.33	0.574
AD	1	5.65	5.65	8.84	0.009
Error	16	10.22	0.64		
Total	23	62.44			

contrary to Lillie (1951). The final ingredient in Schiff's recipe is metabisulfite. Based on these results, freshly manufactured metabisulfite has an adverse impact on absorption. This may be counterintuitive and may require further investigation to determine how the fresh and aged ingredients may be different.

I find the impact of formaldehyde rather curious. It shows that its concentration had a reverse impact on absorptance; in other words, the higher the concentration, the lower the absorptance. Recall that this material is not used in the main formulation of the Schiff but as a tool for testing it. Clearly, this conclusion does not hold across any range of concentration because at zero levels of formaldehyde, the solution does not change color. The lesson learned from this aspect of the DOE is that test methods need to be scrutinized and upper and lower bounds need to be understood.

Minitab provides the following regression equation for absorptance (α) in the uncoded units:

$$\alpha = 2.143 + 1.268A - 0.020B - 0.146C - 0.544D + 0.079AB$$
$$-0.094AC - 0.485AD$$

This equation may be considered as the transfer (or characterization) function for the Schiff reagent. Recall that A stands for fuchsine dye, B for activated charcoal, C for metabisulfite, and D for formaldehyde. Analysis of variance provided by Minitab is presented in Table 11.6. Note that R^2 associated with this analysis is 83%.

Summary

- Often, governing equations derived on the basis of physics, chemistry, and engineering constitute transfer functions.

- The transfer function enables the design team to set various parameters and helps the team understand the interaction of one component (or assembly) with its neighboring components.
- A DOE may be used to develop the response surface and ultimately the transfer function.

Exercises

1. Continue the design work for the concept selected in the Chapter 10 exercise.
   a. In the conceptual design, identify the elements that are CTQs.
   b. Identify the factors that impact the CTQs. Would you be able to develop the associated transfer functions?

Notes

1 Scientists may use the term "characterization" to mean the same set of activities.
2 Any review of these differences is beyond the scope of this work.

12

The Virtual Product

Introduction

In my years as a design engineer, I can say that to me this step, to actually work on the modeling, was the most exciting step in product development. This is the step to actually design the so-called nuts and bolts of a product and then see it materialize. It brings a sense of elation to me.

In Chapter 6, a product development roadmap was presented. Based on this map, by developing the transfer functions and assembly requirements, nearly one third of the road has been traversed. Developing the virtual product and the associated documentation constitute the middle third, and design transfer, verification, and validation activities make up the last third as shown in Figure 12.1.

Before discussing the details of the design approach, it should be mentioned that the outcome of design activities is *paperwork* and not a *product* – as popularly believed. For this reason, it is important that ample time is given to documentation and providing information for the individuals who would be involved in not only manufacturing and servicing a product but also in sustaining it and providing support. Should proper documentation *not* be maintained, product knowledge and information become "tribal knowledge." This knowledge is lost when people move about either by retirement, different career paths, or attrition. A Systems approach to product development always advocates preserving and maintaining product knowledge and information. Thus, it is appropriate to review needed documentations that will be released into a product's design history file (DHF) to ensure that they provide the mindset and the rationale for the specific design decisions made.

The outcome of design activities is *paperwork* and not a *product* – as popularly believed. For this reason, it is important that ample time is given to documentation and providing information for the individuals who would be involved in not only manufacturing and servicing a product but also in sustaining it and providing support.

In practice, development of major transfer functions takes place parallel to developing the mechanical "look and feel" of the product and/or subassemblies. However, the detailed design may begin in earnest when various subsystem behaviors have been well understood; i.e., the transfer functions have been developed. This is true not only for electromechanical components but also for any chemical solutions and/or reagents that may be used in combination with the electromechanical subsystems. Consider, for example, a color inkjet printer. The behavior of inks of different colors and their interactions – not only with each other but also with their container and delivery system – must be clearly understood before the design of the product may advance to its next logical stage.

Figure 12.2 provides an overview of the activities associated with developing the virtual product's components, assemblies, subsystems, and finally the finished system. In short, Figure 12.2 is a summary of the material covered in this chapter.

DOI: 10.1201/9781003564508-16

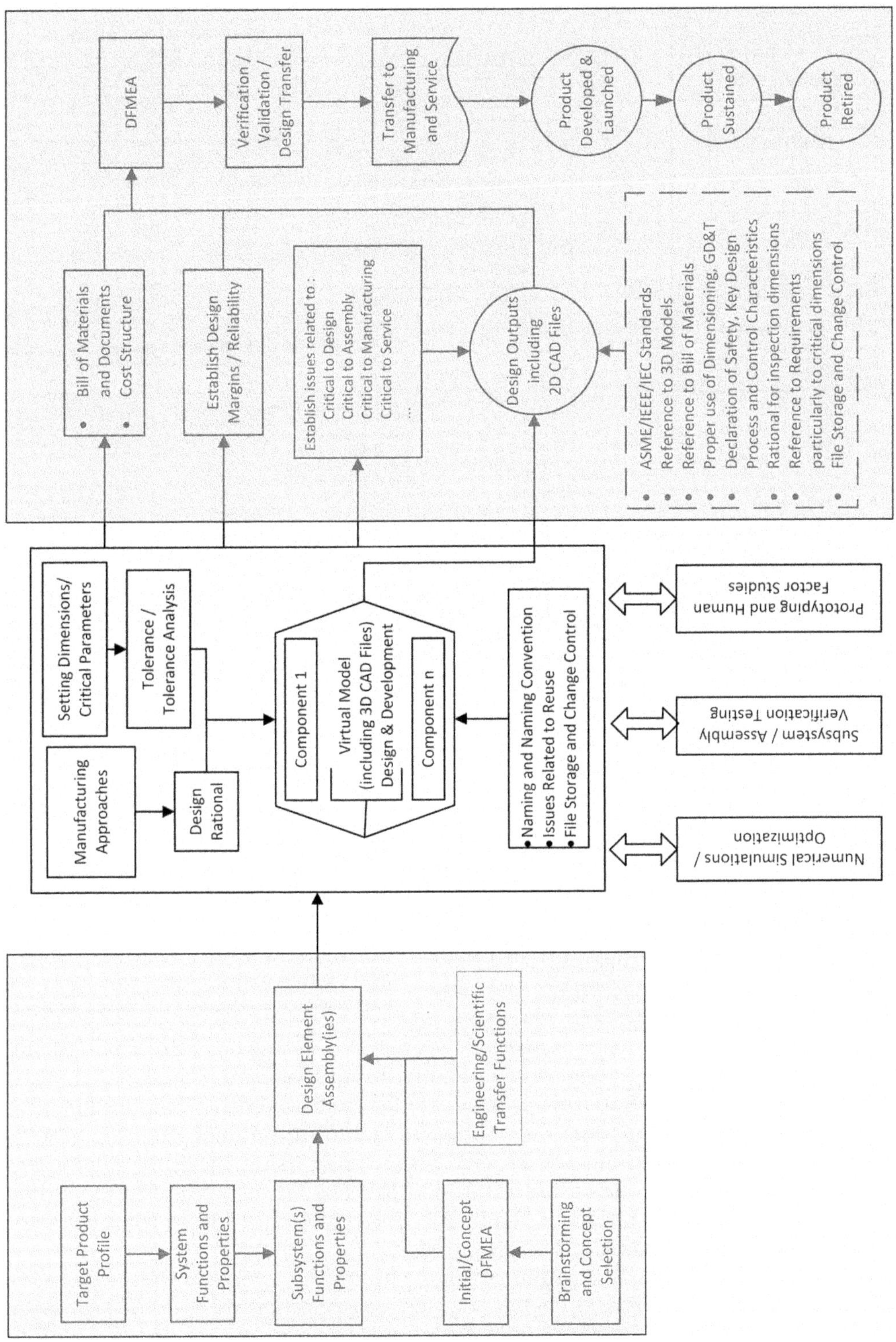

FIGURE 12.1
Product development roadmap highlighting relevant sections in Chapter 12

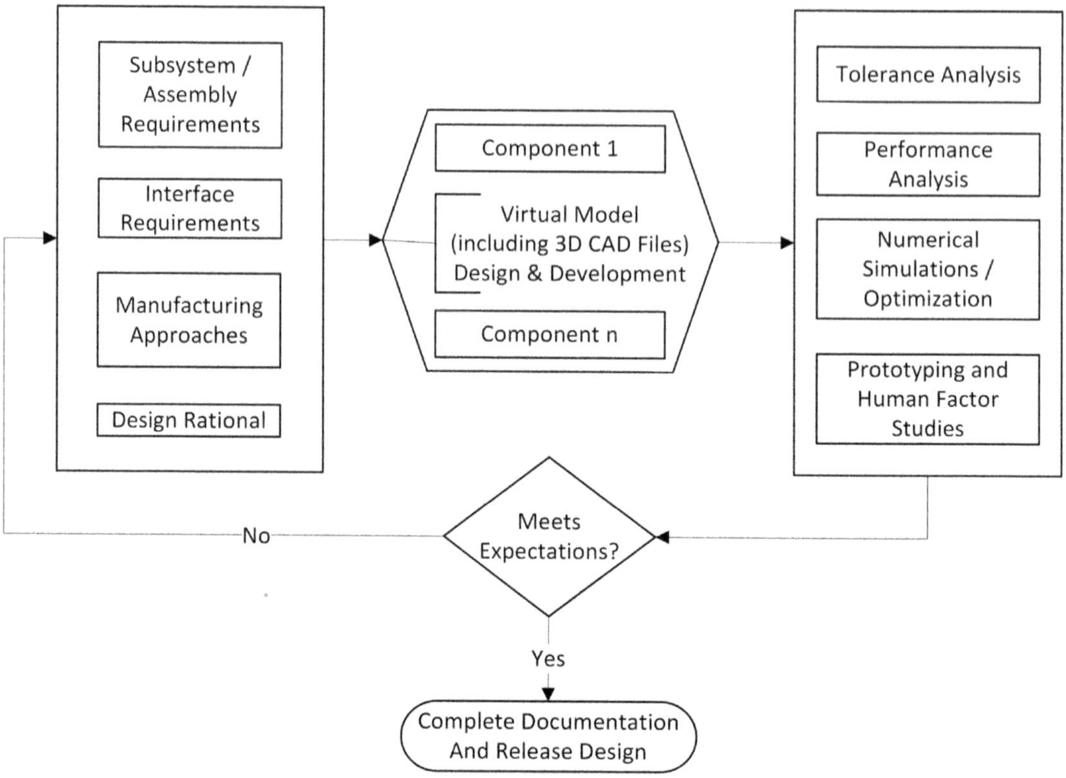

FIGURE 12.2
Steps in developing a virtual model and its verification

> To develop a comprehensive set of specifications, inputs from other functions within R&D as well as from a variety of organizations outside of R&D should be taken into account and incorporated.

This design stage, by its rather creative nature, is iterative. It begins by examining the subsystem and/or assembly along with interface requirements. Next, manufacturing approaches along with design rationale are selected. In other words, if plastics are selected as the material over metal and injection molding over machining, the rationale should be clearly documented. The same is true for chemical compositions. This is needed because, depending on the manufacturing and selected material, the component and assembly designs (as well as associated costs) may be very different. To illustrate this point, consider the bracket shown in Figure 12.3. They both provide the same function. Part A may be manufactured using injection molding or machining; however, it may not be readily fabricated using sheet metal bending. Should sheet metal fabrication be of interest, the design should be modified to look like part B. Having a documented rationale maintains design intent for future technical support teams and helps any potential product upgrades.

Once (as shown in Figure 12.2) requirements have been decomposed into design specifications and the manufacturing approach is selected, components and assemblies may be designed using a variety of computer-aided design (CAD) software.[1] Depending on the assigned function of the component or assembly, there may be a need for tolerance and

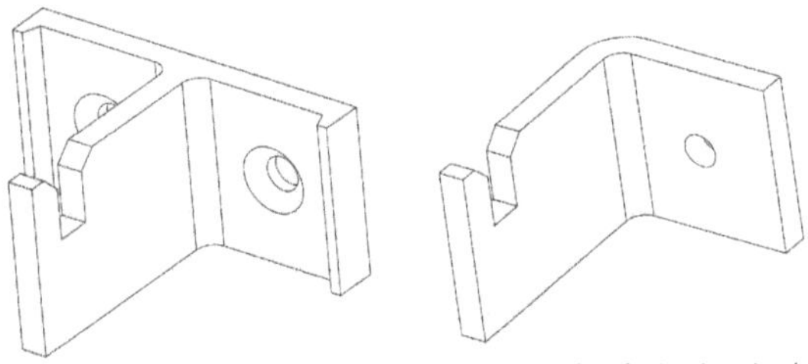

FIGURE 12.3
Component design and manufacturing interdependence example. (a) This design is suited for injection molding or machining, and (b) this design is suited for stamping and sheet metal fabrication

performance analysis, numerical simulation of either functions or creating prototypes for human factor considerations and studies. Typically, this aspect of the design is an iterative process to converge to the desired configuration and, ultimately, lead to releasing the design for manufacturing tooling. It should be noted that any requirements verification testing of assembly or subsystem requirements takes place using production tooling – and not using prototype units. If verification of requirements fails, this aspect of design must be repeated at least one more time. This broader process is demonstrated in Figure 12.4.

The process as represented in Figure 12.2 or Figure 12.4 is somewhat deceptive – or should I say oversimplified. It correctly suggests that the design team begin by decomposing assembly (or subsystem) requirements and develop design specifications. The reality is that to develop a comprehensive set of specifications, inputs from other functions within R&D as well as from a variety of organizations outside of R&D should be taken into account and incorporated. In some organizations, the outcome of these activities and the decisions made as applicable to the manufacture and service of the end product are documented in a *device master record* document. The outcome and decisions concerning interfacing various functional inputs from within R&D are documented in a set of *interface documents.*

As shown in Figure 12.5 (and in no particular order), organizations that communicate design intent with R&D and their category of inputs are listed below:

1. Marketing as well as compliance (including regulatory) teams are *labeling* stakeholders. Labeling refers to any markings on the product (or shipped along with it) that communicate a message or a series of messages. An example of a label is a decal that provides information required by a regulatory body such as UL (Underwriters Laboratories), CE (Conformité Européenne, or European Conformity), or TuV (Technischer Überwachungsverein, or Technical Inspection Association). Another form of labeling is any marketing decal, such the product brand, that may be printed or otherwise placed on the product. Product and shipping containers and any manuals also fall within this category. Often, the marketing team designs these documents; however, these designs are treated, maintained, and controlled as any other component design document.
2. Business team is focused on the financial aspect of the products and their impact on the so-called bottom line. Hence, one of the major and early requirements for any product is its projected cost of manufacture. This cost requirement should

be properly cascaded to each subsystem and ultimately to each component. This factor influences the choice of manufacturing and production to a large extent. Associated with this cost sensitivity are influencers such as reliability. Clearly, the higher the product reliability, the lower the service and warranty costs. In addition to reliability, risk also plays an important role. No product is without risk; it so happens that some products have higher risks than others. It is the responsibility of the design team to identify risk factors and develop a risk benefit analysis of the product and identify safety-related components within the design.

3. Manufacturing team provides input and feedback on design for assembly (DFA) and design for manufacturability (DFM) concerns. Should the rules of DFA and DFM be properly followed, the overall cost of the product – both direct and indirect – will be lowered and optimized. However, while R&D receives information and input on DFA and DFM, it provides leadership and guidelines on what appropriate manufacturing approaches should be selected and works closely with manufacturing to establish needed processes and set needed factors so that manufactured components meet design intent. An example of this R&D input is

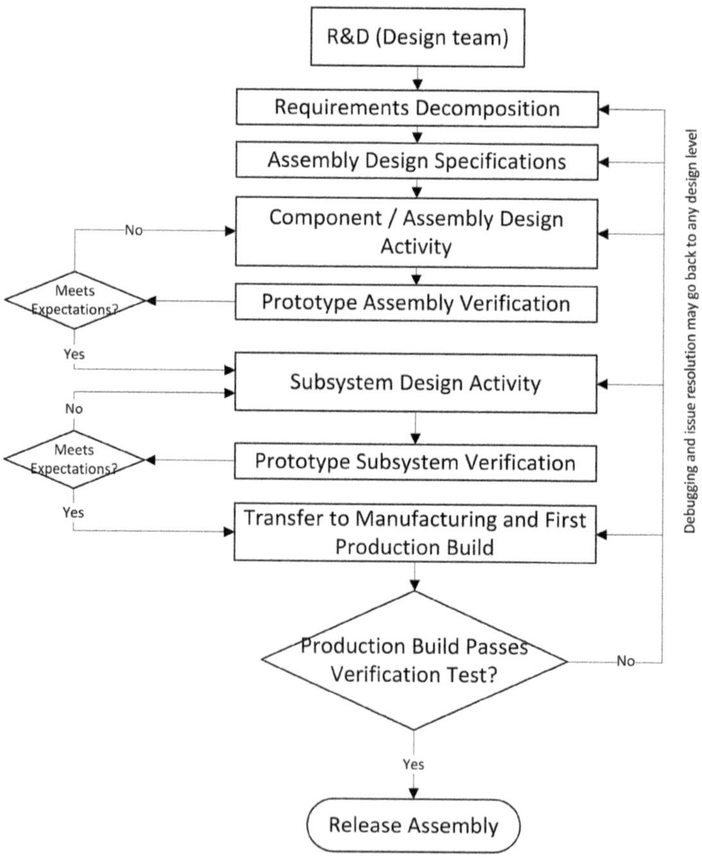

FIGURE 12.4
Steps in developing a model, including its production build and verification

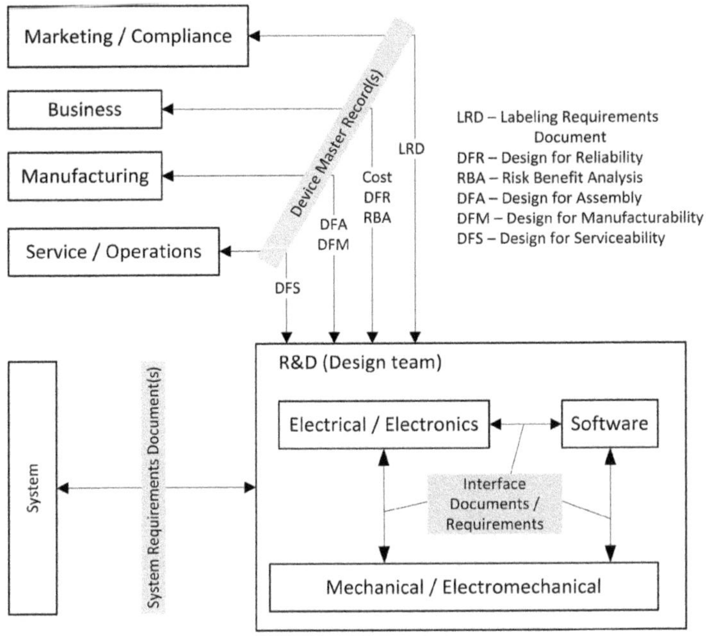

FIGURE 12.5
The interaction of cross-functional teams and their influence on design

in injection molding components. Often, by reducing mold hold-time and pressure, an injection molding operation can save the cost of fabricating components; however, the price to be paid for this saving is higher part-to-part variability. R&D and manufacturing should partner to identify optimum process conditions that meet both design intent and manufacturing goals. Once identified, these process parameters should be captured in a related Device Master Record.

4. Finally, the service team has a stake in the design process as well. They bring to the table their expertise and know-how in service to arrange assemblies so that the service process is straightforward and time is not wasted to open the product and replace failed components. Similarly, R&D works with service organizations to provide effective diagnostic tools and codes such that service personnel could easily and effectively identify a faulty component. As an example, many of the new computers may be considered. Once a malfunction is experienced, these newer units are equipped with diagnostic software that may be executed at the startup of the computer. The system conducts a test of various critical components (such as memory, hard drive, or display) and reports if any issues are observed. Another example is an engine electronics module that is installed on many of the newer model automobiles. To identify any malfunctions, the mechanic attaches a diagnosis device to this module, and within a short period of time, the issue is identified and located.

In addition to showing the interrelationships between R&D's design team and other stakeholders, Figure 12.5 points to the interrelationships between various functions within R&D. This inter-functional communication and collaboration is important even at some mundane levels. For instance, PCBAs typically designed by the electronics teams have to be mounted in enclosures typically designed by mechanical engineers. The simplest issue is the placement of mounting holes and their size; who gets to dictate where these holes should be? Should it be the mechanical team, based on enclosure concerns, or the electronics team, based on circuit design and board layout? Other interface issues may involve heat generation and removal, vibration or EMI, electromagnetic compatibility (EMC), or ESD concerns. In one situation, the decision may be based on the electronics concerns; in another, the mechanical engineers make the decision; and yet, in another situation, it is the scientists who have to dictate. The point that I am trying to make here is that while design decisions are made, they should be captured in *interface documents*. Documentation ensures that decisions and design intents go beyond individuals' memory spans and "tribal" knowledge.

Requirements Decomposition

Requirements decomposition is simply translating input requirements into usable specifications. It is simple, though the challenge is to ensure that functional lines are not crossed. For instance, the mechanical decomposition should not involve elements of electronics or software functions. For instance, suppose that a particular electrical power consumption requirement is 10 watts. It may be tempting to decompose the mechanical function requirements as "the trace shall be able to carry a power load of 10 watts." It is more appropriate to specify "Power traces shall have a width of x mm and a copper weight of y ounces."

Requirements decomposition is simply translating input requirements into usable specifications. To put it in a different way, design specifications are the outcome of decompositions.

Figures 12.6 and 12.7 depict the inputs and outputs of the decomposition process and the information that needs to be specified (or fleshed out) for a mechanical requirements decomposition as well as an electronics requirements decomposition. A few points to consider: first, on the input side, one assembly may be impacted by a number of cascaded requirements. For instance, there may not only be mechanical requirements such as weight and volume but also concerns such as vibration and shock as well as heat dissipation.

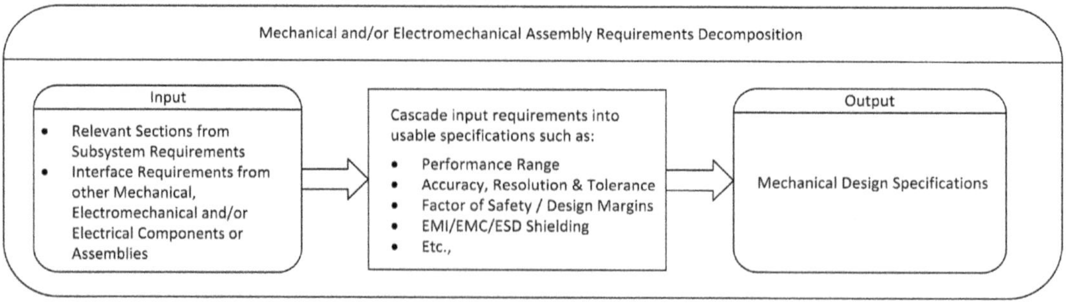

FIGURE 12.6
Inputs and outputs of mechanical/electromechanical

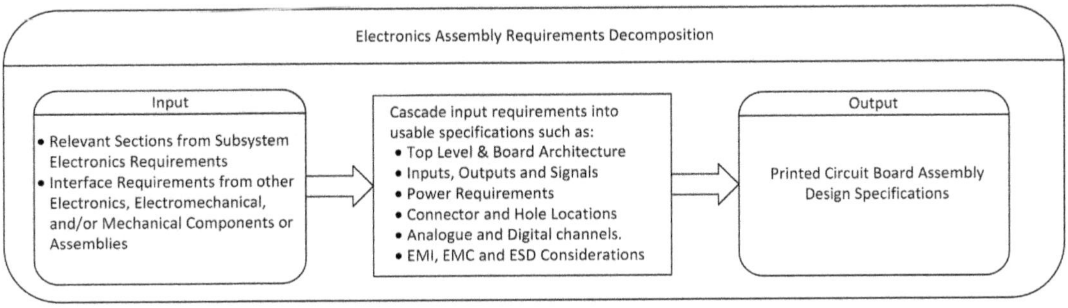

FIGURE 12.7
Inputs and outputs of electronics requirements decomposition

Also, on the electronics side, requirement cascade may include power consumption as well as ESD or EMI/EMC issues. If we are developing a chemical reagent, the decomposition may need to include environment, health, or safety concerns. In a way, various system and subsystem requirements converge on the assembly and components. Thus, their designs should be traceable back to all their various and potentially conflicting system-level requirements.

To illustrate the decomposition process, consider the air-in-line sensor example of Chapter 11.[2] At the product level, there are two requirements that impact the air-in-line sensor subsystem and design. The first – and the obvious – is the requirement that the product should sense air bubbles greater than a particular size. The second – and more subtle – requirement is that the product has a size and weight limitation. These two high-level requirements are decomposed in Tables 12.1 and 12.2.

Design specifications are the outcome of decompositions. In developing these specifications, every effort should be made to phrase them in such a way that they are clear, viable, and measurable. For instance, suppose that a leaf spring exerts pressure on a shaft, and hence, it keeps the shaft from moving through friction forces. The specification for the spring may be developed as follows: "The leaf spring shall be capable of preventing shaft movement at a minimum of 1 N force applied laterally to the shaft at a rate of 0.2 N/sec, in dry conditions at 38 degrees C."

Once design specifications are developed, several concepts are proposed and discussed among the design team for feasibility. The combination of specifications and the chosen concept(s) are from a portion of a set of inputs required to develop the design details and the three-dimensional mechanical CAD models and schematic files for electronics. Other inputs into developing detailed designs may be enumerated as follows:

1) Interface issues and functional tolerance concerns
2) Manufacturing approaches, techniques, and rationale for selection
3) Make versus buy decisions
4) Re-Use and naming convention
5) Design for manufacturability, assembly, service, reliability, etc.
6) Prototyping and human factor studies

TABLE 12.1

An Example of Mechanical Requirements Decomposition

Requirement	Description			
Product	The product shall have a mass less than 10 kg and a volume less than 0.012 cubic meters, including packaging and manuals.			
System	The device shall have a mass less than 7 kg and a volume less than 0.002 cubic meter.			
Relevant Requirement at Sub-system	The air-in-line sensor subsystem shall detect air bubbles with the following parameters			
	Range	From	To	Units
	Weight	N/A	150	grams
	Total Sensor Assembly Envelope	10 x 10 x 35		mm
	Total Board Envelope	75 x 50 x 10		mm
Mechanical Components	The ultrasonic sensor must meet the minimum parameters in the table below			
	Range	From	To	Units
	Weight	N/A	150	grams
	Connector	Locking 3-pin		N/A
	Operation Temperature	-15	55	°C
	Tube Size	5.5	6.5	mm
	Sensor Housing Wall Thickness	nominal – 2		mm
	PCB Thickness	nominal – 1.6		mm
	PCB Mounting	4 x 2mm through holes		N/A

Similarly, the list of the final outputs may include:

7) Two-dimensional drawings that are used to control fabrication and manufacture of the product
8) A list of critical design components and/or dimensions as well as design margins
9) A traceability matrix that links critical components and/or dimensions to requirements or risk
10) Bill of materials and cost structure
11) Theory of operation for each assembly, subsystem, and finally the system
12) Any special design considerations that are needed for manufacturing
13) An understanding of service and/or failure conditions experienced by the end user
14) An understanding of reliability concerns and preventative maintenance concerns.
15) Bill of documents

This chapter will focus on various design inputs and considerations, that is, items 1 through 6, and Chapter 13 will focus on outputs (items 7 through 15).

Typically, once the initial draft of components has been developed, the first of the detailed design activities is to ensure that all parts fit together at their nominal specified

TABLE 12.2

An Example of Electronic Requirements Decomposition

Requirement	Description			
Product	The device shall detect air-in-line while pumping			
System	The pumping mechanism shall detect air bubbles larger than 10 mL			
Relevant Requirement at Sub-system	The air-in-line sensor subsystem must be able to detect air bubbles with the following parameters			
	Range	From	To	Units
	Bubble Size detected	10	N/A	mL
	Accuracy	N/A	5	%
	Resolution	N/A	0.1	%
Electronics Component	The ultrasonic sensor must meet the minimum parameters in the table below			
	Range	From	To	Units
	Bubble Size detected	10	N/A	mL
	Supply Voltage	N/A	5	VDC
	Supply Power	N/A	0.25	W
	Response Time	N/A	20	µSec
	Digital Output (Fluid)	4	4.5	VDC
	Digital Output (Air)	0.1	0.5	VDC
	Output Tolerance	3 at high	5 at Low	%
	Zero Balance	N/A	0.3	mV/V_{ex}

dimensions. Once the design team is convinced that the design meets this criterion, depending on the complexity of the product, other simulations, such as FEA, may be conducted to evaluate localized or overall stresses under the applied loads. This is to examine the margin of safety for part or product failure. It should be noted that by loads and stresses, I mean mechanical, electrical, or even chemical loads and their corresponding stresses.

These topics lead to the first design consideration, namely, that of interface issues and functional tolerance concerns.

Interface Issues and Functional Tolerance Concerns

In my previous work (Jamnia 2025), I had pointed out that failures typically occur when design team members overlook certain (and possibly critical) factors. One of these factors is an investigation of how the delivered function of an assembly or subsystem is affected by various part-to-part variations. Often, engineers conduct worst-case tolerance analysis on both mechanical and electrical components; however, they do not investigate these variations beyond stack-up. The same may be true for solutions and reagents, as (slight) variations and their impact on chemical compositions are often overlooked. In this section, I like to explore this concern by way of an example and to highlight issues that need to be evaluated at interfaces and the resulting functional tolerances. I should add that the design provided here is not representative of a working or a commercial unit.

> One of the overlooked factors in design is the investigation of how the delivered function of an assembly or subsystem is affected by various part-to-part variations.

Having said this, now suppose that the ultrasonic sensor whose requirements were composed in Tables 12.1 and 12.2 was designed based on a transmitter/receiver set as characterized in Figure 12.8. The mechanical configuration, as shown in Figure 12.9, consists of an upper housing and a lower housing. One requirement for this sensor is that it should accommodate at least two different-sized tubes: a smaller 4.5 mm tube diameter and a larger tube diameter of 5.5 mm. To achieve this design objective, the upper housing has been equipped with a spring to close the gap between the upper and lower housings. In order that this spring does not crush smaller tubes, the lower housing is equipped with a feature to ensure that the gap between the two sensors remains at a nominal value of 4.0 mm. By squeezing the diameter of the tube by 0.5 mm, two flat surfaces on the tube will be formed that will interface with the transmitter and receiver surfaces. This will ensure proper sonic path and transfer of energy. For the larger tube, however, squeezing the tube by 1.5 mm (from a 5.5 mm diameter down to 4.0 mm) will constrict the flow and hence is considered to be excessive. As such, the spring constant is chosen in such a way to be balanced with the elasticity of the tube, and hence a nominal transmitter/receiver gap of 5.0 may be maintained for the larger tube.

Another sensor requirement was that it exhibit no more than a 10% variation (i.e., output tolerance). As such, the signal strength for a large tube with water would be approximately 2.09 to 2.31 (nominal 2.20 – as shown in Figure 12.8); similarly, it would be 2.42 to 2.67 for a smaller tube. Conversely, the range of signal strength for air-in-line is 0.91 to 1.01 for the larger tube and 1.23 to 1.36 for the smaller tube. The nominal signal strengths and their maximum and minimum are summarized in Table 12.3.

Hence, it may be conceivable that through the use of embedded software and a look-up table, one may conclude that for any signal strength above 2.0, there is water in the tube and no air bubbles to be detected. On the other hand, for signal strengths between 0.9 to 1.0 and 1.2 and 1.4, air bubbles are detected, and an alarm should be sounded. Furthermore, it may be possible to identify the tube size from the signal strength values; that is, any air bubble signal belonging to the 0.9 to 1.0 belongs to the larger tube.

Functional Tolerance Analysis

In the foregoing analysis, design decisions were based on the manufacturing supplier's datasheet, the sensor's tolerance limits, and the expected distance between the transmitter and the receiver. However, an initial DFMEA was introduced in Chapter 10 to identify failure modes and potential causes of failures. For the ultrasonic air-in-line sensor, the causes of failure were associated with the distance between the transmitter and receiver as well as their proper lateral alignment.

In Chapter 11, DOE was introduced as a tool employed to identify critical factors in a given study. This tool may be employed effectively to find the relationship between various factors involved in the interface between components and/or assemblies and to

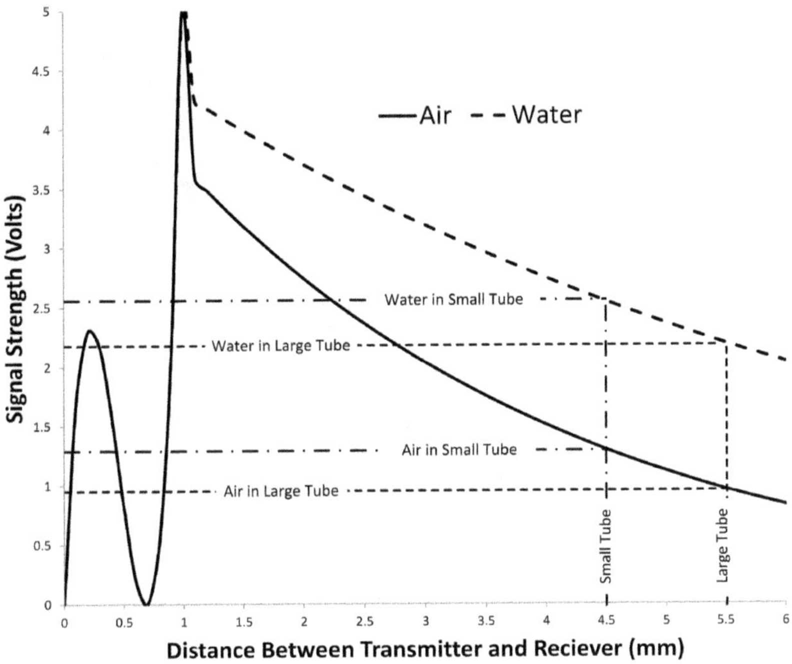

FIGURE 12.8
Hypothetical ultrasonic sensor response based on available datasheets

develop the transfer function. Once this relationship is established, it can be used to study the impact of part-to-part variations on the overall response.

For the ultrasonic air-in-line sensor example, a design of experiments was developed for the transmitter/receiver set whose characteristics are depicted in Figure 12.8. The resulting transfer functions associated with signals passing through air and passing through water are as follows:

$$Y_{\text{Air}} = 3.83305 - 0.50238\ d - 1.x + 0.15407\ d\ x$$

and,

$$Y_{\text{Water}} = 4.5567 - 0.42224\ d - 0.66755\ x + 0.09532\ d\ x$$

Where, d and x are shown in Figure 12.10.

Now that the transfer functions have been developed, let's look at the detailed design and various specified feature dimensions and their associated tolerances as depicted in Figure 12.11. Again, the goal is to study the impact of part-to-part and assembly variations on the response signal strength.

The top and bottom housings slide laterally relative to each other; however, they can also slide transversely relative to each other due to tongue and groove tolerances. This transverse movement (x) is a maximum of 0.6 mm $\left(\text{gap on either side} = (3+0.1)-(2-0.1)\right)/2$ on either side.

Recall that requirement decomposition indicated that any developed design shall be used for two different tube diameters: a smaller 4.5 mm and a larger 5.5 mm. As a

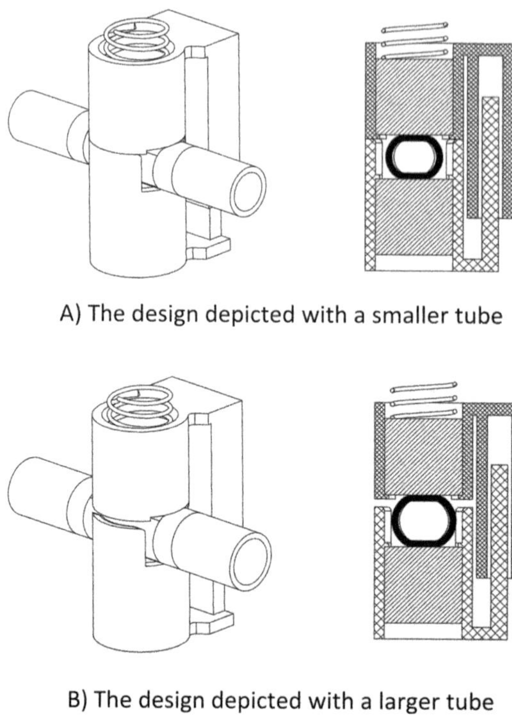

FIGURE 12.9
The design embodiment depicting both tube sizes. In this design the spring acts to close the gap between the top and bottom housing: (A) Force exerted by the spring is counteracted by the lower housing's walls; (B) Force exerted by the spring is counteracted by the elasticity of the tube

TABLE 12.3
Approximate Signal Strength Variations Based on Figure 12.8 and 5% Allowed Tolerance

	Signal Strength in Air			Signal Strength in Water		
	Min	Nominal	Max	Min	Nominal	Max
Small Tube	1.23	1.30	1.36	2.42	2.55	2.67
Large Tube	0.91	0.96	1.01	2.09	2.20	2.31

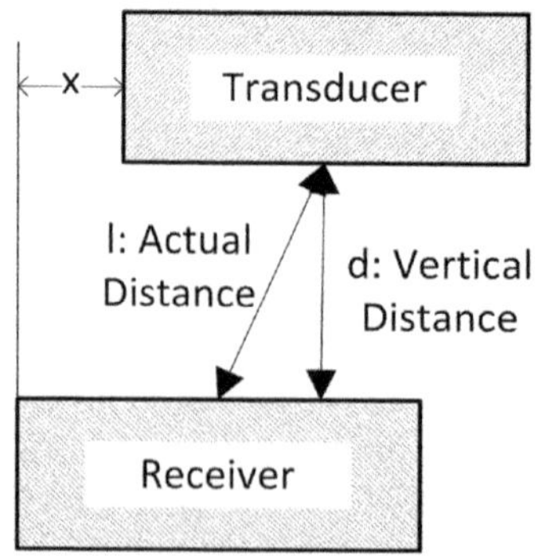

FIGURE 12.10
Two factors d and x affecting signal output

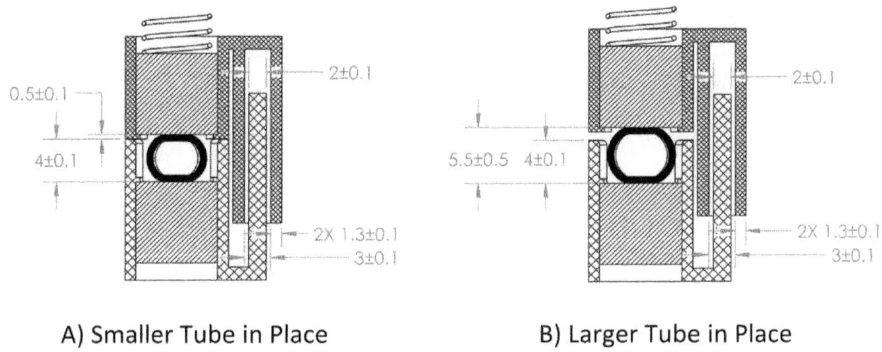

FIGURE 12.11
Various dimensions and tolerances affecting the signal output

result, there are two problems that should be solved here; one to consider the conditions at 4.5 mm and the other to evaluate the 5.5 mm; however, the design is such that it squeezed the tube by 0.5 mm. Hence, for the small tube, the distance d is as follows: $d = (4 \pm 0.1) + (0.5 \pm 0.1) \rightarrow d = 4.5 \pm 0.2$ (Figure 12.11A). For the larger tube, $d = 5.5 \pm 0.5$ as shown in Figure 12.11B, the larger tolerance value is due to the combined variations in both the spring constant, and tube diameter variations and its elasticity, which would counteract the spring force. Now the question that needs to be answered is this: How do the variations in d and x impact the response signal strength?

Just a reminder: why do we need to know this? Because we will determine – through software and look-up tables – whether there are air bubbles in the tubes or not. And there has to be a large degree of confidence in this determination.

Monte Carlo Technique

A relatively simple technique to study the impact of input variability on the output is to use the Monte Carlo technique. This approach, which was initially developed to study (or model) gambling outcomes (Bethea and Rhinehart 1991), can be used effectively to model the outcome variations based on input changes and variations.

> Monte Carlo is a simple yet powerful technique to study the impact of input variability on the output. This technique enables the design team to play "what-if" games and explore design modifications.

Briefly, in this method, nominal values for each input variable (here d and x) are specified along with specified distribution curves for each variable. Then, based on the nominal value of each variable and its expected distribution curve, random numbers are selected and the outcome calculated. As an example, consider the signal strength transfer function:

$$Y_{\text{Air}} = 3.83305 - 0.50238\,d - 1.x + 0.15407\,d\,x$$

Now, assume that d has a normal distribution and x has a uniform distribution and the following variation:

$$4.3 \le d \le 4.7$$
$$0 \le x \le 0.6$$

The Monte Carlo technique chooses a random number for each of the two variables and calculates Y_{Air}. In this instance, suppose that the randomly selected numbers are $d = 4.7$ and $x = 0.2$, then

$$Y_{\text{Air}} = 3.83305 - 0.50238(4.7) - 1.(0.2) + 0.15407(4.7)(0.2)$$

$$Y_{\text{Air}} = 1.416685$$

Now imagine that this process is repeated 10,000 times, each time using different random values of d and x. The outcome will be a distribution curve for Y_{Air}.

Signal Response Variations

Figures 12.12 and 12.13 provide the results of the Monte Carlo simulations using a, commercial software (Apogee 2011) for the small and large tubes, assuming a normal distribution for d and a uniform distribution for x using the following data:

For the small tube:

$$4.3 \le d \le 4.7$$

$$0 \le x \le 0.6$$

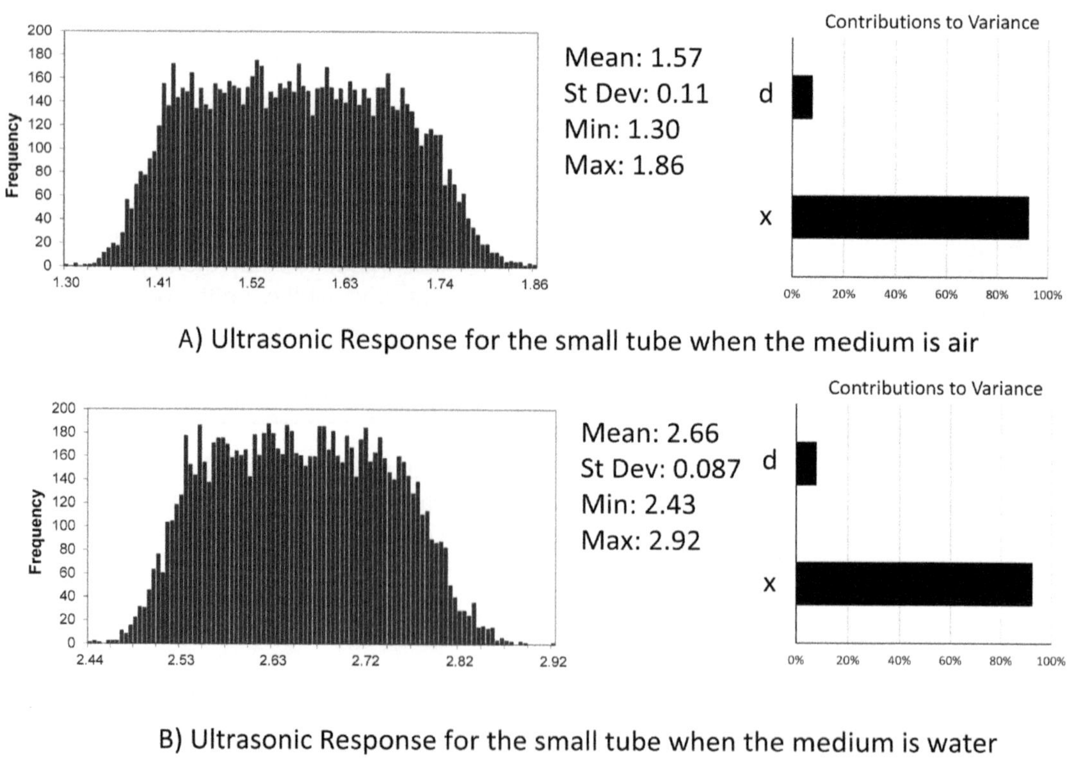

FIGURE 12.12
Signal response distribution for small tubes

For the large tube

$$5.0 \le d \le 6.0$$
$$0 \le x \le 0.6$$

A normal distribution for d was assumed for the following reason. This variable depends on features that are controlled through a manufacturing process; as such it reasonable to assume that production variability is controlled. A uniform distribution for x was assumed because there is no control on how the top and bottom half of this sensor close and how the parts may slide sideways relative to each other.

The results are tabulated in Table 12.4 for ease of reference. Clearly, these results are very different from the ones presented in Table 12.3. The reason is that in Table 12.3, the interaction of the variables d and x are ignored whereas the use of the transfer function along with the Monte Carlo technique takes this interaction into account and replicates what will be experienced in any potentially fielded product.

The first conclusion that may be drawn here is that there are no distinct bands of response that may belong to one tube size or the other. Furthermore, the maximum signal value for a small tube with air bubbles (1.86) is close to the minimum signal value of a large tube filled with water (1.92). This may cause either some false alarms or allow air when it should not.

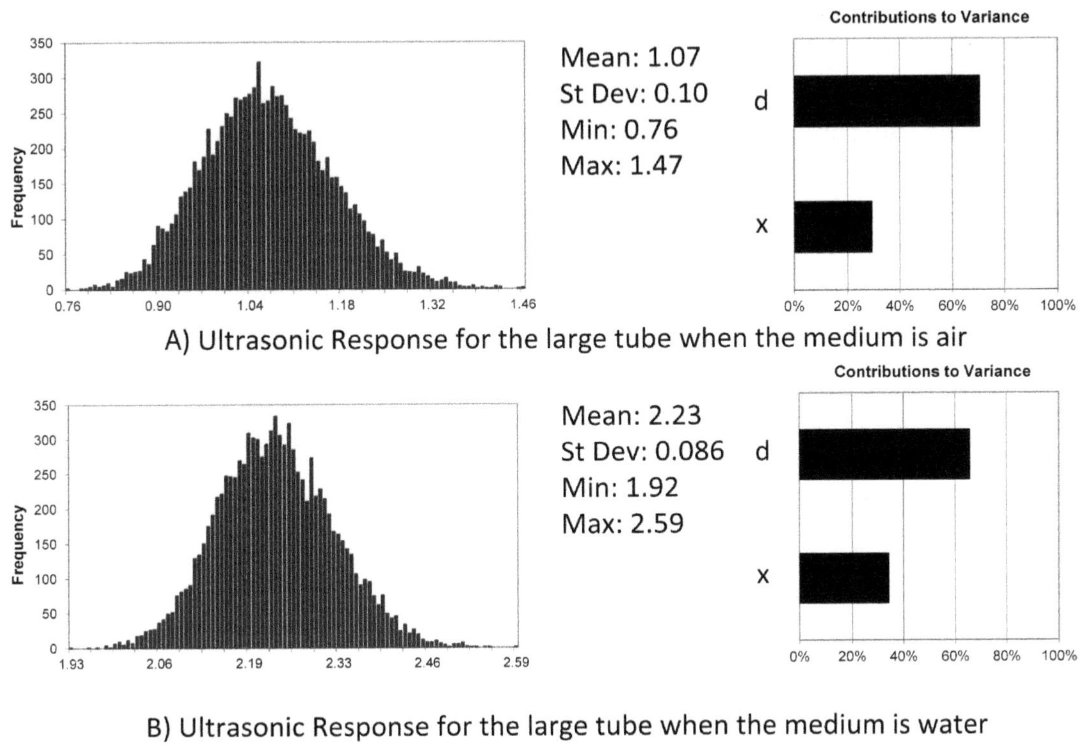

FIGURE 12.13
Signal response distribution for large tubes

TABLE 12.4

Signal Strength Variations Based on Monte Carlo Simulation

	Signal Strength in Air			Signal Strength in Water		
	Min	Nominal	Max	Min	Nominal	Max
Small Tube	1.30	1.57	1.86	2.43	2.66	2.92
Large Tube	0.76	1.07	1.47	1.92	2.23	2.59

The second conclusion that may be drawn is to identify what factor contributes to the output variability. As shown in Figure 12.12, for the small tube, x variability is the major contributor; whereas, d is the major contributor to variability for the large tube, as shown in Figure 12.13.

This approach enables the design team to play "what-if" games and explore design modifications. For instance, in this example, the variability of d may be reduced if two distinct configurations (i.e., assemblies) are used, one for each tube size, as opposed to one configuration to accommodate both tube sizes. The other question may be whether the variability in x may be reduced by shrinking the gap between the tongue and groove in the top and bottom housings. Here is the concern: depending on the material and the manufacturing process used, it may not be possible to assemble the product should the variability of x be reduced. For instance, if the material of choice is a commodity polymer – say, nylon – then, it is possible that the tongue is not molded straight and is slightly warped or bent. Having said this, it may be possible to machine the housings to a large degree of precision. In this case, it may be possible to reduce the variability in x to negligible levels but at a higher cost.

In short, it is prudent to evaluate expected assembly and subsystem functions by evaluating interfaces and the impact of part-to-part or mating variabilities. The design should be simulated using any appropriate number of analysis tools to understand the impact of change on the outcome.

Polymer Disposables and Chemical Consumables

The material discussed up to now is heavily focused on electromechanical systems. However, the basic principles still apply to polymer disposables and chemical solutions. In Chapter 11, I introduced a design of experiment applied to a chemical reagent and developed its transfer function. Monte Carlo analysis can easily be applied to this transfer function to understand that the impact of component variations on the final outcome.

Stack-up Tolerance Analysis

A stack-up tolerance analysis is performed in order to review the effects of dimensional variations within a mechanical (or electric/electronic PCBA) assembly to produce a product with the lowest manufacturing cost possible while meeting all functional requirements of the product.

For mechanical systems, tolerance analysis is thought of as a process that studies relationships between the dimensions of various components within a given assembly. Relevant feature dimensions and their associated tolerances are added in a prescribed sequence in order to determine the upper and lower dimensional values of a target feature. Using this information, nominal gaps or interferences along with their associated tolerances are determined. Typical tools that are used in these calculations are worst-case analysis, root sum of squares method, or Monte Carlo analysis. Chapter 24 provides an in-depth review of tolerance analysis

Similarly, tolerance analysis may be conducted on electric/electronic assemblies. Here, however, the concern is no longer with the physical dimensions of a component but whether voltages and currents exceed the capacity of each component to withstand the load or not. In addition, a number of electronic components show sensitivity to temperature ranges; as a result, knowledge of component thermal dissipation and/or coefficient of thermal expansion play a role in the electronics tolerance analysis as well. Additionally, tolerance analysis on electronics may mean a study of component electrical property variations on the output variability. This study is effectively a functional analysis, as explained in the previous section.

Theory of Operation, Manufacturing Approaches, and Selection

> The theory of operation document contains the design intent as well as how the assembly works, along with a rationale for any selected manufacturing approach.

It goes without saying that once the design team completes the functional tolerance analysis, they develop a solid understanding of how the design works and where the potential pitfalls may be. Furthermore, as it was briefly mentioned earlier, the choice of material and manufacturing approach has a detrimental impact on specified dimensional tolerances and part-to-part variability. What I am advocating here is that the design team commit their learned lessons to a document called *theory of operation*. This document contains the design intent, how the assembly works, and the rationale for any selected manufacturing approach. Such a document enables support teams to make correct decisions when supplier notices of changes are evaluated or cost savings proposals are considered long after product launch.

By way of an example, consider a hypothetical surgical blade and its carrier (i.e., container) as shown in Figure 12.14A. The function of the carrier is to hold the blade in place as it is transported from one surgical station to another. The component that provides the function is a retaining spring, as shown in Figure 12.14B.

The retaining spring may be fabricated using a number of different manufacturing approaches; however, the most commonplace and economical is wire-forming. This is the same operation whereby springs are formed, a well-established technology. As shown in Figure 12.14, the critical dimension is at the mouth of this clip. Wire-forming is only capable of maintaining a 0.03 inch (0.76 mm) tolerance in this design – with a uniform distribution. In other words, this dimension may be any value between 0.11 and 0.17 inches (2.79 mm to 4.32 mm). This excessive variation causes large deviations of the force exerted by the retaining ring on the blade in order to hold it in place during transport, leading to the blade falling out. An alternative manufacturing approach, however, may be a wire electro-discharge machining (EDM) operation. In this approach, a stack of metallic strips is cut by means of electric discharge through a very thin wire electrode. This operation is able to provide tolerances as small as 0.0005 inch (0.013 mm).

If the design team makes the decision to wire EDM and not document it, it may be quite possible that a future cost-savings activity replaces the more expensive operation with the cheaper one, only to increase field failures, cause customer dissatisfaction, and company embarrassments.

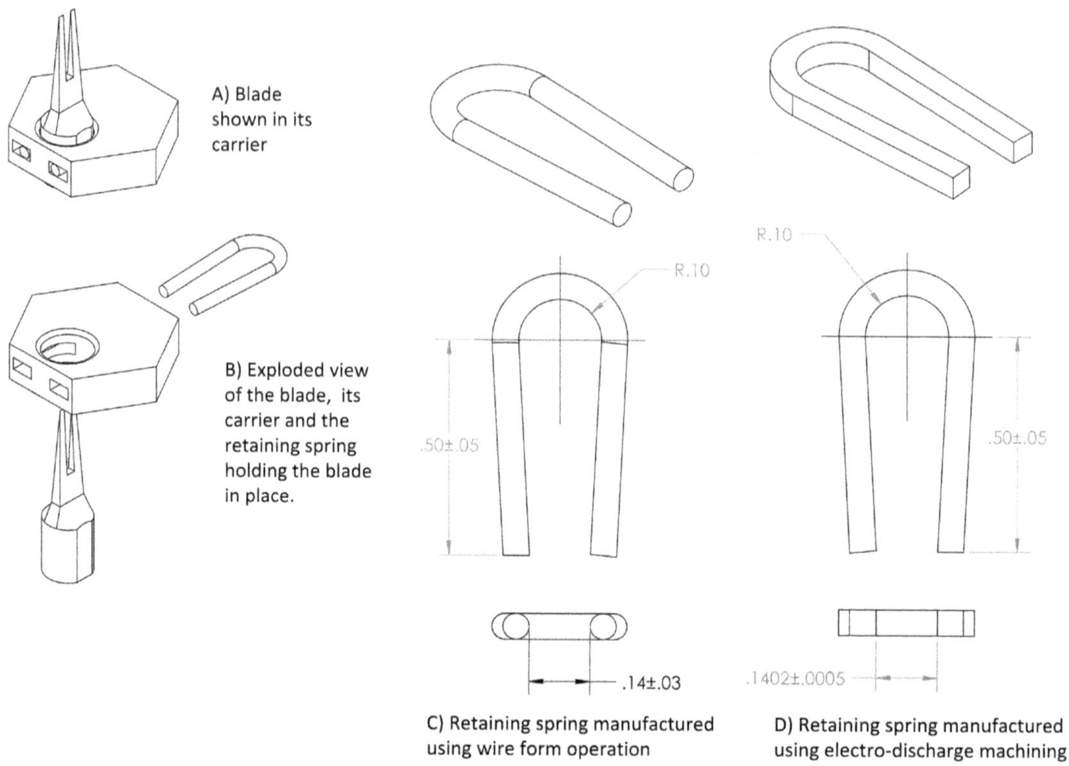

FIGURE 12.14
A hypothetical surgical blade and its carrier tool dimensions shown are in inches

Theory of Operation

As was mentioned, documentation is important to communicate design intent to the sustaining team. It begins by keeping the back-of-the-envelope sketches and designs to documenting the reason why certain decisions were made. For instance, to incorporate a particular purchased part into the design, a bracket may need to be designed. This dependency should be captured in a design rationale document. Once the design is matured, verified, and validated, this design rational document will morph into a theory of operations capturing not only how a subassembly works but why it works as well.

Choosing Manufacturing Approaches

Even though the design team depends primarily on the input of manufacturing engineers to propose a cost-effective manufacturing approach, it is important for the design team as a group to understand a variety of manufacturing and fabrication techniques. An aspect of the design rationale documented in the *theory of operation* is to discuss manufacturing approaches and the underlying reasons. Furthermore, if information is available, there should also be the reasons why other possible manufacturing methods are not appropriate.

As an example, suppose that a fluid passage manifold is machined from a particular plastic. From a cost reduction point of view, unwittingly, a sustaining team member may suggest molding the part, not knowing that the in-mold residual stresses would cause warping of the component, which would lead to leakage. Furthermore, the reason that plastic is used as opposed to (say) aluminum is because aluminum is heavier on the one hand and, on the other, may potentially interact with some of the fluids and/or chemicals passing through.

For completeness sake, I would like to briefly review some of the variety of manufacturing and/or fabrication methods that exist today and how they can potentially impact a design team's product- and process-related decisions. Ultimately, the team needs to decide either to fabricate a part in-house or buy it off-the-shelf. The decision of make or buy is an important design decision.

Mechanical Components

Fabrication of a mechanical component is typically done through one of the following means – though the list is not exhaustive:

1) *Machining* – This operation involves the removal of material through a variety of means. The modern machining units are not only computer numerical control (CNC) driven but also have the ability to interface with common solid modeling software to read three-dimensional models. Grinding, turning, milling, and electro-discharge operations are also considered as sub-branches of machining. Machining has the capability to produce and maintain very precise dimensions and tight tolerances.
2) *Molding* – A very common manufacturing operation is molding. Typically, molding is used in the context of plastics and polymers; however, this technique is also used with metals and ceramics as well. In the case of metals, metal-injection-molding may be used to create complex metallic geometries with metal properties as high as 98% to 99% of wrought metals. For polymers, injection or blow molding is typically used in conjunction with thermoplastic materials, whereas compression and transfer moldings are used to process thermoset polymers. This is not a hard and fast rule, as a class of silicon materials (a thermoset material) is processed in injectable molds – though at reduced mold pressures.
3) *Metal Casting* – There are two common types of metal casting classifications based on whether the mold is expended with each operation or kept and maintained. Sand and investment casting belong to the first category, and die casting belongs to the second. In sand and investment casting, molten metal is poured into the mold and then allowed to cool down. To remove the part, the mold has to be broken.

   In die-casting, material such as aluminum or zinc turns into a semi-liquid state under high pressure and then is moved into a die or mold. In die-casting, the mold has a relatively long life – possibly on the order of 100,000 to 150,000 shots, depending on component size and material. Magnesium thixomolding belongs to this category as well. Thixomolding allows relatively thin-walled metallic components to be manufactured.
4) *Metal Bending* – This term is generically used when flat metal sheets or strips are manipulated into formed shapes. This includes operations such as stamping, bending, drawing, etc.
5) *Additive Manufacturing* – An emerging technology is additive manufacturing, or 3D printing. In this technology, the finished product is made one layer at a time

by placing a fine layer (often as thin as 0.001 in) of material in the desired shape. In this technology, tooling is not required, and the parts are made directly from the 3D solid model. This manufacturing approach has matured to the level that both plastics and metals may be printed, and the parts are able to withstand realistic loads. Although currently this is an expensive fabrication option, it is quite attractive for manufacturing parts with extremely small production runs.

6) *Other Operations* – The last approach worth mentioning is that of combining smaller components to create larger, more complex geometries. Typically, this would involve welding-type operations. Note that with certain materials, terms such as "brazing" and/or "soldering" are more appropriate.

Electronics PCB and PCBA Production

The manufacture of electronic assemblies involves first the selection and manufacture of the boards, followed by the choice of conductive material, and finally the assembly and soldering process:

1) *FR-4* – This is the most common type of material in printed circuit boards. It is a combination of woven glass fabric impregnated with epoxy resin and a brominated flame retardant (hence the acronym *FR*).
2) *G10* – This material is a predecessor of FR-4 without the flame-retardant additives.
3) *CEM 1* – Another type of material is called *Composite Epoxy Material.* This is a low-cost material made of one layer of woven glass fabric and paper-based laminate. It is used only with single-layer PCB designs, as it is not suitable for plated through holes.
4) *Ceramic* – In some applications, such as automotive under-hood applications, ceramic boards are used as the base material. Of the four options presented here, ceramics are the most expensive option.

To create the traces on a PCB, typically, FR4, G10, or CEM1 boards are copper plated first, and then, through a screening process, the desired trace areas on the board are covered. Then, by placing the boards in an acid solution, the exposed copper is removed. Once the boards are washed of the excess acids, what remains are the required copper traces and the dielectric material. If the selected material is a ceramic board, typically, conductive ink (containing silver particles) is printed on, and then the board is allowed to be cured. The curing process may be as simple as drying or done through the use of heat or *UV* light. Once the ink is cured, laser trimming is used to adjust the resistivity of the pad.

Once the PCB is ready, fabricating the PCBA may begin. The following are common steps:

1) *Component Placement* – there are two different ways to place components as described below.
   a. *Manual Component Placement and Soldering* – This approach is used when production runs are very small and there are large and bulky components. Depending on the labor market, this approach is both expensive and error-prone.

b. *Pick and Place Machine* – It is an automated means of picking various electronic components from a reel tape and placing them on a PCB. This machine works with surface mount technologies.

2) *Soldering* – Again there are two means of soldering the component in place
   a. *Reflow Oven* – Once surface mount components are placed on a PCB, they are taken through a reflow oven, which then heats the boards to a temperature causing the solder that exists at the pads – both on the boards and the components – to melt and "reflow" so that components are soldered in place.
   b. *Solder Wave* – This manufacturing technique is often used in association with leaded components whereby the PCBA sits at a short distance above a pool of molten solder. A wave of solder is generated in such a way that the crest of the wave comes into contact with the bottom of the PCBA and solders the leads in place.

Electromechanical Component Productions

By definition, electromechanical devices either convert electricity into movement;or convert movement into electricity. These are typically called motors and/or actuators. This energy conversion is done either by the use of piezoelectric ceramics or by the use of electromagnetics. *Wire winding machines* are used to wind wire around a core using a specified geometry. In contrast, piezo-ceramics are mechanically assembled together, and the wires are soldered to their pads.

Polymers and Chemical Consumables

Earlier in this section, I mentioned the significance of the design team's understanding of the manufacturing process in generating robust designs. This understanding becomes even more important in the design of polymers and chemical consumables. The reason is that developing computer models for this product – unlike their electromechanical counterparts – is significantly more challenging. Many scientists rely on lab techniques and experiments to generate prototypes and drive their formulation and product synthesis. Factors such as heat transfer, blend mechanism, and/or process time will potentially be significantly different in a lab and small-scale environment than on the production floor. Many organizations that rely on the manufacturing of chemical agents do in fact have a dedicated environment where products developed in a laboratory environment are scaled up before being released for manufacturing.

Make-versus-Buy Decisions

In the last two sections, I provided a (very) brief overview of some of the manufacturing approaches and techniques that may be commonly used in the process of developing a new product. I purposely sidestepped the manufacturing of electronics products in part because most design teams, while developing specifications, never really get involved in fabricating them. This is probably the most insignificant *make-or-buy* decision that the design team makes, because, in this instance, the decision is almost always to *buy*. Most teams make this decision almost subconsciously; however, for argument's sake, let's review the reasons.

Firms that purchase electronic components are in the business of developing end products. As such, they typically do not have the needed expertise and the equipment in-house to manufacture the specified components. Thus, the first question to be answered in the *make-or-buy* decision is to ask whether the needed in-house expertise or competence exists or not. Now, there is a business decision to be made here: if the in-house expertise does not exist, is there a reason to develop the needed in-house expertise? If the answer is "no," the decision is to *buy* – just as it is made in buying commodity components. While a large number of "buy" decisions fall in the category of obtaining standard parts such as hardware (screws, connectors, structural components, etc.) and electronic components, other decisions may involve components that may be designed in house but fabricated by external suppliers. Examples of this may be injection molding of enclosures, stamping of brackets, or fabrication of PCBAs. The higher level of *make-or-buy* decision is to consider whether to purchase the technology or develop it in-house (assuming that what is delivered meets the specified requirements). For instance, a corporation with expertise in electronics design and manufacturing may outsource the development of its embedded software.

The decision tree for *make* versus *buy* is depicted in Figure 12.15. There are instances when the in-house expertise may exist to produce a component (including software); however, if production lines are running at full capacity, the choice may be made to outsource the work. This resolution may also depend on whether the outsourced technology is of a proprietary nature or not. Other associated elements are cost of *make* versus *buy*, supply chain management, and quality control; and last but not least, the company culture. There are organizations – particularly in the eastern hemisphere – that pride themselves on being vertically integrated with no (or very little) dependency on external suppliers. The rationale is that if the company is vertically integrated, it has the highest level of control over quality and cost.

Not shown here are cost considerations.

In general, the *make-or-buy* decision is not a purely technical decision to be made, particularly when the capacity may exist in-house. Shorten et al. (2006) speak of three "pillars" to be considered in making this judgment:

1) *Business Strategy Factors* – Considerations should be made to competitive positioning and how the core competencies may be impacted and the rate at which design changes may need to be implemented.
2) *Supply Chain Factors* – Risks such as shipment continuity need to be evaluated. Interruptions may be due to inclement weather such as earthquakes and tsunamis as well as political instabilities.
3) *Economic Factors* – The concern in this is whether or not a supplier may be able to provide the same component (or possible service) at a lower cost and with higher reliability and quality.

Three factors impact *make-or-buy* decisions:

1) Business strategy concerns,
2) Supply chain concerns, and
3) Economic concerns.

All in all, the decision to make or buy at a corporate level should be made with care and diligent considerations; the impact may be felt throughout the entire organization. The greater implication of *make or buy* is whether competencies should be

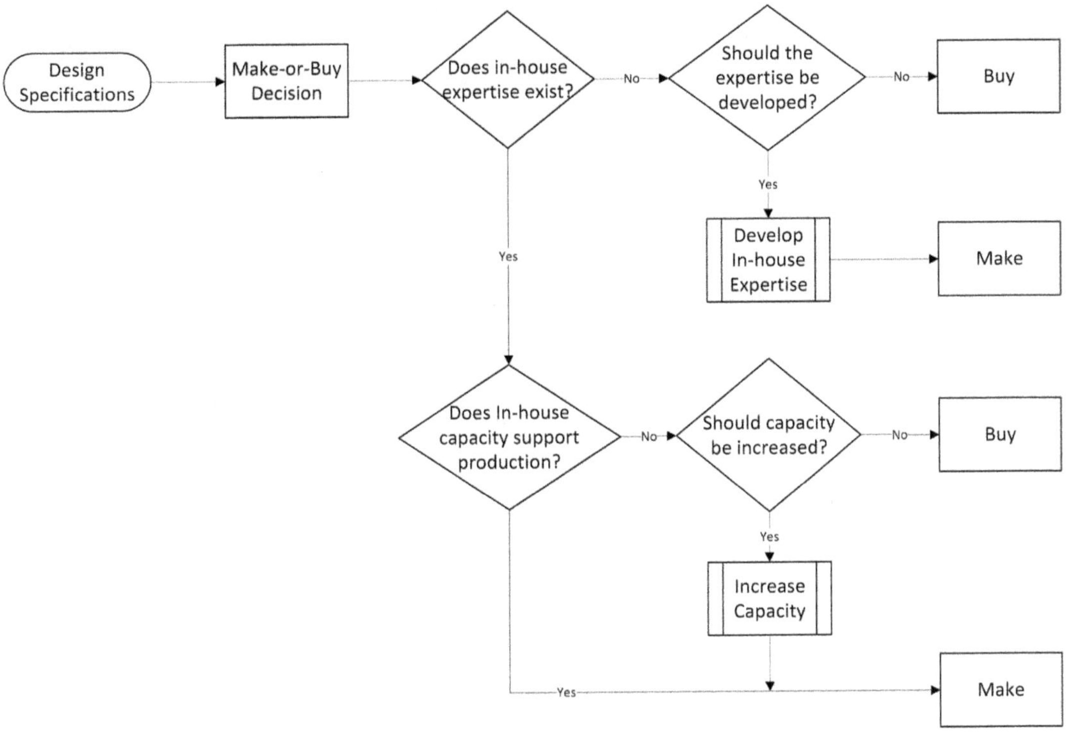

FIGURE 12.15
The *make-vs-buy* decision matrix

kept in-house or outsourced. For instance, an organization may decide that outsourcing a particular operation – say machining of a component – may be more economical. Hence the decision may be to outsource the operation and lay off the operators and subsequently other personnel involved. The outcome of this decision is that the specific expertise has also left the organization.

Standard Parts, Reuse, and Naming Convention

There are two lenses for looking at new products being developed. First is the lens of a manufacturer developing a product in-house. New products being developed are rarely new in a way that none of the components being used or specified have ever been used by the producer. What I am trying to say is that new product development is typically developing a new variation of an existing older product. The question from a financial point of view is how can the organization take advantage of economies of scale when a number of components are the same or extremely similar between these product variations.

The second lens is that of a contract manufacturing house that develops turnkey solutions for a variety of customers. Superficially, there may be no commonalities between these products. Is there any way for them to take advantage of the same or similar economies of scale?

The answer lies in how the Document Control department and its database is set-up and whether or not a *Configuration Management* protocol has been established. In Chapter 21, I will discuss configuration management in some detail. As for the document control department, unfortunately, there are no standard practices that are used for naming and assigning part numbers. Some organizations tend to use what are called *significant* part numbers. In other words, one, several, or all the digits in a part number signify an attribute. For instance, consider a possible part number for a pressure transducer as shown in Figure 12.16.

This naming convention works particularly well for ordering components from catalogs. It may not be as efficient of a tool for part numbers that are used in-house and for production. One could argue that an unstructured number may work as efficiently if the number is generated from a database and that Document Control has assigned attributes to this unstructured part number within the database. Now, assigning attributes to a part number will encourage a conversation between the design team member requesting a number and the document control personnel assigning it. For instance, for discrete electronic components, the attributes may be as follows:

Part number, Product Index, Packaging, Voltage, Tolerance, Max Power, Operating Temperature

Quite frequently, the design team member may learn that the component of their choice may already be used on a different project. Although this could happen by chance and word of mouth, or a more advanced database may be in place that would use a set of attributes as inputs prior to providing a part number for a component that might already exist. Whatever the approach, the debate on naming conventions and part number assignments will not (in my opinion) come to an end any time soon.

File Storage and Change Control

File management, storage, and change control are main aspects of configuration management, which I will discuss in Chapter 21 as an independent topic. For now, I would like to bring a few points to our attention for file storage and management once component design reaches a certain level of stability and is ready to be integrated into the assembly.

It is important to utilize a product life cycle management (PLM) database effectively. For small projects, it may be "easier" to manage files through the use of spreadsheets. However, as the size of the project and the team associated with it grow, a manual approach becomes cumbersome at best and, at the worst, creates confusion. A common side effect of not having a centralized file storage system is that multiple copies of the same document come to exist, and soon, no one will know which contains the latest copies of changes.

The process of checking files out of a database and then back in enforces that only one active copy of any document exists and is available to everyone. Furthermore, once the file is checked back in the database, the reasons for change are (or at least should be) documented. This enables change traceability that can be used to identify root causes of failures should anything go wrong.

One factor that many people in product development often overlook is the impact of change. At times changes are made to component design, but the impact on the assembly is not considered. Consider the following scenario. Earlier, the surgical blade in Figure 12.14 was discussed, and the impact of the fabrication technique on the function of the retainer was explained. Now assume that someone new to the project changes the design of the retaining clip to save cost without conducting a proper impact assessment. The outcome could be disastrous.

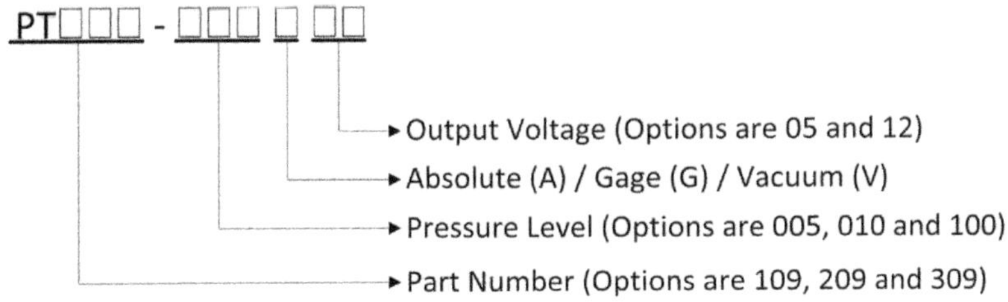

FIGURE 12.16
An example of significant part numbering

Numerical Simulation

> There are several applications for the use of computational methods in the design process. One is a numerical "what-if" game-play. A second application is to evaluate if the intended components or assemblies meet particular design intents. A third application is failure analysis.

Once a component and assembly concept has been chosen and the three dimensional models developed, the design team should inquire whether the virtual product meets the intended design. Traditionally, this question is answered by fabricating a prototype and testing (or evaluating) it. In the days when the paper design was flat; in other words, designs were two-dimensional projections of someone's mental image, it was understandable that a design-build-test-redesign was justifiable. In today's fast-paced product development cycles and the advances in computational fields and 3D solid modeling software, this traditional approach is no longer efficient.

In today's world, much of this old approach is (and should be) replaced by proper engineering and/or scientific calculations and evaluations. Later, in Chapter 17, I will point out in some detail that it may be next to impossible to remove hidden design flaws through testing alone. However, "what-if" scenarios may be conceived and their outcome evaluated via numerical simulations.

There are several applications for the use of computational methods in the design process. The first, as I mentioned, is a numerical "what-if" game play. For instance, what if I used aluminum as opposed to steal, or what if I made the rib thinner? A second application of numerical simulations is to evaluate if the intended components or assemblies meet particular design intents. Questions such as "how much deflection?" or "how hot?" or "what frequency?" are also answered. Failure analysis is another application of numerical simulation. The system or subsystem is modeled and subjected to field stresses in order to identify components that may fail. Other types of analysis may focus on the future performance of the product or the performance of the entire population. Here, the simulation attempts to answer questions such as "How long will this part last?" or "what percent of the population may fail?" Conducting simulations of this type helps the design team to understand possible failure modes while the design is still on paper and changes are inexpensive.

Before providing more details on these three classes of simulation, I would like to suggest that numerical simulations may be divided into three types. The first is the closed-form solutions that may be solved longhand. These are simple applications of engineering "handbook" formulas. The solution may be developed by using a handheld calculator with advanced functions. The second type of numerical simulations involves a set of equations that may not have readily accessible closed-form solutions. Up to the late 1990s, some simple software programs were needed to solve these equations and provide solutions in tabular or graphic format. In today's environment, a program such as MathLab, or Microsoft Excel may easily be used to solve a number of these types of equations and converge to a solution. The third type of numerical simulation is when systems are discretized, and techniques such as FEA are used to solve complex technical problems. Now, I would like to explore these three classes of simulation in some more detail.

What-If Scenarios

One use of numerical simulation is to answer the *what-if* questions. As depicted in Figure 12.17, two different design configurations of a clip are subjected to the same loading conditions, and the resultant Von Mises stresses are compared. This type of analysis helps the design team make decisions on the direction that the product design and/or its configuration may take.

Similarly, numerical simulations may be used to develop the transfer functions between the input values and the expected output(s). Figure 12.18 depicts the magnetic scalar potential and the flux intensity fields. By employing a Hall-effect sensor, the magnetic field strength may be measured. Considering that the magnet and the magnetic paths and leakages must be designed so that the output signal is monotonic and relatively linear, a trial-and-error approach may prove extremely ineffective in bringing such a sensor to market in a timely manner. The transfer function may be developed based on input value variations of geometric and material properties. The numerical simulation would provide an insight into magnetic leakages, and their impact on output values may be easily evaluated.

Another aspect of *what-if* scenarios is to compare the expected results under one set of conditions to another set of conditions. Figure 12.19 shows fan and impedance curves (used in electronics packaging) at sea level and at an altitude of 5000 feet. Note that details of this analysis are not provided here.

Failure Analysis

Many factors may lead to component and/or assembly failures; some may be due to sloppy design and a lack of verification activities, whereas others may simply be due to the designer's lack of insight into potential factors affecting the system; yet, others may be

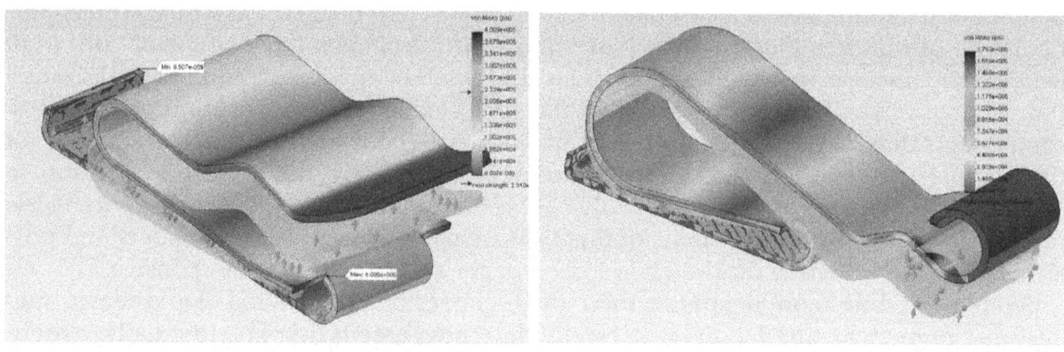

A) Von Mises Stresses For Configuration A B) Von Mises Stresses For Configuration B

FIGURE 12.17
Numerical simulations are used to set design and/or configuration decisions

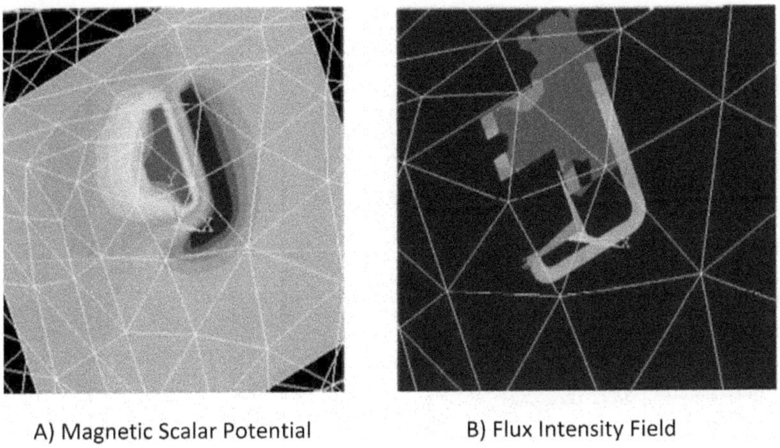

FIGURE 12.18
Numerical simulation of a magnetic field to develop transfer function

contributed to misuse of the product. In Chapter 10, I offered FMEA as a tool to identify sources of failure, potential causes, and possible mitigations. In general, failures of electromechanical systems may be categorized as reversible failures, irreversible failures, sudden failures, progressive failures, and chemical failures.

Reversible Failures

From a mechanical point of view, these failures are caused by elastic deformation of a member in the system. In general, once the failure-causing load is removed, the system would function normally once again. In electronics systems, often some excess heat will interrupt functionality, but when the unit cools down sufficiently, operations are resumed. Another example of this type of failure is resonant vibration of relays or chatter of PCBAs – as shown in Figure 12.20; once vibratory excitations are removed, the assembly works as expected.

Irreversible Failures

These failures are caused by incremental application of loads beyond a certain (reversible) limit. This behavior is typically observed in systems with nonlinear (softening) behavior. In mechanical components, the applied loads and the associated stress factors create stress fields that are beyond the proportional (yield) limit and in the neighborhood of the yield point. In electronic systems, there is component damage, say, due to arching.

An example of this phenomenon in an electromechanical device may be the PCBA shown in Figure 12.20. Should the magnitudes of vibration pass a certain limit, the PCBAs' chatter may lead to an electric short, causing an irreversible failure. Another example is overloading a motor; some overload conditions may be tolerated, and others may cause significant degradations.

Sudden Failures

Sudden failures are caused by excessive application of loads where loads have surpassed their ultimate values. A blown fuse is an example of this type of failure. Cracking of

Openning	Area	N	K		Flow Rate	Pres. Loss	Fan Curve	Pres. Loss @ 5000 ft	Fan @ 5000
inlet	9.31	0.50	1.30E-03		0	0.00	0.360	0.00	0.31
Inital Stage	10.20	0.50	1.09E-03		5	0.01	0.348	0.01	0.30
Fan Tray Area	49.28	1.00	9.31E-05		10	0.02	0.335	0.02	0.29
Flow Thru PCB's	15.75	2.50	9.13E-05		15	0.06	0.323	0.05	0.28
90 deg bend (right)	7.50	1.50	6.03E-03		20	0.10	0.310	0.09	0.27
90 deg bend (top)	37.50	1.50	2.41E-04		25	0.16	0.298	0.13	0.26
90 deg bend (left)	7.50	1.50	6.03E-03		30	0.22	0.285	0.19	0.25
Outlet (right)	2.93	0.50	1.32E-02		35	0.30	0.273	0.26	0.23
Outlet (top)	10.08	0.50	1.11E-03		40	0.40	0.260	0.34	0.22
Outlet (left)	2.93	0.50	1.32E-02		45	0.50	0.248	0.43	0.21

Total K = 0.042				
Air Density = 0.0765	at sea level or	0.0659	at 5000 feet	
Total Flow = 33.000	at sea level or	33.00	at 5000 feet	

number of fans = 3

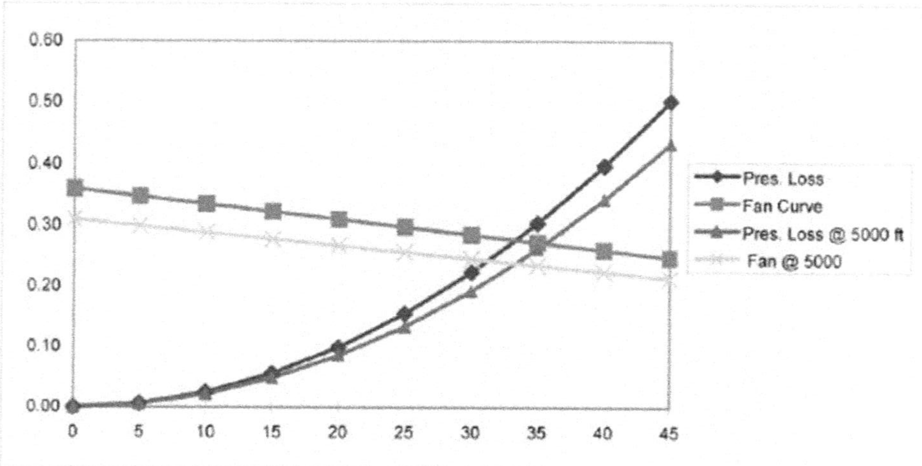

At Sea Level

Delta T =	17.33	deg. F
Input CFM =	33.00	CFM
VA =	245.96	Watts
Power Factor =	0.75	
Barrometric Pressure =	29.92	inches Hg
Ambient Temperature =	60.00	deg. F

At 5000 Feet Alititude

Delta T =	20.03	deg. F
Input CFM =	33.00	CFM
VA =	245.96	Watts
Power Factor =	0.75	
Barrometric Pressure =	24.90	inches Hg
Ambient Temperature =	40.00	deg. F

FIGURE 12.19
Comparison of outcome under two different environmental conditions

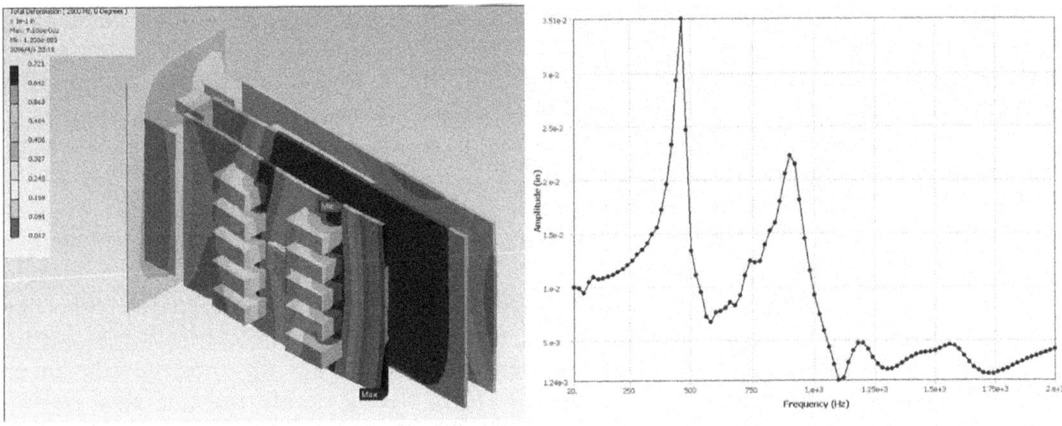

A) PCBA Displacements Under Vibration

B) PCBA Vibration Response

FIGURE 12.20
An example of PCBA chatter and its response

ceramic boards is another example. Numerical simulations of sudden failures are rather simple: once the load distribution is calculated, load (or stress) values at desired locations are compared to permissible levels. This type of analysis is also known as the worst-case analysis.

Progressive Failures

These failures are the most serious of failures because initially a device passes all its verification requirements, yet after some time in the field, it begins to fail. Mechanical creep and fatigue belong to this category. Typically, failures of electrolytic capacitors or lead-acid batteries may also be due to this class of failures.

Prediction of progressive failures is more challenging than other types of failures. On the one hand, the expected failure mode should be known and well understood. On the other hand, failure time is not deterministic, and a certain level of uncertainty and probability exists; hence, along with an understanding of the physics of failure, a knowledge of statistics and probabilities is also needed to analyze progressive failures and create predictive models.

Chemical Failures

The most common chemical failure in an electromechanical product is of an electrochemical nature and is generally known as corrosion. It occurs in the interconnects where the design engineer has not been careful in specifying the right material for the contacts or in disposable components that is operated near salt solutions or chemical assays. It is also observed in equipment where moisture condensates and forms small pools of water.

The modes of failure of chemical attacks on plastics are generally different than those of metals; however, with the advent of engineered plastics and various fillers such as carbon or steel fibers, galvanic cells may form between a fiber-filled plastic and an adjacent, adjoining metal. The corrosion process may even be exacerbated because many resins absorb and retain moisture, leading to a process that is referred to as dry corrosion.

Aside from the possibility of galvanic cell formation, specifying plastics as a packaging material must take into account the impact of the service environment. If a plastic material is in contact with a corrodent (or solvent), it will either resist it or will fail very rapidly. However, there are other environmental factors, such as ozone or *UV* light, that may have a detrimental impact on the reliability of a plastic part. The causes and/or mechanisms of this failure are beyond the scope of this book.

Migration is the process by which material moves from one area to another area. If migration happens between two adjacent metals, such as copper and solder, it is called diffusion. If it happens due to internal stresses and in the *absence* of an electric field, it is called whiskers. Finally, if it happens in the presence of an electric field and between *similar* metals, it is called dendritic growth.

Diffusion may potentially pose yet another problem. While the intermetallic layer formed between copper and solder is needed to form strong bonds, the thickness of this bond may have a detrimental effect on the ductility of the joint; the thicker the layer, the more brittle the joint. Thus, the choice of solder and the pad material are quite important.

The growth of whiskers is related to the materials and processes used to plate and solder the components on the PCBA as well as the environment to which the equipment is subjected. It does not depend on the components, used and a variety of components such as diodes, transistors, ICs, microcircuit leads, and even PCBAs, have been affected. Many metals have been known to whisker, and as such, designers and engineers should be mindful to avoid components plated with materials as well as processes that would facilitate whisker formation.

Migration and electromigration have traditionally been very important failure mechanisms in the electromechanical combination products. The presence of chemical, or biological agents or pollutants along with an electric field provides an ideal environment for electromigration and dendritic growth. Dendritic growths are leaf-like structures that grow in the presence of an electrolyte and an electric field. They are aggravated as temperatures and voltage levels increase. Eventually, dendrites can cause electrical shorts and failures.

Single Point Solutions versus Distributions

In conducting numerical simulations, often single-point calculations are made; in other words, loading and material properties under a given set of conditions are used as inputs for numerical simulations, and a single deterministic output is calculated. However, in reality, these input values are subject to variations. Should these variations be ignored, a great deal of information may be lost.

An example may provide more clarity: in the design of cabinets and racks for electronics systems, a design criterion that may be overlooked is buckling. A simple approach to account for buckling is to identify components under compressive loads and then to calculate these components' load-carrying capacities. For this example, assume that a rack system is made of a number of slender truss members. A simple slender member would be able to carry a compressive load so long as it is below P_{max}.

$$P_{max} = \frac{\pi^2 EI}{L^2} \tag{12.1}$$

where E is young's modulus of elasticity, I is second moment of inertia, and L is the column's length. Should the compressive load in a given member exceed this value, it would lose its load-carrying capacity and buckle. For the sake of this example, assume that the member has a $0.5\text{in} \times 0.5\text{in}$ cross-section $(12.5\text{mm} \times 12.5\text{mm})$ and a length of 45 inches (1.143 m). Furthermore, assume that the material is steel with a modulus of elasticity of 30,000,000 psi (206.8 GPa). Based on these values $P_{\max}$ is calculated to be 761.5 lbs (or 3.39 KN).

Now suppose that the material property used in this design has a uniform distribution with a tolerance of $\pm 1.5 \times 10^6$ (in other words $E = 30 \times 10^6 \pm 1.5 \times 10^6$. The component is provided by two different suppliers with the following nominal and tolerance dimensional values:

$$(0.5 \pm 0.01\ \text{inch}) \times (0.5 \pm 0.01\ \text{inch}) \times (45.0 \pm 0.5\ \text{inch})$$

The difference between the two suppliers is that the dimensional distribution by supplier A is uniform (i.e., any component may have a dimension within the minimum and maximum values), and the distribution of component dimensions by supplier B is normal (i.e., about 68% of the dimensional values cluster around the nominal value).

Figure 12.21 depicts the result of two Monte Carlo simulations conducted using an MS Excel spreadsheet using Equation 12.1. It shows the distribution of the expected column buckling loads for the components delivered by the two suppliers. Note that the minimum load-carrying capacity of the components provided by supplier B is greater than those by supplier A (690 vs. 671.8). Even though the difference is small, it is possible that parts fabricated by supplier A would fail, whereas parts fabricated by supplier B would not.

Now consider a different example. Assume that an injection-molded box as shown in Figure 12.22 is to be sourced from two different suppliers. This time, the molding process at supplier A is such that a normal distribution of various geometric dimensions may be maintained. The process at supplier B, however, only allows a uniform distribution.

The material selected for this enclosure is a PC/ABS blend with a Young's modulus of elasticity of 392 (± 50) ksi and a yield strength of 4900 psi. The maximum stress induced at the base of the snap-fit feature is calculated based on the assumption that the snap-fit feature is approximated as a beam.

$$\sigma = \frac{3Ec\delta}{L^2} \tag{12.2}$$

Where E is Young's modulus of elasticity, c is half the thickness $(c = h/2)$, δ is the maximum deflection, and L is the length of the beam at the point of maximum deflection. Table 12.5 summarizes single-point calculations of maximum stress at nominal dimensions as well as the maximum and minimum values.

Clearly a nominal value of stress has been calculated along with upper and lower bounds. Furthermore, this material may likely fail at a stress level of 4900 psi. Considering that the upper stress bound is 6882 psi, is a redesign necessary? The fact is that the upper stress bound is only realized if the maximum values E, h and δ are aligned with the minimum value of L in the same part. What is the probability of this occurring? Or, more appropriately, what is the probability of stress values being greater than 4900 psi? Will the parts from supplier B have more failures than supplier A? To answer these questions, two Monte Carlo runs were conducted as shown in Figure 12.23.

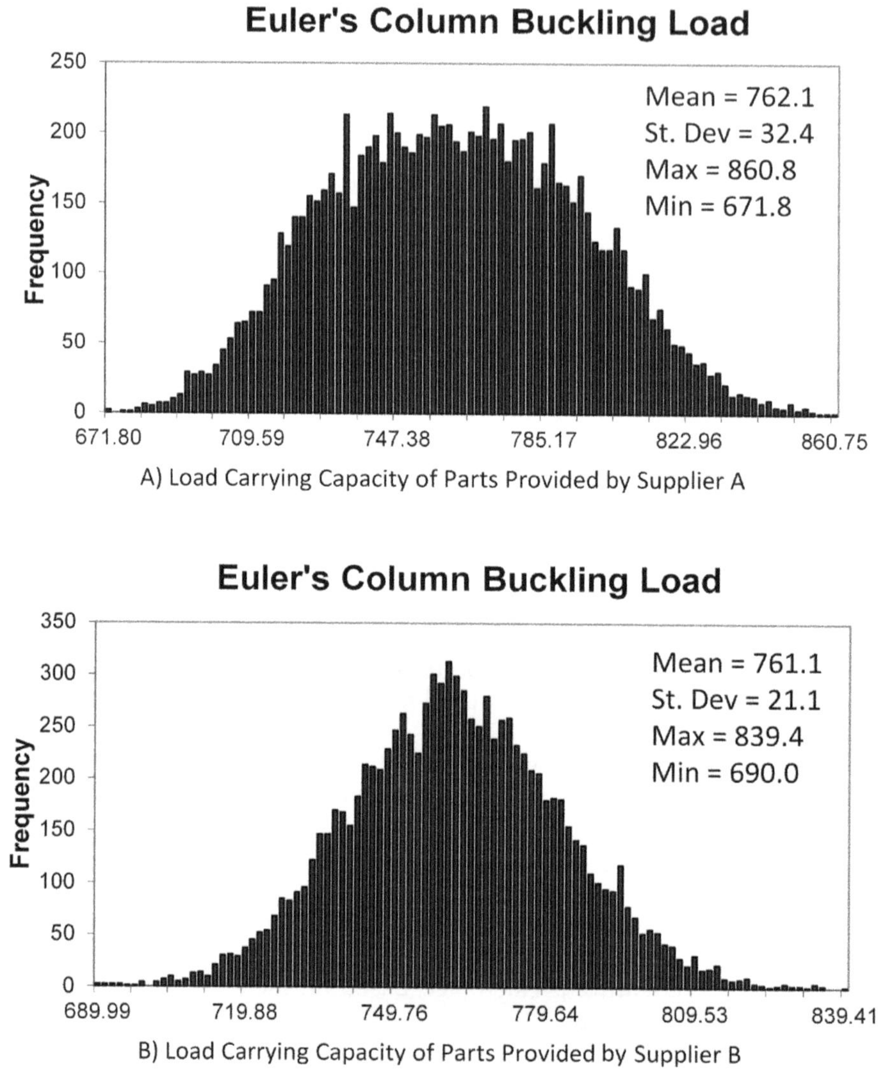

FIGURE 12.21
Impact of input variation and distributions on expected outcome

There are some interesting factors to be noted here. First, the average stress value of the Monte Carlo population is lower than the calculated nominal value (3759 or 3753 compared to 4212). This is due to the random combination of feature dimensions. Second, the failure rate of parts (i.e., parts that fail due to stress levels above 4900 psi) produced by supplier A is only 0.2% because of the process control and a normal distribution of various feature dimensions. This failure rate increases to 2.86% for supplier B, who does not have the same process controls and, hence, can only provide parts that meet the "print."

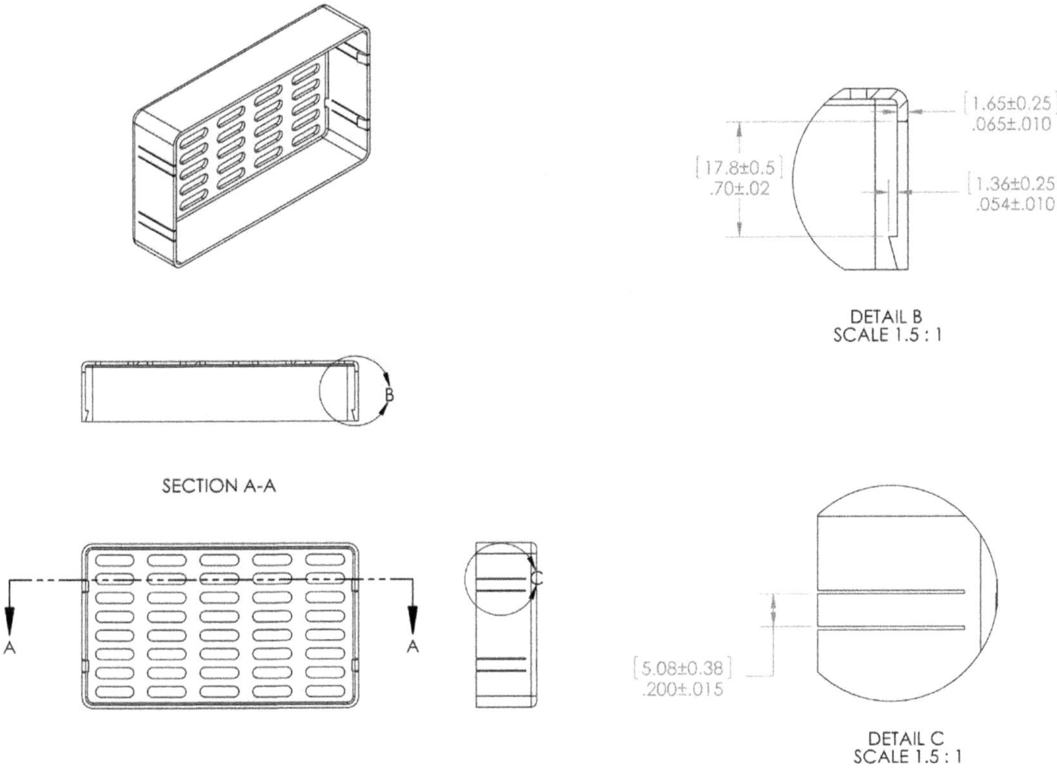

FIGURE 12.22
An injection-molded enclosure with snap-fit features

TABLE 12.5
Single Point Stress Calculations

	Nominal	Tolerance	Min	Max
δ	0.054	0.01	0.044	0.064
h	0.065	0.01	0.055	0.075
L	0.7	0.02	0.68	0.72
E	392000	50000	342000	442000
σ	4212		2395*	6882*

* Care must be exercised in calculating maximum and minimum stress values.

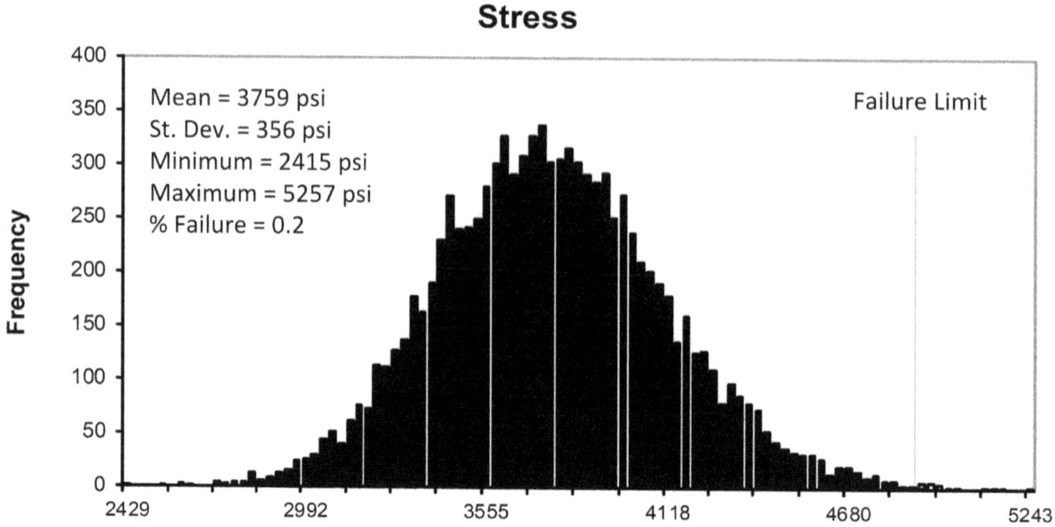

A) Stress Distribution in The Snap-Fit area of Parts Provided by Supplier A

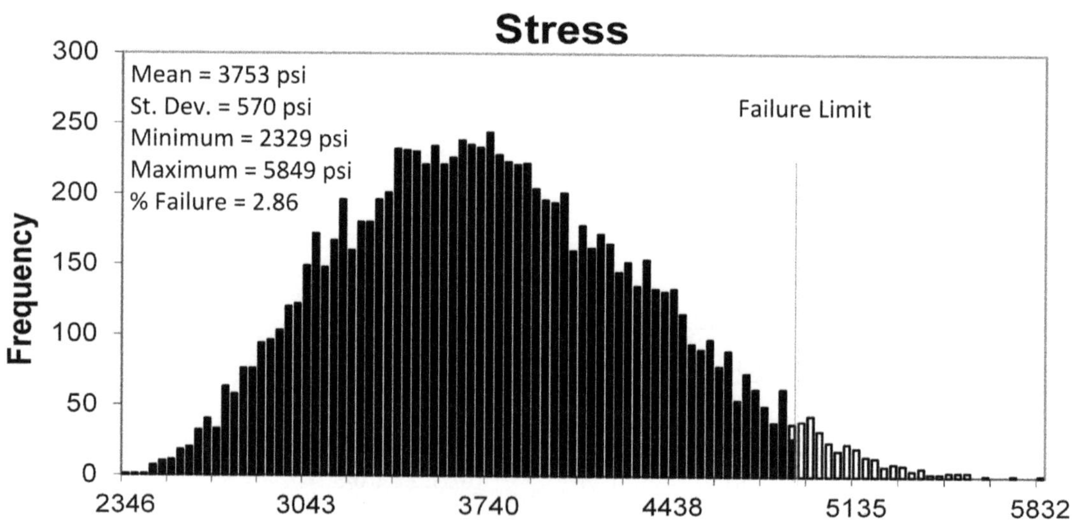

B) Stress Distribution in The Snap-Fit area of Parts Provided by Supplier B

FIGURE 12.23
Impact of dimensional variation and distributions on part stresses

This type of analysis provides insight into two other factors. First, essential or critical dimensions that impact either the functionality of a component or an assembly may be easily determined and marked to be controlled in manufacturing. Second, a cost-based decision may be made to use a component with higher variability if it is not an essential or critical component, and higher failure rates may be tolerated.

These exercises go to demonstrate that a great deal of information may be lost should the numerical simulations only focus on single-point calculations and ignore the impact of component geometric and material variations.

Prototyping and Human Factors

In my career, I have seen a number of different uses for "prototypes." In the days that products were designed flat on drawing boards, the design engineer had no other choice but to rush to machine (mill or otherwise fabricate) prototypes in order to identify his (or her) own errors and correct them. These errors could have been as trivial as misplaced (or forgotten) hole locations. In the same era, wooden (and scaled-down) models were often developed for user studies and feedback. Often, these prototypes were not functional because adding functionality introduced a much higher level of complexity.

Nowadays, with advances in 3D solid modeling, the need for design-error-proofing is eliminated to a large extent. With the advent of rapid prototype machines such as stereolithography (SLA) or fused deposition modeling (FDM), the idea of developing proof of concept with working prototypes has gained some currency. This was an important step in product development because a number of different industrial designs could be developed with relative ease. It was also possible to incorporate some electronics or electromechanical elements, albeit with care. This pseudo-working unit could be evaluated for feasibility by potential users. Feedback in these early stages of design was important as it would prevent costly design modifications at later stages.

As the quality of material used in fabricating prototypes improved, the produced units began to resemble manufacturing equivalent units. Today, direct metal laser sintering (DMLS) has reached such a high degree of quality that in certain low-quantity production runs, it is more economical than the traditional manufacturing methods.

Finally, the high quality of prototype materials and the ease with which dimensions of components may change, enable the design team to develop DOEs. These DOEs may be used to develop transfer functions when developing a mathematical or numerical function may prove difficult or seemingly impossible. Similarly, information may be gained by careful testing to provide input (or feedback) into the DFMEA. For instance, accelerated testing may be conducted to develops further understanding of design margins and envelopes. A word of caution: as tempting as it may be, testing prototypes for verification of requirements is not the same as testing manufacturing-equivalent parts, assemblies, and subsystems. Design requirements verification is an activity that should take place using parts and components that will eventually be in the hands of the end users.

Assembly or Subsystem Verification Testing

Once the design team's confidence in the design grows to the point that the design is considered frozen, the team should begin tooling and prepare for manufacturing. Recall the V-model for product development shown earlier in Chapter 9 (Figure 9.11). Once the detail design is developed, the process nears the apex of the V in the model as shown in

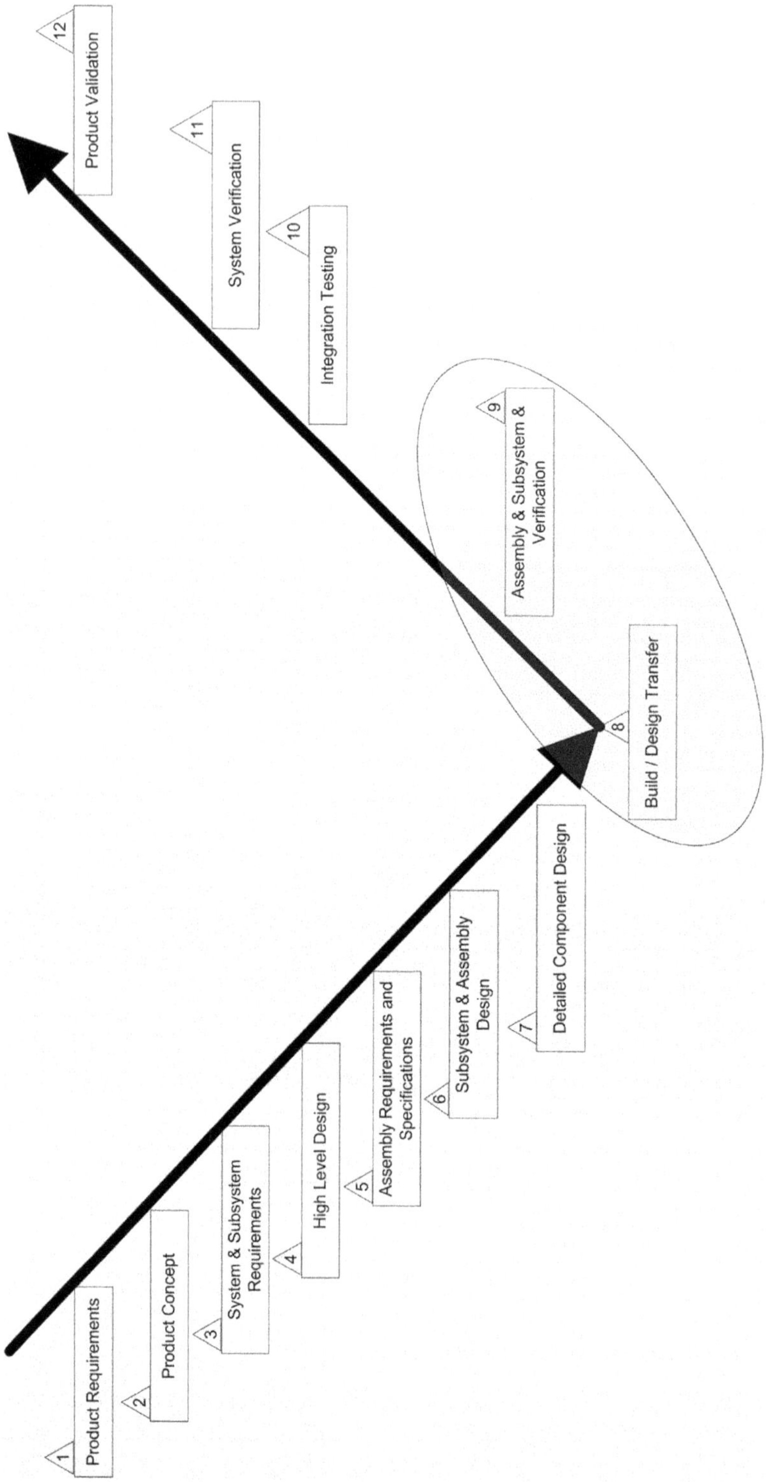

FIGURE 12.24
The stage of assembly and subsystem verification in the V-model for product development

Figure 12.24. The last step of detailed design is developing the 2D engineering drawings, which will be discussed in Chapter 13. It should be noted that although the steps are shown in a sequential manner in Figure 12.24, in reality, there are a number of iterations between these various steps.

In the build and design transfer step leading to assembly and subsystem verification, manufacturing and service organizations develop a more intimate understanding and knowledge of the assembly. Manufacturing and service-related documents may be initiated and written while the design team focuses on working with the manufacturing process engineers to create manufacturing-equivalent components and assemblies.

The first steps of verification involve testing whether the component and/or assembly was fabricated as presumed in the paper design stage. This reveals whether any hidden pitfalls exist that have been overlooked.

The second and more important step is to ensure that the assembly performs as expected and meets its requirements. In Chapter 17, I will review testing verification and validation methods and will review potential pitfalls. For now, I suffice to say that one cannot verify a requirement if it does not exist. By exist, I mean it is documented and approved by the team *a priori*.

Summary

- The outcome of design activities is paperwork and not a product; hence, ample time should be given to documentation.
- Detailed design should begin in earnest when various subsystem behaviors have been well understood.
- Design specifications are the outcome of requirements decompositions.
- Delivered function of an assembly or subsystem is affected by various part-to-part variations of its components. This relationship should be quantified and optimized.
- Two tools for developing this relationship are design of experiments and Monte Carlo simulation.
- *Make-or-buy* decisions are not purely based on technical options. They also depend on business strategy, supply chain and economic factors.
- The greater implication of *make or buy* is whether competencies should be kept in-house or outsourced.
- One use of numerical simulation is to answer the what-if questions. Another is to develop the transfer functions between the input values and the expected output(s).
- In conducting numerical simulations, often single point calculations are. However, in reality, these input values are subject to variations leading to variations in the outcome. Should these variations be ignored, a great deal of information may be lost.
- Once the design team's confidence in the design grows to the point that the design is considered frozen, the team should begin tooling and prepare for manufacturing.

Exercises

1. Continue the design work for the concept selected in the Chapter 11 exercise.
   a. Develop the 3D model of the selected concept.
   b. If capabilities are available, create prototypes of components and assemble them.
   c. Do the parts assemble as expected? What changes need to be made and why?

Notes

1 Agreed, this statement is mainly applicable to the electromechanical aspects of a product, but the general argument holds for nearly any product.
2 For the sake of being brief, I will not provide an example of decomposition of a chemical regent here, but for the sake of completeness, I am providing it in Appendix C.

13

Design Details and Documentation

Introduction

Once the design concepts of components belonging to an assembly[1] mature, design engineers and scientists begin to develop associated design-related documents such as two-dimensional engineering drawings, chemical formulation recipes, and other specifications. These documents are tools that are used to communicate what the component is and the aspects of the part that are important to its proper functioning within the assembly, subsystem, and possibly within the system. It is therefore essential that the cascaded requirements that have been translated into the design specifications are properly communicated in engineering drawings and/or chemical formulation instructions, primarily for the sake of manufacturing the solutions, parts (or assembly).

This means of communication enables an easy way of relating engineering and scientific needs to the members of the purchasing organization, who are the liaisons to any potential suppliers. The sooner these drawings/documents are developed in the design process, the sooner supplier feedback may be solicited to identify alternative manufacturing approaches that may reduce cost or to ascertain primary and alternative processes.

As shown in the product development roadmap (Figure 13.1), once the virtual product is developed, the following may be established:

1) A bill of materials (BOM) that dictates what components have been specified, which ones are custom designed and require fabrication, and which ones are purchased (at times these are called COTS – commercial off the shelf). The BOM will enable a cost structure.
2) Through the use of numerical simulations, DOEs, or both, a sense of design margins and product reliability is developed. This assessment is further enhanced as manufacturing equivalent parts are fabricated and used in testing.
3) Similarly, through simulation and/or DOE, factors critical to design, assembly, manufacturing, service, etc. are identified and captured.
4) Finally, the 2D engineering drawings and other design documents, such as recipes for chemical formulations, are created to communicate not only the physical configuration(s) of a component but its design intent as well.

There are two aspects of developing a detailed design of components and assemblies of a product. The first aspect is technical and deals with the *what* that appears on the drawings or other documents to communicate design intent. The second aspect is behavioral and deals with the *how* of keeping everyone informed of the changes that may be taking place on these last steps of design and development.

DOI: 10.1201/9781003564508-17

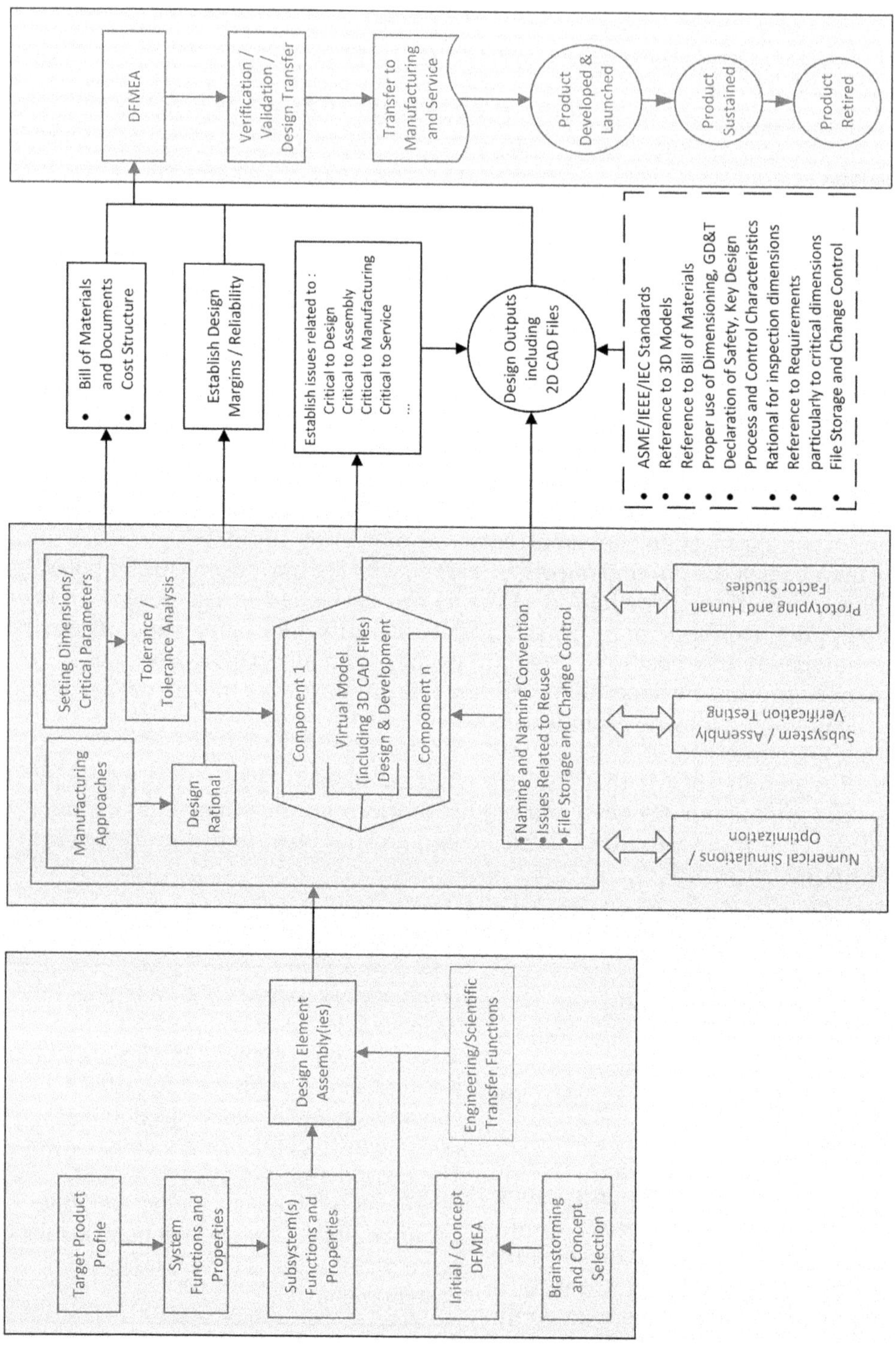

FIGURE 13.1
Output of developing the virtual product

> These documents are used by a variety of support organizations (such as purchasing and external suppliers) to enable product development and ultimately launch into the market. Internal customers of engineering drawings include (in addition to manufacturing and service) sales and marketing, purchasing, operations, and quality (both assurance and control). At times, external customers also require or request copies of the engineering drawings. Finally from a systems point of view, these documents provide rational on how factors that were called CTQ cascaded down to design specifications and ultimately the final product.

In this chapter, the focus is on developing the 2D engineering drawings, whereas issues related to BOM and costing (along with a related design for X topic) are discussed in later chapters.

The What and How of Detailed Design

There are two aspects of developing the detailed design of components and assemblies of a product. The first aspect is technical and deals with the *what* that appears on the drawings to communicate design intent. The second aspect is behavioral and deals with the *how* of keeping everyone informed of the changes that may be taking place on these last steps of design and development. Let's discuss these behavioral issues first, and then the technical details can be reviewed.

Earlier, I suggested that although the V-model shows various steps in a linear fashion, the process is iterative in nature. Therefore, it may be that design specifications and requirements change. Should these changes touch any of the drawings, affected suppliers should be notified and kept in the loop. Should fabrication of a part be impacted, cost may change; hence, the program manager should be aware. Needless to say, holding periodic meetings with program management, purchasing, suppliers. and internal manufacturing helps to expose any potential issue that may later become a major roadblock and to highlight critical paths for development. Once all issues are resolved, the production tooling design is approved and ordered. Now that I have covered the behavioral aspects, I can turn to the technical aspects of 2D engineering drawings. However, before I do so, let me say a few words about scientific design documents.

Scientific Design Documents

The rest of this chapter focuses heavily on creating engineering drawings based on ASME standards. At the moment, there are no standards for conveying chemical formulations and/or recipes. Safety data sheet (SDS) may be the only common standard used to communicate hazards associated with a chemical. Having said this, the need still remains that such a document should be created in each organization and be given a standard format for every scientist to follow for sake of clear communication to other functions and teams.

Format of a Detailed Engineering Drawing

For completeness sake, I like to review some elemental features of engineering drawings. The first feature to discuss is the size of the drawing. In the ASME standards, These are the smallest A-size (8.5 in x 11.0 in) paper through the largest E-size (34.0 in x 44.0 in) paper. International standards may use A0 (the largest size, 841 mm x 1189 mm) through A4 (the smallest, 210 mm x 297 mm). Each size may be used in either portrait or landscape settings. Table 13.1 summarizes various paper sizes.

Figure 13.2 depicts a typical engineering drawing and its various other features. Although the content of the engineering drawing frame may vary greatly from one corporation to another, there are certain commonalities:

1. All engineering drawings have an information block. This information block contains the name of the corporation owning the design, the design engineer, the approver, etc.
2. Legal description to identify the corporation as the legal owner of the design, part, etc. and copy-right issues related to the drawing.
3. A revision block containing a description of changes made.
4. A zone identifier, which is generally in the form of numbers horizontally and letters vertically.

There are two other features of an engineering drawings that is common place, although they are not part of the drawing frame in a strict sense. These are

5. A status note indicating whether the depicted design is preliminary or final.
6. A "Notes" section where specifications and requirements relative to fabrication of the part are communicated.

Finally, at times there are tables that are placed on the drawings. The most common of these (across all industries) is a BOM table associated with assemblies. This table is often placed on a drawing to identify the part numbers of the components of an assembly. An example of this is shown in Figure 13.3. Other types of tables are to specify hole locations, welds, bends, and punch specifications.

The BOM may be as simple as the one shown in Figure 13.3 or as elaborate as providing further details as make or buy items or other information. In my experience, more detailed

TABLE 13.1

A Comparison of U.S. and European Drawing Sizes

US Designation	Paper Size in x in	European Designation	Paper Size mm x mm
E	34.0 x 44.0	A0	841 x 1189
D	22.0 x 34.0	A1	594 x 841
C	17.0 x 22.0	A2	420 x 549
B	11.0 x 17.0	A3	297 x 420
A	8.5 x 11.0	A4	210 x 297

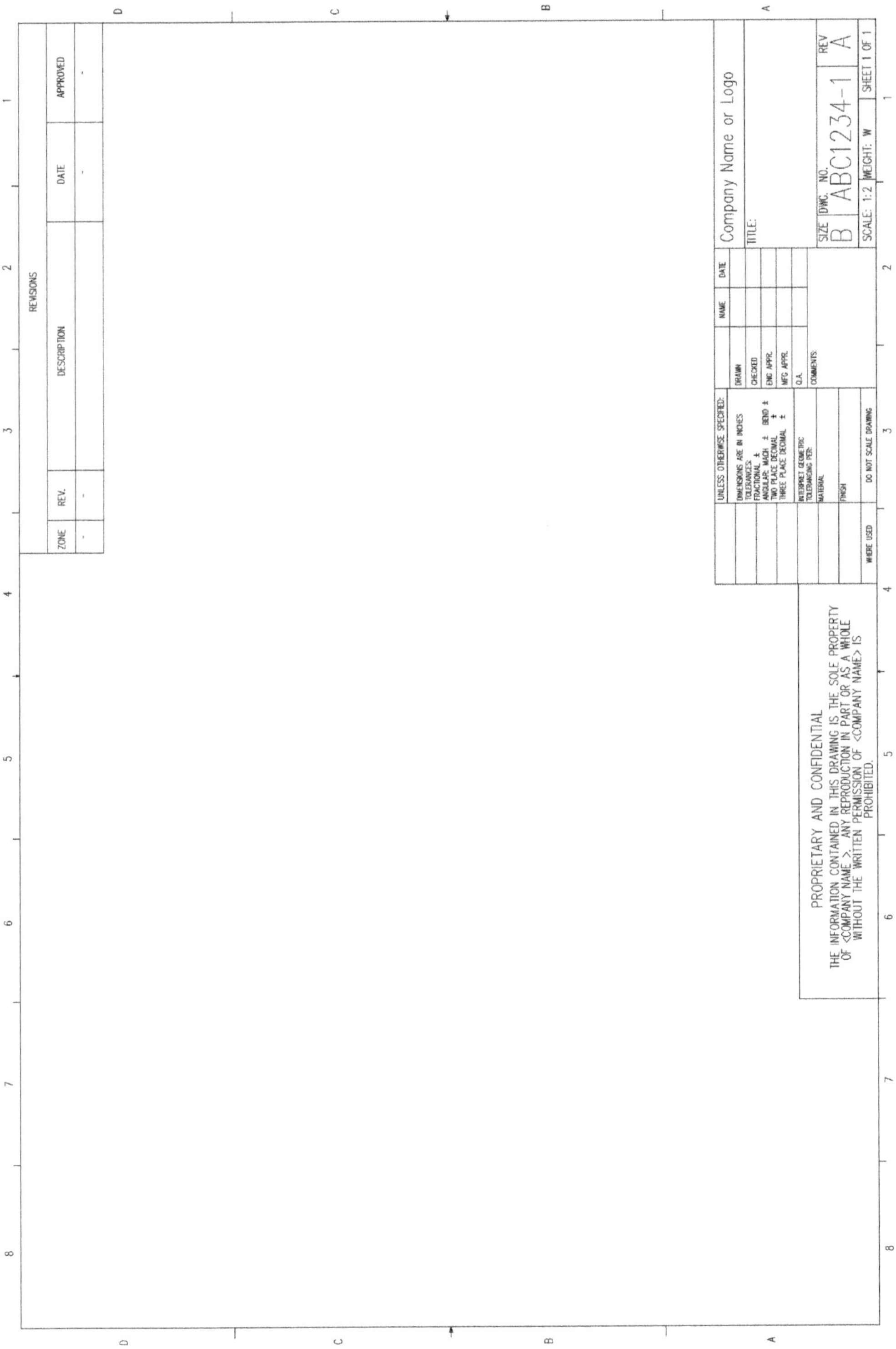

FIGURE 13.2
A common engineering drawing frame

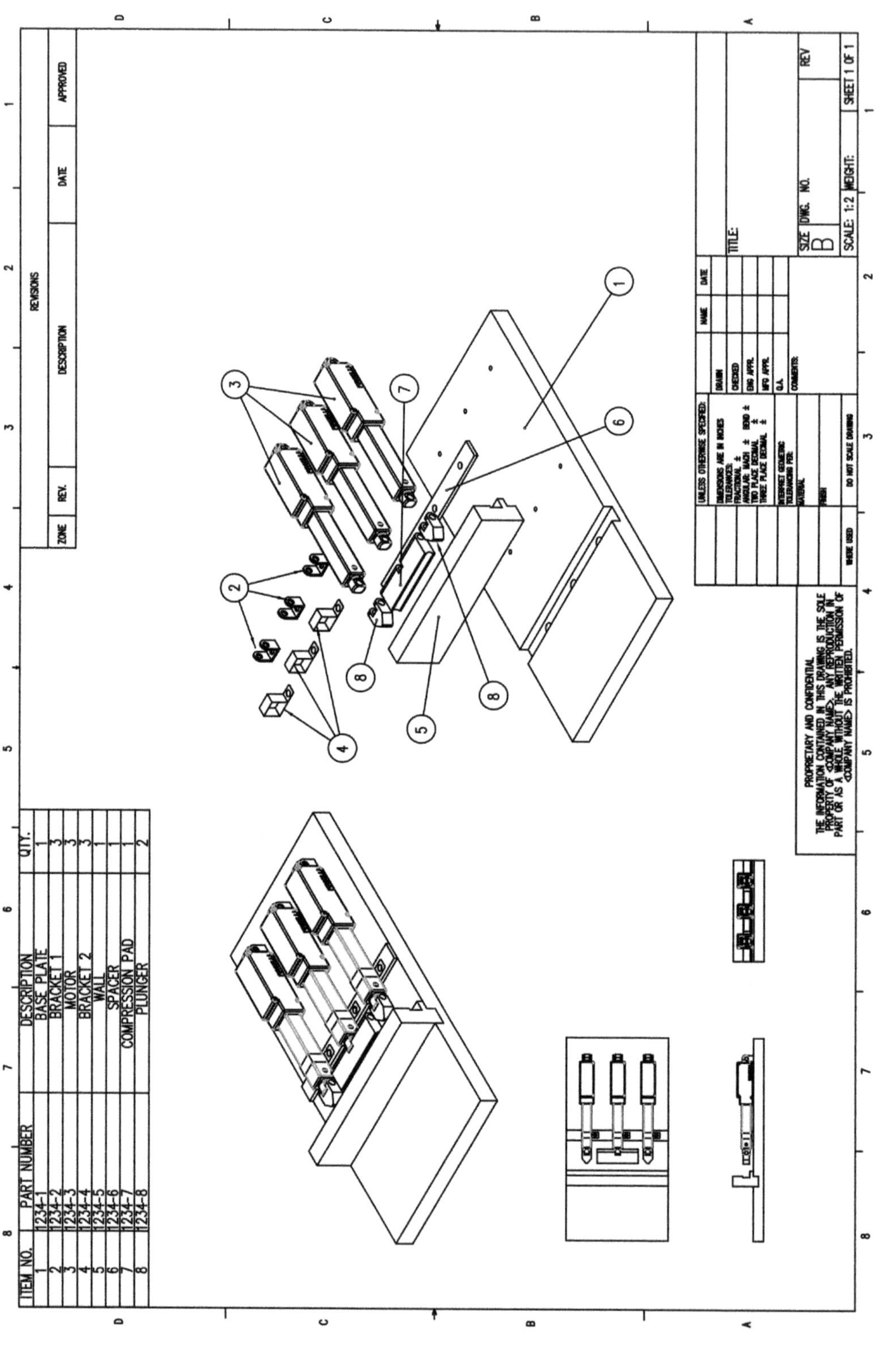

FIGURE 13.3
An engineering drawing of an assembly depicting various components in the BOM

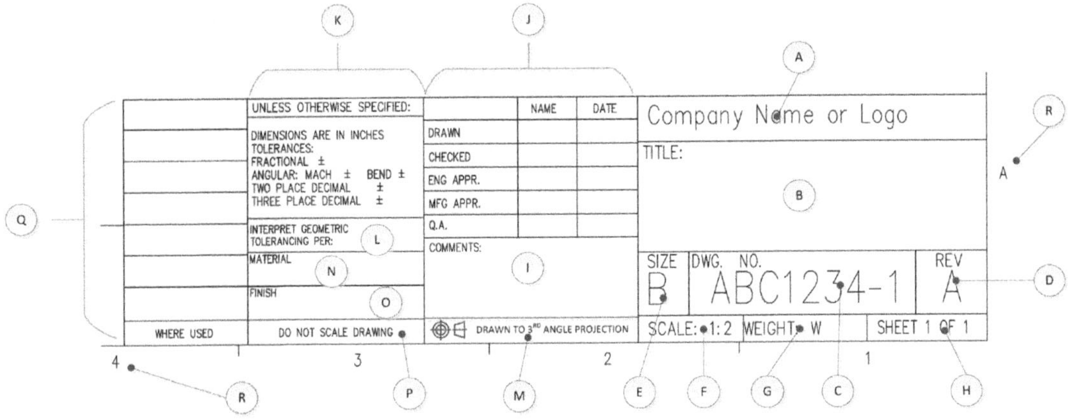

FIGURE 13.4
Details of the information block

information of a BOM is maintained in a separate file. The information on the drawing is only maintained for ease of part identification and is only shown to the detail of the first level down, otherwise the amount of detail will be overwhelming and defeats any practical purposes.

Before reviewing various aspects of engineering drawings, I like to comment that for traditional reasons, the texts provided in engineering drawings are all in upper case block letters with a text height of 0.187 in. Historically, this text format was needed to easily read and understand the information, although I am not sure if this rule serves any more practical purposes anymore.

Information Block

The information block contains the essential information for an engineering drawing and is often found on the lower corner on the right-hand side of the drawing. Common elements of this block are depicted in Figure 13.4 and described below.

(A) Company Name Block: The simplest format of this block provides the organization's name and/or its logo along with a city or a branch name is also provided. In addition to establishing legal ownership, this information facilitates identifying subject matter experts who can answer questions that might be raised at any given time.

(B) Title Block: This section should contain a description of the part or assembly being represented. Commonly, there are between two and four lines of titles. In situations where a component may be common to a number of different products, having a naming convention may be the difference between taking advantage of economy of scale or not. For instance, suppose the drawing depicts a 0.5 in 6–32 fully threaded zinc plated cap screw. Now, imagine that one engineer calls this a bolt instead of a screw, one engineer does not mention the zinc plating, and a third thinks being fully threaded is not important. The purchasing agent entering this data in a database could potentially consider them as three different and distinct components and subsequently issue three different part numbers. This means that an opportunity to take advantage of economy of scale is lost. I will explore this issue in more detail in Chapter 20 "Configuration and Change Management." Needless to say, naming convention is of less concern for *make* components.

(C) Drawing Number Block: This is the part identifier or part number. I strongly recommend that this number be the same as the associated 3D CAD file unless there is a compelling reason not to. Should there be multiple pages, the part number should appear on all pages.

(D) Revision Block: An alphanumeric character(s), this is the identifier of the latest revision of a part and at times is used with a time stamp (or block – not shown in Figure 13.4). It is used in association with a revision table (see the corresponding section below). This block is best used in early stages of product design and development where there are rapid changes that need to be communicated with different team members. Once the product is released, it loses a great deal of its use. There are differing views on whether the revision naming should be the same at pre-production versus post-product release. I will discuss the preferred approach in configuration management in the chapter discussing that topic.

(E) Drawing Size Block: Table 13.1 provided common engineering drawing sizes. The alphabet associated with the physical size of the drawing is captured in this block.

(F) Scale Designation Block: This block reflects the scale used in creating the drawings. Typically, the CAD software chooses a value that may be modified by the design engineer or the draftsperson. It is also common to have drawings with a different scale than the block scale. These should be noted near the specific area.

(G) Weight Block: This section provides either a calculated or a measure of how much a component weighs.

(H) Sheet Number Block: This portion identifies the number of sheets used to describe a part and the specific location of a sheet (i.e., SHEET three of five total pages).

(I) Comments Block: This area is reserved for any relative comments that do not fit into other blocks.

(J) Drawing Approval Block: This area identifies individuals who have developed and approved the drawing. Typical entries are as follows:

- DRAWN: Name of the individual who completed the drawing and date of completion
- CHECKED: Name of the individual who checked the drawing and date of completion
- ENG APPR: Name of the individual who approved the drawing from a design point of view and date of approval
- MFG APPR: Name of the individual who approved the drawing from a manufacturing point of view and date of approval
- Q.A.: Name of the individual who approved the drawing from a quality assurance point of view and date of approval

(K) Dimension Specification Block: In this area, the unit of dimensions (in, mm, or others) is specified. In addition, associated tolerances with the number of used significant places are provided. The following are some examples:

.XX = ±.01 in or ±.25 mm

.XXX = ±.005 in or ±.125 mm

Fractional = +1/32 in

Angular = +1/2 degree

(L) Drawing Standard Block: In this area the standard that is followed is provided. In the United States of America, this standard is typically ASME Y14.5M-1994; however, it was recently updated to ASME Y14.5–2009. It may also refer to British standards, German (DIN) standards, or ISO standards. Though in many respects these standards are similar. Henzold (2006) provides a description of their difference.

(M) Angle Projection Block: Here, the first or third angle projection is specified. In this book, I do not mention the basics of engineering graphics and projection planes. This is partly because of my assumption of the readers' familiarity with the subject and partly because of the use of modern CAD software and engineers' reliance on 3D solid modeling and the automation that CAD entails. There are a number of references on this subject; see, for example, Simmons et al. (2009).

(N) Material Block: This area is used to specify material.

(O) Finish Block: This area is used to specify component finish; at times, color is also specified here.

(P) Scaling Block: Typically this block warns against scaling the drawing.

(Q) Where Used Block: This block is not very common but serves an important purpose. This table identifies all the assemblies where this specific component is used. This information helps with assessing the impact of a proposed change to the form, fit, or function of a part. I will discuss the concept of where-used in some detail in Chapter 20 "Configuration and Change Management."

(R) Zone Designation: Zone designators are used to help locate details of a design or changes made. They are pairs of alphanumeric characters that are placed on the outside of the drawing frame. Typically numbers are placed on the top and bottom of the frame and letters on the sides.

Tables in Engineering Drawings

There are a number of tables that may be incorporated into an engineering drawing. The most common of these are the *revision* table and *bill of materials* table. Others are *hole, weld, bend,* and *punch* tables. There could also be a general table specific to a family of parts or components.

Revision Table

These tables are used to identify the latest design or revision, which are typically designated by alphanumeric characters. Figure 13.5 is a depiction of such a table.

The most common elements of a revision table are the revision letter, a brief description, date of change, the approver, and, at times, a zone identifier. Often a symbol such as a triangle with the revision number in its center is placed near each change for ease of change identification. Some corporations require that the engineering change order (ECO) number be recorded on the revision table as well. Typically, this information is only captured on the first page of the drawing, with subsequent pages referring to the first page.

REVISIONS				
ZONE	REV.	DESCRIPTION	DATE	APPROVED
-	A	INITIAL RELEASE	DATE	Name
2B	B	WHAT CHANGED	DATE	Name

FIGURE 13.5
An example of a revision table with its elements

ITEM NO.	PART NUMBER	DESCRIPTION	QTY.
1	1234-1	BASE PLATE	1
2	1234-2	BRACKET 1	3
3	1234-3	MOTOR	3
4	1234-4	BRACKET 2	3
5	1234-5	WALL	1
6	1234-6	SPACER	1
7	1234-7	COMPRESSION PAD	1
8	1234-8	PLUNGER	2

FIGURE 13.6
An example of a BOM table with its elements

Bill of Materials Table

An example of a drawing with a BOM was provided in Figure 13.3. Figure 13.6 focuses on details of the table area as shown above.

Typical elements of a BOM table are item and part numbers along with a brief description and quantities. Some add the element of make-or-buy to indicate which items are purchased. I like to be clear that this BOM is for ease of reading the drawing and should only identify the next level down on assembly. For instance, if a system is depicted, only subsystems should be shown. Should a subsystem be placed on the drawing, then the next lower level to identify will be assemblies. And so on, down to components. A detailed manufacturing bill of materials should be maintained elsewhere.

Statement of Proprietary and Confidential

Corporations protect their intellectual properties by placing a proprietary and confidential note on all engineering drawings. The wording is similar to what is shown in Figure 13.7. Some teams choose to incorporate this notice into their information block, although others may place it on the drawings as a separate note.

Status Notes

There are two notes on drawings that should be placed separately from other notes. In addition, I suggest that a relatively large, bold, red-colored font be used to draw attention.

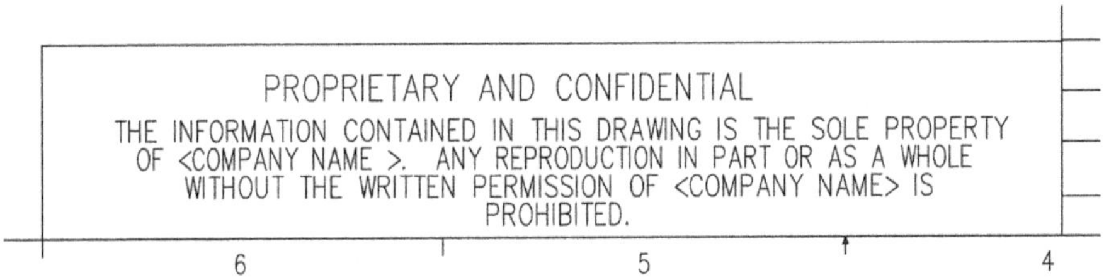

FIGURE 13.7
An example of a bill of a proprietary and confidential statement

The first is a "PRELIMINARY" status note. The designation should be used while a drawing is being routed for quotation or review purposes. When a preliminary drawing is sent to a potential supplier, it is highly recommended that a second note be added: FOR QUOTATION ONLY. This designation alerts the supplier that the requested quote is budgetary because the final design may be altered.

Drawing Practices

Often, when there is a mention of engineering drawing, the conjured image may be that of a so-called blueprint. In fact, the term "blueprint" (or just "print") is used interchangeably with the engineering drawing. Although for the purposes of this book, the focus is on the drawings that convey design data, there are other uses and, in fact, types of drawings that need to be mentioned. Earlier, I suggested that there are both internal and external customers of these drawings. As they are customers (and not suppliers), they may have their own set of requirements on what needs to be communicated to them.

For instance, customers may require a copy of the assembly drawing. A special drawing should be prepared in order to simplify the final design and only provide the essential size of the product. Should the product require connections for installation and/or assembly, the drawing would provide pictorial guidance for these purposes. Customer-facing drawings should not contain any references to internal documents and/or procedures. In other instances, some manufacturing corporations require full-scale drawings in order to review the accuracy of the design dimensions and how various parts assemble.

American Society of Mechanical Engineers Standards

Although a history of the evolution of practices relative to developing engineering drawings is beyond the scope of this work, it is noteworthy that the United States Department of Defense (DoD) played a significant role in the development of American Society of Mechanical Engineers (ASME) standards. It began when DoD sought to standardize engineering drawing practices and to convert MIL-STD-100 to general standards used by industry and government alike. The conversion of MIL-STD-100 has not taken place in its entirety; however, the effort led to the development of ASME Y14 series standards. It began with developing ASME Y14.100, founding standard practices governing the development of engineering drawings and their related documentations. Other documents followed that completed the ASME Y14 standards set. The lists that comprise the basic engineering drawing standards are as described below.

ASME Y14.1, Decimal Inch Drawing Sheet Size and Format – This standard provides the size and format for engineering drawings. Most files generated by CAD software conform to this standard.

ASME Y14.1M, Metric Drawing Sheet Size and Format – This is similar to ASME Y14.1; however, it is developed for the metric units of measurement.

ASME Y14.2, Line Conventions and Lettering – This document focuses on creating a uniform means of using lines (thickness and length) and lettering (font and size) in engineering drawings.

ASME Y14.3, Orthographic and Pictorial Views – The requirements for developing orthographic and pictorial views – particularly as applied to constructed views – are developed in this document.

ASME Y14.5, Dimensioning and Tolerancing – This document which will be reviewed in more detail in a subsequent section of this chapter is the topic of much discussion. It is considered to be the essential standard for creating and maintaining uniform practices for writing and interpreting *Geometric Dimensioning and Tolerancing* (GD&T).

ASME Y14.24, Types and Applications of Engineering Drawings – This document established the requirements for various types of general engineering drawings – though specialized fields such as marine or construction are not covered.

ASME Y14.34, Associated Lists – ASME recommends that this standard that provides a uniform approach to developing lists (e.g., index, parts, or wire lists) be used in conjunction with ASME Y14.24, ASME Y14.35M, ASME Y14.41, and ASME Y14.100 documents.

ASME Y14.35M, Revision of Engineering Drawings and Associated Documents – This document focuses on drawing change and change identification.

ASME Y14.36M, Surface Texture Symbols – The focus of this document is to provide a standard set of symbols to communicate how to control surface properties such as roughness and waviness.

ASME Y14.38, Abbreviations and Acronyms for Use on Drawings and Related Documents – This document is a compilation of abbreviations used in engineering drawings.

ASME Y14.41, Digital Product Definition Data Practices – This document provides exceptions and additional requirements as applied to the development of digital product data not covered by other ASME standards.

ASME Y14.100, Engineering Drawing Practices – This document provides the essential requirements applicable to the development of engineering drawings (computer-generated or manual). ASME recommends that this document be used in conjunction with ASME Y14.24, ASME Y14.34, ASME Y14.35M, and ASME Y14.41.

All other ASME Y14 standards are considered specialty types of standards and contain additional requirements or make exceptions to the basic standards as required to support a process or type of drawing.

I like to emphasize that the value of these standards is to bring uniformity and similarity to an engineering form of communication. In a way, to enable everyone to speak the same language and to reduce "misunderstandings" and, hence, reduce variability. Using the allegory of languages, certainly local dialects develop by necessity to express or address specific needs or requirements.

For an extensive treatment of engineering drawing standards, see Genium Drafting Manual (2009). For a treatment of European and international standards see Henzold (2006).

The Need for Standards and Its Relationship with Interchangeability

In the centuries past, when master builders and craftsmen designed and built a structure or a machine, components were fabricated and fitted together for the final assembly; one component from one structure could not be used in a similar structure without some minor or significant modifications. Since antiquity, engineers have searched for a means of developing interchangeable parts – specially for developing war machines; however, since precise ways of measurement did not exist, developing consistent and uniformly similar parts was almost impossible.

Eli Whitney has been credited with being the first person who developed the tools and fixtures needed to develop the first muskets with interchangeable parts. This work was done under contract for the U.S. government in the early 1800s. By 1940s, the idea of positional tolerancing was developed and adopted in MIL-STD-8A. This military standard ultimately evolved into the widely used geometric dimensioning and tolerancing (GD&T) ASME Y14.5M-1994 (Wilson 2001) and now ASME Y14.5-2009.

It should be noted that interchangeability requires that components be made precisely. In other words, they have to be duplicates of one another. It is of little importance if they are made accurately relative to a unit of measure. A musket barrel was considered interchangable if all the production lot had the same length. It mattered little if the length was 28, 29, or 32 inches. The requirement of "same length" necessitated the principles of tolerancing.

As the concept of interchangeability matured and was applied outside of military circles, an idea began to take root that parts could be produced in different locations, and yet, the end product could be assembled without having to "custom-fit" the components together. More importantly, products could be serviced easily with interchangeable parts. To achieve this goal, it became apparent that not only did components have to be precise (the same geometry and shape), but also they had to be accurate (the same dimensions). This may be the beginning of applying two basic principles: first, the principle of dimensioning and tolerancing; second, the principle of measurement systems. Clearly, the concept of accuracy and precision is meaningless if a measurement system is not in place. Next, I will focus on coordinate system dimensioning. I will cover measurement systems at the end of this chapter.

A Common Engineering Language

Accuracy and precision affect the size, form, location, and orientation of features of components and assemblies. These components, along with design intent, are communicated on engineering drawings. There are two languages for this communication: a local language and a universal language. The *local* language uses coordinate system dimensioning and positional tolerancing, whereas the *universal* language uses GD&T. As with any language, there are syntax and grammar. It so

There are two languages for communicating form, location, and orientation of component features. The first is a local language, and the second is a universal language. The local language uses coordinate system dimensioning and positional tolerancing, whereas the universal language uses GD&T.

happens that the syntax and grammar of the local language are only known and appreciated by a particular company and are subject to wide variation and interpretation outside of its local influence. On the other hand, the universal language is based on a syntax that is widely known. Furthermore, it reflects both design intent and function (for more details on GD&T, see Krulikowski 2007 or Henzold 2009). To demonstrate this point, consider the following example.

CASE STUDY

Consider an assembly as shown in Figure 13.8. There are four components in this assembly: a baseplate, pivot pin, a rotor, and a lock clip. The requirement is that the rotor moves freely around the pivot pin with minimal lateral and longitudinal play. This minimal play has been determined to be 0.02 in. side to side laterally and 0.01 in. longitudinally. Furthermore, the pivot pin is not to protrude from the distal end.

Figure 13.9 provides the dimensions and tolerances associated with the *made* parts. The lock clip is a *buy* item that is not shown here. The pivot pin has to be heat treated to a Rockwell Hardness index of HRC 48 to HRC 52 (not shown here). Parts are fabricated and assembled in-house for the initial launch, but they will be outsourced should production needs exceed in-house capacities. The assembly process and inspection dimension for the subassembly are provided in Figure 13.10. As shown, the pivot pin is press-fitted onto the baseplate.

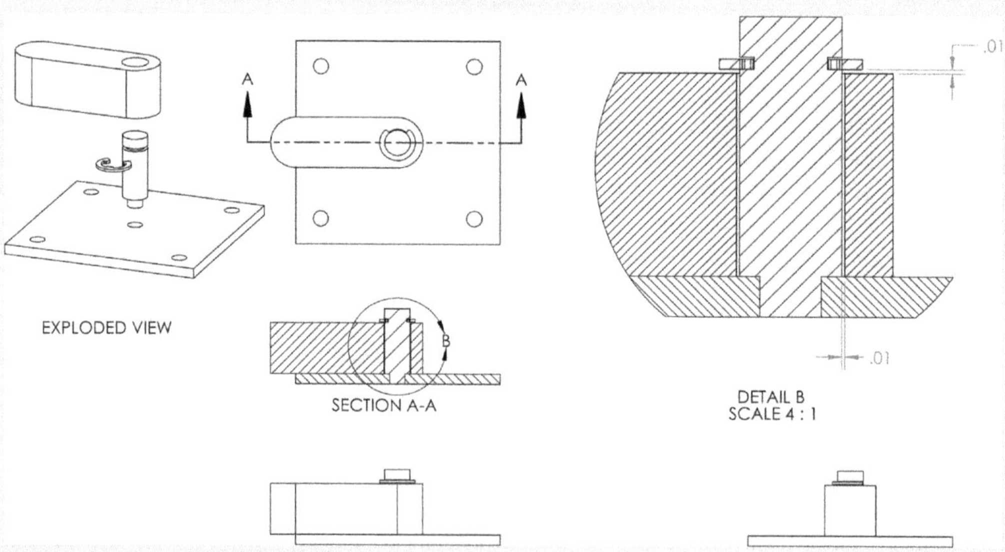

FIGURE 13.8
An example for interchangeability

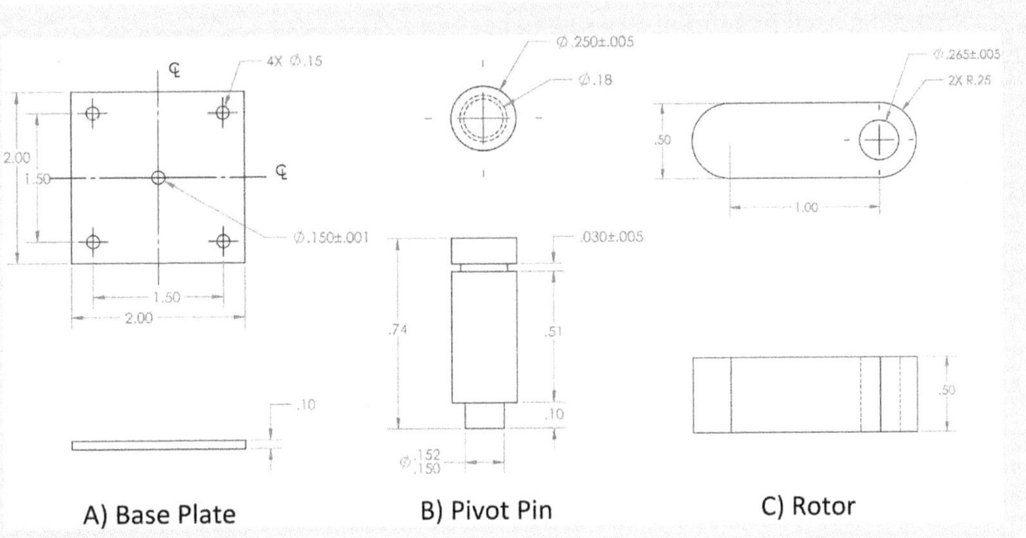

FIGURE 13.9
specified component dimensions and tolerances
(Implied general tolerance is ± 0.01 in)

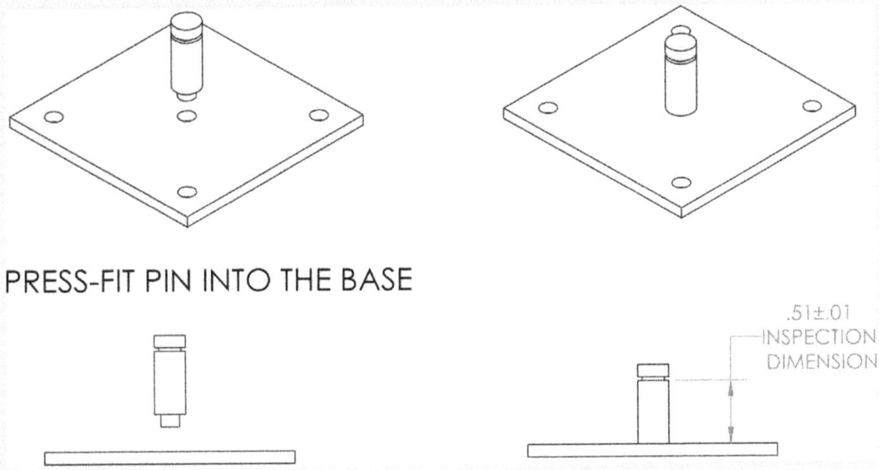

FIGURE 13.10
Subassembly specifications
(Implied general tolerance is ± 0.01 in)

In this scenario, for the product launch, all manufacturing activities were done in-house, and yield was in excess of 95%. However, as the production volumes increased, fabrication and subassembly were outsourced to meet both volume and a new lower cost target. However, final assembly was kept in-house. Soon after, manufacturing yield began to drop at alarming rates. The first reported problem from the manufacturing floor was that the rotator does not freely move in the rejected assemblies. The second reported problem was that some rotators could not be easily assembled onto some subassembly pivot pins. In other words, the pivot pins would not easily go through the provided holes on the rotors. Finally, it was observed that while some rotors could not be assembled on some pivot pins, they could be assembled on others.

In order to prevent a line-down, a 100% inspection of the final assembly was instituted, and the rejected parts were analyzed. The surprising factor was that all rejected components were within the specified tolerances and passed the inspection criteria (Figure 13.11). Further investigation revealed that the angle of the pivot pin press fit into the base plate varied between 87° and 90°. Depending on the severity of this variation, the rotor could move freely or jam.

Additional investigations showed that a large majority of the pivot pins were bent, as shown in Figure 13.12. Again, all the bent components had passed their inspection

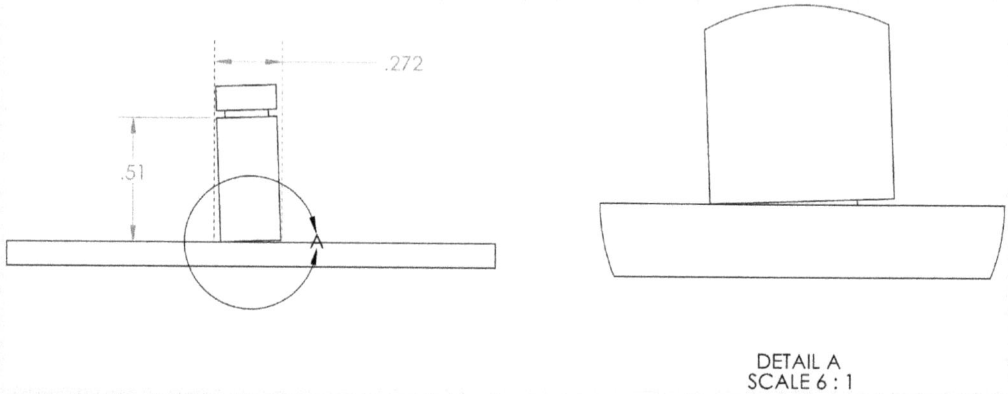

FIGURE 13.11
Unacceptable subassembly passes inspection

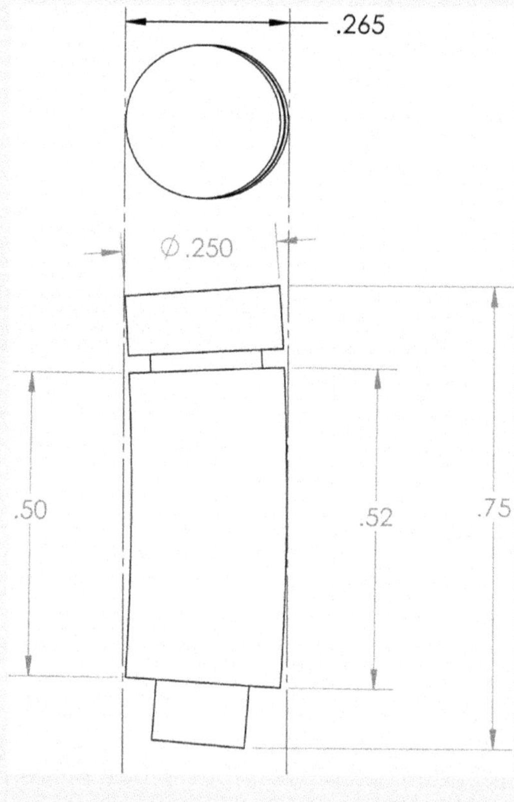

FIGURE 13.12
Bent part passed inspection

criteria. The root-cause investigation then focused on the measurement system and its suitability. It was learned that these parts were inspected with a caliper.

Often the engineering reaction, to control assembly failures and to increase yield, is to reduce tolerances. The realistic impact of this reduction is to make manufacturing more difficult and the part to cost more. In this given case study, however, the engineering department suggested that a Go/No-Go gage be developed for the inspection of the incoming sub-assemblies. Furthermore, a "binning" process was put in place to separate the rotors into different size groups. Although this activity increased manufacturing yield, a relatively large number of parts were scrapped, and the intended cost savings by outsourcing were never realized.

From a broader point of view, two interrelated conclusions may be drawn in this case study. First, interchangeability, which is the cornerstone of cost savings, was not fully recognized. The cause of this loss is because the design intents were not clearly communicated to the suppliers. Internally, the engineers knew what they meant through their drawings. However, their intentions were partially communicated through the use of positional tolerances and coordinate system dimensioning. This communication might have been vague at best or conflicting at times. As an example, consider the way the rotor was dimensioned as shown in Figure 13.9. For clarity's sake, this portion of Figure 13.9 is shown separately in Figure 13.13.

Often the engineering reaction, to control assembly failures and to increase yield, is to reduce tolerances. The realistic impact of this reduction is to make manufacturing more difficult and the part to cost more.

If I were to ask what the width of the rotor is, the answer may be 0.5in nominal. This is a correct "design" answer, but it is a wrong inspection response. The nominal 0.5″ may come from the summation of the two radii (0.25″ plus 0.25″) or the specified 0.5in. However, as the tolerances are applied, should one apply the ± 0.01″ tolerance to the radius dimension, resulting in a ± 0.02″ variation, or apply ± 0.01″ to the specified 0.5in thickness. In other words, should the inspection dimension be 0.5in ± 0.02in or 0.5in ± 0.01in?

Another confusing dimension is this: How would one measure the dimension marked as 1.00 in? One answer may be with a caliper and an approximate location of the center of both semicircles (using the less scientific "eyeballing" approach). Another answer may be to use an optical comparator and pin-point the center of the hole as well as the left-hand side. Is that really measuring the 1.00 dimension? Needless to say, this assumes that the hole and the right-hand side semicircle are concentric. What if, in practice, they are not? Should one measure from the center of the semicircle or the hole?

Often, design engineers make use of dimensions during detailed CAD design to develop a component. Later, as these dimensions are transferred onto the print, they may not readily make sense for manufacturing and/or inspection purposes. As I mentioned earlier, detailed engineering drawings are tools for communicating what essentially is an abstract concept and idea into a physical reality. The more clear this transfer of concept, the more accurate the produced part.

If we liken this communication to an engineering language, then as in any language, abstract concepts require a set of symbols and a syntax for the use of symbols

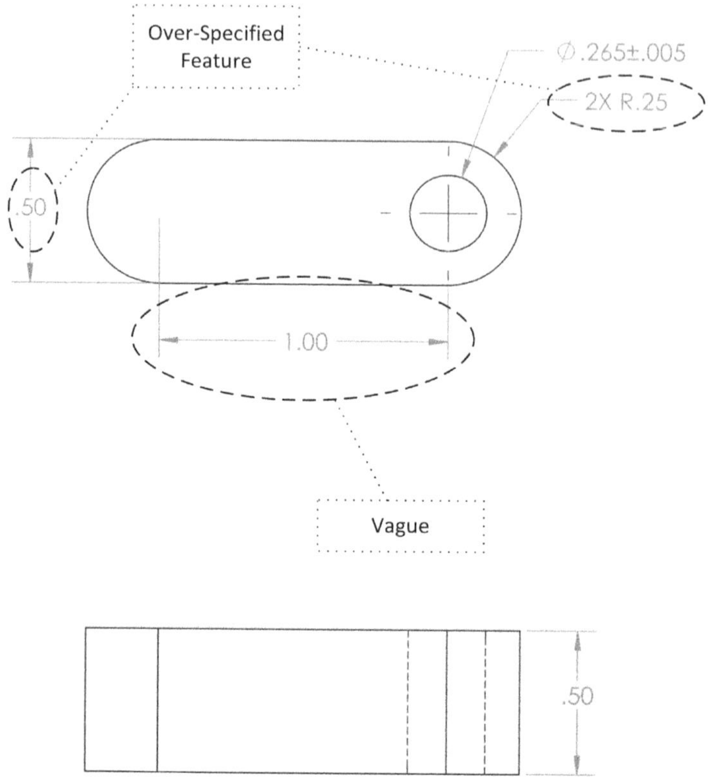

FIGURE 13.13
An example of potentially vague dimensioning
(Implied general tolerance is ± 0.01 in)

(i.e., grammar) for clear and meaningful conversations between parties. These symbols and their use convey how parts come together for assembly or need to be measured so that the design intent is ensured.

For instance, consider the subassembly as shown in Figure 13.14. The ⊥ and the boxed "A" characters underneath the pivot pin diameter dimension signify that the pivot pin has to be perpendicular to the surface that is marked "A." Later, I will explain that this surface will be called a *datum* and it is a reference plane or surface. The number 0.01 next to the perpendicularity symbol (⊥) along with the letter "M" indicates that a maximum pin tip-to-tip tilt variation of 0.01 in relative to the datum A is permissible.

Figure 13.15 depicts the pivot pin with GD&T. The two boxed letters "B" and "C" signify two datum surfaces. They are indicative that these surfaces will be used to make connections to other surfaces in other parts. In this situation, datum "B" will interface with the baseplate and datum "C" will interact with the rotor. The two concentric circles above the datum "B" convey that the bottom cylindrical feature has to be concentric with the main body of the part. The positional value of 0.005 in provides the allowable variation from a perfect concentricity. Finally, the letters "CF" (new in ASME Y14.5-2009) tell the manufacturer that the 0.250 in diameter and its tolerance are continuous features. Thus, the same dimension does not need to be repeated across the groove.

Additionally, GD&T's convention is that the form of the feature may be imperfect so long as it fits within the envelope defined by the largest tolerance value. In other words, the

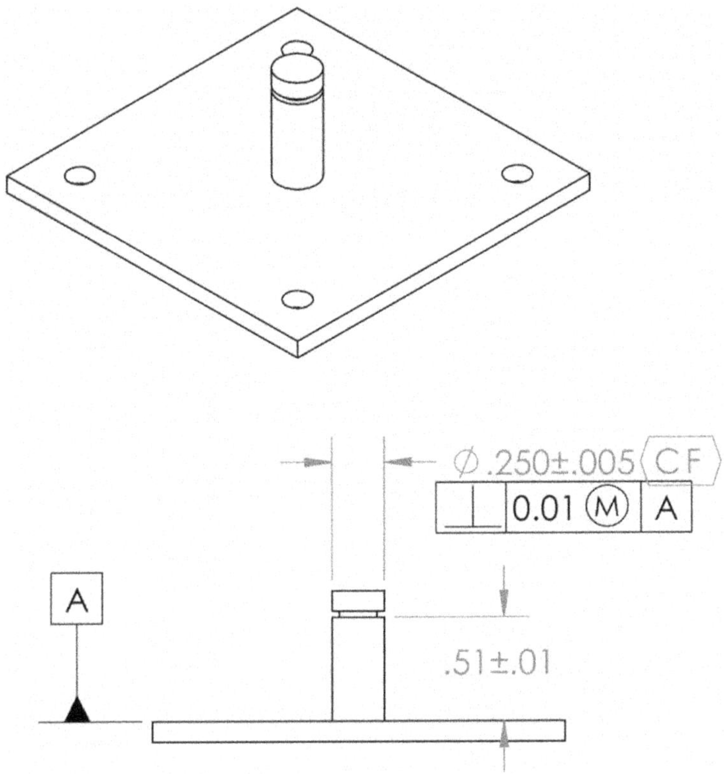

FIGURE 13.14
Subassembly specifications based on GD&T
(Implied general tolerance is ± 0.01 in)

pivot pin may have the largest form imperfection (e.g., bent) when its diameter is 0.245″. As the diameter grows toward the maximum 0.255″ diameter, there has to be closer to perfect. Thus, at maximum material condition (i.e., diameter of 0.255 in), the part has to be perfectly straight.

Principles of Geometric Dimensioning and Tolerancing

Earlier, I suggested that there may be different customers for the engineering drawing(s). This necessitates that the engineer or draftsperson be mindful of the level of detail or information that is provided on the drawing(s). Having said this, I like to add that in the context of the developing drawings for the purposes of manufacturing and subsequent inspection, nothing should be left to guesswork. I do not mean to be disrespectful of the individuals whose responsibility it is to create these drawings, but modern CAD software enables one to place any dimension that they desire with little thought or consideration. Furthermore, at the push of a (virtual) button, the entire drawing may be automatically filled with a variety of dimensions – albeit, dimensions that were used to create the model. This is not how dimensioning and tolerancing should work.

A system-based approach to developing detail drawings would ensure that cascaded requirements and functions are reflected on the drawings.

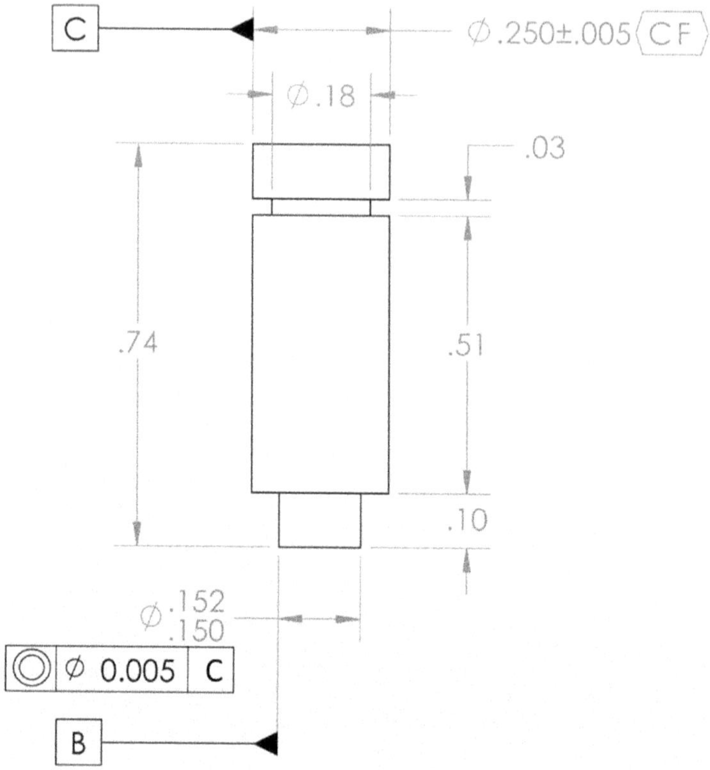

FIGURE 13.15
Component specifications based on GD&T
(Implied general tolerance is ±0.01 in)

Considering that detailed drawings describe a part or component, they should provide accurate information to communicate part geometry for purposes of fabrication and manufacturing, assembly, and ultimately the delivery of the intended function. In short, a system-based approach to developing detail drawings would ensure that cascaded requirements and functions are reflected on the drawings. In addition, the drawings will specify exactly how the components come together. GD&T is the standard for accomplishing this.

Features

In the previous rotor case study, as I discussed the rotor and its detailed drawing, it became clear that dimensions should be placed on items that properly define the part being described. When parts are defined in such a way that the relationships between points, lines, or surfaces are shown, and at the same time, the interrelationship of these elements to one another or to other mating components is provided, the goals of clear communication have been achieved. In engineering terms, this simply means that one should place dimensions on elements of a part that one can see, touch, and feel (Foster and Dadez 2011a, 2011b). These elements are called part *features*. In Figure 13.13, the 1.00″ dimension does not reference a feature. For this reason, it cannot be measured and/or inspected. ASME Y14.5 refers to surfaces, holes, pins, tabs, etc. as examples of features. In the language of ISO standards, the two terms *single feature* and *compound feature* have been used. A single feature has been defined as an entity that is formed by one point, line, or surface. With this definition, a hole is considered to be a single feature, whereas a slot (or tab) is defined as a compound feature because it is formed from a combination of a number of surfaces and lines (Henzold 2006). While this distinction may have some merits, I tend to follow ASME's simpler definition.

There are two types of features. The first type contributes to the function of the part, such as a keyhole. Called *features of size*, these are the elements of a part that require having dimensions and tolerance values. The hole on the rotor shown in Figure 13.13 is a features of size because its function is to allow assembly onto the base subassembly and enable free (or unhindered) rotation of the rotor. Classically, holes or pins, slots or tabs, and balls or sockets define features of size. Another aspect of features of size is that they have opposite elements (i.e., they can be "picked-up" by a gripper or two fingers). Per ASME Y14.5, Rule # 1[2] (implies that there has to be perfect form at maximum size) applies to features of size.

The second type of feature is called (unimaginatively) as *features of non-size* (Foster and Dadez 2011a, 2011b). The important difference is that features of non-size do not contribute to part functionality. As such, Rule #1 does not apply. This allows relaxation of tolerances and possible use of lesser manufacturing techniques for cost savings.

Feature Control

Features may be specified by their size, form, orientation, or location. It is clear why the location or orientation of a feature should be controlled; however, as size controls form (Rule #1), why should there be independent size and form conditions? The answer to this question is two-fold. First, Rule #1 applies to *features of size* as defined earlier. If the element at hand presents itself as nonessential to the functionality of the part, it may be specified with a basic dimension and a form (e.g., profile) specification. Second, even if the element is a feature of size, there may be functional requirements that may specify a tighter control on the feature's form.

In the GD&T language, size controls appear in the form of a nominal dimension along with a tolerance band. In contrast, form, orientation, or location appear in what is called a *feature control frame*. Figures 8.14 and 8.15 provide examples of orientation (⊥ symbol for perpendicularity) and location (◎ symbol for concentricity) controls.

As mentioned earlier, form controls are specified to provide tolerances for the shape of features. The most common control of form is its profile; however, other forms of control are flatness and straightness, among others.

It stands to reason that orientation controls specify the orientation of one feature relative to another. This "another" is almost always a feature called a *datum* (Foster and Dadez 2011a, 2011b). For now, consider that orientation has three aspects: perpendicularity, angularity, or parallelism.

The word "location" in location control not only suggests controlling the *position* of one feature relative to a second feature, but also it can be used to control *concentricity* or *symmetry* of two features. It applies exclusively to features of size and frequently references a datum.

Datums

The *Merriam-Webster* online dictionary definition of the word *datum* is: "something used as a basis for calculating or measuring." In the context of engineering drawings, a datum is a reference point, line, or surface. Theoretically, a datum is a perfect geometric (not physical) shape; that is, point, line, or plane. The reason datums are important is because they enable the design engineer to align and assemble various components together. By providing a sequence of datums and how various features of components relate to these datums, the design engineer conveys to manufacturing what the design intents; and, how the parts would assemble once fabricated. On the basis of this knowledge of parts assembly, incoming quality inspection may take place to release parts for assembly. Developing accurate datums across engineering drawings is the cornerstone of interchangeability.

The reason datums are important is that they enable the design engineer to align and assemble various components together.

From a practical point of view, datums should be placed and identified with elements of a part. For instance, in the previous example, the pivot pin has to be pressed into the hole on the base plate. We can appreciate how the position (and any possible variation) of the hole on the plate may impact the placement and rotation of the rotor. So it would be logical that this position as well as the pivot pin's diameter are used as references (or datums). These references would enable part measurements in such a way as to ensure proper assembly every time.

Once a feature of a part is associated with a datum, it is called a datum feature. It is this datum feature that is referenced in the feature control frame.

Requirements and Critical Design Outputs

The following is a recap of all that has been discussed in this chapter so far: engineering drawings are meant to be a vehicle through which product requirements and design intent are communicated to manufacturing and their suppliers. The two tools for this communication were the drawing format and GD&T. The format sets the tone of this communication, and GD&T provides the language. There is yet a third aspect of communication: *emphasis*. How do we emphasize that one component (or a feature of a component) plays a more important role than another?

From a systems point of view, engineering drawings should provide linkages to design specifications, thus providing traceability back to upper-level requirements.

From a systems point of view, engineering drawings should provide linkages to design specifications, thus providing traceability back to upper-level requirements. Emphasis stems from identifying those requirements that have been identified as CTQ. This provides "objective evidence" that a CTQ has not been ignored. Another more important reason for having traceability is for change control and its impact. Should a drawing dimension be subject to change, the product support engineer(s) can readily assess the impact of change on the overall product (i.e., its requirement and ultimately user needs and voice of the customer) and accept or deny the change. Unfortunately, this traceability does not exist in the majority of cases. Once the original design team disassembles, so does the product knowledge. It is important that engineering drawings provide an accompanying table or reference a document that provides the linkage between drawing specifications and requirements.

Although CTQ identifiers are generally associated with requirements, there is another term that is widely used. That is a *critical dimension.* These dimensions are often identified on engineering drawings with a symbol – sometimes a solid triangle (▲), other times with a star (★) or asterisks (*). I am not aware if there is a symbol more widely used for this purpose. Designating a dimension as a critical dimension generally signifies that excessive variations lead to an inability of the parts to be assembled.

In theory, if the manufacturing process is capable of producing parts to the required tolerances and, through GD&T, a proper relationship between components and their assembly has been established, there would be no need to identify critical dimensions for the sake of manufacturing. In practice, chaos is at work, and often processes slip, and nonconforming parts are made. That is why incoming and outgoing inspections are needed to ensure some degree of quality assurance. Identifying critical and CTQ dimensions works to raise awareness to ensure that any major deviations are prevented. The rationale for this identification, however, should be documented, preferably in the drawings themselves or in an associated document.

In theory, if the manufacturing process is capable of producing parts to the required tolerances and, through GD&T, a proper relationship between components and their assembly has been established, there would be no need to identify critical dimensions for the sake of manufacturing

Sources for Tolerance Specifications

In my days as a junior design engineer, it was not uncommon for me to ask a supplier what type of tolerance they could maintain, and that I would use their number as my tolerance specification. I even recall a senior engineer advising me as such: "For glass-filled engineering plastics with 30% or higher glass, use a ± 0.005″ tolerance. For engineering plastics with no glass, use a ± 0.010″ tolerance. For commodity plastics, use ± 0.015″ tolerance or more." This was not bad advice, and for the industry that we served at the time, it worked. But there are better ways.

Typically, three factors influence the value of a specified tolerance. These are:

1. Function Requirements
2. Assembly Requirements
3. Manufacturing Capability

Function Requirements

The function of an assembly should set the initial ranges of the required tolerances for its components. As an example, consider the ultrasonic sensor described in Chapter 12. In that chapter, I suggested an approach whereby the dependence of the output on physical variations may be determined through establishing the transfer function. Figure 13.16 depicts the configuration for the small tube and its resultant variation.

Based on Figure 13.16B, lateral variation (i.e., x) has the most influence on the sensor function. Therefore, the design engineer should control this variation. In this case, an initial tolerance band of ± 0.1 is indicated in the drawing. Based on the discussion on CTQ, dimensions affecting "x" should be identified as critical to ensure that sensors perform within the desired range.

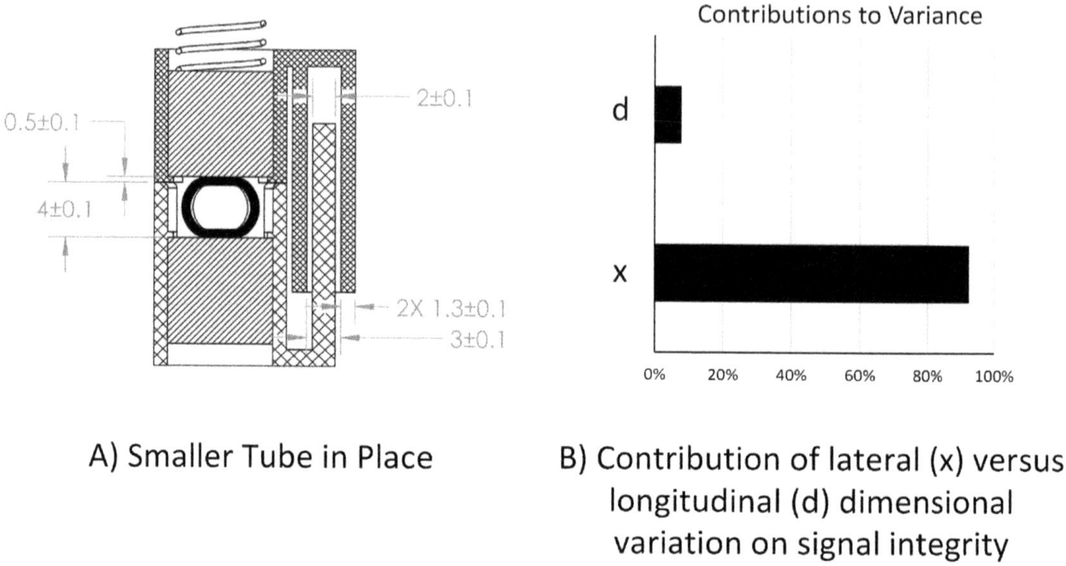

FIGURE 13.16
Transfer function as a source of data for tolerance specification
Refer to Figures 12.11 and 12.12 in Chapter 12 for details of these drawings

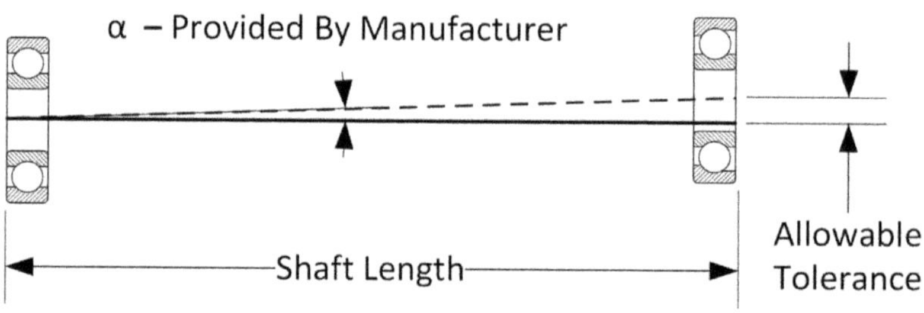

FIGURE 13.17
Component specifications as a source of data for tolerance values

Assembly Requirements

There are several aspects of assembly requirements. First, it has to do with standard-size holes, pins, and their fits, whether clearance or press-fits. Many standards (as well as some part catalogs) provide the required tolerance for proper assembly.

Similarly, there are components that are used as a coupler between two different components, such as shafts. Or, a bearing on a shaft. The manufacturer of these components provides the information to specify the needed tolerance. As an example, consider the two ball bearings as shown in Figure 13.17 (walls and the shaft are not shown for clarity). In this example, the two bears are to be mounted on opposing walls, and a shaft will pass through them. The manufacturer of the bearing provides what cone angles will be permissible as a part of the component specification. This information may be used to set the required positional tolerances – in this case, for the bearing holes located on the wall on the right hand side.

A last resort source for setting tolerance values is from tolerance stack-up calculations. This approach enables a balancing of various tolerances from an overall assembly point of view to ensure certain areas of the assembly are not "starving" for room.

Manufacturing Capability

Each manufacturing process may be optimized to provide the least amount of variation. The source of variation is different from process to process. For instance, in machining a part, tool wear has a large influence. It is not uncommon that a machinist would program the machine at one end of the tolerance to get the maximum life of the tool. As the tool wears, the specified size moves from one end of the tolerance to the other. Another influencing factor is the part deflection under machining loads. Needless to say, if tighter tolerances are required, either cutting tools should be replaced more quickly or machines with different technologies should be commissioned. These actions generally lead to higher component costs.

In injection molding, influencing factors may be the temperature, and pressure of the machinery along with the part dwell time in the mold. Another source of variation in molded parts may be the presence of sink marks and warping. The root cause of these imperfections may be found not only in the process factors but component design as well.

As the design team specifies various tolerances, having open communication with their manufacturing counterparts will ensure that correct expectations for the manufacturing capabilities exist and are reflected in the drawings through the specified tolerances.

Measurement Systems

The study and qualification of measurement systems is called *Measurement Systems Analysis* (MSA).

Recall the earlier case study of interchangeability. It was a rotor that was assembled on a pivot pin and a baseplate as depicted in Figures 13.8 and 13.9. In this example, I suggested that one reason that the bent pins were not discovered at incoming inspection is because a caliper was used to measure the diameter and the overall length of the pin. The point that I did not raise at the time was this. Would another operator have realized that the pins are bent? There was nothing on the drawing to tell the inspector how to measure the part. If an operator noticed the bent pieces, it was not a systemic observation but a personal awareness.

The point that I am suggesting now is that even if GD&T were used on the print, it is still very possible that one inspector measures a component differently than another. Furthermore, an inspector may measure a part with an instrument (say, a caliper), put both down on a table, pick them both up to make a second measurement, and report a slightly different number. This is called repeatability and reproducibility, and it is related to measurement qualifications. The study and qualification of measurement systems is called *measurement systems analysis* (MSA).

On the basis of this rationale, two sources of measurement variation may be identified; namely, manufacturing variations and measurement variations. The question is which source has a larger variation. Needless to say, to be viable, measurement variations and the resulting error should be much smaller than manufacturing variation.

For this reason and to reduce measurement time, many inspections are done on the basis of Go/No-Go gages. In general, these are very simple tools with very precise dimensions that measure the simple geometries such as holes, slots, pins, and tabs. For a hole, for instance, a No-Go-gage is slightly (typically about 0.0005") larger than the high tolerance dimension. If this gage enters the hole, it is too large and should be rejected. The Go-gage is slightly (typically about 0.0005") smaller than the low-tolerance dimension. Similarly, if this gage does not enter the hole, it is too small and should be rejected.

There are two drawbacks to Go/No-Go gages. First, there are those borderline cases when the gages almost go (or don't go) in the hole. Acceptance or rejection becomes an inspector's choice (i.e., a source of variation). The second drawback is that on complicated geometries where there are feature interactions, a simple gage cannot determine if the accepted part will be functionally acceptable. To overcome these shortcomings, one may take advantage of the lessons of GD&T to develop functional gages if maximum material conditions are specified.

Functional Gages

The essence of developing a functional gage is to design a tool that mimics the virtual envelope of all the possible part variations. Because this gage is developed to assure part interchangeability, it is important that the components that are tested follow the same order of assembly as they would in actual production. This suggests that datums need to be properly (and correctly) placed on the drawing to begin with. Subsequently, the inspector should also have the same knowledge and understanding of the datums and the sequence in which they are engaged. Once these two steps are taken, then the gage would work as designed.

The essence of developing a functional gage is to design a tool that mimics the virtual envelope of all the possible part variations.

The second important notion in developing a functional gage is GD&T's Rule #1 (that at maximum material conditions, form has to be perfect). As an example, a functional gage for the subassembly depicted in Figure 13.14 is shown in Figure 13.17. In this simple case, the gage is a block with a hole in its center. The dimension of the hole is derived from the maximum material conditions of the pin diameter and its orientation tolerance $(0.255 + 2 \times 0.01 = 0.275 \text{ diameter})$. Note that the dimensions of the hole on the gage is shown with four (4) significant figures. The reason is that the gage must be fabricated with more precision than the part being measured.

One last note: functional gages test for orientation and form; they are not meant to check for size. As such, to ensure that the correct subassembly has been fabricated, the size of the pin must be measured, then it must be subjected to the gage.

Now that GD&T is covered, the last item associated with this segment of the roadmap is file storage and change control.

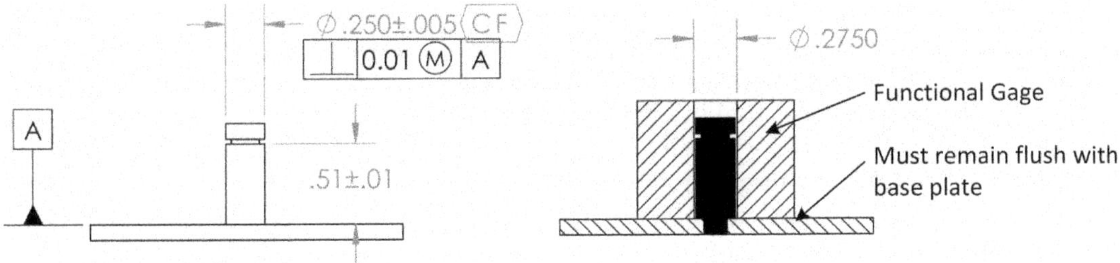

FIGURE 13.18
An example of the use of a functional gage

File Storage and Change Control

The period through which engineering drawings are being developed and released is indicative of stages of design maturity. Shortly after the initial release of drawings, the first building activities begin. The level of scrutiny that the release process receives depends on what may be built. Initial prototypes may be built directly from the 3D solid model. These hardly require any degree of over-sight.

Once drawings receive a preliminary stamp and are ready to be communicated to external teams, there should be a design review by a subset of the full design team. The individual who would generally conduct this initial design review would be the subsystem or assembly owner, the responsible design engineer, and the draftsperson if different from design engineer. In addition, a member of quality engineering should be considered for attendance as an independent reviewer. The first design review of the drawing should cover areas such as requirements traceability, and identification of critical dimensions either for manufacturing or for quality. Completeness and comprehensiveness of drawing notes. Any generated action items should be closed prior to releasing the drawings into a database.

Preliminary drawings are used to communicate with internal customers as well as suppliers. It should reflect the design intent at the moment of release. As these drawings are reviewed by others, comments and suggestions for change are frequently offered. It is important that these changes are not made unilaterally by the design engineer and a change management process be followed. Through this process, the requests are evaluated and their impacts are considered. It would not be surprising if two or more conflicting change requests were made at the same time.

Documentation of requested changes, rationale for accepting or rejecting them, and their impact on the design and manufacturing provide a basis for having a complete file of how design had evolved through its initial stages.

Emerging Concepts and MIL-STD-31000A

It is not hard to imagine why, in this day and age, when parts are fabricated or printed directly from the 3D solid models, we may question the need. to have 2D drawings. I could never get a convincing answer when I used to ask this question. Later, I learned that 2D drawings are required to convey an understanding of allowable dimensional variations. Also, I have learned that it was also a requirement of the U.S. Department of Defense, which was extremely influential in setting trends.

In 2013, MIL-STD-31000A was released, which may hold the promise of removing the need for having 2D drawings altogether. This standard introduces an organizational schema standard for what is called model-base definition (MBD). It provides a set of reference standards and guidelines for CAD users. It is intended to be the foundation for design development efforts.

Historically, 3D solid models had accompanying 2D drawings and were represented by them. However, with advances in solid modeling software architecture, all product definitions previously shown on a drawing may now be defined and displayed directly in the 3D model. The standard suggests: "This makes the 3D model a single master source for obtaining product definition data and eliminates the need for a 2D drawing."

The scope of this standard is to "Provide requirements for the deliverable data products associated with a Technical Data Package (TDP) and its related TDP Data Management Products." Of particular interest from the point of view of this chapter is Appendix B of this standard. The purpose of this appendix was stated as,

> "This standard is necessary to establish a common method to facilitate access to the digital product definition data by downstream users. While this document was initially focused on mechanical piece part focused but the intent is for it to provide a foundation for use in any discipline.
>
> The Schema document is designed to be compliant with ASME Y14.41. It is defined to work with any CAD system.
>
> The goal of this schema is to define a common practice to improve design productivity and to deliver consistent data to the customer."

Although exploring MIL-STD-31000A is beyond the scope of this chapter, it is important to be aware of emerging concepts in regard to the evolution of engineering and technical documents and their communication means to other organizations. Currently, efforts are underway to translate the CAD 3D data into portable data file (PDF) format for ease of electronic transmission. For more information, one may visit www.milstd3000pack.com.

Summary

- The two aspects of developing the detailed engineering drawings are to communicate design intent and to communicate any design changes.
- The US DoD played a significant role in the development of ASME standards.
- ASME Y14.5 is the U.S. standard for GD&T.
- Part interchangeability requires that components be made precisely. In other words, they have to be duplicates of one another.
- Accuracy and precision affect size, form, location, and orientation of features of components and assemblies.

Exercises

1. Continue the design work for the concept used in the Chapter 12 exercise.
   a. Develop the 2D engineering drawings.
   b. What materials have you selected and why?

Notes

1 Here, I am using "assembly," "component," and "part" in a generic sense.
2 There are only two rules: Rule #2 indicates that unless a material modifier is applied, the term Regardless of Feature Size (RFS) applies. A more in-depth description of this rule requires details of GD&T that are beyond the scope of this book.

14

Design for X (DfX)

Introduction

In a way, the term *Design for X* (abbreviated as DfX) stands for a concurrent engineering mindset, where the design team has recognized that product design is greater than the sum of its elements. "X" may stand for manufacturing, assembly, service and reliability, and at times, other concerns such as cost, environment, or international markets. Some even say "X" stands for Excellence.

One may argue that, with a few exceptions, just about anything with a realistic design can be manufactured, assembled, and serviced. For this reason, I should say that DfX stands for *design for ease of* (or *optimum*) X. The implication of interjecting the word *ease* (or *optimum*) is that both the time and costs associated with the process (i.e., manufacturing, service, etc.) are minimized. Hence, the element of cost plays a major role in DfX. This is the reason why if labor-intensive processes may not be optimized, they are moved to geographical regions and places where labor is much cheaper.[1]

> *Design for X* stands for a concurrent engineering mindset, where the design team has recognized that product design is greater than the sum of its elements.

DfX should not be an afterthought. " 'D'" in DfX stands for design. Without the "D," f(X) remains an unknown.[2]" (Baniasad 2015). It is important that manufacturing and service as well as reliability (as the three major stakeholders) should be invited as principal stakeholders to the table at the early stages of design. In the early chapters of this book, I had alluded to this collaboration when I had emphasized the role that support organizations would (and should) play in product design and development.

In this chapter, I will review elements of manufacturability and assembly a long with serviceability and reliability. In a way, the purpose of the entire chapter may be summarized in the following. Design for X activities should

1. identify the most appropriate materials and processes to deliver design specifications,
2. identify which components/assemblies have a shorter life span than others, and
3. arrange and locate these components within subsystems such that they can be replaced easily and quickly.

Design for Manufacturability

I have heard the term *design for manufacturability* (or *manufacturing*) (DfM) used by some, while others use the term *design for assembly*. Still, I have also heard the term *design for*

DOI: 10.1201/9781003564508-18

manufacturing and assembly being commonly used. To my knowledge, a unified approach and process to DfM, DfA, or DfMA does not exist; however, most people involved in manufacturing best practices do speak of similar mindsets and approaches, albeit slightly different from place to place.

> Design for manufacturability (DfM) should properly focus on producing the special-purpose parts in the most economical and cost-effective means while delivering the design intent.

Poli (2001) uses the terms "standard" versus "special purpose" parts – much the same way that *make*-versus-*buy* components were defined earlier. He argues that a design engineer has no or very little control over buy items. Hence, DfM should properly focus on producing the special-purpose parts in the most economical and cost-effective way while delivering the design intent. Using this argument, it is easy to see that design for manufacturing is more focused on choosing materials and processes that are appropriate to the designed component. In this mindset *design for assembly* is focused on how various components are integrated into an assembly, assemblies into subsystems, and the subsystems into the system and the final product.

A Few Case Studies

Let's explore this idea a bit further. Consider two instruments as shown in Figure 14.1. One is a common household tool, and the other is a surgical instrument for extracting teeth. From a design point of view, they are very similar. They both have large ergonomic handles along with a working end comprised of a cylindrical shaft. The working ends have been fabricated to deliver the required function. This is where the similarity ends. From a cost point of view, the screwdriver may be purchased for about $10, whereas the surgical instrument's price may be as much as $200. What drives this price difference? Would it be possible to bring the price of the surgical tool to similar levels as the screw driver?

Figure 14.2 depicts the working ends of the two devices. The screwdriver shaft has been tailored for a high-volume production. It is primarily based on two stamping operations, one on each end. The blank shaft may be cut to size from stock material and standard

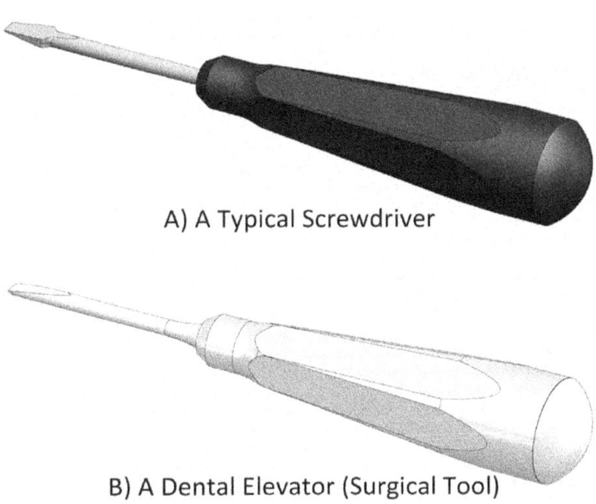

FIGURE 14.1
Comparison of two similar devices from a manufacturing point of view

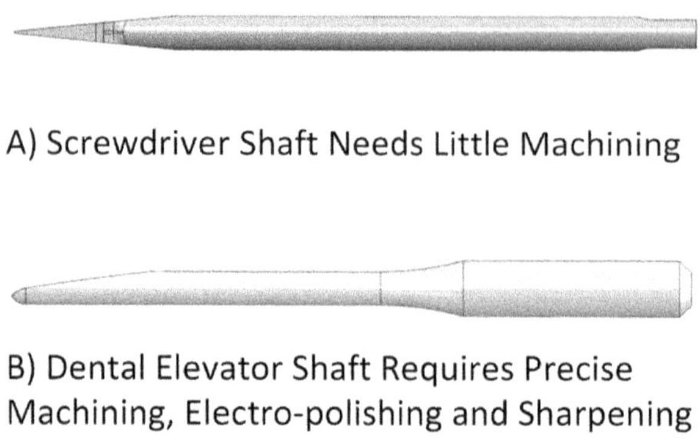

FIGURE 14.2
Comparison of the working ends from a manufacturing point of view

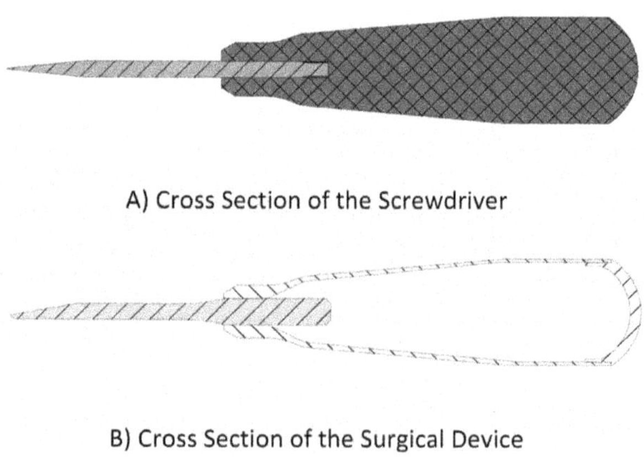

FIGURE 14.3
Comparison of the handles from a manufacturing point of view

gages (sizes). Once the blanks are stamped to shape, there may be minimal grinding operations and heat treating to harden the tip. Hundreds of thousands (if not millions) of these parts are made on an annual basis.

The surgical tool, on the other hand, requires extensive machining to provide the step-down shaft transition as well as the shape of its tip to provide a sharp instrument. Once machining is completed, parts are cleaned, followed by heat treatment for precise hardness measures. The tips may need to be honed by hand if there are sharpness requirements. Typical production quantities of these surgical tools are in the hundreds or possibly thousands of units annually, not hundreds of thousands.

Finally, the handles of these two instruments may be compared. As shown in Figure 14.3, the screwdriver handle is (typically) solid, whereas the surgical instrument has a hollow handle. There are a couple of simple ways of making the screwdriver handle. It may be over-molded directly on its shaft. For large productions, a robot may load the working ends onto a multi-cavity mold, remove the molded parts, and place them in shipping packages.

By contrast, the surgical device handle has a much more complicated fabrication process. First, it is not plastic but metallic. The choice of material only allows two approaches. One approach is to swage a tube to the approximate shape, weld a bushing on the neck, and weld a cap on the larger end. Once this is done, the welds are ground, and the handle is cleaned and buffed. Finally, the tip is pressed in, and the fabrication process is complete. The second manufacturing approach may be to replace the swaging process with a draw of a flat sheet into the shape of a half handle. Once two halves are welded (along with the bushing) and the welds cleaned, the handle is ready.

The foregoing explanation provides an answer to the question "What drives this price difference?" The answer to "Would it be possible to bring the price of the surgical tool to similar levels as the screw driver?" is that it is highly doubtful if the design intent of the surgical instrument is to be met.

The major manufacturing difference between these two instruments was primarily based on production volume. Typically, high production volumes require (and enable) investment in large capital equipment and automation. Large capital investments, particularly in automation, are more difficult to justify for low volume productions.

As another case study, consider the cassette shown in Figure 14.4. In Chapter 9, I presented this product and its role in orthopedic surgery or dental procedures. Note that the design on the right is slightly different from the design on the left. This difference is particularly obvious around the corners. The reason for this difference is that there are two ways to manufacture this product. One way is to take the path of injection molding, and the other is to use sheet metals. Let us consider the merits of each approach.

To have a plastic part made, a mold is first fabricated, and subsequently, hot or molten polymer is inserted into the cavity that exists within the body of the mold. Once the polymer has solidified, the mold is opened for the part to be removed. Any formed part or geometry that does not interfere with mold open/close operation is easily created. For this product, consider the holes at the bottom of the cassette. They are in the same direction as the mold opening. The same, however, cannot be said about the side openings. To create these side features on the part, a number of slides must be designed into the mold. These slides move into place once the mold is closed to form the opening when polymer flows into the mold. When the mold begins to open, these slides have to be retracted to allow the part removal from the mold. In other words, the slides need to move out of the way. The challenge with the particular design shown in Figure 14.4 is that there need to be slides on all four walls. This would make for a complicated and costly tool. On the basis

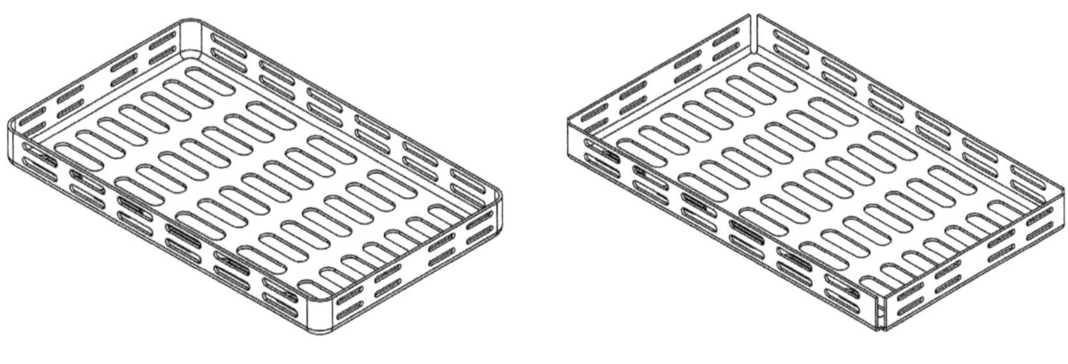

FIGURE 14.4
Similar designs fabricated by two different means

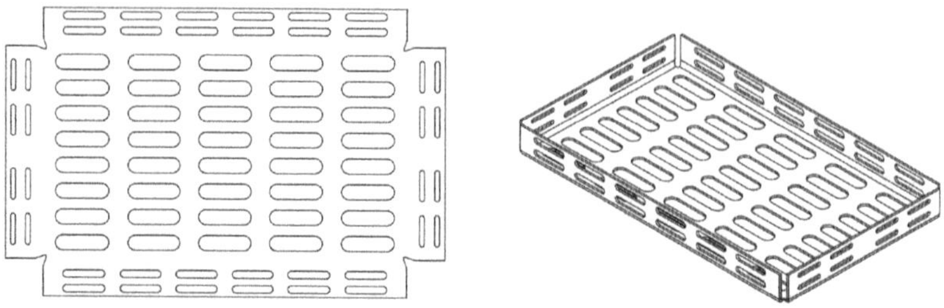

FIGURE 14.5
Sheet metal fabrication from flattened state to finished State

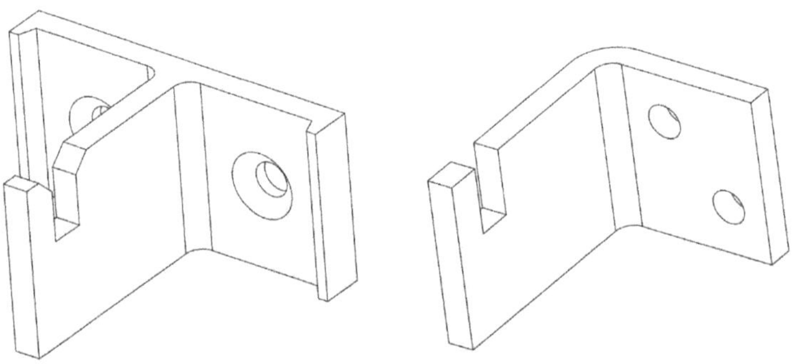

FIGURE 14.6
Two different brackets with the same design intent

of this complication, one may say that this approach is not easily *manufacturable*! Unless a high initial tooling cost is acceptable to the project. High tooling costs may be justifiable if the production volumes are high and profit margins allow a timely return on investments.

In contrast, the sheet metal approach requires a stamping tool to create the blank as shown in Figure 14.5A. This blank is then bent into the final shape (Figure 14.5B). Compared to fabricating molds with side actions (slides), tooling is much less costly. Furthermore, this process has the flexibility to change the dimensions of the cassette to create different configurations. Thus, one may suggest that this design is easily *manufactureable*.

Figure 14.6 shows an example of a bracket with similar functions that may be fabricated by either machining a block of metal (Figure 14.6A) or using a stamping and bending operation (Figure 14.6B). Clearly, machining is the less efficient means.[3]

The question is this: What if the bracket in Figure 14.6A was needed and not its cheaper alternative? In this case, alternative means need to be found to reduce machining. One solution for manufacturability is to use an extrusion process to create what is called a *near-net shape*. Then, the holes and the slot may be machined for a fraction of the original cost.

A design feature that often causes manufacturing headaches is an undercut. If this issue is not resolved by a combination of design relief and manufacturing process, secondary operations such as welding may be required to provide an undercut. An example

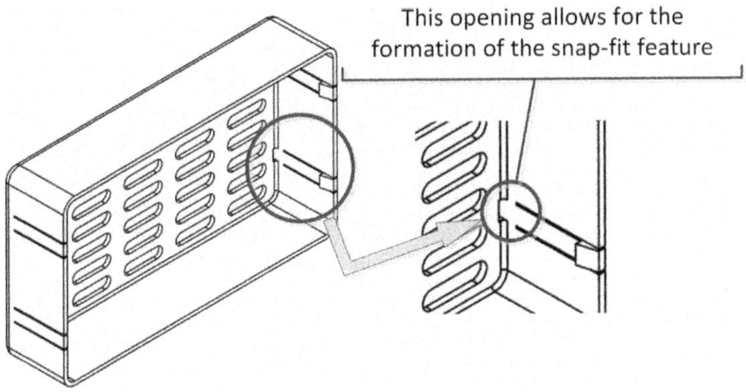

FIGURE 14.7
Manufacturing challenges of undercuts

of an undercut, as shown in Figure 14.7, is the tip of a snap-fit design. The box design follows all the required rules for manufacturability using injection molding, i.e., all features are in the direction of mold opening. All with the exception of the snap-fit feature. A slide action slide may add substantial cost to the tool. Instead, an opening below each snap-fit feature is provided. This allows for a small bar in the mold to help shape the snap-fit and remove the need for side-action slides.

> There is a strong link between the design of a component and its manufacturing process. DfM is a dual interaction of design, materials, and process interaction.

Materials and Processes

The last few case studies demonstrated the strong link between the design of a component and its manufacturing process, and subsequently the class of specified materials. Some suggest that design for manufacturability is a dual interaction of design, material, and process interaction. Any two may be specified; however, the third will be driven by the first two. Others may argue that design for manufacturability is really a design-material-process triad relationship, as shown in Figure 14.8. In a way, I would suggest that if these links are ignored and violated, then a component may not be manufacturable.

Scalability

This is an area of process and its selection where the design team needs to pay attention, particularly in chemical products. We can formulate a product (or assemble it) in a controlled environment of a laboratory on a small scale and develop a false sense of success; yet, experience a high failure rate (as experienced in low manufacturing yield rates). The reason is in a manufacturing environment there are different sets of noise factors than in a laboratory setting with significantly less control. A best-in-class approach is to set up a location where the product may be manufactured or fabricated at larger volumes than a lab but smaller than a full manufacturing rate. Through this intermediate stage, any ramp-up challenges may be identified and resolved.

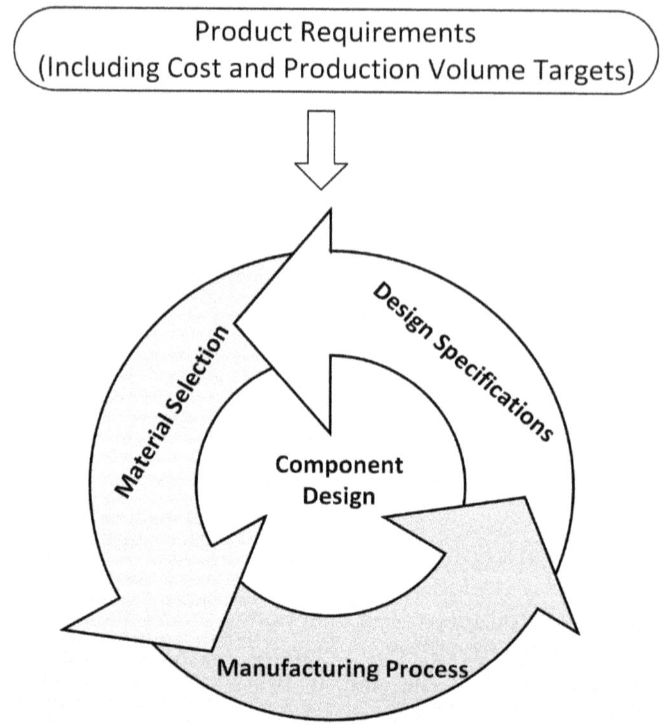

FIGURE 14.8
The interrelationship of design specifications, manufacturing process, and material selection and the flow-down from product requirements

Poli (2001) emphasizes a material-process interrelationship and provides a process-first or a material-first approach. In a material-first approach, he suggests a systematic four-level hierarchy to specify a particular material. For instance, in his first level, a design team decides on using metal or plastic[4] in his part. In the second level, the design team identifies a general class of processing that best fits the required cost and production volume. For instance, if the selected material is metal, the general class of processing may be casting. If the selected material is polymers, the second tier will be to choose between thermosets and thermoplastics. In the third level, the design team identifies an associated process such as sand casting or drawing, and finally, the material most appropriate for the process is specified, say 440C stainless.

In the process-first approach, Poli (2001) suggests that the design team first select an appropriate process type based on part information. Then, a material class may be identified. For instance, injection-molding is first selected and then thermoplastics as the material class by considering factors such as part design complexity and production volumes.

It may be that in practice, as shown in Figure 14.8, all three factors (design specifications – including cost and production volume – manufacturing and material selection) need to be considered at the same time. This concurrency would then drive a detailed component design to the pinnacle of optimum simplicity of configuration, quality, and cost – including material, labor, and overhead.

Design for Assembly and Serviceability

I have to admit that I have trepidations about separating *design for assembly* (DfA) and *design for serviceability* (DfS) from DfM. The reason is that on the one hand, these areas are interrelated, and on the other hand, depending on how DfM is defined, DfA and DfS are its subsets. Regardless of how these areas are categorized, the physical reality is that once a series of components are first manufactured, they need to be assembled to create assemblies, subsystems, and finally the product that is marketed. It thus makes chronological sense to talk of DfM first, followed by DfA and DfS.

> Design for assembly (and serviceability) is really a set of guidelines that design engineers need to follow to enable ease of assembly (or service).

To a large extent, DfA is really a set of guidelines that the design team needs to follow to enable ease of assembly. Similarly, design for serviceability is also a set of design guidelines to enable service personnel to easily reach and replace a faulty component or subassembly. The basis of DfS is founded on the design engineer's acceptance that the product that he (or she) is developing will at some time fail and the faulty component(s) must be replaced. This requires, first, a knowledge of the sequence of components that fail, and next, placing the ones with the highest failure rates at an accessible location in the assembly. This necessitates a knowledge and understanding of the product's reliability, which will be discussed later in this chapter.

A Few Case Studies

CASE STUDY # 1

Let's review the design of a few products from a DfA and/or DfS point of view. Consider the automotive module as shown in Figure 14.9. This module is designed to be used around the engine compartment. It is a relatively simple design with a heavy heatsink, two thermal pads, a PCBA, and a backplate that is glued to the heatsink. Once the module is assembled, it is mounted using the four holes on the heatsink.

This module is easy to manufacture in an assembly line using proper fixtures as assembly aids; however, as the components are bulky, the assembly process will not lend itself to automation unless volumes and quantities justify substantial investments in robotics. There are no screws, and the part count is kept to a minimum. Note that there are no hidden holes or screws placed in hard-to-reach areas.

From a serviceability point of view, this design is a bit more challenging. Should the PCBA fail, its removal requires prying the back plate because it is glued to the heatsink. In addition, the entire module has to be removed from the vehicle before it can be repaired. Depending on the cost structure, this module may be considered disposable as opposed to repairable unit.

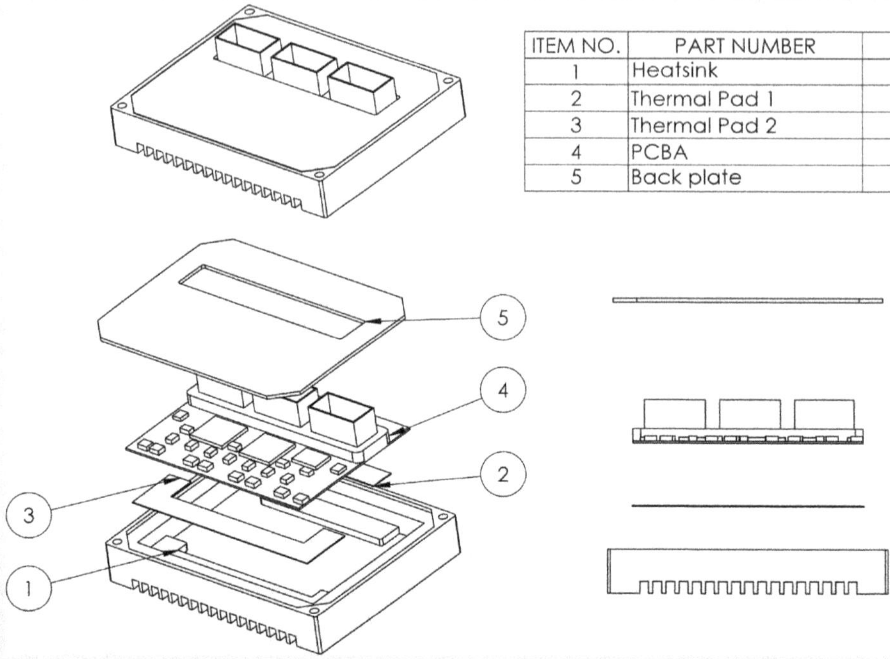

ITEM NO.	PART NUMBER	QTY.
1	Heatsink	1
2	Thermal Pad 1	1
3	Thermal Pad 2	1
4	PCBA	1
5	Back plate	1

FIGURE 14.9
An automotive module

CASE STUDY # 2

In contrast, consider an avionics telecommunication box as shown in Figure 14.10. This box is mounted on a rack with the connectors to plug in on the back of the unit. The front handles are used to carry the unit as well as position it on the mounting rack. These devices are designed to be field replaceable, and the failed units are serviced by opening the front plate.

Once the enclosure is assembled, the process for assembling the boards is identical in either manufacturing or service facilities. PCBAs are designed to have their connectors placed on the back. The enclosure has built-in rails that position the PCBA to properly align with the back connectors. Once in place, PCBAs are locked in. Built-in diagnostic software allows the service personnel to identify the failed board. Its replacement is as easy as opening the front plate, unlocking the failed board, and replacing it. While there are screws used in this design (not shown), their number is kept to a minimum, and they are all the same size.

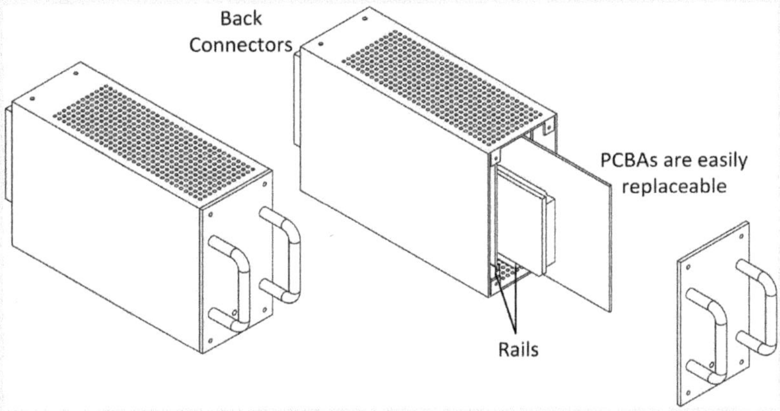

FIGURE 14.10
An avionics module

CASE STUDY # 3

Figure 14.11 depicts an early style of a telecommunication box. Similar to the avionics module, it is simple to assemble both PCBAs once the box has been fabricated and the baseboard (not shown) is placed and bolted to the chassis. The connectors are all horizontal to receive vertical boards. The lid opens from the top, and once removed, there is full view of and access to the entire assembly. There are two components with relatively high failure rates. The first is the power supply. It is placed in front of the unit for ease of access. The second component with a relatively high failure rate is a tray with three fans. This tray slides in and out of place, though it is secured in place by two screws.

A microprocessor controls the speed of the individual fans. This has the benefit that the fans run only when the system temperatures exceed a certain level. This control lowers the duty cycles of the fans, which adds to their useful life. Should one of the three fans fail, the processor runs the other two at elevated speeds in order to compensate for the lost airflow. This allows for higher device availability until a service person replaces the faulty fan tray.

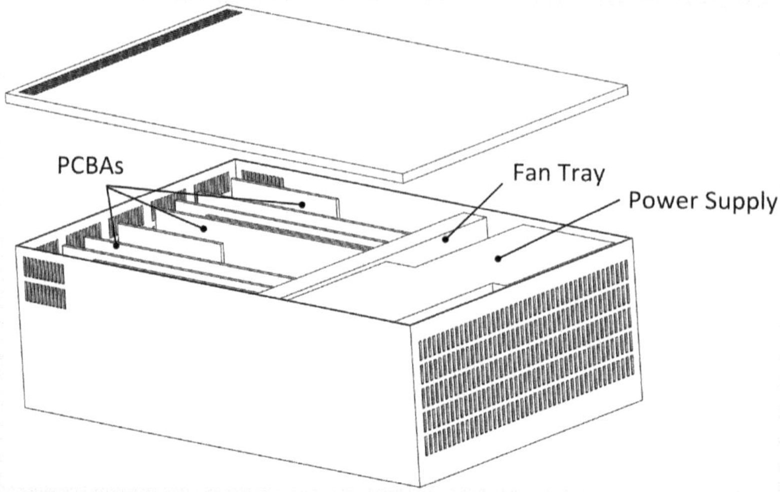

FIGURE 14.11
A telecommunication module

CASE STUDY # 4

In Figures 14.12 through 14.15, several configurations of an enclosure and their impact on assembly and serviceability are reviewed and discussed. Figure 14.12 shows a basic design where a printed circuit board is assembled onto an enclosure using five screws. This is a basic design. From an assembly point of view, all assembly is done in a vertical line; components do not require manipulation to be positioned properly. There are, however, five screws to contend with. From a service point of view, five screws are taken out, the board is replaced, and then the screws are placed back in.

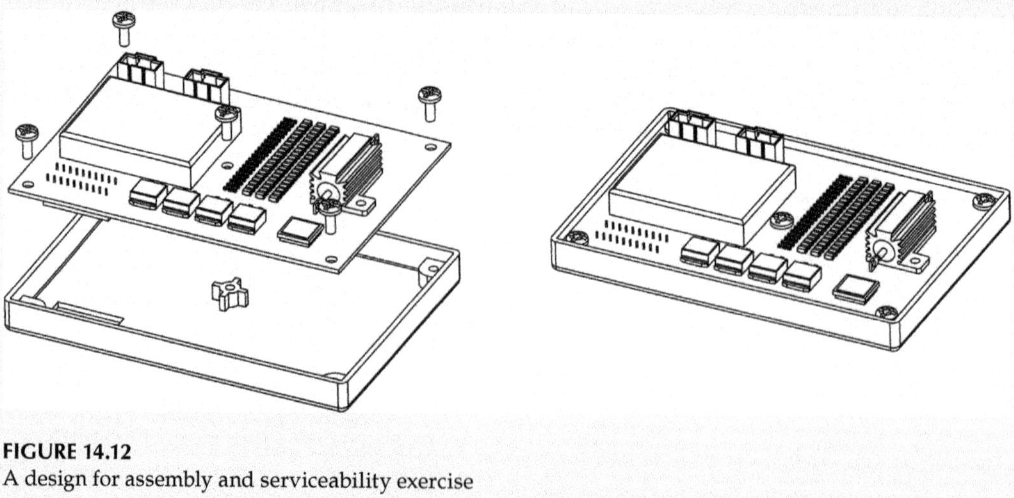

FIGURE 14.12
A design for assembly and serviceability exercise

One aspect of DfA or DfS is to reduce part count. In this example, the number of screws may be reduced down to two, as shown in Figure 14.13. This is accomplished by providing two tabs on the opposite side of where the two screws will be used. Note that by doing so, the vertical line of assembly is no longer true. This is because the board must slide underneath the two tabs.

From a serviceability point of view, there are no major drawbacks. By undoing the two screws, the board may be lifted at an angle and then removed. One may note that there is a component on the backside of the board. The practical challenge is to ensure that this method of assembly does not adversely impact the performance of this back-side component.

One may argue that the screws may be eliminated altogether. Figure 14.14 depicts a configuration where there are studs on the enclosure that would position the PCBA onto the enclosure in a vertical line of assembly. Then, by means of a heat-staking operation, these studs are melted down to secure the PCBA in place. This by far is the easiest and fastest manufacturing approach. The drawback is that this design is no longer serviceable.

A screw-less design that still maintains serviceability may be accomplished by designing a snap-fit into the enclosure, as shown in Figure 14.15.

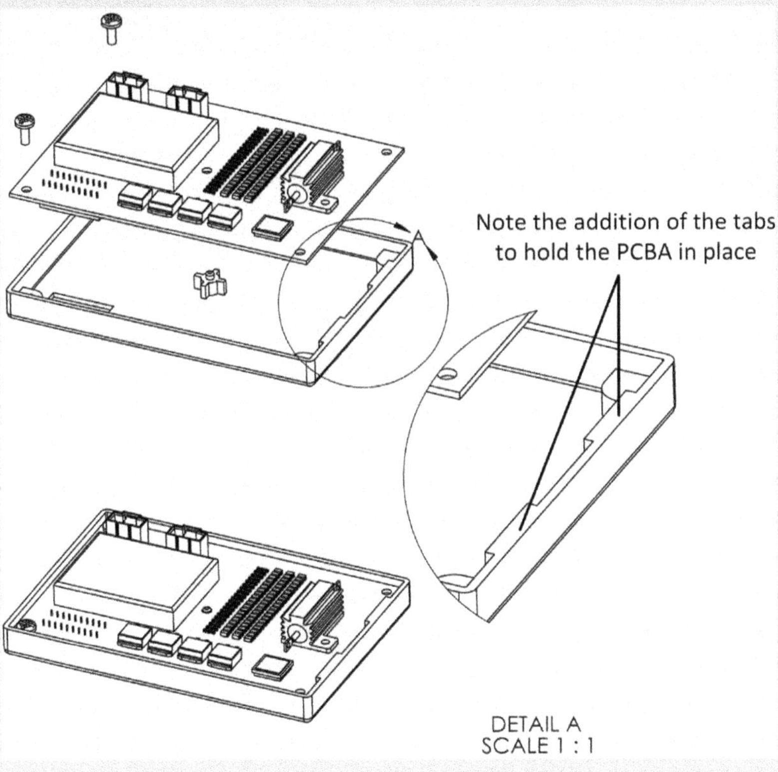

FIGURE 14.13
A design for assembly and serviceability exercise – number of screws reduced to two

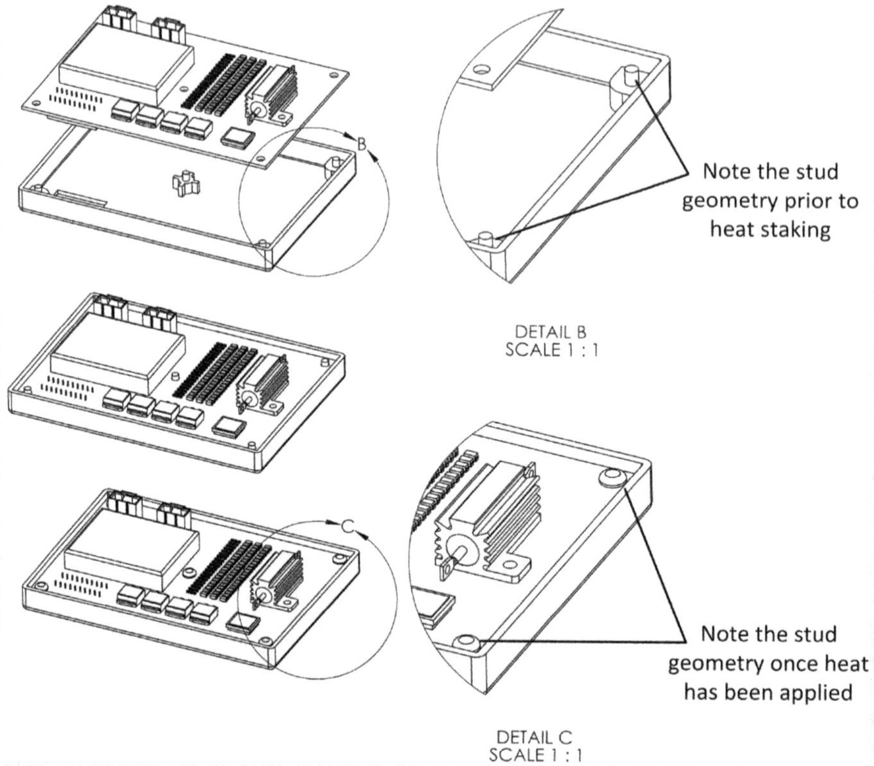

FIGURE 14.14
A Design for assembly and serviceability exercise – screws are eliminated completely by means of heat-staking.

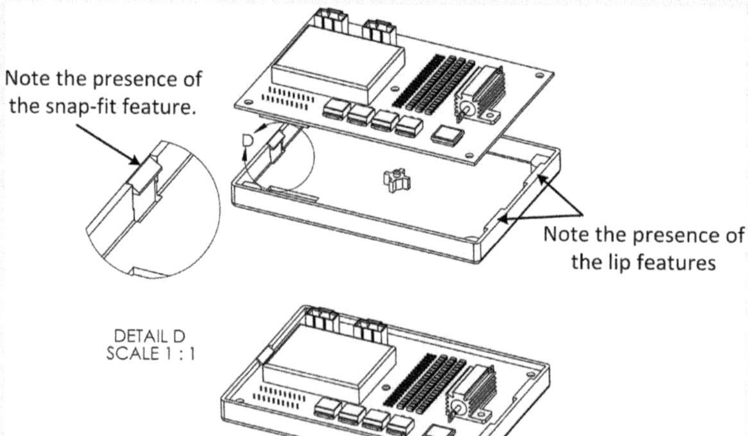

FIGURE 14.15
A design for assembly and serviceability exercise – screws are eliminated completely by means of a snap-fit design.

In my opinion, the 80/20 rule also prevails in developing an optimum DfX product. A product design may improve greatly from a manufacturability, assembly, or serviceability point of view by incorporating only a few changes. However, beyond a certain point, compromises between various competing factors need to be made to bring improvements only in specific and localized areas.

For instance, consider once again, the configuration shown in Figure 14.15. By designing a snap-fit and a tab feature, an opening (not shown here) in the enclosure had to be made to make the enclosure manufacturable. For the purposes of this example, the assumption is that this opening is acceptable. However, such an opening might not have been acceptable if it was a customer-facing feature. A trade-off may be that the openings caused by the presence of these two features may be covered by a decal. Another approach to achieve optimum assembly is the heat-stake approach, as depicted in Figure 14.14. The trade-off is that once the PCBA is secured, it may not be easily removed; thus, serviceability is lost; hence, the assembly becomes disposable. These design trade-offs are real and almost always impact cost one way or another.

Printed Circuit Board Assemblies

Bralla (1996) suggests that DfX might have started with the electronics industry, partly due to the rapid advances that this field has experienced in the last few decades. The fundamental principles of DfX are the same in electronics as they are in other fields: the PCBA should be easily manufactured, assembled, and serviced. With the advent of automatic pick-and-place automation, this implies that a board with surface mount components can be assembled and manufactured more easily than leaded components – even though some automatic devices can even trim and bend leads. Typically heavy components such as transformers present a challenge to automation and have to be soldered in place manually. An example of this type of product is a power supply, as shown in Figure 14.16, where almost all components are placed by hand.

In a way, design for serviceability does not apply to PCBAs – with a large number of components on the majority of these assemblies, it is not economical to identify and then try to replace a faulty component on a board. Instead, the entire board is replaced. Design

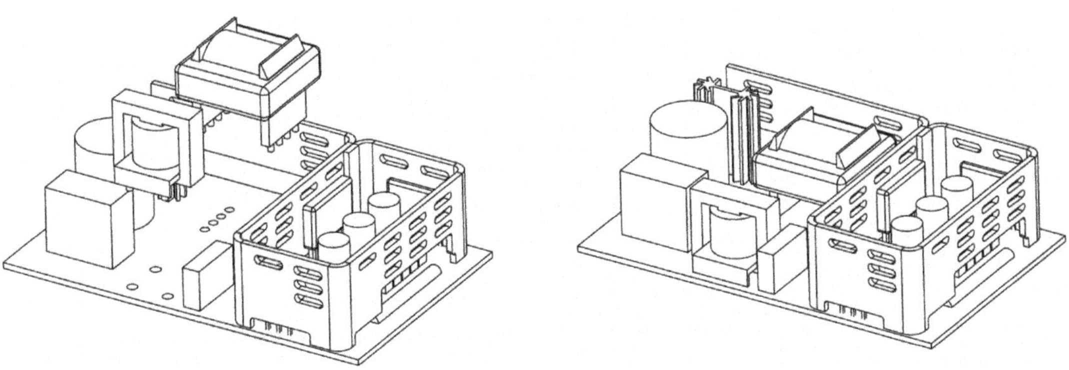

FIGURE 14.16
A typical power supply PCBA

for manufacturability of PCBAs is a different matter entirely. Components should be selected to a large extent based on the capability of the pick-and-place machine that is available on the manufacturing floor. Furthermore, as the part density of a board increases, so does the role of panel and component layout. A technical difficulty with the manufacturability of high-density PCBAs is the inability of solvents to properly clean and remove the flux. Flux contamination is among the top reasons for low manufacturing yields. Additionally, factors such as the size tolerance of components and their assigned space become critical (Bralla 1996).

Design for serviceability does not apply to PCBAs. It is not economical to identify and then try to replace a faulty component on a board. Typically, the entire board is replaced. Design for manufacturability of PCBAs is a different matter entirely. DfM for PCBAs needs to focus not only on the capabilities of the pick-and-place machine but also on the board layout and the panel design.

An important influence on the serviceability of PCBAs is the location and design of the interconnects. In a way, interconnects to PCBAs are like screws and holes to mechanical components. If these features are placed out of direct line of sight, then reaching them becomes cumbersome, time-consuming, and costly. In electronics, there is always the chance that the pins are bent, leading to larger than expected scraps.

Mistake-Proofing (Poka-Yoke)

An aspect of design for assembly and serviceability is to assess what may go wrong during the assembly or service process and to anticipate the mistakes that operators can make. Today, *mistake-proofing* or *poka-yoke* is commonly used and is another example of Japanese influence on increasing production throughput and quality (Beauregard et al. 1997).

An aspect of design for assembly and serviceability is to assess what may go wrong during the assembly or service process and to anticipate the mistakes that operators can make.

Mistake-proofing is based on a zero-defect mindset and a field of study in and by itself. This subject is being reviewed here as a reminder that the best place to provide a mistake-proofing mindset is at the design stage.

For instance, consider the torque wrench as shown in Figure 14.17 – adapted from a Jamnia et al. 2007 patent assigned to Hu-Friedy Corporation. In this configuration, the spring element provides resistance to rotation until a certain level of torque value is reached. The torque is applied through the housing and is transferred to a workpiece (not shown here) through the rotor. Because of the symmetry of the spring element, it is possible that it is assembled onto the rotor and eventually the wrench in a wrong orientation. This mistake may not potentially be discovered until the end user tries it for the first time.

The solution to this problem is to introduce a degree of asymmetry that would allow assembly of parts only for the correct orientation, as shown in Figure 14.17B. A small tab on the spring element and a corresponding groove on the rotor solve the design for assembly issue here in this configuration.

A more common (and well-known) approach to mistake-proofing takes place on the manufacturing floor. This form of mistake-proofing often involves fixtures. For instance, consider the sheet metal with five holes as shown in Figure 14.18. Note that the two sides are nearly the same length. The hole patterns are the same, but the distance between them is not the same. Thus, the possibility of visual mistakes is high. This example is an adaptation of a similar example provided by Beauregard et al. (1997). If the holes on this sheet are

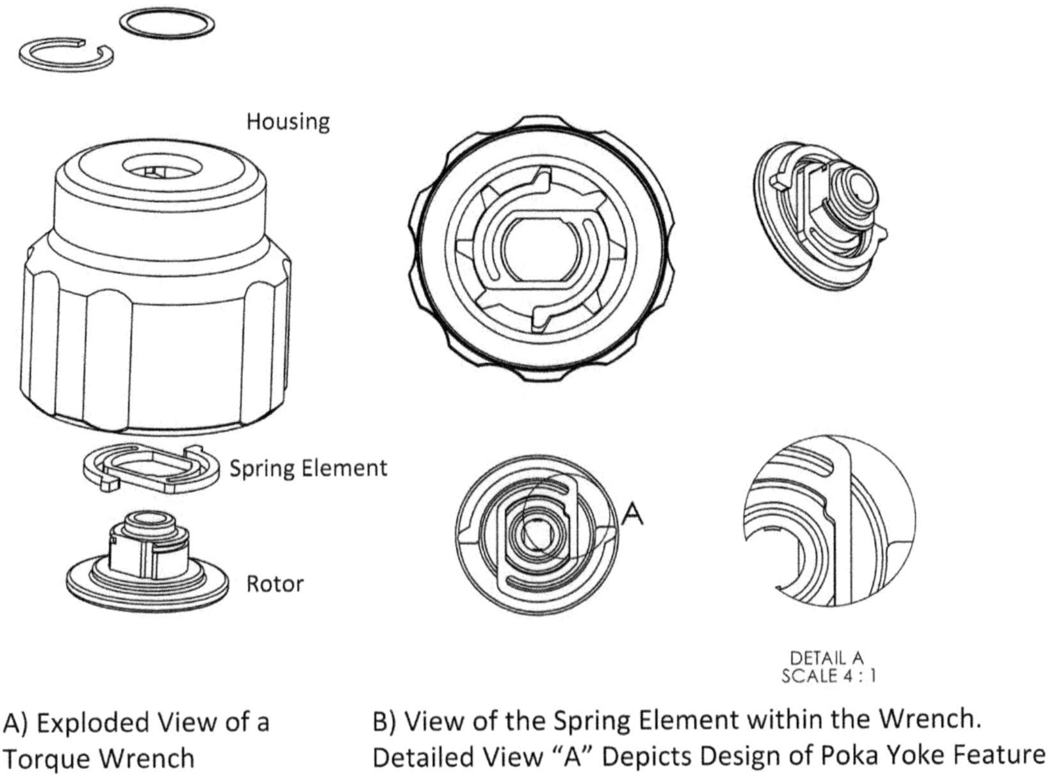

FIGURE 14.17
An example of mistake proofing in design
Adapted from Jamnia et al. 2007

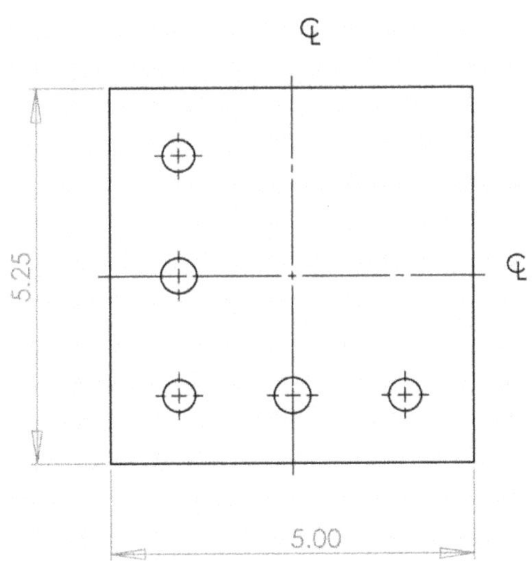

FIGURE 14.18
An example of mistake proofing in manufacturing
Adapted from Beauregard et al. 1997

stamped out without any jigs in place, the part-to-part variations will be at a maximum. So, a fixture is needed to ensure that the holes are punched at the correct location on the sheet every time. The second challenge with this design is that the two sides are nearly the same but not exactly: one edge is at 5.25 in and the other at 5.00 in. It is very easy for the operator to present the wrong edge to the punch press. The mistake proofing solution to this design is to develop a fixture in such a way as to prevent the wrong orientation of the plate from being used.

As Beauregard et al. (1997) point out, poka-yoke strives to either prevent an error from taking place or to detect it as soon as it has happened. Prevention takes place in the design phase of the product (or even a process). As the design team meets to develop the assembly process failure modes and effects (PFMEA), they need to be cognizant of operator errors and capture them in these PFMEA documents and to provide effective mitigations, possibly in the design itself – similar to Figure 14.17.

Detection, as opposed to prevention, is an aftereffect condition, and its discussion is beyond the scope of this work.

A Few Guidelines

An aspect of mistake-proofing in assembly and service is to consider the ergonomic factors associated with assembly or disassembly of the unit. This is to ask how fabrication or assembly operators, as well as service personnel, would handle and manipulate various parts and components. Are there a mix of similar but different parts? Are there blind spots?

> Mistake proofing relies on standardization of manufacturing processes, engineering designs based on self-help, and finally ergonomics for assembly.

I consider the following as seeds that should be used to stimulate brainstorming discussions for design for manufacturing, assembly, and service:

Standardization

By maximizing off-the-shelf components and minimizing special-purpose parts, two goals may be achieved. First, by taking advantage of economies of scale, there are cost-savings opportunities that may be realized. Second, personnel training time is reduced because of familiarity with the component and its handling.

Furthermore, by simplifying the design and reducing the part count, the number of steps is reduced, and by its very nature, the probability of making an error is reduced. Consider, for example, that a design configuration uses 3 different screw sizes and a total of 18 screws. Every time an operator attempts to pick up a screw, there is a probability that the wrong screw is picked up even though they are in different bins. Furthermore, every time a screw is actually picked up, there is a probability that it is dropped. The combination of these two probabilities leads to wasted time and the potential for poor product quality. Part reduction leads to removing these probabilities of error and mistakes.

Principle of Self-Help and Component Design

Though the *principle of self-help* may sound like a legal or psychological term, in engineering, it refers to design configurations in which a natural coupling exists between its components and the assembled configurations provide a path of least energy expenditure. An example of this principle in everyday life is a funnel used to fill a bottle with a liquid from another container. It is not hard to imagine the results of pouring a fluid from one bottle into another should a funnel not be available and the associated waste. Examples in engineering design include chamfers on the edges of features that are assembled together. For instance, machine screws frequently have a chamfer at their tips, and the receiving hole is also provided with a corresponding chamfer. This would allow for easy of part to part insertion and eventual assembly.

Other features that may be categorized in this group are components with clear axes or planes of symmetry or, otherwise, a clear asymmetry. The example shown in Figure 14.18 is a clear case for potential confusions and assembly error. It appears that a line of symmetry exists, but that is not truly the case. Furthermore, the middle holes have different, slightly larger dimensions than other holes.

I suppose the most important aspect of the self-help principle is to ask the question, how easily would the assembly process be automated? Keep in mind that although in assembly, automation may be utilized, the same may not be said for service.

For an automated assembling process, components must have certain characteristics. For example, they should not tangle with each other. An example of tangling components is soft, small springs. Reducing component handling time is another facet of design for manufacturing, assembly, and service.

Ergonomics for Assembly and Service

Often, ergonomics is mentioned in terms of the design of a product; however, ergonomics applies just as readily to the components, assemblies, and subsystems as well. In a way, one may say that the mistake-proofing principles that were discussed in the previous section as well as the other places are all aspects of ergonomics as applied to assembly and service. I like to make a disclaimer here. I do not have any expertise in ergonomics and human factors; however, experience has taught me a few lessons. Elements of good ergonomics in assembly and service include minimizing the number of different hand tools required and eliminating special-purpose tools if at all possible. Another element is having a line of sight for the operator for aligning components and placing/tightening their fasteners.

DfM/A/S in the Design Process

Figure 14.8 provides the interrelationship of design specifications, manufacturing process, and material selection and their flow-down from product requirements as it applies to design for manufacturability. Similarly, the same cascade may be presented in terms of the V-model.

As shown in Figure 14.19, design for manufacturing (or assembly or service) begins at the high-level design of the product. Once this step is completed, an analysis of the proposed configuration must be completed. Should there be concerns, a redesign is suggested, and the high-level/concept design is revised and updated. The development activities then moves onto developing assembly requirements and specifications on the basis of this redesigned concept (stage 5, Figure 14.19). This is followed up by subsystem and assembly design. Once this step is completed, the output is evaluated for material and process and is subjected to parts and process cost analysis (Collis and Clohessey 2015). Again, should adjustments need to be made, a redesign is proposed and steps 4 through 7 are repeated.

It is common for the traditionally minded engineers to raise the question that the development process is prolonged as a result of these activities. That is quite correct – in

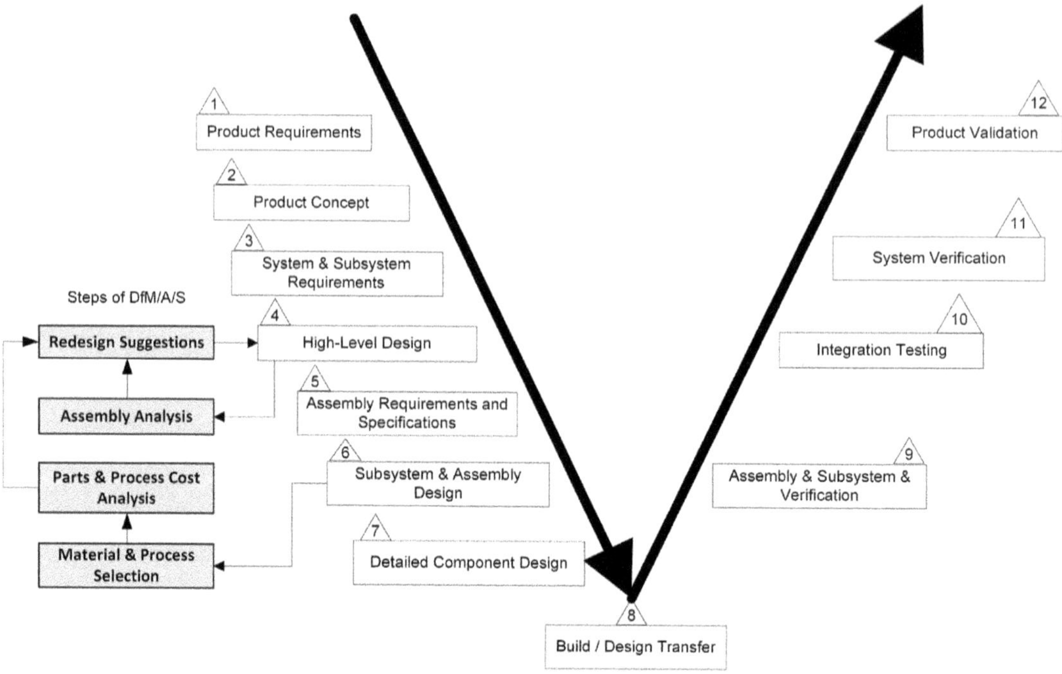

FIGURE 14.19
DfM/A/S and the V-model

some way! The initial design time increases; however, the overall development time may decrease by as much as 45%, as shown in Figure 14.20 (Collis and Clohessey 2015). DfM/A/S at the initial design provides feedback at an early stage so that the need for multiple iterations of build-test-rebuild is reduced substantially. This, in turn, reduces the need to change and update the detailed design drawings and documentation. Furthermore, the paperwork and documentation associated with writing test protocols and results are also reduced, which results in substantial savings.

Design for Reliability

> Reliability is defined as the probability of a product or device performing adequately for the period of time intended under the operating conditions encountered. Reliability is different from robustness. As we are designing for excellence, we need to keep the difference between these two terms and the means of achieving each in mind.

At an early stage of the development, the design teams should ask questions such as, "How reliable will my product be?" "How reliable should my product be?" And, "How frequently does the product need to repaired/maintained?" To answer these questions, the design team needs to develop an understanding of how and why their products fail. They should develop a means and

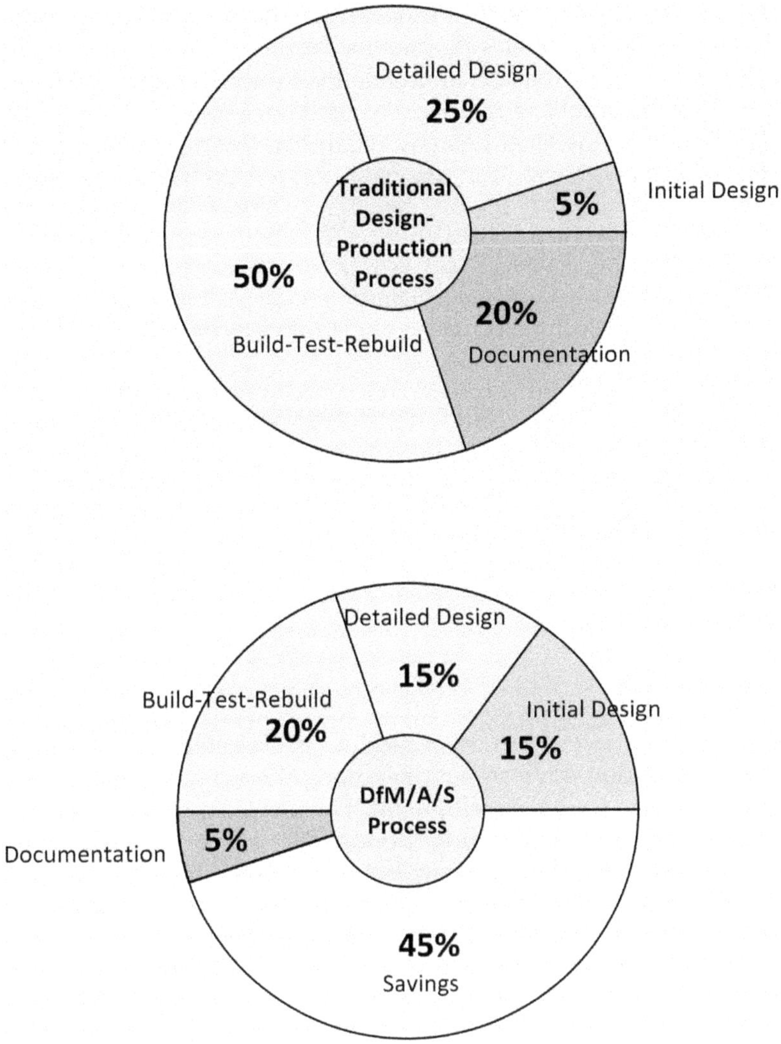

FIGURE 14.20
Estimated cost comparisons and the savings associated with DfM/A/S

understanding to predict these failures and their frequencies in future. The set of activities used to accomplish this is called *design for reliability* (DfR). It involves both mathematical modeling as well as testing and analysis. Reliability should not be mistaken for robustness (Jamnia and Atua 2020). Robustness is the ability of a product to function as expected under the influence of noise factors, whereas reliability is defined as the probability of a product or device performing adequately for the period of time intended under the operating conditions encountered. Typically, for consumable products, robustness is a more significant element than reliability, which is usually discussed in the context of electromechanical products. In this section, I will review elements of DfR. I will review robustness later in this chapter.

Reliability concerns those characteristics that are dependent on time: stability, failure, mean time to failure/repair, etc. If harm or injury comes as a result of poor reliability, the manufacturer or distributor may have a responsibility to compensate for these losses and/or injuries. This is a matter for the legal system to settle.

Putting the law and litigation aside, another, albeit unofficial, definition of reliability is "the willingness and the ability to maintain a customer's good faith and business." So, it may be suggested that if a product has low reliability, the distributor and the manufacturer still suffer – even if no harm or injury has occurred. As Taguchi and Clausing (1990) have explained:

> When a product fails, you must replace it or fix it. In either case, you must track it, transport it, and apologize for it. Losses will be much greater than the costs of manufacture, and none of this expense will necessarily recoup the loss to your reputation.

Design for reliability is an extensive field concerned with understanding and managing failures in equipment and systems (MIL-HDBK-338B 1988; IEEE Std. 1624 2008). There are, in general, three aspects of reliability that concern electromechanical design engineers, namely, mechanical, electrical, and chemical issues.

Additionally, shelf life, chemical reactions, solderability, moisture, aging, etc., have important influences on reliability as well. Furthermore, the reliability of the software controlling the device should not be ignored. A discussion of these other factors, however, is beyond the scope of this book.

Product Use Profile

A use or mission profile describes how a product is expected to be used. Typically, use profile along with operation environment is among the very first items discussed when a new concept product is under development. For instance, suppose that a company is developing a new minivan. Furthermore, suppose that a technical team is discussing the design of the brake system for this minivan. Before deciding on what the shape, configuration, and materials used in the system, the team should understand the driving behaviors of minivan driver. Based on market research, they may recognize that 50th percentile of the driver population applies their brakes just over 320,000 times in a ten-year period; whereas the 90th percentile of the same population apply their brakes about 530,000 times during the same period (Dodson and Schwab 2006). The implication of this comparison is to contemplate which of these two usages should the design support. Clearly, there are lower levels of stress associated with the lower use. A product developed for the brake at the 50th percentile use will probably demonstrate higher failure rates at higher stresses.

Typically, power consumption, environmental conditions, customer usage, and even transportation conditions are included in the description of a product's usage.

Failure Modes and Mechanisms

Reliability is, in a way, a study of failure; hence, it is important to first define failure. Failure is defined as the inability of a system/subsystem/assembly/component to meet its design specifications. It may be that it was poorly designed and never met its objectives, or that it initially met its design objectives, but after some time it failed. In these two scenarios, clearly, there have been certain overlooked factors.

> Failure is defined as the inability of a system/subsystem/assembly/component to meet its design specifications.

Many factors lead to failures of a fielded product: some may be caused by a lack of proper design or verification activities, whereas others could be attributed to misuse, and still others may be a result of the designer's lack of insight into potential influencers. FMEA helps the design team to clearly identify sources of failure and indicate whether the solutions may be sought in mechanical, electrical, or other causes. In general, there are four causes of mechanical failures. These are as follows:

1. Failures by elastic deflection – These failures are caused by elastic deformation of a member in the system. Once the load causing the deformation is removed, the system functions normally once again.
2. Failures by extensive yielding – These failures are caused by the application of excessive loads where the material exhibits a ductile behavior. Generally, the applied loads and the associated stress factors create stress fields that are beyond the proportional limit and in the neighborhood of the yield point. In these scenarios, the structure is permanently deformed, and it does not recover its original shape once the loads are removed. This is generally a concern with metallic structures such as chassis and racks.
3. Failures by fracture – These failures are caused by the application of excessive loads where the material exhibits a brittle behavior or, in ductile materials, where stresses have surpassed the ultimate value.
4. Progressive failures – These failures are the most serious because initially the system passes most, if not all, test regiments, and yet after some time in the field, it begins to fail. Creep and fatigue belong to this category.

In electronics and in general, PCBA and electronics (sub)systems exhibit three modes of failure. These are open circuits, short circuits, and intermittent.

1. Open Circuit – When an undesired discontinuity forms in an electrical circuit, it is said that the circuit is open. This is often caused by solder cracks due to a variety of thermomechanical stresses. Another contributor to developing this failure mode is chemical attack (such as corrosion).
2. Short Circuit – When an undesired connection forms between two electrical circuits or between a circuit and ground, it is said that the circuit is shorted. Major contributors to circuit short circuits are based on chemical attacks such as dendrites and whiskers or contaminants left behind from the manufacturing process combined with humidity and high temperatures. Short circuits may also occur under vibration if the PCBAs are not properly secured.
3. Intermittent Failure – An intermittent failure is a microscopic open (or short) circuit that is also under other environmental influences. For instance, a crack in solder may be exacerbated under thermal loads, leading to an open circuit failure; however, as the board is removed from its environmental conditions and cools down, the gap closes, and it would work as intended.

Failure mechanisms in electronics packages have been well studied over the years. Failure mechanisms may be attributed to

1) design inefficiencies,
2) fabrication and production issues, and

3) stresses such thermal, electrical, mechanical, and environmental factors such as humidity.

In addition to the two main mechanical and electrical influences on failure, a third and, to some extent, less considered influence is chemical and electrochemical factors. This may be divided into two broad categories:

1. Corrosion is a natural two-step process whereby a metal loses one or more electrons in an oxidation step, resulting in freed electrons and metallic ions. The freed electrons are conducted away to another site where they combine with another material in contact with the original metal in a reduction step. This second material may be either a nonmetallic element or another metallic ion. The oxidation site where metallic atoms lose electrons is called the anode and the reduction site where electrons are transferred is called the cathode. In electromechanical systems, corrosion is often caused by galvanic cells, which are formed when one of the following two criteria is satisfied: two dissimilar metals exist in close proximity, or the metal is a multiphase alloy in the presence of an electrolyte.
2. Migration is a process by which material moves from one area to another area. If migration happens between two adjacent metals, such as copper and solder, it is called diffusion. If it happens as a result of internal stresses and in the absence of an electric field, it is called whiskers. Finally, if it happens in the presence of an electric field and between similar metals, it is called dendritic growth. Two factors have given this topic relevance in PCB and system package design. The first factor is that as the PCBs are more densely populated, it becomes exceedingly difficult to clean and wash away all the processing chemicals. The presence of the pollutants along with an electric field provides an ideal environment for electromigration and dendritic growth. The second factor is more related to legislature and a lead-free environment. Pure tin solder, which is a natural replacement for a leaded solder, has a tendency to grow conductive needles, known as whiskers, in an out-of-plane direction (z-axis).

Common failure modes and mechanisms in electromechanical devices have been reviewed to some extent in Jamnia (2025).

Life Expectancy, Design Life, Reliability, and Failure Rate

Earlier, I suggested that reliability is, in a way, a study of failure. Although this statement is true, it is incomplete. Reliability is also about product useful or effective life.

Often, *life expectancy* is equated with *design life*. Design life is defined as the length of time during which a product/system/component is expected to operate within its specifications. However, this definition appears to suggest that failures do not happen, although that is not the case at all. For this reason, some are inclined to use *operating life expectancy*, or *operating life* with the definition that it refers to a length of time that a product/system/component operates within its *expected failure* rate.

> Life expectancy or design life is defined as the length of time during which a product/system/component is expected to operate within its specifications.

In its most general term, the relationship between reliability, failure rate, and time for a population of fielded products may be defined as

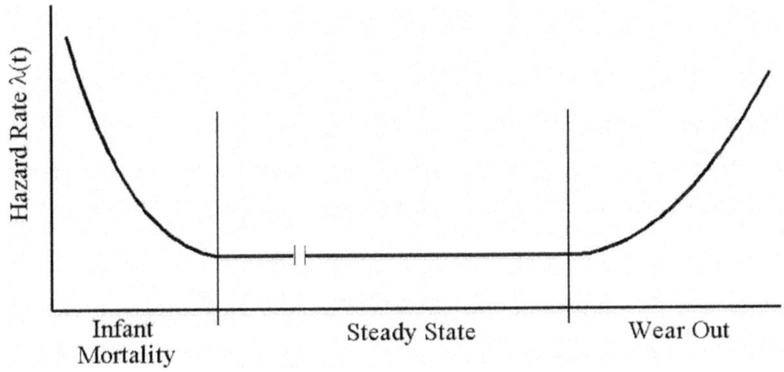

FIGURE 14.21
Hazard rate versus operating time

$$R = e^{-\lambda t}$$

where R is defined as reliability of a component or a part, λ is defined as the failure rate, and t is time. This represents one equation and three variables. If λ is to be calculated, both R and t should be known. There are two ways to interpret this equation. First, for a single device, as time goes by, its reliability drops. Thus, for a given constant failure rate, there is not a single value that denotes the life expectancy of a single product. For a given time, there is a certain probability that the product functions as expected. The next time the same product is used, that probability is somewhat less. For a population of products, life expectancy refers to a time period when a substantial increase in the failure rate is observed; thus, the failure rate is no longer a constant value. This is shown in Figure 14.21. *Hazard (or failure) rate* [λ(t)] is the instantaneous rate of failure of a population that has survived to time t.

There are three regions in this curve:

1. Infant mortality – The initially high but rapidly decreasing hazard rate corresponding to inherently defective parts. This portion is also called early mortality or early life.
2. Steady state – The constant or slowly changing hazard rate.
3. Wear out – This generally occurs in mechanical electronic parts (such as relays) or when component degradation and wear exist. In this stage reliability degrades rapidly with ever-increasing hazard rates.

Associated with failure rate, *mean time to failure (MTTF)* is also a term that is commonly used in reliability calculations. It may be shown that for a constant failure rate (λ), we have

$$\text{MTTF} = \frac{1}{\lambda}$$

Product Reliability Modeling

There are two aspects of modeling the reliability of a product. First, the reliability of a product may be modeled using theoretical values based on feedback from various industries and generic databases. The second is through information obtained from test data and predicate devices.

Theoretical Modeling

There are a number of ways to model the reliability behavior of a product. Three common models are the *exponential distribution*, the *lognormal distribution* and the *Weibull distribution* models. Unfortunately, there is no simple answer to the question of what is the right distribution to choose. The chosen distribution model should represent the failure behaviors of the product of interest. This information is not typically available at the onset of the project; however, it should be obtained systematically and collected as the design is developed and progresses from its conception to its completion.

> Three common distributions to model the reliability behavior of a product are *exponential, lognormal,* and *Weibull.*

There are reference documents and software that may provide expected (or typical) values for active and/or passive electronic components – similar to MIL-HDBK–217. There are also databases that provide similar information for PCBA assemblies based on industry-reported data. Based on this information, it is possible to develop a theoretical reliability model for the device under development. In many instances, a product's failure rate is an algebraic sum of its components failure rates (including certain stress factors). For a series network, the failure rate of the system may be calculated as follows:

$$\lambda_{SS} = \pi_E \sum_{i=1}^{n} N_i \lambda_{G_i} \pi_{Q_i} \pi_{S_i} \pi_{T_i}$$

where π_E is the environment factor, and N_i is the number of each component in the system. λ_{G_i} denotes the generic value of the failure rate, which is generally provided in handbooks and supplier published data, π_{Q_i} is the quality factor as defined previously; and π_{S_i} is electrical stress factor. π_{T_i} is the thermal stress factor. This approach provides the strictest sense of reliability calculation, in which the failure of any component will flag the failure of the entire system.

Any detailed explanation of reliability modeling is beyond the scope of this book. For a simple description and more details, see Jamnia 2025 or Jamnia and Atua 2020. Other good resources include *Practical Reliability Engineering*, 5th edition (O'Connor and Kleyner 2012), *AT&T Reliability Manual* (Klinger et al. 1990), *Practical Reliability of Electronic Equipment and Products* (Hnatek 2003), *Reliability Improvement with Design of Experiments* (Condra 2001), *Accelerated Testing, Statistical Models, Test Plans, and Data Analysis* (Nelson 2004) and MIL-HDBK-338B (1988).

Reliability Testing

It is possible to develop a best-guess estimate for both the expected reliability of the product and a preventive maintenance schedule. These estimates will be based on the reliability block diagram of the product and whether there are redundancies, the assumption of the exponential failure distribution, and the use of generic and published failure rates or predicate devices. However, as mentioned, this would only be an estimate. It is through systematic reliability testing that we can verify whether or not these estimates are acceptable and the made assumptions appropriate.

> The purpose of a *highly accelerated limit test* (HALT) is to identify all the potential failure modes of a product in a very short period of time. The purpose of *highly accelerated stress screening* (HASS) is to identify faulty units in manufacturing.

Categorically, disproportionate failures are rooted in one or a combination of the following areas: poor design or design elements, defective materials, weak manufacturing techniques (i.e., excessive variations) and/or poor service procedures. Proper reliability testing exposes these deficiencies.

In general, there are two types of reliability testing, namely, those tests that seek to induce failures and those that seek to identify problematic products. Typically, in the design stages of the product development, the reliability engineer tries to gather as much information from testing to failure as possible. The HALT is used to push components, assemblies, and eventually the finished product well beyond their stated limits and usage. Once the product is launched and manufacturing becomes the primary activity in the product's life cycle, batches of finished products are subjected to HASS. The purpose of this test is to separate flawed products from robust units.

The purpose of a HALT is to identify all the potential failure modes of a product in a very short period of time so that the design team would take one of the two actions: either to improve the design by replacing the failed component with a more robust one or to develop a service maintenance plan to replace a less reliable component on a regular basis.

The purpose of highly accelerated stress screening (HASS) is to identify faulty units in manufacturing. This is done in an effort to eliminate the early (or infant) mortality failures. The applied stresses in HASS are similar to HALT, but their levels and duration are much shorter. The reason is that good units should not be damaged during the screening process, while faulty units should become noticeable. These defective units may have faulty components or have been assembled with subpar manufacturing processes. HASS enables engineers to understand whether excessive manufacturing or process variations exist.

A major difference between HALT and HASS is that in HALT, the product is tested in small quantities, whereas in HASS, 100% of production quantities are tested. Therefore, larger equipment is needed.

It is important now to emphasize that although both HALT and HASS are tools within the reliability tool kit, they do not provide the information needed to make predictions. The primary reason is that the induced failures are the results of accumulated stresses at various levels. For this reason, life and/or reliability predictions should be based on other accelerated life tests (ALT). In a simple ALT, generally two or more levels of one stress (e.g., temperature) are applied as follows. One stress is applied at a constant level until failure is induced. The time of failure and the level of stress are recorded. Then, the same stress at a different level is applied until failure is obtained. Again, the time of failure and this different level of stress are recorded. By fitting the data into an appropriate model, life expectancy at nominal conditions may be calculated.

Reliability Allocation

Once the theoretical model prediction as well as product testing is completed, it may be possible to identify the contribution of component failures to assembly failure and the contribution of assembly failures to subsystem failures. This would constitute an apportionment table.

It would then be a straight forward calculation to cascade the system-level reliability requirements into subsystems using the apportionment table. The steps for this flow-down are as follows.

1) First, calculate System failure rate based on the System mean time to failure (MTTF) requirement:

$$\lambda_{System} = \frac{1}{MTTF_{System}}$$

2) Next, multiply the percentage of apportionment for each subsystem with the system failure rate to get the subsystem failure rate:

$$\lambda_{subsystem} = \%\ Apportionment \times \lambda_{System}$$

3) Finally, calculate each subsystem's MTTF as the inverse of the subsystem failure rate:

$$MTTF_{subsystem} = \frac{1}{\lambda_{subsystem}}$$

As an example, consider Table 14.1. The reliability profile for a hypothetical system with seven subsystems is presented. The first data column in this table is representative of data collected from either reliability testing or information collected from predicate devices or field data. The middle data column is the information that is based on predictive theoretical models. The last column (on the right) is a hybrid of the two. In this example, it is an average of the other two; however, in many cases, other techniques such as gamma conjugate may be used to combine these values. Why combine the two? Simply because the theoretical is representative of possibly an ideal state and the test/field data representative of the current state. By choosing a middle ground, realistic goals for improvements may be established.

Note that all columns should add up to 100%.

Once a budgeting cascade is determined, it may be used to develop subsystem reliability allocations. For this example, suppose that the system-level MTTF requirement is 50 months. Thus, the system failure rate is

TABLE 14.1

Example of Failure Apportionment for Subsystems

Subsystem	From test/Field Data	From Prediction	Cascading Values
Controls	30%	33%	31.5%
Wireless	25%	12%	18.5%
User Interface	23%	34%	28.5%
Hydraulics	10%	10%	10%
Receiving Module	7%	8%	7.5%
Wiring	3%	2%	2.5%
Caseworks	2%	1%	1.5%
Total	100%	100%	100%

Note: Note that all columns should add up to 100%.

TABLE 14.2

Example of System to Subsystems

Subsystem	Cascading Values	Subsystem λ	Subsystem MTTF
Controls	31.50%	31.5% x 0.02 = 0.0063	159
Wireless	18.50%	18.5% x 0.02 = 0.0037	270
User Interface	28.50%	0.0057	175
Hydraulics	10.00%	0.002	500
Receiving Module	7.50%	0.0015	667
Wiring	2.50%	0.0005	2000
Caseworks	1.50%	0.0003	3333

$$\lambda_{\text{System}} = \frac{1}{\text{MTTF}_{\text{System}}} = \frac{1}{50}$$

$$\lambda_{\text{System}} = 0.02$$

Based on this value and the apportionment in Table 14.1, the MTTF (in months) for subsystems may be calculated as shown in Table 14.2.

Reliability Requirement Cascading

The simple reason for using this approach is that all requirements cascade from system to subsystems should be traceable using a transfer function. In Chapter 9, quality function deployment was used as a means of cascading from high-level user needs to system and subsystem requirements. Later, the application of engineering transfer functions was demonstrated, and now the same methodology is used for cascading reliability requirements.

Other Aspects of Reliability

MIL-HDBK-338B (section 7.1) defines reliability engineering as

> the technical discipline of estimating, controlling, and managing the probability of failure in devices, equipment and systems. In a sense, it is engineering in its most practical form, since it consists of two fundamental aspects:
>
> (1) Paying attention to detail
> (2) Handling uncertainties

However, merely to specify, allocate, and predict reliability is not enough.

This handbook goes on to recommend that each organization should develop a series of design guidelines that design teams may access and use to ensure product reliability at the design stage because once the design is frozen, making changes is difficult and costly; and, once the product is launched and marketed, making any meaningful changes is nearly impossible due to associated high costs.

Design teams typically develop the initial reliability assessment inputs, such as operational and use profiles, followed by FMEA, fault tree analysis (FTA), and other similar tools. As the design and development progress, the team focuses on analyzing the details of the design by utilizing FEA for component mechanical stress analysis or circuit design and part derating analysis. Eventually, the team's focus shifts toward the initial build and verification testing.

The reliability aspects that may be missed by the team are component management and obsolescence. Unfortunately, both of these areas are often treated haphazardly by the sustaining team and within a change management process – if one is in place. Thus, it is highly recommended that as the detailed design progresses, three documents be developed. These are the preferred components list, critical and/or safety components list, and finally preferred supplier list. These documents need to be developed collaboratively among components, reliability, design, and manufacturing engineering alongside purchasing and supplier quality teams (MIL-HDBK-338B 1988).

In Chapter 12, in the discussion of naming convention, *parts reuse* was mentioned, and its impact on purchasing ability was discussed. Another motivation for establishing a parts database is reliability concerns. A parts database enables the design team to identify component reliability measures along with functional and physical requirements. Furthermore, part equivalency may be established early on that would ease the burdens of proving part replacements should a component become obsolete.

The critical and/or safety components list is a reflection of the DFMEA file, where component failures may have a determinable impact on the function or safety of the product. In regulated industries where listings with agencies such as Underwriters Laboratories (UL) or TUV are required, any critical component changes require notification and possibly technical file updates. Typically, the compliance team owns this list; however, by having a clear understanding of this list, the design and reliability teams select these critical components not only based on the part's useful life failure rate but also on its infant mortality as well as wear-out behaviors.

Finally, suppliers should provide quality and reliable components. In larger corporations, supplier quality teams audit and approve vendors prior to order placement and purchasing. In smaller corporations, purchasing or quality teams may conduct this audit. Figure 14.22 proposes a typical process that includes reliability as a component of selection criteria.

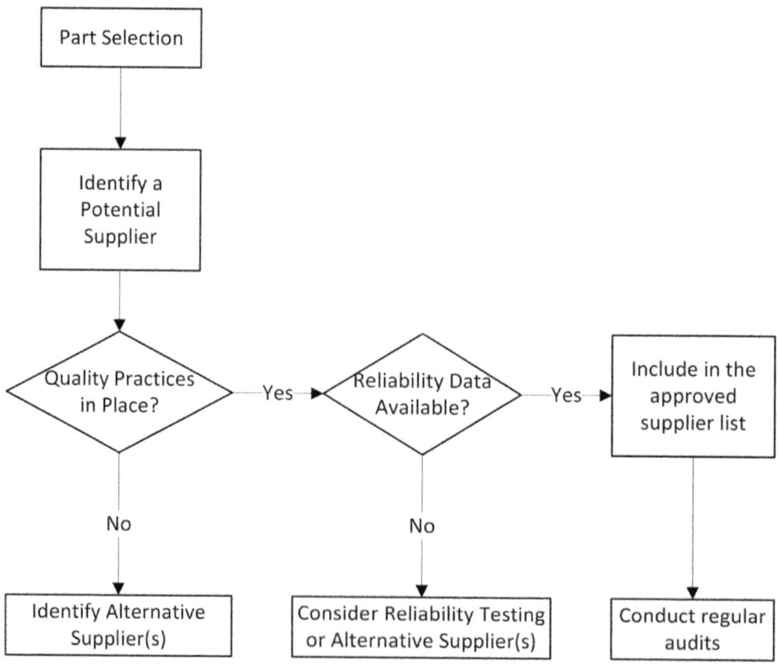

FIGURE 14.22
Vendor selection criteria including reliability

Reliability Planning and Execution in Design Process

Figure 14.23 provides a pictorial view of the reliability planning process and its relationship with the V-model. As depicted here, design for reliability begins at the product requirements level when the operational and use profile as well as customer expectations are reviewed, discussed, and approved. Next, reliability requirements based on customer expectations as well as past performances of competitive products are established. It is common that initially, reliability goals are set. Once the design is established and these goals demonstrated, then documents are updated and goals converted into requirements. Furthermore, a reliability growth typically within 24 or 36 months is also anticipated.

Once components have been selected or designed, theoretical reliability models may be developed based on the bill of materials and industry standards such as MIL-HDBH-217 or Telecordia. At this step, a deeper understanding of system and subsystem behavior is developed by allocating the system-level reliability requirements to each subsystem. This information may dictate a redesign of a part or an entire assembly.

The next steps are to develop a reliability test design and include the number of needed samples in the first (or subsequent) builds. By conducting HALT, expected failure modes may be verified and any potential new ones uncovered. As a result of these HALTs, DFMEAs may need to be updated, or components or assemblies may need to be redesigned.

Once these iterative steps are completed, accelerated life testing may be conducted to establish product life expectancy and evaluate service maintenance schedules. Finally, reliability demonstration testing is conducted to provide objective evidence that reliability requirements can be met.

Design for Robustness

Up to this point in this chapter, our focus has been on electromechanical products. However, as the product development world evolves, we see an integration of either consumable or disposable products into electromechanical products. A rudimentary example is the ink cartridge used in printers. Once the ink is consumed, its container is disposed of. Yet, as a product, it is a subsystem to the printer and is hence subject to the same DfX mindset as the rest of the system. The question is how does the design team evaluate design for reliability for a product that is either disposable or is consumed over a short period of time? To answer this question, we need to look at reliability versus robustness and understand their difference.

Robustness versus Reliability

Before diving into the details of robustness, it is important to know of the interrelationship between robustness and reliability. The classic definition of reliability is the probability of a design performing its intended function under defined operating conditions for a specified period of time. This definition implies that reliability is a function of time, and by merit, the reliability of a product is expected to degrade and decline as the product or device ages. Eventually, the product loses the capability to support the applied stresses, and it fails.

Robustness is different from reliability in that the element of time does not play a role. A Google search of this word gives the following definition: "the ability to withstand or

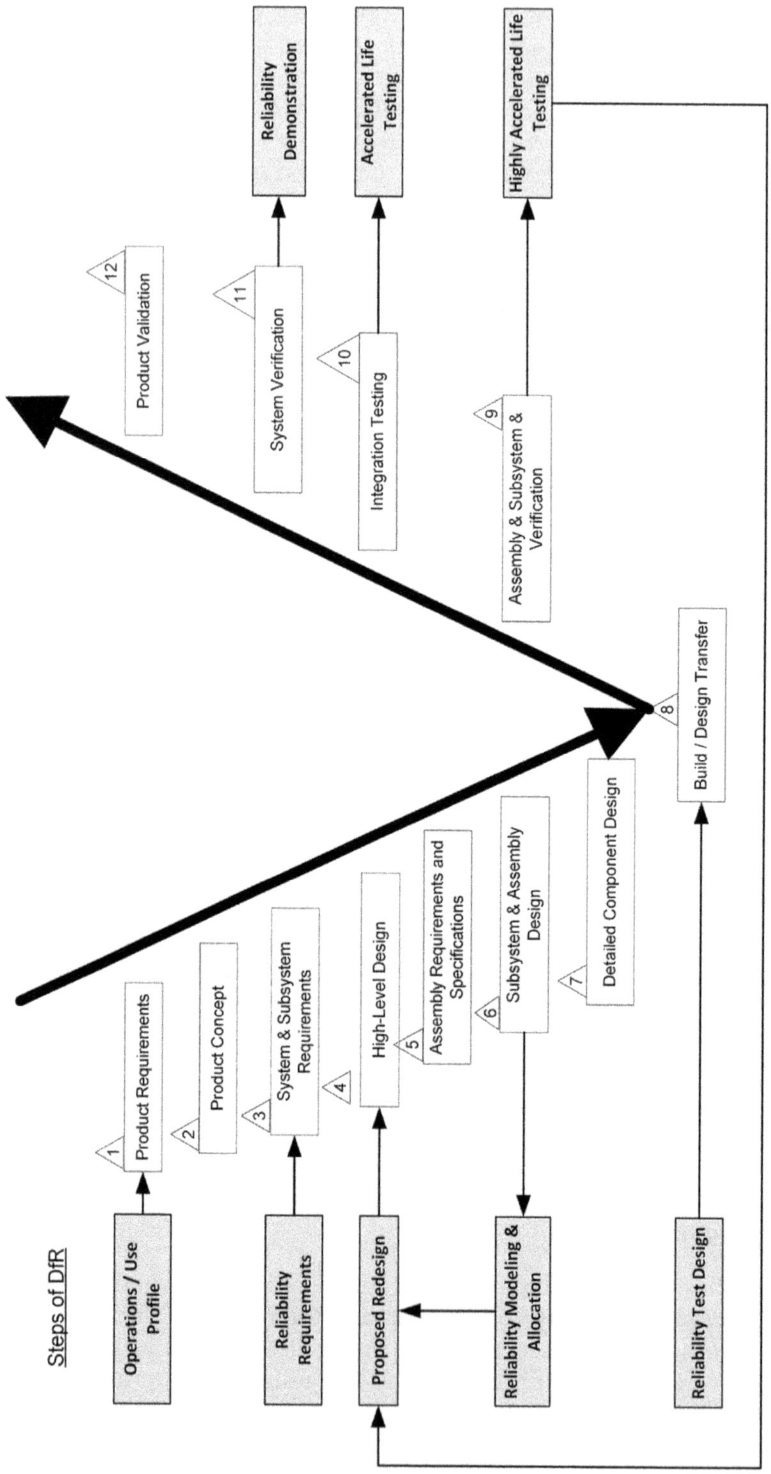

FIGURE 14.23
DfR and the V-model

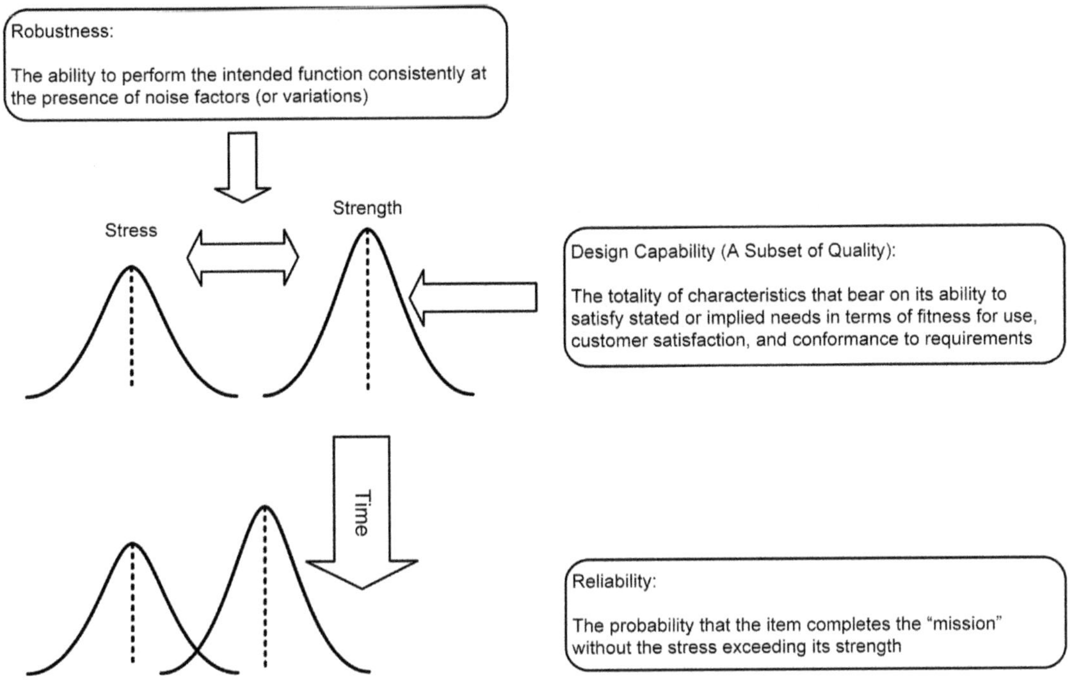

FIGURE 14.24
Relationship between design capability, robustness, and reliability in terms of stress, strength, and time

overcome adverse conditions." In the context of product development, robustness is a product's ability to perform as intended without eliminating variations that influence it. For instance, adhesive left in its bottle in a hot summer sun does not thicken and enables a carpenter to securely glue two pieces of boards together.

In simple engineering terms, robustness is having adequate *design margins*. Design margins are the threshold settings that define the design, and yet the design remains impervious to them. Also called *noise factors*, they include factors such as piece-to-piece manufacturing variations, customer usage, and interaction with other systems. A design is not robust if it is inherently sensitive to variations and uncertainties. In essence, robustness is *the ability of the design to perform its intended function consistently in the presence of noise factors.*

In Figure 14.24, reliability and robustness are depicted graphically, and a related concept called *design capability* – a subset of quality – is introduced. Design capability is the totality of a product's characteristics to meet its intended use and requirements. The relationship between capability and stress defines robustness. As time goes by, due to a variety of factors, a design's capability degrades, and eventually, the stress distribution overlaps a product's capability distribution. The common area between the two defines the probability of failure or unreliability. As the degradation increases, so does this probability.

As product developers and designers, our goal is to bring to market a product that is both robust and reliable. Uncovering robustness issues during the design process requires that relatively large samples be evaluated under a variety of use and environmental conditions. Whereas, discovering reliability issues necessitates testing the product over a period of time – in presence of noise factors – to allow for increasing the chances of lowering a product's strength and its interference with stresses. In essence, reliability testing is testing robustness over time.

Chemical Stability and Shelf Life Studies

As I mentioned previously, a design's capability degrades over time. This degradation is more significant for disposable and/or consumable products as compared to electromechanical products. In particular, three questions need to be answered:

1) How long will my consumable (i.e., chemical or biological) product remain stable in a closed container? In other words, the container has not been opened.
2) Will my consumable interact with its container? And if so, how long will the combination maintain its design's capability?
3) How long will my consumable product remain stable in an open container? In other words, the container has been opened and partially used.

To answer these questions, accelerated aging studies are conducted to evaluate the design capabilities during shelf life. Depending on the study being conducted, samples may be removed at specific time intervals to be assessed for degradation and shelf-life determination. It is commonplace to use a form of the Arrhenius equation called Q_{10}. Simplistically speaking, Q_{10} is an acceleration factor for a reduction of shelf for every 10°C increase in temperature. In other words, life decreases by a factor of Q_{10} for every 10°C increase in temperature. For instance, assume that a product has a shelf life of 12 months at a room temperature of 25°C and a Q_{10} value of 2. At a temperature of 35°C, its shelf life will reduce to 6 months; and, at a temperature of 45°C, its shelf life will reduce to 3 months. Mathematically, this is expressed as

$$L_A = \frac{L_S}{Q_{10}^{\frac{T_A - T_S}{10}}}$$

L_S is life under storage conditions, L_A is accelerated life. T_A and T_S are accelerated and storage temperatures (°C), respectively. Q_{10} is often assumed to be 2. For a more detailed treatment of Q_{10} and implications for its value, see Hemmerich (1998).

I like to note that temperature is not necessarily the only accelerant for aging polymers and chemicals. Other factors such as moisture and light may play significant roles as well, though treatment of these topics is beyond the scope of this work. For more information on the impact of moisture, see Thor et al. (2021), and for the impact of light see Feller (1994).

Summary

- Design for X should not be an afterthought. Manufacturing, service, and reliability, as the three major stakeholders, should be included in the design process as principal stakeholders.
- There is a strong link between the design of a component and its manufacturing process, and subsequently the class of specified materials to be used.

- DfM is really a design-material-process triad relationship.
- The concurrency design specifications, manufacturing, and material selection would drive a detailed component design to the pinnacle of optimum simplicity of configuration, quality, and cost.
- To a large extent, design for assembly is really a set of guidelines that design engineers need to follow to enable ease of assembly.
- Mistake-proofing or poka-yoke is to anticipate the mistakes that operators can make and to mitigate them.
- Reliability is defined as the probability of a product or device performing adequately for the period of time intended under the operating conditions encountered.
- Design for reliability is an extensive field concerned with understanding and managing failures in equipment and systems.
- The purpose of a HALT is to identify all the potential failure modes of a product in a very short period of time.
- The purpose of HASS is to identify faulty units in manufacturing.

Exercises

1. Continue the design work for the concept used in the Chapter 13 exercise.
   a. Conduct a DFMA.
   b. Is the design easily manufacturable?
   c. Is the design serviceable?
   d. Is it possible to reduce part count with redesign?

Notes

1 I do understand that associated with this move, other concerns, such as quality, may be raised. This, however, is the subject of another conversation.
2 This is a play on words: fX becomes f(X). In most basic mathematical equations, X stands for the unknown to be solved.
3 Even injection molding tools are complicated because of the presence of the holes and chamfers requiring side actions.
4 Or a chemical component with a certain purity.

Part V

Preparation for Product Launch

15

Cost Requirements Cascade and Purchasing

Introduction

At times, I believe design teams are kept between a hammer and an anvil. On the one hand, we are told that as engineers and/or scientists, we should not worry about cost; and, in the same breath, we are told that we should keep the design within budget. Which is it?

The answer to this question depends very much on the stage of product development. In the ideation stage, where new technologies are created, costs that are within the overall budget are of little concern. The economical and long-term strategic goal is to situate the business at a competitive advantage at a future date. Once a product concept advances beyond ideation and is placed in the product development track, cost is placed front and center.

An appreciation of cost issues begins at the market and what the end users are willing to pay for a particular product. In a general sense, there are three categories of price points, as shown in Figure 15.1. These may be called economy, market, and premium prices. Effectively, these levels of pricing refer to the customers' expectations and their willingness to pay for particular goods or services. What the majority of a target market for a business is willing to pay for the offered product is called the *market price*. This population tends to set the expectation not only for a product's price but also for its perceived quality, reliability, and other performance metrics. Associated with the target market, there are two other distinct groups. The first may be categorized as the *bargain hunters* and are drawn to the economy side of pricing. This group is characterized by their willingness to sacrifice some features of a product for a price reduction. For instance, instead of having cosmetic features such as touchscreens, they may be willing to settle for knobs and keys if they could be given substantial savings.

> An appreciation of cost issues begins at the market and what the end users are willing to pay for a particular product.

The second group is on the premium side of the market. They may be characterized as the people who are willing to pay extra to ensure that the product that they purchase is a *quality* product. For this group, there have to be distinguishing features between what they select and the rest in the market. Ironically, these expectations are not static (or fixed) and may vary from geographic region to region.

Cost, Price, and Revenues

The marketing team's role is to set the direction and provide the VOC insofar as the price point for the product to be developed. Associated with the price, there are projected sales

DOI: 10.1201/9781003564508-20

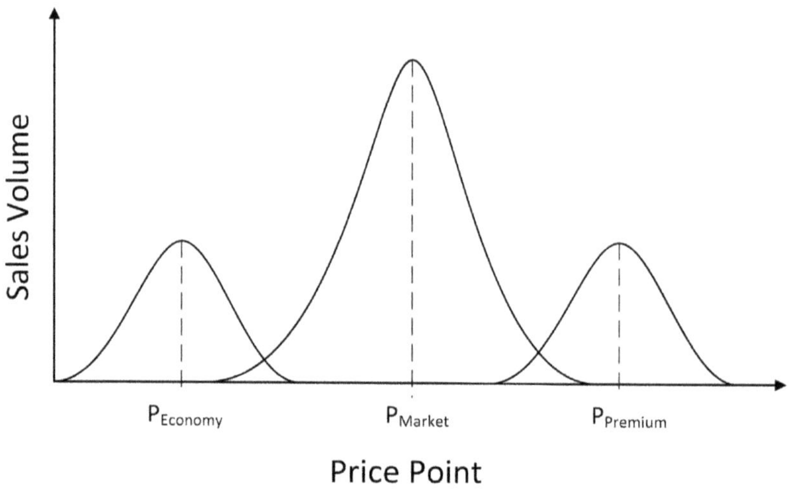

FIGURE 15.1
The relationship between price point and sales volume

volumes typically in the first three or five years. On the basis of sale price and product volumes, an estimate of monthly, yearly, and five year revenue may be developed.

Once the revenues are calculated, the overall costs are estimated.[1] A large portion of the costs are allocated to the distribution channel and sales teams. Next, the impact of development, fixed overhead, warranty, and other costs are evaluated and subtracted. Finally, two other costs are also calculated and subtracted from the revenue. These are the expected profits, and the second is the cost of investment in the product. This last one is somewhat difficult to explain: suppose that an investor has an initial capital of $1,000,000. This investor may be able to invest in a safe investment and have a rate of return of 5% a year. Conversely, the same investor may invest in a high-growth stock and anticipate a return of, say, 20%. By investing in the product, however, the expectation is that a greater than 20% return may be realized. Once these calculations are done, one may realize that the cost of a product is often 10% to 20% of its sales price. Figure 15.2 provides a pictorial (albeit simplistic) view of various factors that contribute to the profitability of an organization. The simplicity of the model allows it to be applied as equally to an organization, enterprise, or even a product line.

Cost and profit modeling can be a rather complex and dynamic field. There are scenarios in which the cost of a product may not be reduced sufficiently for profitable sales; however, the same product may facilitate the sale of lucrative ancillary products.

Figure 15.2 is only a simplistic view of a rather complex and dynamic field. There are scenarios in which the cost of a product may not be reduced sufficiently for profitable sales; however, the same product may facilitate the sale of lucrative ancillary products. In this case, the overall financial gains may warrant giving away the device itself at lower than the cost of manufacture or even for free or allow for other creative marketing approaches to ensure that the total package, that is, the device and the ancillary products, is profitable. The so-called razor and razor-blade model is a well-known example of this mindset, where razors are given away freely in order to secure the sales of the blades.

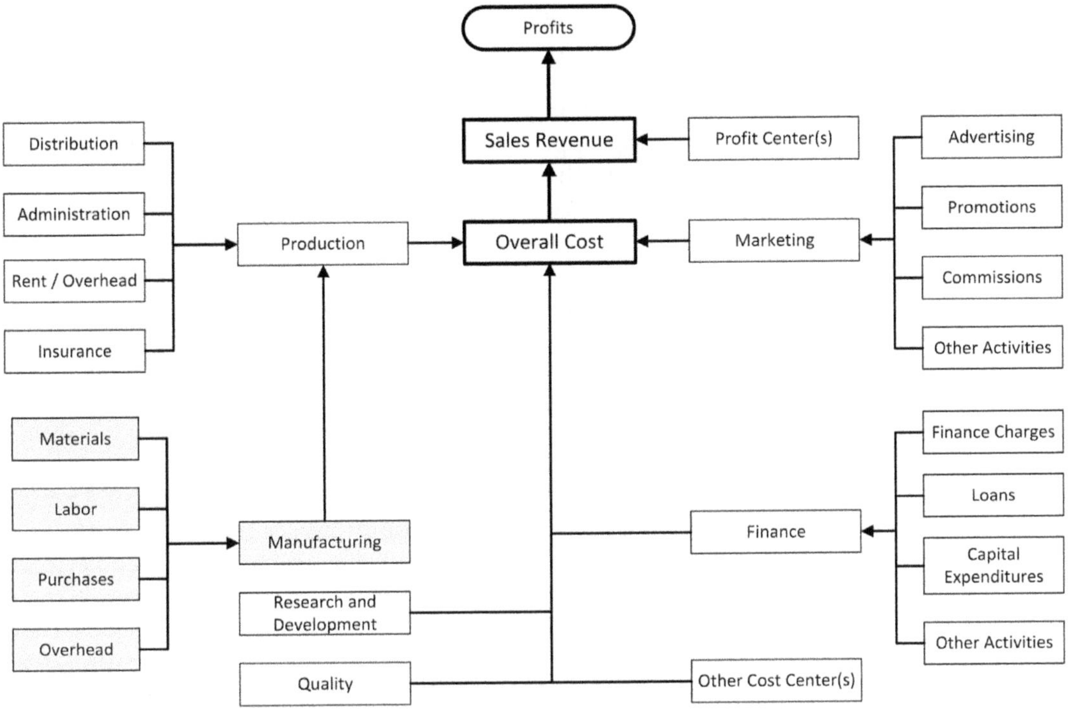

FIGURE 15.2
A simplistic view of various cost points of an organization and their relationship to profits

Cost of Manufacture and Its Structure

> Cost ceiling should be considered as a product requirement and be cascaded into various subsystems, assemblies, and components.

The point of the previous argument is that the relationship between the price of a product and the cost of its manufacture can be very complex (Nagle and Holden 2002). To a large extent, setting the selling price of a product depends not only on the cost of its manufacture but also on other market factors. Although a discussion of these market factors is beyond the scope of this book, an understanding of product cost is one of the major elements of product development activities.

In Chapter 14, topics of design for manufacturability and/or serviceability were discussed, and the relationship between cost and design – as expressed in manufacturing and service – was explored. Because of the potentially complex price-cost relationship, in addition to providing the VOC, the role of marketing is to determine an acceptable price along with a ceiling for manufacturing cost. This price ceiling helps the development team make decisions on the direction that the design of the product takes.

Figure 15.3 provides a pictorial view of when and where cost factors should be considered within the product development cycle. Just like any other requirement, there need to be cost ceiling requirements at the product level. Having this ceiling enables the product team to stop the program and financial bleeding early in the program when cost estimates developed at later stages exceed this upper threshold.

When product concepts are developed and considered, the cost ceiling may be allocated to each subsystem and evaluated based on predicate devices and prior experience. At high-level design (step 4 of Figure 15.3), even though the details are not clear, a direction for device development design and concept is set. This means that the design team makes decisions in regard to manufacturing approach; for instance, will injection molding or casting be used to create components with complex shapes? What roles may other manufacturing approaches, such as stamping play? On the basis of these estimates, the approximate number of new tools may be identified and their associated costs. This information would then enable a budget determination during capital expenditure budget planning. As more details become known, estimates of component costs may be developed, and eventually exact numbers may be established, and the product moves toward the first build and the eventual transfer to manufacturing. Needless to say, this is an iterative process where the interaction of various factors may influence changes in costs. As an example, consider that a selected component design may not pass a particular design and/or, say, reliability requirement. In this case, a design change may necessitate a cost update. These changes are expected and should not be discouraged. However, the impact on the cost ceiling must always be evaluated, and if necessary, the appropriate compromises should be made.

For clarity's sake, I would like to take a step back and review the difference between the cost of a program and the cost of a product. New program development may be canceled due to budget overruns on either of these two. As shown in Figure 15.4, a program's cost depends initially on design and development activities, fabrication of prototypes, and expenses associated with testing and verifying them. As a product launches into the market, the source of expenses shifts from development costs to maintenance activities. The expenses due to sustaining a product focus primarily on distribution and operations challenges in various markets, followed by warranty repairs along with the labor costs of completing design change.[2] Program costs are typically associated with R&D budget.

As shown in Figure 15.4, the cost of a product, which is typically allocated in manufacturing budgets, depends on the total number of tools required, the number of purchased parts (buy items on the bill of materials), material costs for make items, labor, and overhead (or fixed) costs. Labor costs may be further subdivided into activities that are directly related to the product, such as assembly activities, or indirectly related, such as inspection and other processing activities.

It is noteworthy to consider that the only nonrecurring item on the list contributing to product cost is the total tooling cost. In other words, as the number of manufactured products increases, the contribution of tooling to the individual product cost decreases. As an example, consider that a product requires $1,000,000 worth of tooling. If only 100,000 units are sold in the first five years (the time typically used to depreciate capital expenditure), the tooling cost contributor to the product is $10 each. On the other hand, if 1,000,000 units are sold, the cost contributor drops by a factor of ten. Similarly, and for the same reasons, the price of purchased parts decreased substantially when large quantities were ordered.

Figure 15.5 depicts the same contributors to cost at a component level. For a buy item, cost is primarily driven by quantities when all other factors (tolerance, quality, and/or reliability requirements) are the same. This is also true for components that are not off-the-shelf but require an external supplier to manufacture them. For instance, an injection-molded component may be farmed out. The quoted price depends heavily on the required quantities. For make items, other factors contributing to component cost are material as well as processing and handling needs of the component.

Finally, the last topic on cost contributors to each component is the tooling required to produce the part. As depicted in Figure 15.6, and for the sake of argument, I have

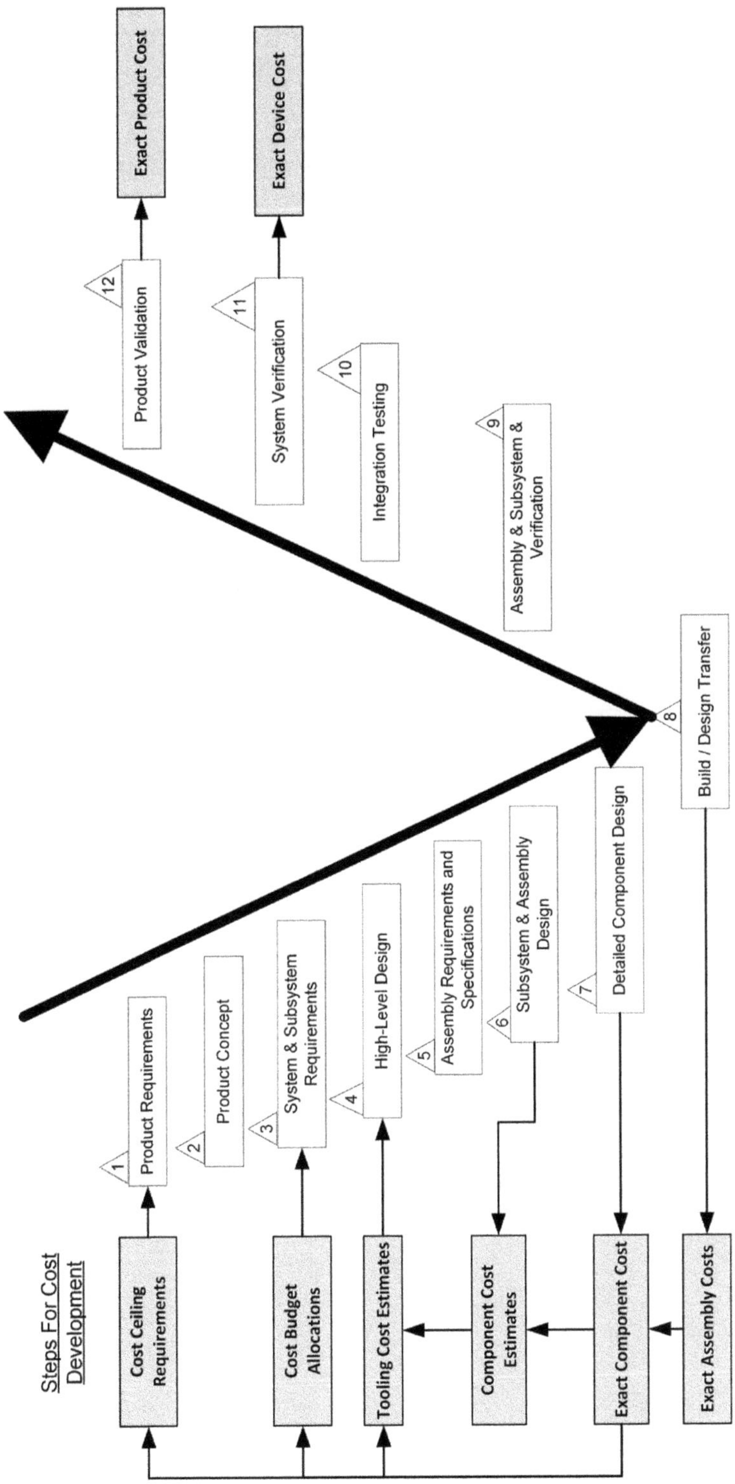

FIGURE 15.3
Cost structure within the V-model

suggested two main branches in tooling. For simplicity's sake, let's suggest that parts are either machined or formed.[3] If machining is the path for fabricating the component, then, the cost of a machining center will be based on the required number of milling axes, needed fixtures, required tolerance, and finally consumable items such as milling bits and cutting oils.

If forming is the path for fabricating the considered component, then the two major categories in forming operations are molding and stamping. Similarly, associated with this category, mold (or die) material and the complexity of the design are major cost contributors, followed by part tolerance and finally regular tooling maintenance to ensure part quality.

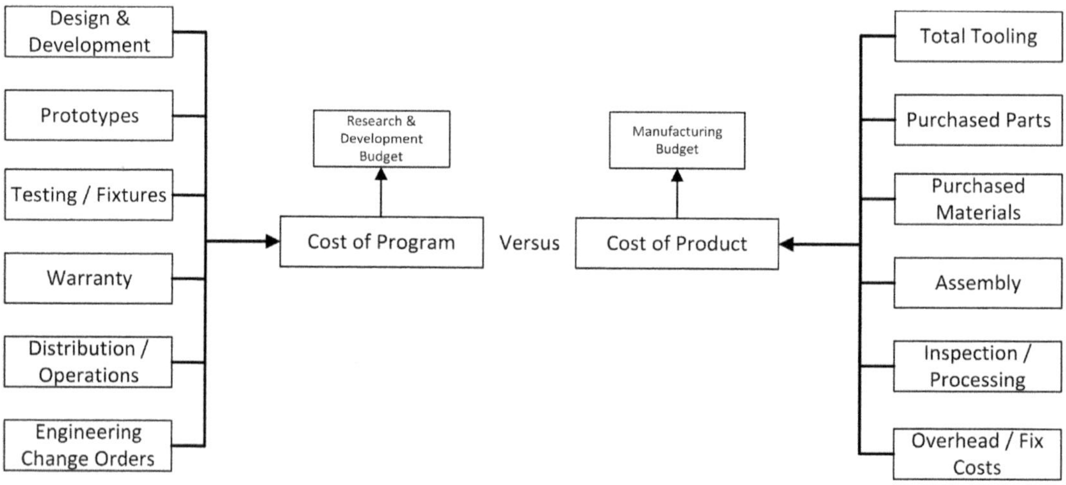

FIGURE 15.4
Program costs versus product costs

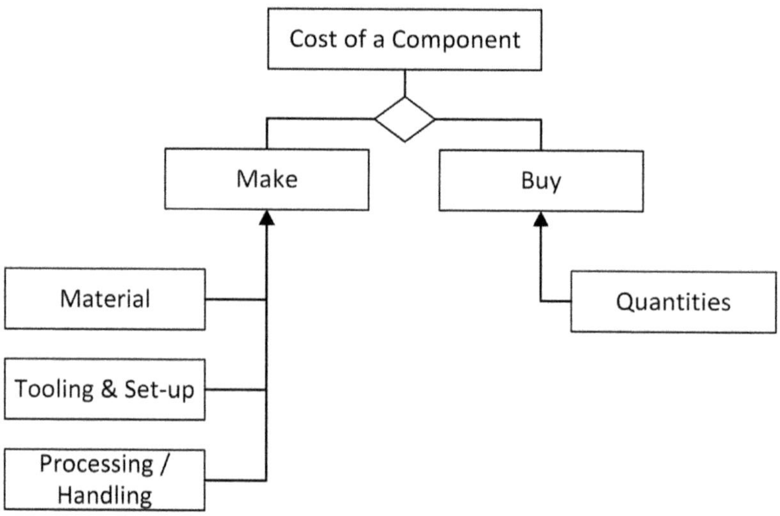

FIGURE 15.5
Elements of component costs

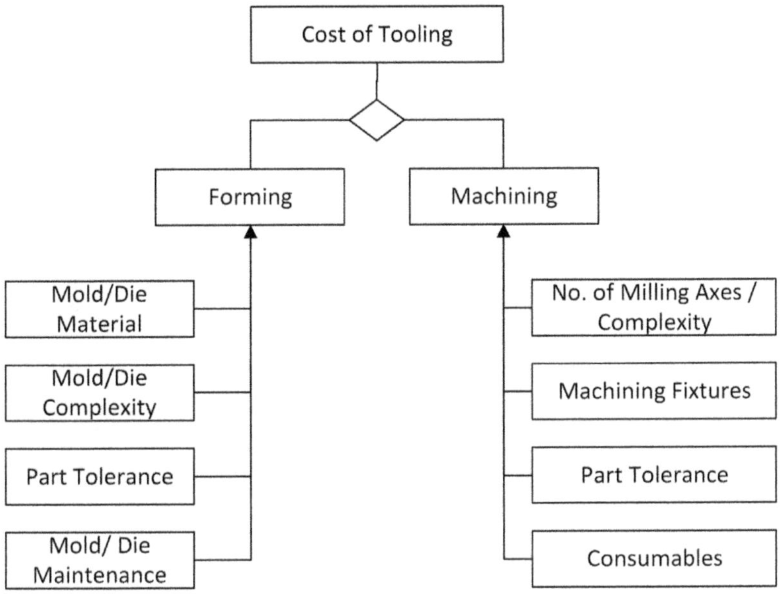

FIGURE 15.6
Elements of tooling costs

In my own experience, I have always outsourced the manufacture of forming operations. In that regard, providing costing for sourced components has been easy. Suppliers provide them. Costing for make components may be complex due to how the finance department tracks internal job centers and their associated costs. Jones (2009) provides an overview of various costing approaches; however, a design team will do well to consult and work closely with the finance team to develop internal cost estimates for part development.

Cost Estimates

> The purpose of developing *budgetary* estimates is to help guide the development direction of a product by offering the opportunity to investigate manufacturing costs and/or design options that may exist for certain parts or assemblies.

I have to admit that estimating costs is part science and part art. It is an art form because it relies heavily on one's previous experience with similar components used in predicate devices. An input to the cost of a component is whether it is a *buy* or *make* item. My own technique of estimating costs for purchased items has been to engage suppliers early on to provide approximate estimates. Depending on the degree of provided detail, even some estimates of tooling costs may be developed.

In lieu of having an estimate from a potential supplier, I have two formulas that have worked for me in the past, and even though they are not accurate, they have put me in the ballpark of the actual cost. For injection-molded parts, I typically calculate (or measure) the component weight. Then, I calculate the cost of material used in the component based on its weight. My estimate of the piece cost is about four or five times the cost of its material.

Having shared this, I would like to add that this formula works better when production volumes are large (in excess of 100,000 units annually) and when components are relatively small (in the range of a few ounces). For machined or formed parts, my rule of thumb is to estimate the length of time that is used to fabricate the component. On this basis, and an estimate of shop floor time, the cost of the part to a supplier may be calculated. Then, by using a realistic profit margin, component cost may be estimated.

The purpose of developing these types of *budgetary* estimates is to help guide the development direction of a product by offering the opportunity to investigate manufacturing costs and/or design options that may exist for certain parts or assemblies. Typically, a cost spreadsheet is developed that captures each component's unique attributes. This tool is used to develop the cost per component, assembly, and ultimately the top-level assembly manufacturing cost. The costs, materials, processes, and outputs from this spreadsheet will then populate values in an enterprise resource planning (ERP) software. In new product development, this spreadsheet can quickly provide costing scenarios to investigate changes in variables, such as manufacturing procedures or changes to an existing process.

Three Case Studies

In this section, a few cases will be reviewed from a number of different points of view. The first example focuses on a new product development effort to reduce cost. The second case will focus on a "job-shop" mindset to estimate tooling and component costs based on prior experience. The final example will examine the cost of a product from assembly and service points of view.

SURGICAL TOOL HANDLE COST SAVINGS

Consider the surgical tool that was presented in Figure 14.1 (shown here once again in Figure 15.7). A hypothetical bill of material is shown in Table 15.1.

For simplicity's sake, I will only consider the fabrication process of the handle as outlined in Figure 15.8.

Now suppose that this product is being evaluated under a cost-saving initiative. Table 15.1 indicates that although the cost of components of the handle (PRT-003, 004, and 005) is relatively low ($1.59), the finished handle costs nearly six times as much ($9.54). Clearly, this is a very labor-intensive process. With this in mind, the cost of the fabrication process is developed with the support of the finance department. For this example, Table 15.2 reflects the cost at various stations.

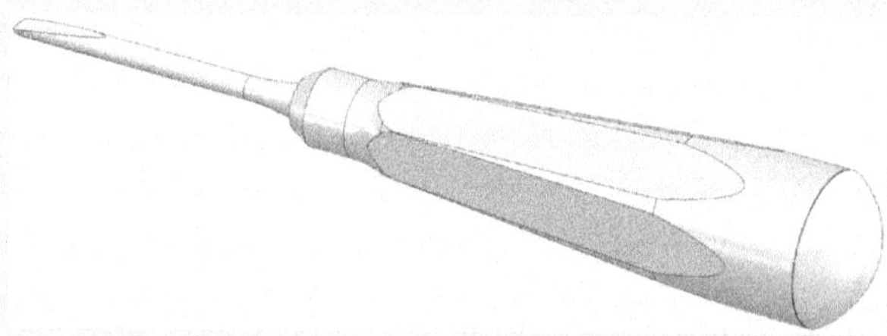

FIGURE 15.7
A handheld surgical tool

TABLE 15.1

A Hypothetical Costed Bill of Material for a Surgical Tool

Item	Part Number	Quantity			Description	Make/ Buy	Process	Material	Cost
1	PRT-000	1			Surgical Tool	Make	In-House Assembly	Stainless Steel	$15.03
2	PRT-001		1		Blade	Make	In-House Fabrication	Stainless Steel	$5.49
3	PRT-002		1		Finished Handle	Make	In-House Fabrication	Stainless Steel	$9.54
4	PRT-003			1	Handle Body	Buy	Stock Material	Stainless Steel	$0.29
5	PRT-004			1	Handle Cap	Buy	Machined Part	Stainless Steel	$1.00
6	PRT-005			1	Bushing	Buy	Machined Part	Stainless Steel	$0.30

Note: Nested quantities refer to components in the assembly hierarchy.

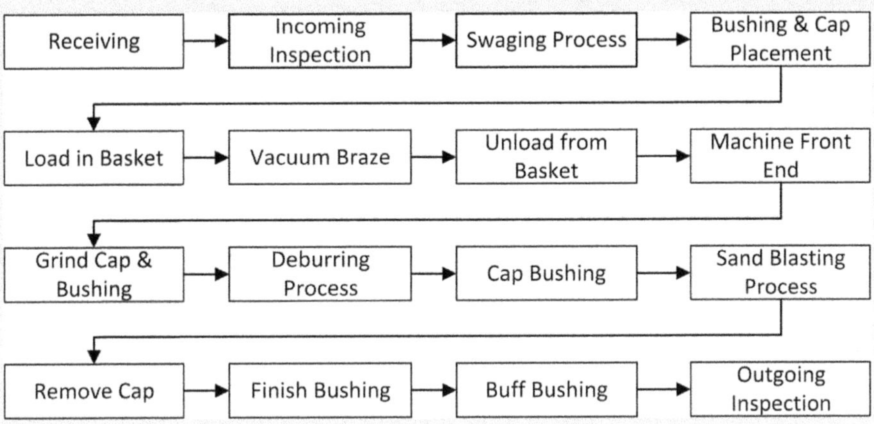

FIGURE 15.8
The process for fabricating the surgical tool handle

A review of Table 15.2, suggests that the highest cost items are the swaging process as well as its associated scrap rate and labor, for a total of $4.35 – highlighted in Table 15.2 for clarity. The question is whether this cost center may be eliminated altogether. The only viable replacement is to develop a metal injection-molded handle. A budgetary quote suggests that a metal injection molding (MIM) handle would cost $8.39 at an annual quantity of 50,000 units and $5.39 at an annual quantity of 90,000 or more. The cost of MIM tooling is suggested to be $95,700. The marketing department has agreed that the market will, in fact, support these quantities.

The initial reaction may be that even at the larger production quantities, the cost of MIM is larger than the cost of swaging and its associated waste. At the same time, how can the cost of tooling be justified? Table 15.3 provides a different picture.

As shown in Table 15.3, in addition to the swaging step, a number of other steps have also been eliminated. Hence, the MIM handle enables a cost savings of $2.60. Based on the marketing group's estimates, an annual quantity of 95,000 units will be needed. This leads to a total cost savings of $247,000 and a net savings of $151,300 once the tooling cost of $95,700 has been deducted. Should this project be approved, the new bill of material will be as shown in Table 15.4.

TABLE 15.2

A Hypothetical Process Cost for a Surgical Tool Handle

Cost Center	Standard Process
Purchased Parts	$1.59
Receiving	$0.02
Incoming Inspection	$0.02
Swaging Process	$3.00
Bushing and Cap Placement	$0.05
Load in Basket	$0.06
Vacuum Braze	$0.70
Unload from Basket	$0.06
Machine Front End	$0.63
Grind Cap & Bushing	$0.55
Deburring Process	$0.29
Cap Bushing	$0.08
Sandblasting Process	$0.10
Remove Cap	$0.08
Finish Bushing	$0.86
Buffing Bushing	$0.08
Outgoing Inspection	$0.02
Subtotal	$8.19
Average Cost of Swaging Scrap	$0.63
Average Cost of Swaging Labor	$0.72
Total	**$9.54**

LESSONS LEARNED

In this particular case study, the new product development effort focused on cost reduction, although a specific target value was not defined. In the V-model (Figure 15.3), cost ceiling is step #1. Considering that this is a product evolution, steps 2 and 3 of the V-model are already in place. At step 4, which is the high-level design concept, MIM was introduced. By consulting several potential suppliers, a budgetary tool cost as well as a part cost based on the existing design of the handle was obtained. Note that the final design of the handle will be conducted at step 7, where particular features unique to the MIM process have to be incorporated. An exact cost of both tooling and each piece will be determined at that time. Presently, the cost estimates indicate that there will be substantial savings should this project come to fruition.

TABLE 15.3
A Hypothetical Cost Comparison between Standard and MIM Processes

Cost Center	Standard Process	MIM Process
Purchased Parts	$1.59	$5.39
Receiving	$0.02	$0.02
Incoming Inspection	$0.02	$0.02
Swaging Process	$3.00	–
Bushing and Cap Placement	$0.05	–
Load in Basket	$0.06	–
Vacuum Braze	$0.70	–
Unload form Basket	$0.06	–
Machine Front End	$0.63	–
Grind Cap and Bushing	$0.55	–
Deburring Process	$0.29	$0.29
Cap Bushing	$0.08	$0.08
Sandblasting Process	$0.10	$0.10
Remove Cap	$0.08	$0.08
Finish Bushing	$0.86	$0.86
Buffing Bushing	$0.08	$0.08
Outgoing Inspection	$0.02	$0.02
Subtotal	**$8.19**	**$6.94**
Average Cost of Scrap	$0.63	–
Average Cost of Temp Labor	$0.72	–
Total	**$9.54**	**$6.94**
Savings Per Piece	$2.60	
Estimated Year 1 Savings	$247,000.00	
Tooling	$95,700.00	
Net Savings Year 1	$151,300.00	

TABLE 15.4
A Hypothetical Costed Bill of Material for a Surgical Tool Using MIM Process

Item	Part Number	Quantity			Description	Make/Buy	Process	Material	Cost
1	PRT-000M	1			Surgical Tool	Make	In-House Assembly	Stainless Steel	$12.43
2	PRT-001		1		Blade	Make	In-House Fabrication	Stainless Steel	$5.49
3	PRT-002M		1		Finished Handle	Make	In-House Fabrication	Stainless Steel	$6.94
4	PRT-003M			1	Handle Body	Buy	MIM Process	Stainless Steel	$5.39

(Nested quantities refer to components in the assembly hierarchy)

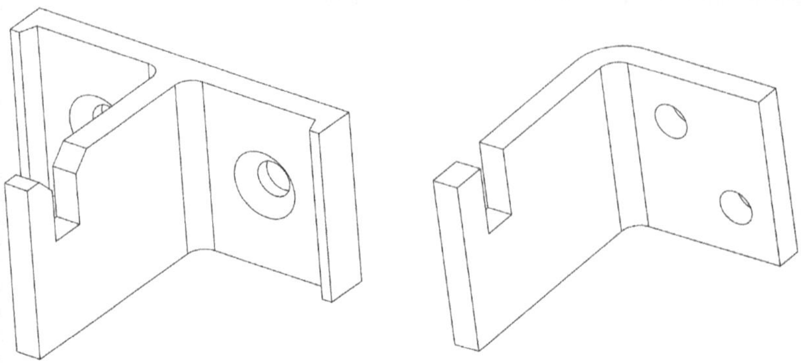

FIGURE 15.9
Two different bracket designs

TABLE 15.5
Three Different Manufacturing Processes to Fabricate a Bracket

Part	Material	Mass	Volume	L	W	H	Process
Bracket A	Al 6061	0.05 lbs (22 gr)	0.49 in^3 (8.09 cm^3)	2″ (50mm)	1″ (25mm)	1.5″ (38mm)	Machined
Bracket A	Al 6061	0.05 lbs (22 gr)	0.49 in^3 (8.09 cm^3)	2″ (50mm)	1″ (25mm)	1.5″ (38mm)	Extruded and Machined
Bracket B	Al 6061	0.03 lbs (15 gr)	0.33 in^3 (5.43 cm^3)	1″ (25mm)	1″ (25mm)	1.5″ (38mm)	Punch and Bend
Al 6061	$5.00	per lb					
Density	0.098	lbs/in^3					

MACHINING VERSUS STAMPING

Consider once again the brackets introduced in Figure 14.6. They are provided here once again in Figure 15.9. If an off-the-shelf bracket is not available, how would one provide a cost estimate for each.

Table 15.5 summarizes three different categories. The first two lines of this table focus on two different ways to produce bracket A and its attributes; the third line provides information on the bracket B component. The assumption is that aluminum is about $5.00 per pound (0.45 kg); however, this price changes with purchased quantities. For simplicity's sake material cost variable is not included in this analysis.

TABLE 15.6

A Hypothetical Cost Comparison of Three Different Manufacturing Processes

Part	Process	Material Cost of Blank Piece	Process Time (Minutes)	Process Cost Per Minute	Total Process Cost	Tooling Cost	Piece Cost 1000 Quantities	Piece Cost 10,000 Quantities
Bracket A	Machined	$1.47	7	$2.00	$14.00	–	$15.47	$15.47
Bracket A	Extruded and Machined	$1.75	0.5	$2.00	$1.00	$4,500	$7.25	$3.20
Bracket B	Punch and bend	$0.15	0.25	$2.00	$0.50	$10,000	$10.65	$1.65

Table 15.6 provides an estimate of processing time for each bracket design. Bracket A may be machined from a block of aluminum. This process involves removing nearly 80% of the material from a blank piece. It is both costly and time-consuming. Alternatively, a custom extrusion die may be made to produce a near-net shape extrusion. These extruded parts may be machined to the net shape at a fraction of the cost. The die cost for this purpose has been estimated at about $4,500. This estimate is based on previous experience and consultation with a potential supplier.

Bracket B is fabricated using a progressive stamping operation. In this operation, a black in stamped to produce the flatten shape. In the next operation, it is bent into the desired shape. The last operation separates the final pieces from the manufacturing panels. The tooling cost for this operation is estimated at about $10,000. Again, this estimate is based on previous experience.

As shown in Table 15.6, one should not assume that bracket B is simply the least expensive approach at all times. Clearly, for very small quantities, a machined part may be the chosen approach. Alternatively, it may be possible to *print* this bracket using a direct metal laser sintering (DMLS) process (otherwise known as 3D metal printing). A design subject matter expert should always consider production quantities before selecting a manufacturing process.

In Table 15.6, piece cost is calculated as follows:

$$\text{Piece Cost} = \text{Material Cost of Blank} + \text{Total Process Cost} + \frac{\text{Tooling Cost}}{\text{Quantities}}.$$

The total process cost is calculated as

$$\text{Total Process Cost} = \text{Process Time (in minutes)} \times \text{Process Cost (Per Minute)}.$$

This simple calculation shows that the impact of tooling costs diminishes rapidly as the production volume increases. Hence, to have a meaningful impact on cost, either the material weight should be reduced or the number of processing steps should be decreased.

ULTRASONIC DENTAL SCALER

In the field of dentistry, scalers are used to remove layers of calculus and scales from teeth. The repetitive nature of hand and wrist movement across the tooth surface often leads to wrist, hand, and arm fatigue. To alleviate this issue, it is possible to develop an electronic instrument that vibrates a scaler tip at ultrasonic frequencies. In addition to this vibration, to flush away the debris, a small pump forces water (and at times medicaments) through the handpiece onto the surface of the teeth. Figure 15.10 depicts this system's architecture.

In this case study, marketing has determined that should a unit be developed at a cost of $550 per unit, the market will support a production of 10,000 units a year for 5 years. Based on the design team's collective experience, the overall device cost has been flowed down to each subsystem and module as shown in Table 15.7. This activity corresponds to step 6 as depicted in Figure 15.3

The approach used in developing this table was twofold. First the *buy* items were identified. Then, by consulting the purchasing department, similar items in inventory were identified, and their purchase price was used as a basis. For items that did not exist in inventory, potential suppliers were identified and budgetary quotations were obtained.

For the *make* items, the first step was to develop a possible fabrication process map. The second step, in collaboration with manufacturing and finance teams, was to allocate timing and cost to each step. Finally, the total cost for each component was calculated based on the cost of each step as well as the needed material.

It should be of no surprise that the first pass at cost estimation exceeded the initial budget by about 15%. On the one hand, the marketing department's judgment of the cost may not be accurate. On the other hand, this is only a budgetary step. Once the final design has been completed, the question of the cost of the finished product must be answered. Should the cost estimates be over the allocated budget, the program must evaluate its financial goals in lieu of the updated financial information. At this stage, the design team should examine their assumption in either part selection or process judgment. At times, these selections are based on the needs of a previous program that may require a higher performance component than may be required in the current project. For instance, the rationale for selecting a $48 pump and a $149 foot pedal should be examined closely. Next, the cost associated with making the electronics generator ($176) should be examined. It may be possible by outsourcing this assembly, to reduce its overall cost be reduced.

Three factors are often overlooked in the initial budgetary discussion. One is the cost of manuals and packaging, the second is the impact of tooling, and the last is the assembly cost of the final device. Have these factors been included in the marketing's cost requirements? In this scenario, packaging accounts for nearly 5% of the total cost.

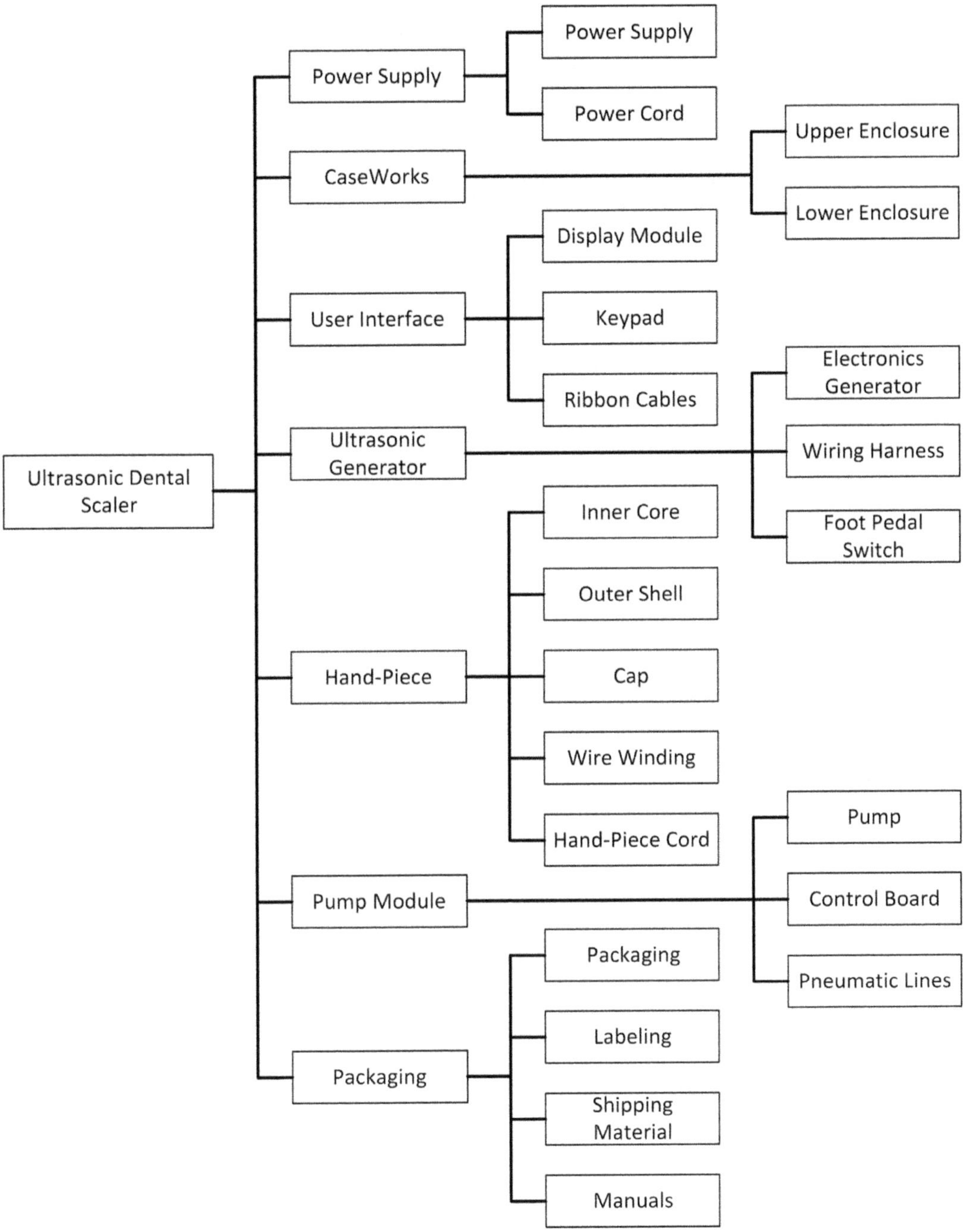

FIGURE 15.10
The system, subsystems, and modules of a dental ultrasonic scaler

TABLE 15.7

A Costed Bill of Materials for an Ultrasonic Dental Scaler

Item	System	Subsystem	Module	Material	Make/ Buy	Process	Piece Price/ Cost	Tooling Cost
1	Ultrasonic Generator	Power Supply	Power Supply	Electronic Assembly	Buy	–	$25.00	–
2			Power Cord	Electric Assembly	Buy	–	$5.00	–
3		Case Works	Upper Enclosure	PC/ABS	Buy	Injection Molding	$48.00	$85,000
4			Lower Enclosure	PC/ABS	Buy	Injection Molding	$36.00	$55,000
5		User Interface	Display Module	Electronic Assembly	Buy	–	$17.00	–
6			Keypad	Electronic Assembly	Buy	–	$7.00	–
7			Ribbon Cables	Electric Assembly	Buy	–	$4.00	–
8		Ultrasonic Generator	Electronics Generator	Electronic Assembly	Make	IR Reflow & Wave	$176.00	–
9			Wiring Harness	Electric Assembly	Make	Automated Wire Assembly	$10.00	–
10			Foot Pedal Switch	Electronic Assembly	Buy	–	$149.00	–
11		Hand-Piece	Inner Core	PPS	Make	Machined	$2.50	–
12			Outer Shell	PPS	Make	Machined	$3.50	–
13			Cap	PPS	Make	Machined	$1.25	–
14			Wire Winding	Copper Wire	Make	Wire Wound	$1.00	–
15			Handpiece Cord	Insulated Wires and Tubes	Make	Co-Extrusion	$0.50	–
16		Pump Module	Pump	Pneumatics	Buy	–	$48.00	–
17			Control Board	Electronic Assembly	Buy	–	$23.00	–
18			Pneumatic Lines	PVC Tubing	Buy	–	$2.00	–
19		Packaging	Packaging	Card Board	Buy	–	$7.00	–
20			Labeling	Printed Material	Buy	–	$5.00	–
21			Shipping Material	Polymeric Material	Buy	–	$12.00	–
22			Manuals	Printed Material	Buy	–	$6.00	–
23	**Sub-total**						**$588.75**	$140,000
24	**Assembly Cost Estimated at 20 Minutes Per Device**						**$40.00**	
25	**Tooling Cost Per Device (a total of 50,000 Units in 5 years)**						**$2.80**	
26	**Estimated Device Cost**						**$631.55**	

Purchasing

In the new product development process time frame, when detailed design is being developed, design team members can almost touch the physical manifestation of what has been a product of their creativity to date. This period has always been an exhilarating time for me, and I imagine it is the same for others as well. There is always a desire to push forward to select a supplier to get to the first samples sooner. The pitfall is that if we are not careful, the *wrong* supplier may be selected. By *wrong*, I really mean a supplier that may not provide what the design, product, and, more importantly, the business need. Figure 14.22 – in Chapter 14 – provided a set of criteria for supplier selection from a reliability point of view. There may be other criteria in the selection process.

Supplier qualification involves several areas; for example, from a regulatory point of view, a supplier must have certain processes in place before the business may engage in commerce with them. From a product point of view, the supplier should be financially viable so that it can support the new product throughout its life cycle. Considering these various factors, it is no wonder why the purchasing department is responsible for the business relationship with a supplier. Thus, R&D reflects only the technical aspect of a supplier assessment. Other areas are quality audits, business relationships, evaluation of the stability and longevity of the vendor, past history, etc. In short, purchasing must be involved with supplier selection to evaluate the business risk of a particular company and to leverage collective purchasing of components for lower prices.

The purpose of this section is to describe a process of working with the purchasing department when selecting a vendor for a new component. If existing qualified vendors cannot meet the specifications for a component being designed, purchasing should be notified. R&D can work with purchasing in order to select a vendor that will meet all of the business requirements as well as the technical requirements. If purchasing is not involved early in the selection process, they may have compelling reasons to disapprove of a selected supplier, and weeks, if not months, of time working with the supplier may be wasted.

Best Practices

In my opinion there are four golden rules that a design team should follow in procuring any elements (or components) of the product under development. These rules are:

1. Never negotiate (or try to) price or terms without a purchasing agent present. Make sure to document all meeting minutes and decisions made.
2. Invite the assigned purchasing agent to any and all meetings where delivery schedules and associated finances are discussed.
3. Copy the assigned purchasing agent on all communications with suppliers and forward any quotes received directly from the supplier.
4. Never sign any agreements with a supplier without the prior approval from the purchasing agent.

Figure 15.11 provides an overall process of working with both purchasing and a potential supplier of either a particular component or consumables on the manufacturing floor. The importance of involving purchasing in the selection of a supplier from the beginning

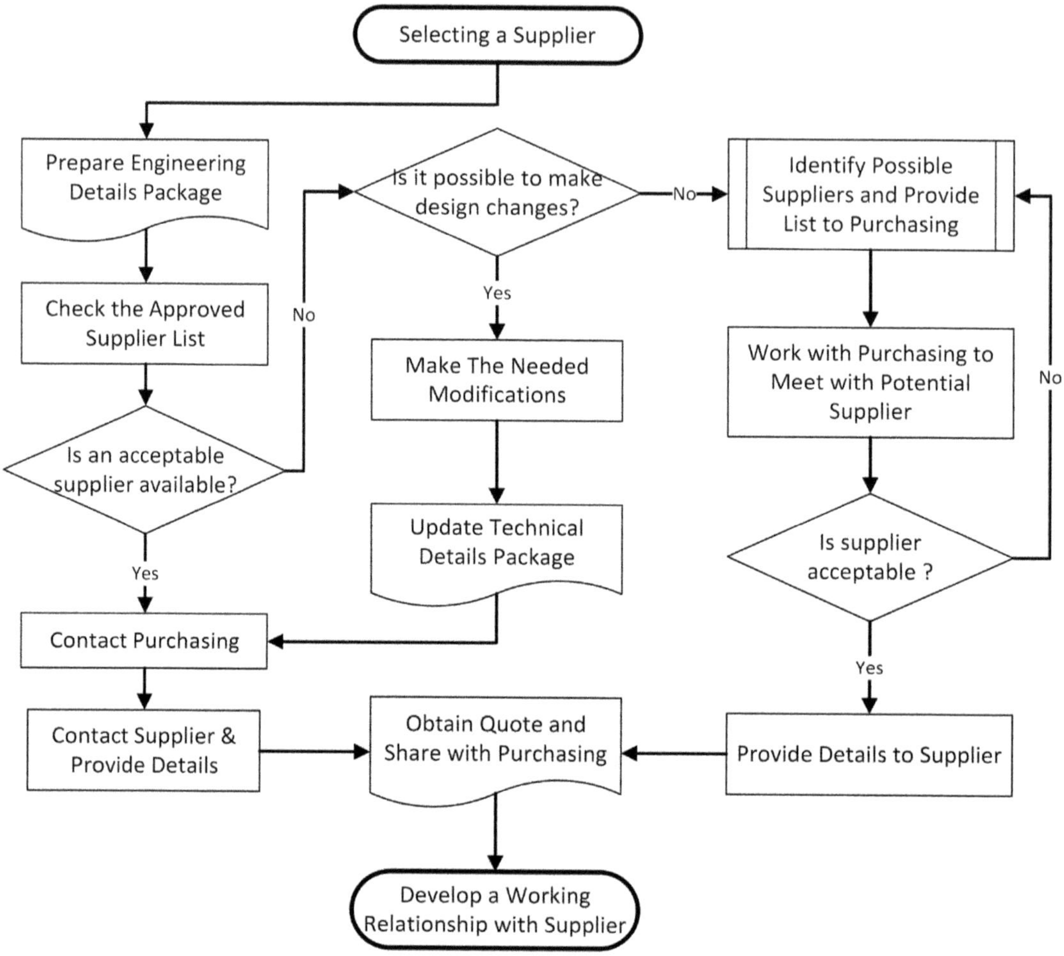

FIGURE 15.11
The process flow to working with both the purchasing department and a potential supplier

of the development process cannot be overemphasized.

It is anticipated that the purchasing department has a list of approved and/or preferred suppliers for the enterprise as a whole and that a qualified supplier exists that may be selected. More often than not the R&D team has specified components that may not have been used in any other product. The question that design team should entertain is this: Is it possible to make design modifications such that a specified component is available from an existing supplier? This mindset emphasized the concept and awareness of reuse and configuration management. Furthermore, it is inline with the ideas of design for manufacturing, assembly, and service.

Four golden rules that design teams should follow are:

1) Let purchasing negotiate the price.
2) Involve purchasing in discussions of delivery schedules and finances.
3) Copy purchasing on quotes received directly from suppliers.

   Never sign any agreements with a supplier without the prior approval from purchasing.

Supplier Selection Process Map

This section provides an explanation of the process map provided in Figure 15.11. The selection process always begins with developing an engineering details package. In other words, what is being asked should be clearly explained and provided in a document. This follows the same mindset of knowing the requirements (inputs) as well as expectations (outputs) at any given segment of product development. Once this is done, the next step is to check the approved supplier list for the appropriate item or service rendered.

Acceptable Supplier Exits

If an acceptable supplier is available, then contact the purchasing agent assigned to the project and obtain the contact person's name. This gives the opportunity to the purchasing department to first learn that a new part is being sourced and to keep them informed. Furthermore, it gives them an opportunity to review the level of the supplier's qualification depending on the component's criticality to the design and possibly safety. If additional qualifications are required, the activity to collect and process this additional information may begin.

The next step is to contact the selected supplier(s) and provide them with the engineering package, including critical information such as estimated volumes. If needed, meet and review component design and specifications with the supplier, making sure that the purchasing agent remains informed of the progress and/or any issues with delivery or quality of the purchased components.

Acceptable Supplier Does Not Exits

If an acceptable supplier is not available, the first question to be asked is whether modifications may be made to the design to enable use of an existing component. This question applies particularly to buying commodity and consumable items such as adhesives and hardware such as screws and nuts. However, it is just as valid when considering components such as motors or electronic components. If possible, make the needed changes and follow the path provided in the previous section.

If it is not possible to modify the design, then the design team should work with the purchasing agent to develop a list of possible suppliers. The next step is to interview the identified suppliers for their capabilities in collaboration with the purchasing agent. If the supplier appears to be capable, the purchasing agent commences the steps for approving the new supplier. If the part under consideration is a critical component, then it is recommended that a minimum of two suppliers be developed and approved. The reason for doing this is self-evident.

In a nutshell, it is critical to the health of the project that the design team to keeps its finger on the pulse of cost as the design is being developed and moves toward completion and transfer to manufacturing. There are numerous examples of budget overruns that are allowed to continue simply because teams have lost sight of product cost due to a variety of reasons. Having a running understanding of the cost will enable the development team as well as management to make course adjustments as needed or possibly even cancel the project if, in the long term, the financial objectives of the business cannot be met.

Summary

- An appreciation of cost issues begins at the market and what the end users are willing to pay for a particular product.
- There are scenarios in which the cost of a product may not be reduced sufficiently for profitable sales; however, the same product may facilitate the sale of lucrative ancillary products.
- An understanding of product cost is one of the major elements of product development activities.
- Cost of a product depends on the total number of tools required, the number of purchased parts (buy items on the bill of materials), material costs for make items, labor, and overhead (or fixed) costs.
- Budgetary cost estimates guide the development direction of a product by offering the opportunity to investigate manufacturing and/or design options that exist for certain parts or assemblies.
- Three factors are often overlooked in the initial budgetary discussions: cost of manuals and packaging, impact of tooling, and finally, assembly cost of the final device.

Exercises

1. Continue the design work for the concept used in the Chapter 14 exercise.
   a. Estimate parts, fabrication, and assembly costs.
   b. Have you considered the impact of tooling or yearly production and service volume?

Notes

1 This approach to setting price and cost is different than the traditional cost-based pricing. For more information, see Nagle and Holden 2002.

2 Often, the fact that technical/engineering change orders are small design projects is ignored. As a design project (albeit very small), they should nevertheless follow the design and development process. Because of this oversight, the expenses associated with the team's time and involvement are ignored, leading to an inaccurate cost estimate of an "engineering" change order.

3 I do acknowledge that many times, components are first formed into a near net shape and then machined into the net shape. This suggests that component cost is influenced by both branches.

16

Detailed DFMEA, PFMEA, and Control Plans

Introduction

FMEA provides a deductive and structured approach to identify problems related to product design, manufacturing, service, and even use. In Chapter 10, I provided some rationale why an early-stage FMEA is an appropriate step in the design and development process. The product development roadmap, as shown in Figure 16.1, indicates that once design details are completed, a second, more detailed set of FMEA files needs to be developed.

The purpose of FMEA studies is to reduce the probability of failures by:

- Focusing on functions and requirements either in design, manufacturing, or service
- Identifying how functions fail (modes) and the impact of these failures (effects) and potential impact on end users (hazards and harms)
- Ascertaining control mechanisms to illuminate or reduce the impact of the effects

The FMEA studies should be conducted as a part of the design and development process; however, FMEA databases are living documents and should be periodically reviewed and updated. In particular, these reviews are important when

> FMEA provides a deductive and structured approach to identify problems related to product design, manufacturing, service, and even use.

- modification are made to component designs or technologies
- modification are made to (or new information is obtained on) manufacturing or service processes
- changes are made to the environment or location of use
- information is obtained on new or unintended customer usage.

xFMEA and Design Process

From a product development process point of view, a system-level DFMEA takes place in the high-level design stage (stage 4 as shown in Figure 16.2). As more details are developed, subsystem and component DFMEAs are further developed and refined – corresponding to the steps near completion of the 2D drawings of the roadmap (Figure 16.1) and stages 6 and 7 of the V-model as depicted in Figure 16.2.

DOI: 10.1201/9781003564508-21

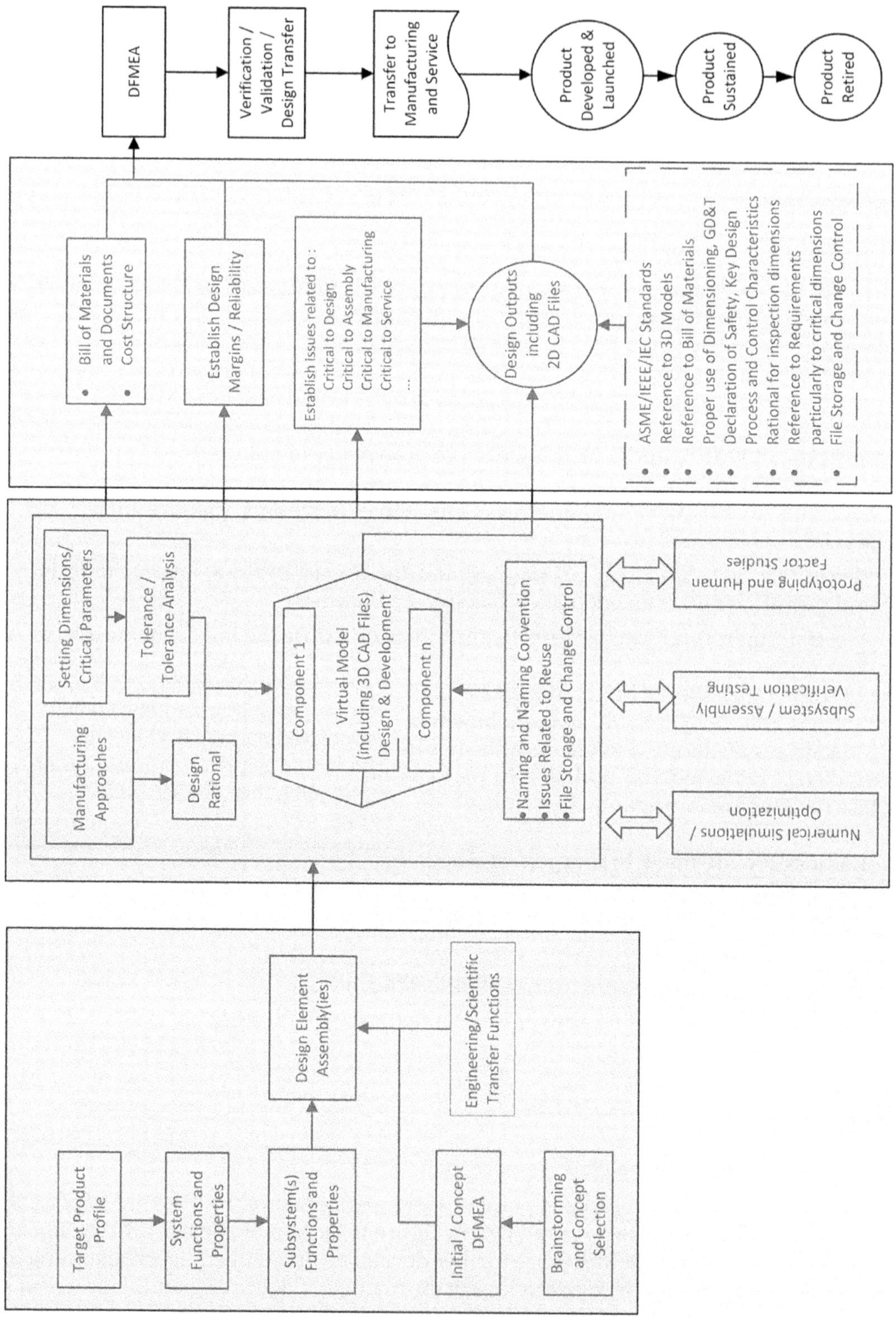

FIGURE 16.1
Post-design details activities

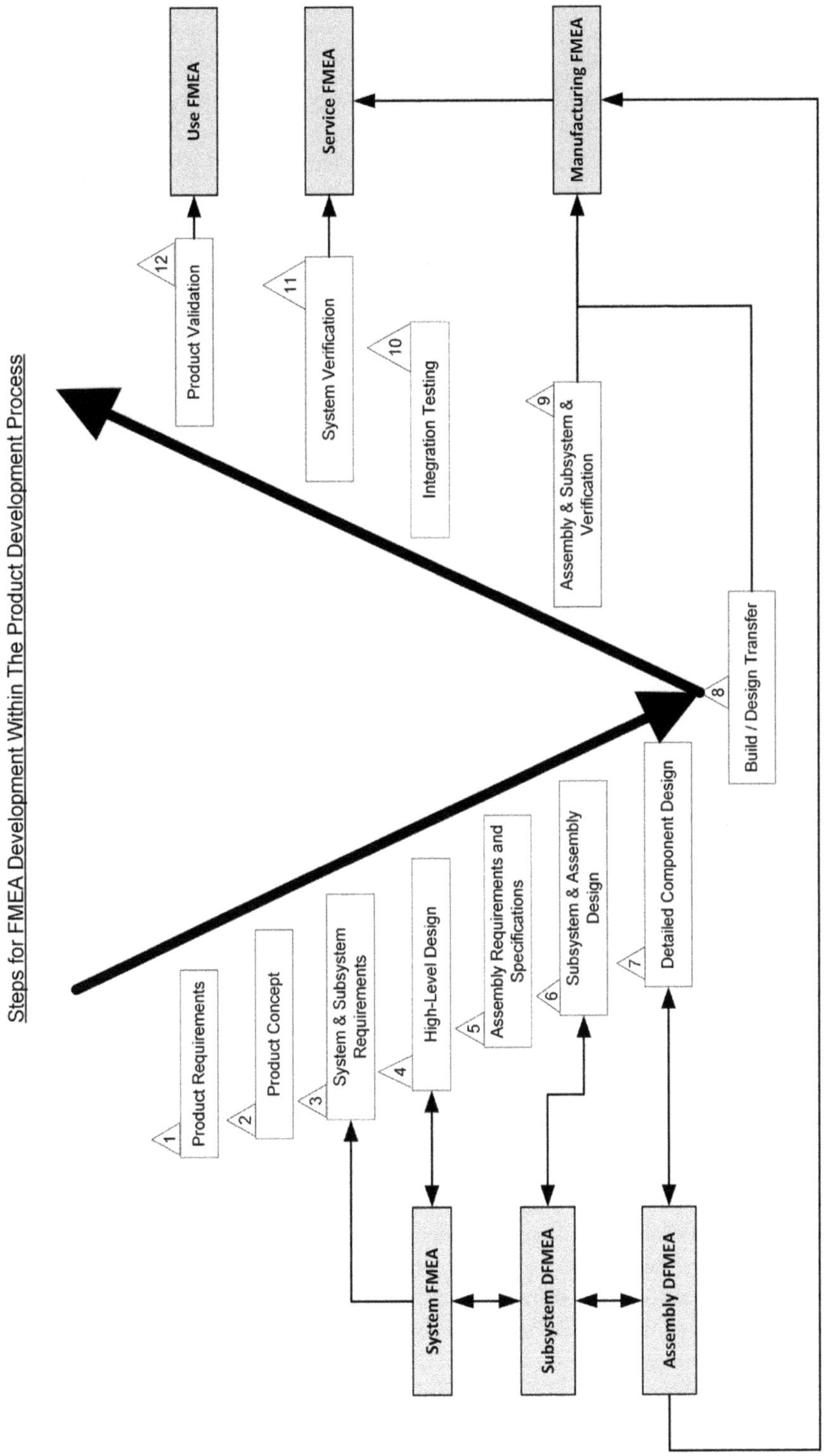

FIGURE 16.2
Steps for failure mode and effect analysis within product development cycle

Once the details of the design are completed, the DFMEA database is refined further. In this refinement, the design team provides evidence of how the initially identified issues were resolved and documents any failure (or risk) elimination or reduction. I will discuss this risk reduction later in this chapter.

Once there is sufficient review of DFMEA results, then process (first manufacturing and then service) failure mode and effect analyses (PFMEA) are conducted and matured during the product development cycle and then continue throughout product life.

The last step in the design and development process – as shown in Figure 16.1 – is product validation. This step studies how the developed product addresses the users' needs. Associated with this step, some design teams, particularly in the medical devices field, develop what is called a use failure mode and effect analysis (UFMEA). This analysis focuses on how an end user may use (or misuse) the product and the resulting failures and associated risks. A review of UFMEA and what it contains is beyond the scope of this work; however, the mindset required is the same as other FMEA studies.

Detailed DFMEA Study

In Chapter 10, I discussed DFMEA and its role in the early stages of the design process. The classic definition of the DFMEA process suggests that it is a bottom-up process. In other words, one starts with detailed design components at the lowest level and builds up to system-level failure modes and their impact on product behavior and the end user(s). The challenge with this approach is that once the design team completes design details and nearly freezes the design, the team's willingness to change all but disappears. At this time, DFMEA loses its ability to impact the design and improve it by identifying weak links and providing pathways to potential redesigns.

Anleitner (2011) suggests that the team begin the DFMEA process when the concept design is developed. He defines the concept design as a design with sufficient details to allow a first-pass cost analysis. At this stage, design details are not yet established, and changes to details may be made easily. Similarly, authors of MIL-HDBK-338B recommend initiating the DFMEA study early in the design process even though details are yet to be developed: "Admittedly, during the early stages, one usually does not have detailed knowledge of the component parts to be used in each equipment. However, one usually has knowledge of the 'black boxes' which make up the system."

DFMEA studies should be initiated early in the design process even though details are yet to be developed.

Therefore, there are, in reality, two aspects of DFMEA. The first approach is a top-down, and the second is the traditional bottom-up formulation. In the first pass, potential design pitfalls are identified. In the second pass, evidence is provided on how these potential drawbacks are either avoided or mitigated (and reduced). Furthermore, DFMEA provides a mechanism by which comprehensive verification test plans may be developed.

The first DFMEA pass begins at the conceptual or high-level design. In Chapter 10, this approach was demonstrated. It was shown how to detect potential design pitfalls based on historical data or current information. In addition, through the examples that were provided, conflicts that might arise as a result of solutions to various design problems were identified. For instance, in high-power electromechanical systems, overheating is typically a failure mode that should be resolved. A second failure mode is excessive electromagnetic

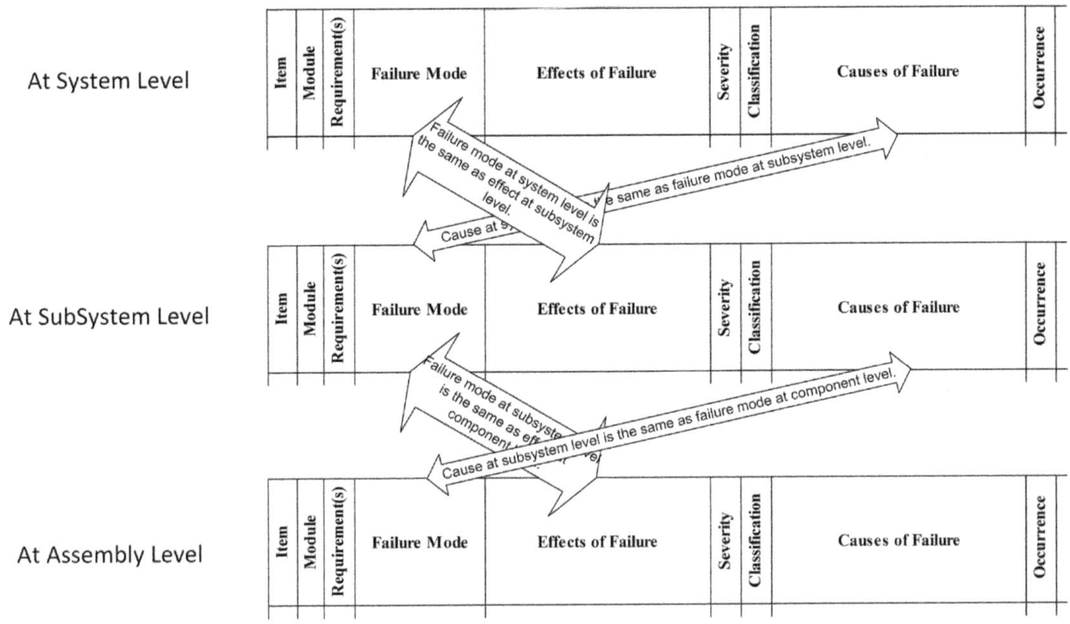

FIGURE 16.3
Failure mode-effect-cause relationships between system, subsystem, and assembly

interference. To resolve the first failure mode, openings in the enclosure are needed. To resolve the second failure mode, the enclosure should not have any openings at all. Clearly, this poses a design conflict. A properly developed early-stage DFMEA enables the design team to identify these conflicts.

Based on the process flow shown in Figure 16.2, once a high level design is developed and the subsystems identified, a system-level DFMEA may take place. At this level, failure modes, effects, and causes are identified based on historical knowledge of predicate and/or competitive devices. This information forms the basis of subsystem DFMEA. As shown in Figure 16.3, system-level *causes* become subsystem-level failure modes. Similarly, system-level failure modes are subsystem *effects*. Through the subsystem analysis process, subsystem causes are identified. In turn, subsystem causes constitute assembly *failure modes*. Subsystem failure modes are assembly level *effects*.

Theoretically, this cascade may extend down to individual components; however, in many repairable electromechanical systems, this cascading down to components may be an unnecessary level of detail. The reason is that, from a service point of view, typically subsystems are repaired by replacing subassembly units. For instance, should a computer monitor have display issues due to a faulty graphics card, the repair-person, does not attempt to restore the card by finding the faulty component. Instead, the repair is done by replacing the entire graphics card without regard to the specific component failure.

Figure 16.3 provides functional links between assemblies, subsystems, and finally the system. Not only the impact of component and assembly failures but also the impact of any change at the component and assembly level may be traced up to the end product. This linkage demonstrates the effectiveness and the power of design failure mode and effect analysis. Design details are developed at assembly and component levels. Hence, at these two levels, solutions may be proposed to either eliminate the failure modes or to reduce their frequency.

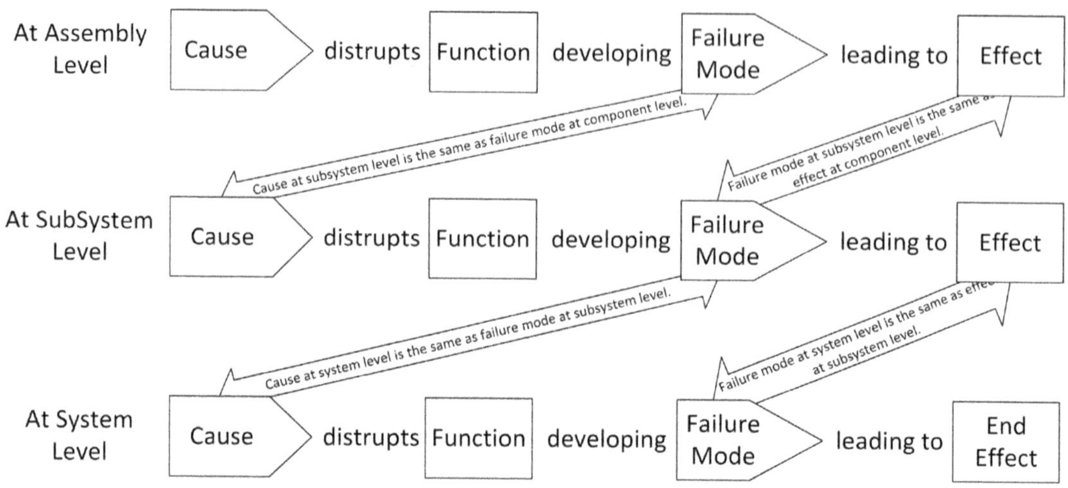

FIGURE 16.4
System failure mode and cause flow-up

Parameter Diagram

The root cause of system-level malfunction (or failure) is always at the level of a single component or the interface of two components. Even in catastrophic failures, one component fails first, which would then cause a chain reaction to rapid failures of other components. The purpose of developing a DFMEA table is to understand how causes of failures affect system behavior or its malfunction. These causes have their roots in variations in the normal conditions that the product was designed for. A parameter diagram (or P-diagram) is a tool to evaluate and understand the impact of variations and noise on the system behavior.

The purpose of developing a DFMEA table is to understand how causes of failures affect system behavior or its malfunction.

Figure 16.4 (a rearrangement of Figure 16.3) depicts a "Cause-Function-Failure Mode." One can then trace the link between Cause at the component level to assembly and eventually the end effect experienced by the user. The conclusion drawn from this figure is that the local impact of disruptions (or causes) must be known and understood in order to appreciate system behavior.

In Chapter 11, I reviewed the function diagram in Figure 11.2. Figure 16.5 depicts the same idea with the addition of *causes* and *failure modes*. In other words, under various causes (also known as perturbations or noises), it is possible to disrupt the intended function and the ideal output.

It should be noted that noises (or causes) may act on the input path to the function, on the function itself, or even on the output path, leading to failures. The following example may provide some clarity.

Suppose that a signal amplifier assembly is subjected to fluid ingress. The liquid may cause corrosion on the input signal leads. In this case, the amplification circuitry would not function properly if it did not receive the right signal. In a different scenario, fluid ingress may cause an intermittent shortage of the output signal lines – the input leads and the amplification circuitry are unaffected – yet failures are present. Finally, the liquid

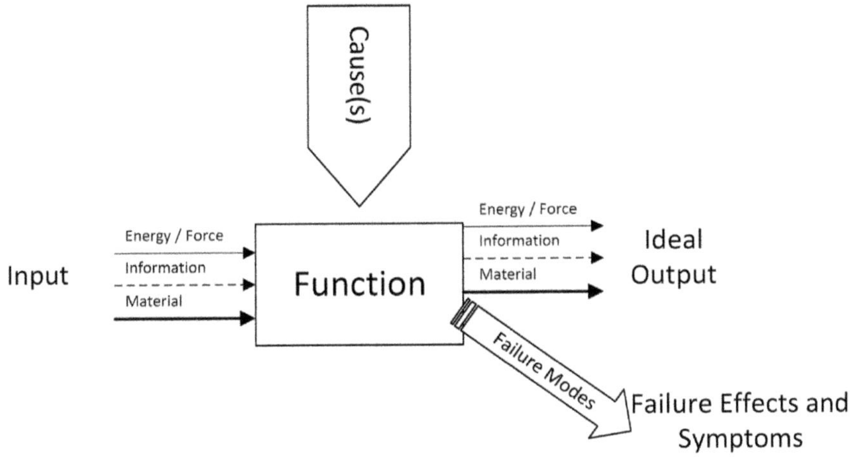

FIGURE 16.5
Cause-function-failure-mode diagram

ingress may impact the actual amplification circuitry by possibly damaging the IC chip's housing.[1]

While noises tend to destabilize function, control mechanisms are put in place for functional stability. For instance, on mechanical components, part tolerances are specified to allow assembly. Similarly, on electronic components, tolerances are specified to ensure expected electrical signals and outputs.

> P-diagrams provide a holistic and bird's-eye view of the factors impacting specific functions of a particular system.

> In P-Diagrams, there are five sources or categories of noise. These are manufacturing variations, changes over time, customer uses/misuse, environmental conditions, and system interactions.

Should these control factors be added to the function diagram as shown in Figure 16.5, the outcome is called a *parameter diagram* (or P-diagram for short) – see Figure 16.6. P-diagrams provide a holistic and bird's-eye view of the factors impacting specific functions of a particular system. In other words, a P-diagram helps the team to fully define expected functions as well as the associated failure modes. This diagram is not only used to complete DFMEAs but also helps with a systematic approach to verification activities.

The transfer function examples provided in Chapter 11 were primarily focused on the impact of dimensional variations on the transfer function. There are, however, other causes or sources of failure. For instance, the failure of the signal amplifier example that I just provided was caused by fluid ingress onto the electronics.

Typically, there are five sources or categories of noise. These are (1) manufacturing variations such as part-to-part variations and assembly variations; (2) change over time, such as material property or shape changes; (3) customer uses, such as storage conditions,

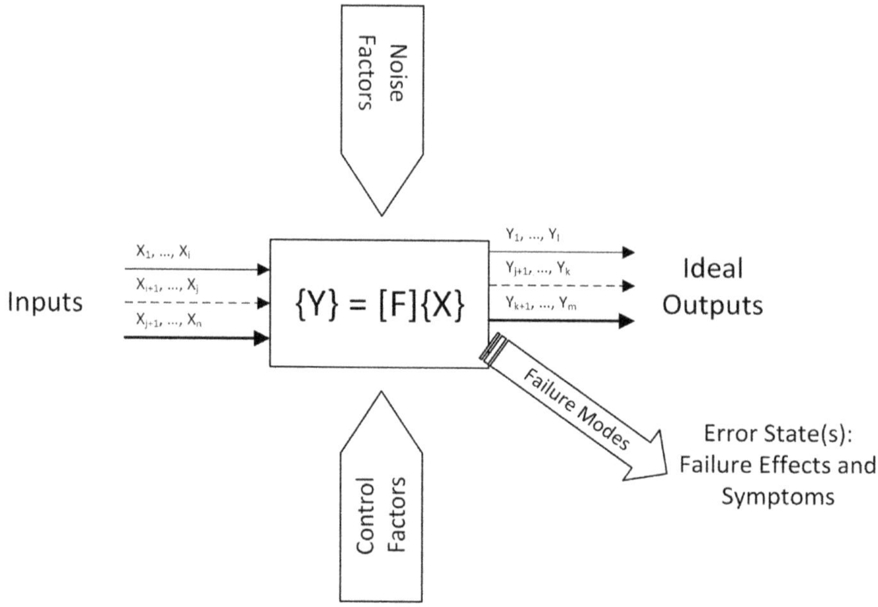

FIGURE 16.6
Parameter diagram

application of different cleaning agents, or off-label use; (4) environmental conditions, such as temperature, pressure, and humidity; and (5) system interactions. Examples of interactions are chatter of electronics boards under vibratory load or warping of a printed circuit board due to heating of an adjacent assembly or part. Unfortunately, of these five categories, system interactions are the most complicated to identify and understand. Furthermore, interactions are often root causes of system failures when individual subsystems pass verification. The classification of noises is summarized in Figure 16.7.

A close examination of these noise factors will reveal that the first two categories (manufacturing variation and change over time) are influenced by properties of the internal elements of the assembly (or component), whereas customer usage, environment, and system interactions are based on the external demands placed on the product.

To counteract the internal noise factors, the product must be designed for manufacturability, assembly, and reliability. Through reliability studies and improvements, the product becomes insensitive to change over time noises. By conducting manufacturability and serviceability studies, the product becomes insensitive to manufacturing and service variability noises.

External noise factors are reflective of the type of demands that are placed on the product by the end users. Many of these factors should be understood based on market studies and historical use data. To counteract these factors, many teams make use of labeling and printed warnings. For instance, medical devices are subject to cleaning after each use. The device manufacturers provide lists of acceptable detergents, meaning that any detergent that is not on the list should not be used. The same may be said about conditions of use. For example, a medical device such as an infusion pump designed for use in a hospital room

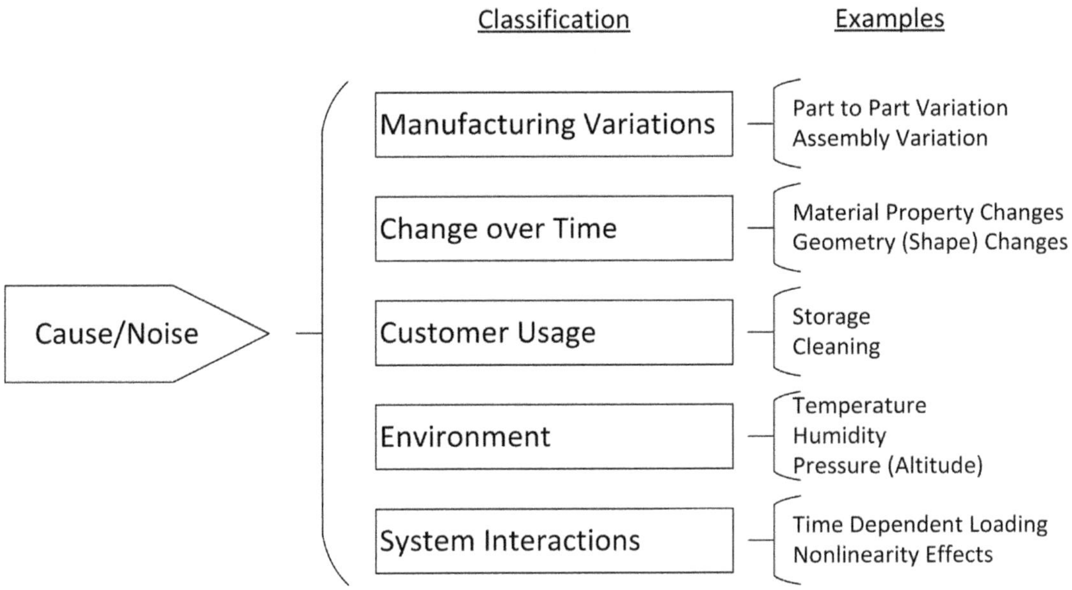

FIGURE 16.7
Noise (or cause) classification

may malfunction if used in an ambulance. To properly understand a product's failure modes, it would be prudent that device manufacturers understand the behavior of their products under some (if not all) off-label uses and provide some degree of warning against misuse.

Although parameter diagrams may be developed for all functions, they are typically developed for the critical components and assemblies in the system.

A CASE STUDY – ULTRASONIC AIR-IN-LINE SENSOR

In Chapter 12, the ultrasonic air-in-line sensor was introduced, and its main transfer function was developed. I like to elaborate on this example here. As shown in Figure 16.8, a mechanical drawing of the aforementioned ultrasonic sensor is depicted (the associated PCBA is not shown). In this embodiment, the spring pushes the receiver down against the IV set. The spring force is balanced either by the elastic forces of the elastomeric tube or the height of the side walls surrounding the IV set. The resulting geometry is a flattened tube. The flat surface of the tube is needed for a straight-path transmission of ultrasonic signals. In this ideal state, the transmitted signals may be received with relatively low noise.

Now, assume that there is warping in the plastic housing in the tongue and groove area, which prevents free movement of the top and bottom housing portions relative

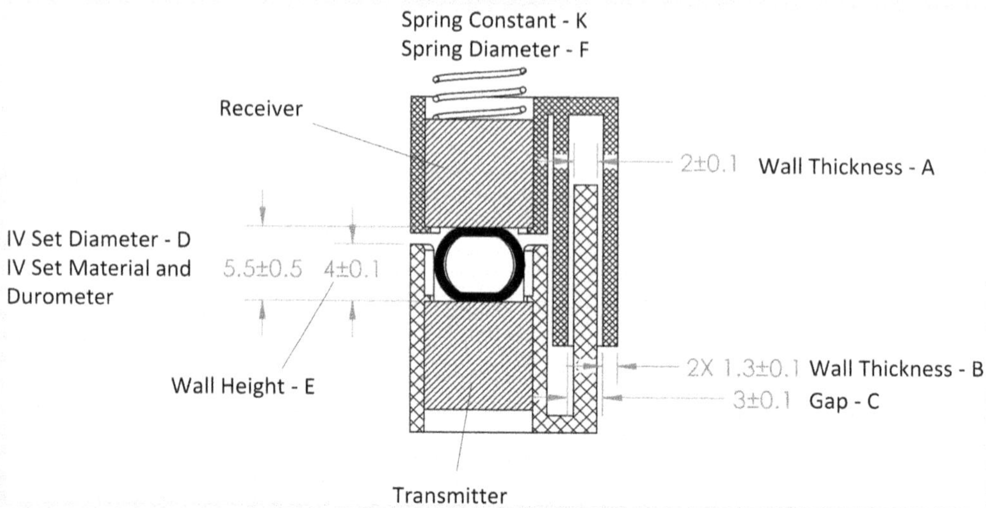

FIGURE 16.8
Ultrasonic air-in-line sensor

to each other. The result may be that the spring force is not sufficient to cause a sufficient flattening of the IV set; hence, a proper ultrasonic coupling between the transmitter and the tube is not formed. These are less-than-ideal signal conditions, and the signal received is noisy and possibly even unreadable.

Another faulty condition may be due to fluid ingress and an IV set with fluid on its outer surface. When this set comes in contact with the transmitter, the ultrasonic waves travel on the surface of the tube and not across it. Again, this results in a false reading on the receiver.

These and other causes may be captured in a P-diagram as shown in Figure 16.9. Developing such diagrams enables the design team to articulate not only various functions within the assembly and components but also to clearly identify failure modes as well. In addition, this information will serve to develop test plans, particularly for verification activities.

I would like to add that there are other techniques to develop a list of appropriate causes. Among these are the interface matrix, fishbone diagram, and brainstorming, among others. Stamatis (2003) provides a variety of tools that may be used for determination of causes. Having an understanding of these tools may ultimately help the design team diagnose issues and anticipate field failures. The knowledge will certainly help build a stronger DFMEA matrix. However, I share Anleitner's (2011) point of view that as design team members, we should not place an over-abundant emphasis on identifying all causes that lead to a relatively small number of failure modes.[2] In a way, we do not need to know all the cause-failure mode relationships (Anleitner 2011); just the major causes. It is just as important to understand the *effects* of the failure mode both locally and globally. It is the effects that propagate from the components up to the system and the end product. Hence, both failure modes and the associated effects must be well understood and mitigated if need be. In other words, if the failure modes (in their entirety) are understood and mitigated (or removed),[3] their causes may all but be forgotten.

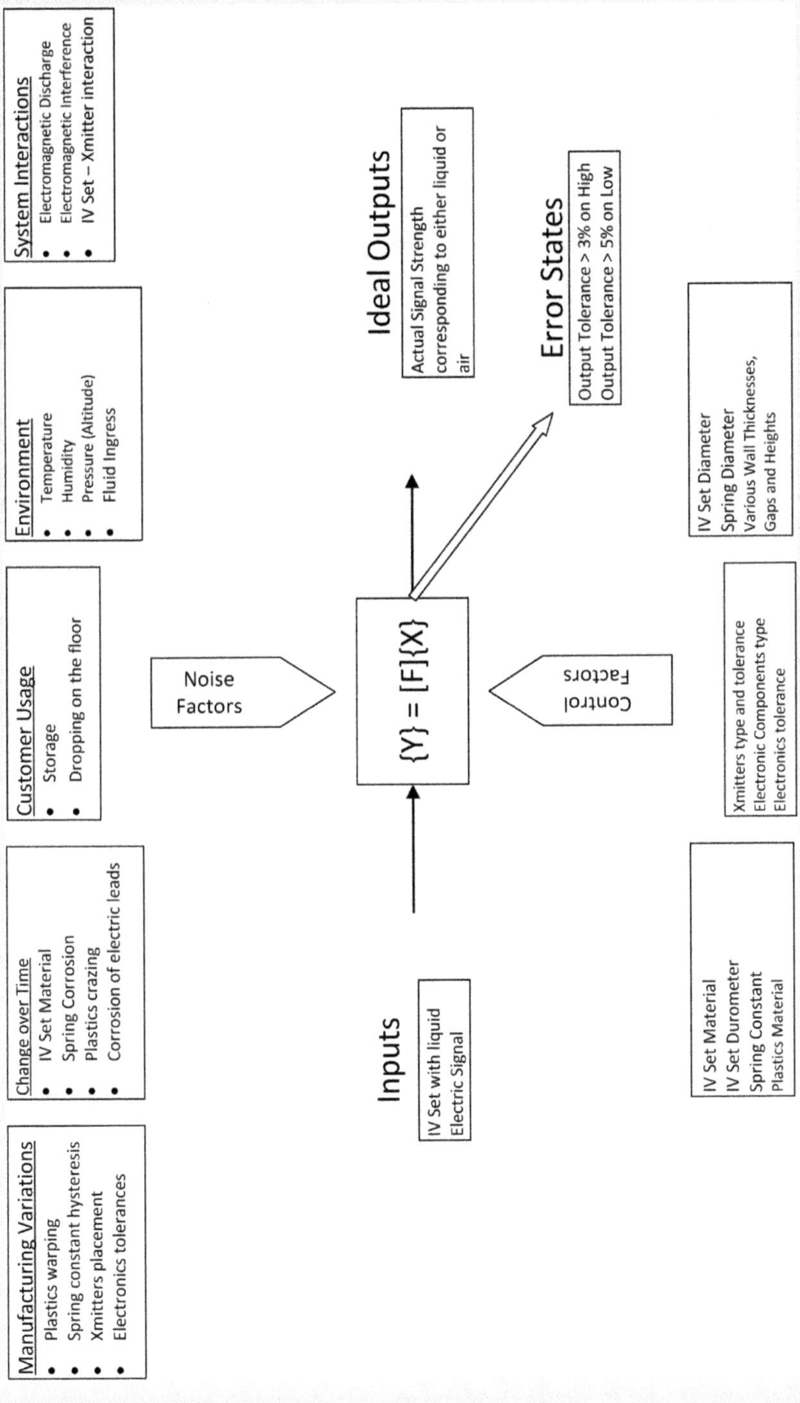

FIGURE 16.9
P-Diagram for the main function of ultrasonic air-in-line sensor

DFMEA Team and Pre-Work

> An FMEA study should not be conducted in a vacuum in the sense that it should not be a "check" on a list of to-do activities.

Up to this point, the focus of DFMEA discussion has been on the so-called *what*-s and developing a proper FMEA mindset. This and the following section focus on the mechanics of conducting a robust failure mode and effects analysis study.

An FMEA study should not be conducted in a vacuum in the sense that it should not be a "check" on a list of to-do activities. In that sense, this activity requires the right team to be assembled. Throughout this book, I have made reference to the *design team*. This team, however, does not need to be a static team. Typically, there is a core design and development team made up of a lead or chief design lead along with program management, manufacturing, service, marketing/operations directors/managers/leads. Other ad hoc teams may be formed to enable completion of various tasks or activities. The xFMEA teams may belong to this category of ad hoc teams that are formed as needed.

Members of a DFMEA typically include a design (or systems) subject matter expert, members of functional areas (typically mechanical, electrical, and software engineers) manufacturing and quality teams. Some also include members of reliability or human factor teams – should these groups exist. At times, it may even be appropriate to include either the customer, suppliers, or both. This should be done in the spirit of discovery of issues and/or problems that might need to be resolved.

> One of the pitfalls of developing an FMEA is an excessive focus on the form and format of the output. Another is the tendency to capture "all" failure modes and their causes.

Once the team is assembled, the roles and responsibilities must be clearly discussed and defined. There are some common xFMEA pitfalls that should be avoided. One of these pitfalls is an excessive focus on the form of the xFMEA output. I recall participating in a meeting when an entire hour was spent on whether the local or the corporate form of the DFMEA should be used. Another pitfall, which I mentioned earlier, albeit briefly, is the mindset that one has to complete an xFMEA because it is required, not because there is value in the work. The third, and possibly the most dire, is a lack of cooperation and engagement between team members. There are those who are extremely vocal, extremely withdrawn, or even have their personal agendas. To remedy or prevent a dysfunctional team, it may be appropriate to review basic ground rules such as, say, the appropriate length of time to discuss or review a specific topic, required participation and commitment by each team member, and, finally, assignment of the final decision to one individual.

Figure 16.10 provides an overview of the steps and the inputs needed to prepare for a DFMEA study.

As suggested, once a team is assembled and the ground rules reviewed, it is important to first develop a problem statement and define the goals that are to be achieved as a result of the FMEA meetings. The problem statement should clarify if the activity is for a new product and, if so, whether the focus is on the initial concept or at a subsystem or lower level. If the work is being done for a legacy product, is it due to a design change, part obsolescence or another reason? Finally, the expected outcome should be clarified as well. As a colleague of mine[4] often says: "What does *done* look like?"

Once the problem statement is defined, the team can focus on collecting the information needed in the study. This includes studying and understanding requirements and functions related to the subject of the DFMEA. Additional design details such as

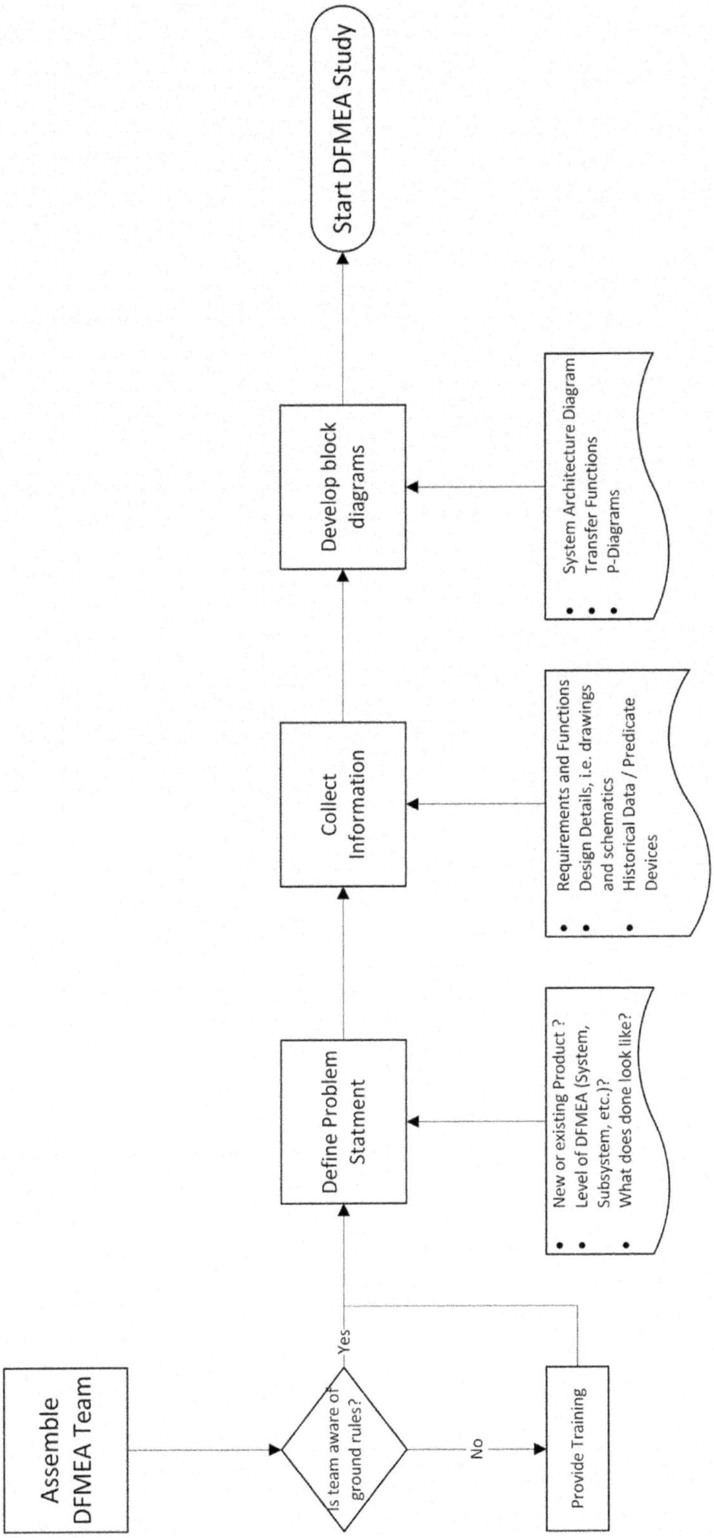

FIGURE 16.10
Steps and inputs needed to prepare for a DFMEA study

drawings, schematics, and even calculations should be reviewed. In addition, historical field failure data collected from service organizations needs to be organized. If a new product is being developed, information on predicate or competitor devices is an important source as well.

The final step prior to the actual FMEA study is to develop block diagrams of areas of interest, preferably of the entire system if one does not exist. In complex systems, it may be easier to develop an interaction matrix[5] that clearly defines the internal elements of what is being evaluated along with external elements affecting it. The output of the interaction matrix is the function and relationship of these various internal and external elements. The last elements that require clarification are various states and conditions of the systems; for instance, how does the operation of the system differ during start-up from steady operation? The environment of operation, customer usage, etc., needs to be known to the extent possible. In other words, the team should collect as much information about the noise factors as possible.

> The final step prior to the actual FMEA study is to develop block diagrams of areas of interest, preferably of the entire system if one does not exists.

Performing DFMEA

Once all the needed information is at the team's disposal, the DFMEA matrix may be completed rather easily. The following is a brief description of each column of the matrix. Here is a word of caution. Various texts and/or standards on developing the DFMEA matrix often provide a description of each column, as I am about to do here; hence, it is only natural that a novice practitioner would tend to complete the DFMEA going from one column to another and effectively fill the matrix row by row. In fact, the analysis flowchart in IEC 60812 prescribes that the DFMEA be developed row by row. In other words, the practitioner is encouraged to consider one function and then one failure mode of that function. Next, a single effect is considered, followed by the first cause, and so on. The next step would be to consider the second cause of the same failure mode, complete the second line, and repeat until all the causes for the first failure mode are exhausted. Then, the team moves on to identifying the second failure mode, and the process repeats itself. Although, theoretically, there is nothing wrong with this approach, from a practical point of view, it leads to rather haphazard DFMEA studies. Haphazard, why? Because the DFMEA team often runs out of steam and has less energy to put into the later lines.

For this reason and to ensure that major factors in a DFMEA are covered, it is highly recommended that the team first identify the major and essential functions of the item(s) under consideration; then, move on to identifying (again) the major failure modes, moving on to the important effects, and so on. Effectively, this is a column-by-column approach to developing the DFMEA matrix. The rational is that in the world of business, there is never enough time and resources to complete this matrix fully, but it is possible to develop a solid foundation in the first pass.

Table 16.1 depicts a DFMEA page with a typical header and relevant information. Note that the format of this table is slightly different from Table 10.5 presented in Chapter 10. This is to be expected, as various organizations and teams may personalize the information and material to fit their needs. The following is a brief description of each column.

TABLE 16.1

An Example of a DFMEA Matrix

DESIGN FAILURE MODE AND EFFECTS ANALYSIS

Part Certification New Product

System	Subsystem	Component	Page	FMEA Number Document Number
Part Number PART NUMBER		Design Responsibility R&D	Prepared by Name	Contact Information
Additional Information		Product Release Date mm/dd/YYYY	Initial FMEA Dat mm/dd/YYYY	
Core Team DFMEA Team members				

Design Item Function or Requirements	Potential Failure Mode	Potential Failure Effect(s)	Severity	Classification	Potential Failure Cause(s) or Mechanism(s)	Occurrence	Prevention	Detection	RPN	Recommended Actions	Responsibility & Target Completion Date	Actions Taken	Severity	Occurrence	Detection	RPN
Function 1	Failure Mode 1	Potential Effect 1			Potential Cause 1		Current Design Prevention 1			Recommended Action 1						
					Potential Cause 2		Current Design Prevention 2			Recommended Action 2						
					Potential Cause 3		Current Design Prevention 3			Recommended Action 3						
		Potential Effect 2			Potential Cause 1		Current Design Prevention 4			Recommended Action 4						
					Potential Cause 2		Current Design Prevention 5			Recommended Action 5						
					Potential Cause 3		Current Design Prevention 6			Recommended Action 6						
					Potential Cause 4		Current Design Prevention 7			Recommended Action 7						

Item Function or Requirement

Define the item and item functions based on requirements, defined functions, and block diagrams. Do this step for all items.

Potential Failure Modes

As defined earlier, a failure mode is the manner in which an item fails to meet its function and/or requirement. This information may be supplied from the P-diagram. Note that each function may have several failure modes. All modes should be listed. Again, this column should be completed in its entirety before moving on to the next column.

Potential Failure Effects

This is the local effect of the failure and is the failure mode of the next higher level in the system architecture. I have observed that in some organizations, the end effect of a local effect is called out. In my opinion, considering the end effect of a local failure mode only leads to confusion and a loss of traceability.

Severity

Severity is a numerical value that is assigned to the degree of the significance of the effect. Chapter 10 provided a categorical severity table adapted from MIL-STD-882E ranging from 1 to 4. Other industries may provide different ranges. In automotive (AIAG 1995), this number ranges between 1 (lowest) and 10 (highest associated with death or serious injury). Stamatis (2003) provides an overview of the governing bodies and standards that a variety of industries are obligated to follow. This reference may be used to identify an appropriate scale.

Classification

Some failures may have critical or safety impacts. These are typically associated with critical or safety (or significant) functions. Should this be the case, the characteristics column is marked with either CC or SC. Verification testing may focus first on areas that have either critical or safety classifications.

Potential Failure Causes or Mechanism

In this column, potential causes of failure are listed. P-diagram may be a good tool for this purpose.

Occurrence

This is the rate at which failure modes occur. Similar to severity, Chapter 10 provided a categorical severity table adapted from MIL-STD-882E ranging from A to F. Other industries may provide different ranges. In automotive (AIAG 1995), this number ranges between 1 (lowest) and 10 (highest associated with death or serious injury) based on expected failure rates. This field should be based on reliability data and calculated or observed failure rates. Some indicate a value of 1 corresponds to a 98% or even 99% reliability and a 10 to a reliability of less than 10%.

Prevention

In this column, prevention methods are listed to ensure that the design is adequate for the function or requirement under consideration. Examples of these design activities are verification testing or analysis. Typically, there are three types of prevention; namely, detect the failure mode, detect the failure mechanism, or finally, prevent the failure mode or reduce its occurrence rate (AIAG 1995).

Detection

Detection is a measure of how well the prevention strategy outlined in the previous column will in fact detect a potential cause before *the design is released for production.* It is important to note that detection of a design shortcoming is meant not as identifying production issues but design shortcomings. This is probably the single point of confusion that DFMEA teams may experience. The tendency is to suggest that any potential issues may be identified during production or post-production inspections. In my opinion, *detection* in DFMEA is a metric indicative of the design team's prior experience with the proposed design concept(s) and the details under development.

As an example, suppose that as an electromechanical design engineer, I incorporate, say, a laser sensing mechanism for air bubble detection in a fluid. The theory is that the laser will identify the boundary between air and the fluid and then calculate the air volume. Clearly, if this is the first time that I am using this approach, there are much higher chances that design flaws are not detected until such time that the product is manufactured and either tested in the lab or found by the end users.

Risk Priority Number

The product of severity, occurrence, and detection is called RPN. This number is a measure of the design risk and is used to determine if any actions are needed. In practice, recommended actions are prioritized based on the highest severity items and then the RPN number.

Recommended Actions

Once the failure modes have been prioritized by the RPN values, to the extent possible, the design should be modified (or corrected) to first reduce the severity and occurrence rates, followed by an increase in detection. The preventive control may include fail-safe designs or tolerance analysis. Alternatively, the design may incorporate detection mechanisms such as on-board sensors or built-in controls to detect failures and alarm the end user. If the DFMEA takes place when the design is frozen, design configuration changes may be difficult to achieve.

Responsibility and Target Completion Date

The purpose of this column is primarily for project management purposes. Frequently, a variety of issues are discussed in meetings, and yet, once the meeting is over, there is no clear documentation on what actions are assigned to whom. This column provides a placeholder so that such decisions are properly documented.

Actions Taken

Once the recommended actions are implemented and verified, a brief description of the actual action and its date of implementation is recorded.

Final RPN

When the recommended actions are implemented, the RPN should be recalculated based on the new estimates of severity, occurrence, and detection. Should RPNs be above a predetermined threshold, steps must be taken once again to reduce it.

Process Failure Mode and Effect Analysis

In the beginning of this chapter, I discussed the place of xFMEAs in the design process. In particular, Figure 16.2 referred to several different types of FMEAs; namely, system, design, manufacturing, service, and use FMEAs. As the development activity moves down the V model, the first FMEAs to be completed are System and Design. Once the build/design transfer stage is completed, the other types of FMEA should be developed. The first of these is manufacturing FMEA, followed by service FMEA. In some industries, such as medical, a use FMEA is also developed.

It is noteworthy to consider that typically, system and design FMEAs focus on the hardware and software alone and do not consider human interaction with the product. However, as the FMEA development moves into manufacturing, service, and use, the impact of human interaction increases substantially. So much so that a Use FMEA considers only the operator's interaction with the product's hardware and software.

Post-design FMEAs typically evaluate the processes of either manufacturing or service. For this reason, often, they are collectively called process FMEAs (PFMEA). One could argue that even the use of a product may be categorized as a type of process. For simplicity's sake, I will refer to them as PFMEA as well.

As the word *process* implies, manufacturing and service (and even use) are really a series of repeatable steps or events whose outcome is a successful completion of manufacture, service, and/or use of a product. Thus, these steps/events need to be clearly defined, and in the process of studying these steps, potential mishaps and their impact on the product or the end user need to be identified. With these mishaps identified, control actions may be instituted to minimize their impact.

> For the PFMEA studies, as the word *process* implies, a series of repeatable steps should be in place whose outcome is a successful completion of manufacture, service, and/or use of a product. Thus, these steps/events need to be clearly defined, and, in the process of studying them, potential mishaps and their impact on the product or the end user need to be identified.

As stated earlier, a PFMEA study should be conducted when:

- modifications are made to component designs or technologies with impact on manufacturing, service, and/or use.
- modifications are made to (or new information is obtained on) manufacturing or service processes.
- information is obtained on new or unintended customer usage

All xFMEA documents are living documents, and no PFMEA is an exception; once established at the onset of manufacturing, these documents should be maintained throughout the product life cycle to reflect the changes mentioned above.

PFMEA Team and Pre-Work

Much of the underlying mindset discussed and described for DFMEA studies holds here as well. A dedicated PFMEA team should be formed that would report to the core design and development team as described previously. The members of a PFMEA typically include design (or systems), manufacturing (and/or service), quality engineer, tooling, industrial, and process engineers. Additionally, the DFMEA team also includes members of purchasing and supply chain management. As before, it may even be appropriate to include external suppliers as well.

Once the team is assembled, the roles and responsibilities must be clearly defined and discussed. Furthermore, the potential pitfalls that derail the study should be reviewed. These common pitfalls were reviewed previously in the DFMEA discussion section.

Figure 16.11 provides an overview of the steps and the inputs needed to prepare for a PFMEA study.

As before, once a team is assembled and the ground rules reviewed, it is important to first develop a problem statement and define the goals that are to be achieved as a result of the PFMEA meetings. The problem statement should clarify if the activity is for a new process for an existing product or an entirely new product.

Once the problem statement is defined, the team can focus on collecting the information needed in the study. This includes studying and understanding the DFMEA developed for this product and identifying any critical components. Additional design details such as drawings, schematics even calculations should be reviewed to identify the relationship between the identified critical components and the assemblies housing these components. The results of associated design for assembly or manufacturing (DFA/M) studies should be reviewed and summarized. Should these studies not exist, this would be the time to conduct them. In addition, historical information on predicate and/or competitive devices should be collected and analyzed.

The final step prior to the actual PFMEA study is to develop the process block diagrams of areas of interest, preferably of the entire manufacturing process if one does not exist. Figure 16.12 depicts a process flow diagram for the manufacturing of the handle of a surgical tool. Notice that primary and secondary areas are identified on this map. PFMEA focuses on the primary stations and what may go wrong.

Performing PFMEA

Once all the information is at the team's disposal, the PFMEA matrix may be completed rather easily. The following is a brief description of each column of the matrix. My earlier caution on the pitfalls of developing the DFMEA matrix holds here as well. It may be more effective to complete the matrix column by column. Table 16.2 depicts a PFMEA page with a typical header and relevant information.

Item Stage Process Requirement

Define the stage and its function based on the process map and workflow diagrams, if any.

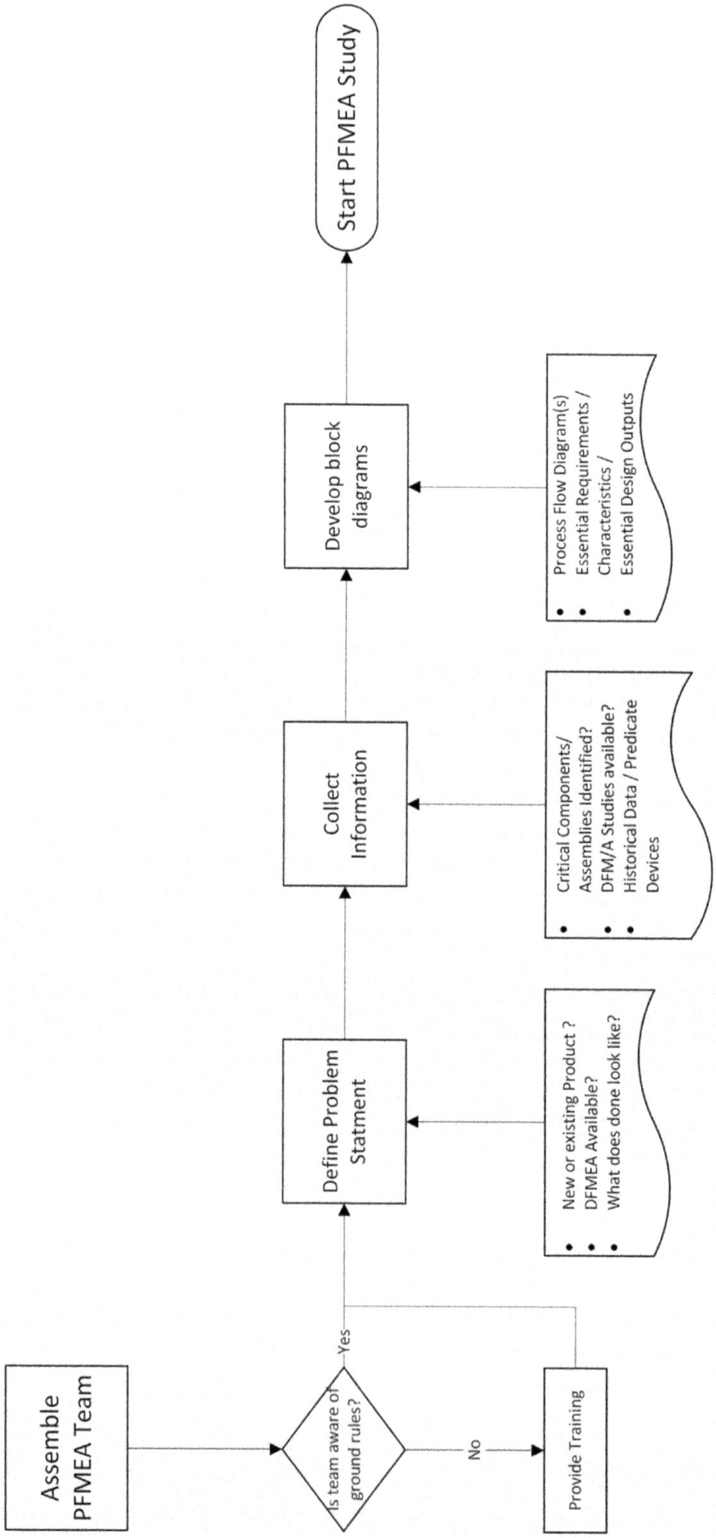

FIGURE 16.11
Steps and inputs needed to prepare for a PFMEA study

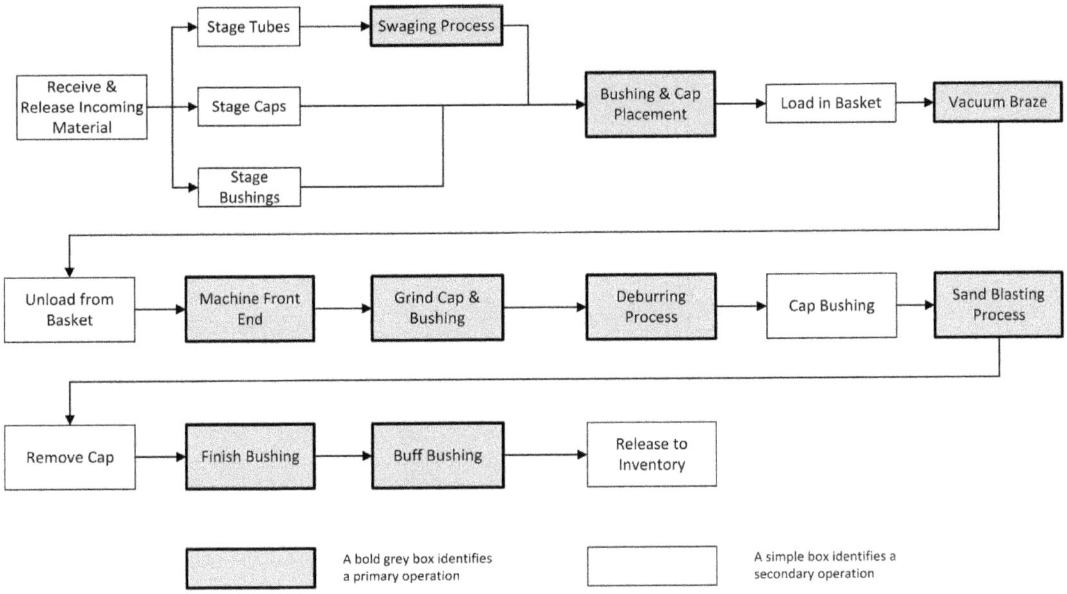

FIGURE 16.12
Process map for the manufacture of a surgical tool handle

Potential Failure Modes

As defined earlier, a failure mode is the manner in which a process step or workflow fails to meet its intended process requirement. Note that each stage process may have several failure modes. All modes should be listed. Again, this column should be completed in its entirety before moving on to the next column.

Potential Failure Effects

This may be both the local effect of the failure on the next stage of the process and the end effect on the product. For example, if the insulation of an electric wire is not properly installed (or not installed at all), it may not have a local effect but the end effect, may be electric shock to the end user.

Severity

Severity is a numerical value that is assigned to the degree of the significance of the effect. Considering a variety of scales may exist, it is important that the proper scale within one's own industry be used.

Classification

Some failures may have critical or safety impacts. These are typically associated with critical or safety (or significant) functions. Should this be the case, the characteristics column is marked with either CC or SC.

TABLE 16.2

An Example of a PFMEA Matrix

PROCESS FAILURE MODE AND EFFECTS ANALYSIS **Manufacturing Service**			
Part Certification	**New Product**		
Process Segment		Page	FMEA Number Document Number
Part Number PART NUMBER	Process Responsibility R&D	Prepared by Name	Contact Information
Additional Information	Product Release Date mm/dd/YYYY	Initial FMEA Dat mm/dd/YYYY	
Core Team xFMEA Team members			

Process Item and what should be completed	Potential Failure Mode	Potential Failure Effect(s)	Severity	Classification	Potential Failure Cause(s) or Mechanism(s)	Occurrence	Prevention	Detection	RPN	Recommended Actions	Responsibility & Target Completion Date	Actions Taken	Severity	Occurrence	Detection	RPN
Stage 1	Failure Mode 1	Potential Effect 1			Potential Cause 1		Current Process Prevention 1			Recommended Action 1						
					Potential Cause 2		Current Process Prevention 2			Recommended Action 2						
					Potential Cause 3		Current Process Prevention 3			Recommended Action 3						
		Potential Effect 2			Potential Cause 1		Current Process Prevention 4			Recommended Action 4						
					Potential Cause 2		Current Process Prevention 5			Recommended Action 5						
					Potential Cause 3		Current Process Prevention 6			Recommended Action 6						
					Potential Cause 4		Current Process Prevention 7			Recommended Action 7						

Potential Failure Causes

In this column, potential causes of failure are listed.

Occurrence

This is the rate at which failure modes occur. Estimates of occurrence should be based on predicate or historical information on similar processes. Process capability indices such as Cpk or Ppk may be used.

Prevention

In this column, prevention methods are listed to ensure that the process step is adequate to deliver the product function or requirement under consideration. Examples of these design activities are verification testing or analysis. Typically, there are three types of prevention; namely, detect the failure mode, detect the failure mechanism, or finally, prevent the failure mode or reduce its occurrence rate (AIAG 1995).

Detection

Detection is a measure of how well the prevention strategy outlined in the previous column will, in fact, detect a potential cause before *the part leaves the stage area for the next step in production*. Some may rely on random inspections to detect failed components. That is not an effective detection method; however, the regular statistical sampling technique is quite effective.

Risk Priority Number

As in the DFMEA, the product of severity, occurrence, and detection is called RPN. This number is a measure of the process risk and is used to determine if any actions are needed. In practice, recommended actions are prioritized based on the highest severity items and then the RPN number.

Recommended Actions

Once failure modes have been prioritized based on the RPN values, to the extent possible, the process should be modified (or corrected) to first reduce the severity and occurrence rates, followed by an increase in detection. The preventive control may include machine functions, assembly procedures, and personnel training.

Responsibility and Target Completion Date

The purpose of this column is primarily for project management purposes. Frequently, a variety of issues are discussed in meetings, and yet, once the meeting is over, there is no clear documentation on what actions are assigned to whom. This column provides a placeholder so that such decisions are properly documented.

Actions Taken

Once the recommended actions are implemented and verified, a brief description of the actual action and its date of implementation is recorded.

Final RPN

Once the recommended actions are implemented, the RPN should be recalculated based on the new estimates of severity, occurrence, and detection. Should RPNs be above a predetermined threshold, steps must be taken once again to reduce it.

Control Plan

Throughout the course of this chapter as well as Chapter 10, I have made repeated reference to critical and/or safety (or significant) characteristics. The xFMEA matrix has a column dedicated to this classification. It should therefore be no surprise that all items associated with either critical or safety characteristics are monitored to ensure that they are in control. As variations are part and parcel of every production and service environment, on the one hand, the design team needs to be aware of how variations impact these critical items. On the other hand, plans need to be developed and implemented to ensure that critical item characteristics are not out of acceptable bounds.

As variations are part and parcel of every production and service environment, the design team needs to be aware of how variations of certain safety (or significant) characteristics of the design may impact the functionality or safety of the product. It is just as important to develop and implement control plans to ensure that critical item characteristics are not out of acceptable bounds.

As mentioned, the process of writing a control plan begins with developing an understanding of process variations. This understanding may come about through a short capability study to ensure that the proposed process is realistic. This requires that a knowledge of measurement systems exists and that it is appropriate. By measurement systems, I mean that the techniques and approaches used to measure the process outcome are appropriate to the process.[6] Often, if a measurement system analysis (MSA) is not available, a design of experiments may be conducted (called gage repeatability and reproducibility or gage R&R) to measure and qualify the levels of variation.

Often a control plan incorporates a statistical sampling method through which, at regular intervals, components are removed, measured, and their results are logged and measured against previous measurements. Once a comparison with previous data are made, production or service personnel should be trained to know what actions are required should negative trends be observed. An example of a control plan is provided in Table 16.3.

Example

To demonstrate the use of this control plan, consider the process map as shown in Figure 16.12. Table 16.4 provides a possible control plan for the first three primary stages of this process. For instance, in the swaging process, the control plan suggests measuring the diameters of five samples using an optical comparator and using the data in a process control chart. The frequency of this check is once per each new lot. The same plan suggests using an in-house test procedure for checking for pinholes in the vacuum braze process. Here the number of samples is increased because the type of data has changed from measured to a pass/fail. I will discuss data and measurements in Chapter 22.

TABLE 16.3
An Example of a Control Plan Template

PROCESS CONTROL PLAN						
Part Number PART NUMBER	Process Responsibility R&D					Prepared by Name
Additional Information	Product Release Date					Initial Control Chart Date
	mm/dd/YYYY					mm/dd/YYYY
Core Team Control Team members						
Process Item and Description	Critical Characteristics	Sample Units	Sample Size	Sample Frequency	Test Method	Reporting Mechanism

TABLE 16.4
An Example of a Control Plan for the Surgical Handle

PROCESS CONTROL PLAN						
Part Number PART NUMBER	Process Responsibility R&D					Prepared by Name
Additional Information	Product Release Date					Initial Control Chart Date
	mm/dd/YYYY					mm/dd/YYYY
Core Team Control Team members						
Process Item and Description	Critical Characteristics	Sample Units	Sample Size	Sample Frequency	Test Method	Reporting Mechanism
Swaging Process	Diameter 1	Each	5	1/lot	Optical Comparator	X Chart
	Diameter 2	Each	5	1/lot	Optical Comparator	X Chart
	Length	Each	5	1/lot	Calipers	X Chart
Bushing and Cap Placement	Cap and Bushing Alignment	Each	3	1/Lot	Visual	U Chart
Vacuum Braze	Pin Holes	Each	10	1/Batch	In-House Test Procedure	U Chart

Summary

- Failure Mode and Effects Analysis (FMEA) provides a deductive and structured approach to identify problems related to product design, manufacturing, service, and even use.
- There are in reality two aspects of DFMEA. The first approach is top-down, and the second is the traditional bottom-up formulation.
- The root cause of system-level malfunction (or failure) is always at the level of a single component or the interface of two components.
- Noise factors tend to destabilize function; control mechanisms are put in place for functional stability.
- There are five sources or categories of noise: (1) manufacturing variations; (2) change over time; (3) customer uses; (4) environmental conditions; and (5) system interactions.
- Parameter diagrams are typically developed for the critical components and assemblies in the system.
- An FMEA study should not be conducted in a vacuum in the sense that it should not be a "check" on a list of to-do activities.
- Control plans are developed and implemented to ensure that critical item characteristics are not out of acceptable bounds.
- Writing a control plan begins with developing an understanding of process variations.

Exercise

1. Develop a DFMEA matrix based on the design of the product used in the Chapter 15 exercise.

Notes

1 This is really an example of failure at interfaces as well as the module itself.
2 Recall that a failure mode is defined as over-delivery, under-delivery, or intermittent-delivery of function.
3 I like to say that removing a failure mode is easier said than done.
4 Kenneth Neltnor, Lead System Designer, Baxter Healthcare Corp.
5 A description and details of developing an interaction are beyond the scope of this work. The reader is encouraged to consult Anleitner (2011).
6 A common grade-school ruler is not a good tool to measure the precision of a machined surgical tool. This example may sound silly, but often we do not stop and think about whether our measurement approach fits what is being measured. Another flaw is measurement in that what is measured may not be relevant to the required function of the component.

17

Design Verification, Product Validation, and Design Transfer

Introduction

I would like to share a personal anecdote to emphasize the need for a process to transfer developed designs to manufacturing and ultimately the market:

A common practice in the field of dental or orthopedic surgery is that innovative tools and techniques developed by thought leaders are supported by manufacturers in these fields. Once there is strong evidence that a new tool is successfully used, manufacturing organizations then collaborate with its inventor(s) to fully develop these tools and bring them to market. I had been a design engineer for a leading dental instrument manufacturer when a periodontist proposed a hand tool to enable general practitioners to place implants in their patients' teeth.

During the course of the development of this instrument, I designed and fabricated an assembly fixture. This fixture was used to construct the initial number of prototypes needed for a variety of reasons, including additional clinical and marketing studies. The project was approved and moved to the manufacturing phase. And I moved on to my next project.

A few weeks later, I came across the assembly fixture and took it to the manufacturing engineer in charge. "Oh, no wonder we have so many failures!" was his response. Fortunately, production had just begun, and the initial production volumes were not large. Here was the miss on my part: I had not documented the fixture anywhere. Here was his miss: he had not asked if any special or additional fixtures were needed!

Although either or both of us can bear the blame for this miss, in reality, the organization was also to blame for not having robust practices in place to ensure a smooth transition from design into manufacturing. Admittedly, the term "design transfer" is often associated with medical device manufacturing and regulatory bodies such as the FDA. In my opinion, design transfer is an aspect of good engineering practices that regulatory bodies have been emphasizing.

It turns out that ISO 13485:2003 (Medical devices – Quality management systems – Requirements for regulatory purposes) does not have a specific section dedicated to design transfer – just that it is a requirement; similarly, ISO 14969:2004 (Medical devices – Quality management systems – Guidance on the application of ISO 13485:2003) only provides a recommendation to have design transfer requirements (Daniel and Kimmelman 2008). Only the FDA's Quality System Regulation has a specific design transfer requirement: "Each manufacturer shall establish and maintain procedures to ensure that the device design is *correctly* translated into production specifications." (21 CFR § 820.30 Section H).

DOI: 10.1201/9781003564508-22

> Design transfer is the process of translating the device design into production specifications.

The emphasis here is on the word "correctly." In the anecdote that I shared, one may argue that the design documents that were placed in the document control database were correct; however, they were not complete. This lack of totality of documentation led to a false sense of completeness for manufacturing personnel. Hence, in a broader sense, "correctly translated" means that the complete knowledge about the product should be encapsulated and explained in production specifications, work instructions, etc. In return, concrete manufacturing instructions ensure that products are produced within process capabilities time and again.

The pressures to bring a product to market generally work against a detailed, meticulous, and documented transfer of knowledge. In simple projects – similar to what I shared earlier – the two sides assume that there is enough familiarity with the new product and its workflows that additional knowledge transfer is not needed. Hence, no documentation is needed. For more complicated systems, a design team member often spends time with the production to iron out issues and provide on-the-fly training that may be needed. However, without proper documentation in place, once this initial period is done and design and production staff decouple, manufacturing relies only on tribal knowledge for successful and problem-free runs. Once key personnel leave, this vital information may be lost as well.

I guess what I am saying is that whether or not a documented design transfer is required by a regulatory body, it is good engineering practice.

Design Transfer in Product Development Model

I may be getting ahead of myself. Let's review the product roadmap and the V-model once again. Consider Figure 17.1. In this figure, the V-model is placed side by side with that of the roadmap. In the V-model, stage 8 is design transfer at the pinnacle of the "V" and prior to any verification and validation testing. The product roadmap places the steps of transferring to manufacturing and service as the last step to the product being considered complete. The apparent discrepancy between the two mindsets is that the V-model is inherently iterative, whereas the roadmap considers only the completed steps.

The iterative nature of the V-model is such that every time that an assembly or subsystem is built for testing, it has to go through the design transfer phase. In other words, once the details of the design and engineering prototypes have provided evidence that the design is feasible, the first manufacturing articles are developed. However, design issues and problems may be discovered that would require a design change or update. These updates would bring design back to the left-hand side of the V-model and back into the design realm. Once the design is updated and closed, the process may not bypass the design transfer phase.

To start the design transfer process, a series of questions need to be entertained and answered. These questions are as follows:

1) Are all *make* component drawings or schematics complete and error-free?
2) Are all *buy* item specifications developed and do they have associated part numbers?
   a. For any custom-designed *buy*-item, has the supplier agreed to meet the critical dimensions and/or features of the part?

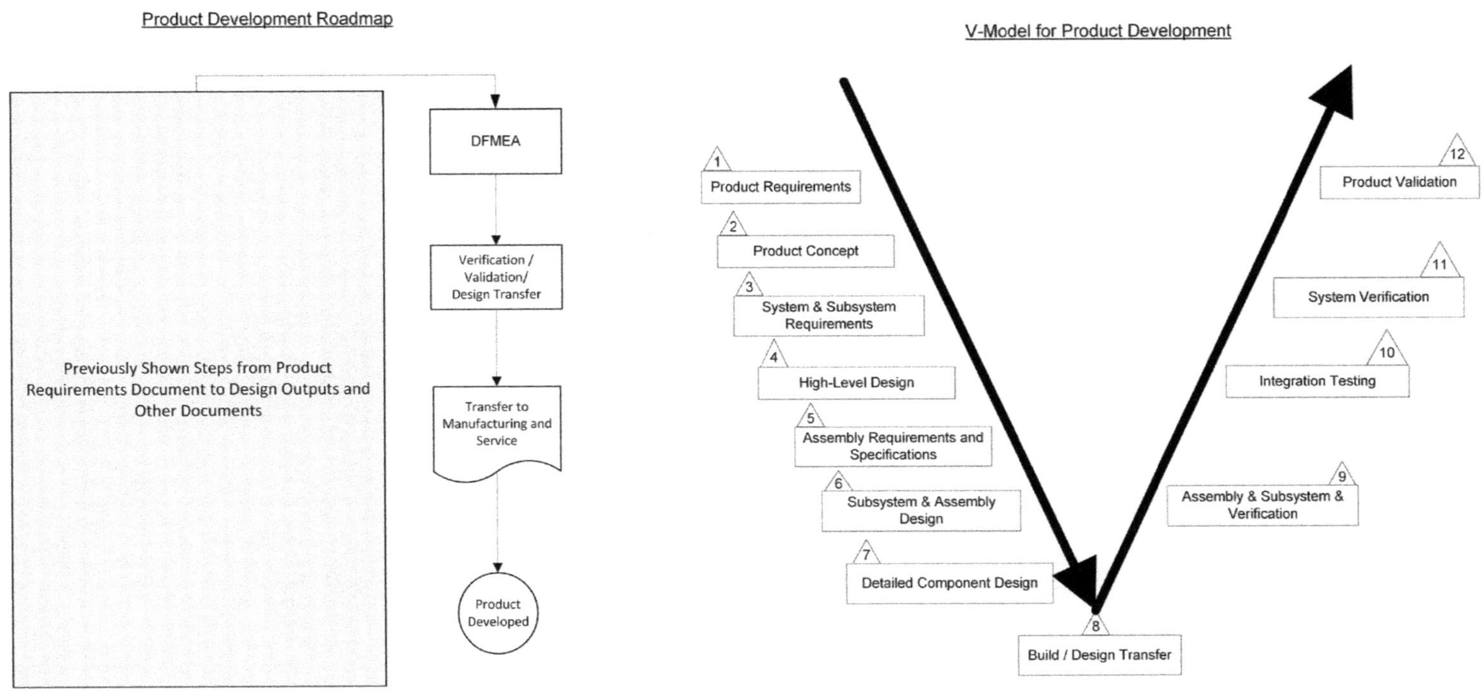

FIGURE 17.1
Design transfer in the product development roadmap versus the V-model

3) Are all *consumable* item specifications developed and do they have associated part numbers?
4) Are all *labeling* item specifications developed and do they have associated part numbers?
5) Is the bill of materials complete and error-free?

Should the response to these questions be lacking, the design must be reevaluated to obtain the right answers. Once the responses to these questions have been properly evaluated and approved, the second tier of inquiries must be made:

1) Have critical (or essential) design outputs been identified?
2) Are assembly procedures available and documented?
3) Are assembly fixtures or jigs needed?
4) Has an error identification code list been created and documented?
5) Are service procedures available and documented?
6) Are service fixtures or jigs needed?
7) Have PFMEAs and control plans been completed?

With the answer to these questions available, an initial build of the assembly and/or subsystem may be made.

A Successful Design Transfer Activity

The goal of design transfer is to test whether the production (and/or service) process is robust and that variations would not deter manufacturing performance and the quality of the product. This end result reflects itself in a few metrics. The first of which is a demonstration that the manufacturing process is both reproducible and repeatable with a high degree of confidence. As an example of a reproducible and repeatable process, the following tool (as shown in Figure 17.2), used in dentistry to mix various cements and/or resins. The main requirements for this tool are (1) its extreme flexibility; (2) its lack of adherence to any resin or cement; and finally, (3) its high resistance to corrosion.

The manufacturing process of this part is quite extensive and costly. A cost savings proposal was made to use MIM instead of the traditional methods. The potential cost savings could be up to 50%. Engineering prototypes provided evidence that the end product is

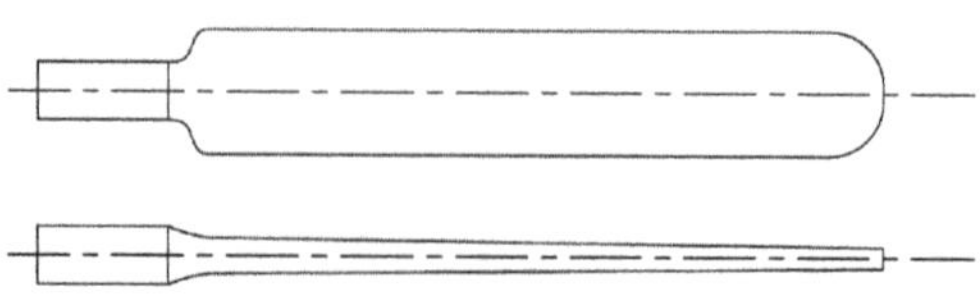

FIGURE 17.2
A simple rendition of a dental spatula
(Handle is not shown)

> The goal of design transfer is to test whether the production (and/or service) process is robust and that variations would not deter manufacturing performance and the quality of the product.

equivalent to parts made using traditional means. With this data, the design team decided to move into the design transfer phase based on a one-cavity mold to evaluate the production environment and process.

The MIM process is a two-step process. In the first step, a slurry of metal particles and binding material is injected into a cavity that forms what is called a green part. This part is about 40% larger than the finished article. In the next step, the green parts are staged and placed into a curing oven where the binding material is "burned" away and the metal particles fuse together. The staging of parts is a crucial step: during the curing process the part volume is reduced. Should this reduction be hindered or be nonuniform for any reason, the final part warps or deforms out of shape.

Although this spatula is a seemingly simple part, it has a very complicated geometry. During the design transfer phase, as process capabilities were studied, the design team learned that the MIM process for this component was not capable. In other words, repeatable and reproducible parts could not be fabricated.

Needless to say, as the capability of a specific process is being studied, various process variables need to be closely monitored, and their impact on the process in general and critical, significant, and/or safety characteristics (CC or SC) needs to be determined. These factors are identified in the DFMEA files. They can now be linked to key process indicators through the process studies. The results of process studies are used as inputs into PFMEAs.

The last aspect of successful design transfer activities is to evaluate received raw materials and components to ensure that they meet product requirements. I recall another incident where a marketing requirement dictated a custom-color engineering plastic. These plastics were used to create the bezel of a high-end product. To ensure color consistency, the decision was made to have a supplier provide a custom blend of the material with the desired color. Again, initial samples were qualified; however, as production capabilities were being examined, the team identified small black spots that would appear in random locations on some of the bezels. This was a violation of cosmetic requirements. The root cause of this issue was traced to the supplied material and an inherent flaw in the production of the custom-blend, custom-colored resin.

Design Transfer Process

As suggested earlier, the design transfer process begins by collecting information about the particular design that is being transferred to production and ensuring that this dataset is complete. It is important that the significant, safety, or critical components of the design be clearly identified. The next step is to identify or develop any assembly or test fixtures that may be needed. By "development," I mean creating and documenting any needed drawings and/or details of operation. Next, production equipment should be identified, followed by a draft of the production process and training of the operators who will be using this equipment.

Once these steps are done, the production team will be ready to analyze their measurement system by running a reproducibility and repeatability study followed by installation, operational, and performance qualification (IQ/OQ/PQ) activities. Qualification studies (IQ/OQ/PQ) answer the following three questions via providing objective and documented evidence:

1) Are the equipment used in the process installed correctly and functioning as expected?
2) Have the factors that impact the quality of the product been examined, tested, and placed under control? For instance, in an injection molding process, mold temperature, clamp and injection pressure, and dwell time are among these factors.
3) Can the process consistently produce a product that meets its requirements in full?

At the end of PQ activities, the production team may answer two concerns: first, whether the process is performing as expected, and second, whether the product (i.e., part, assembly, subsystem, or system) is behaving as expected. If the response is affirmative, then testing may continue to ensure that performance qualifications are also done. For a more comprehensive description, refer to the Global Harmonization Task Force document (GHTF/SG3/N99-10:2004 2004).

The final steps in design transfer are to document lessons learned about the process and the key process variables that need to be controlled in order to ensure the quality of the object undergoing design transfer. A control plan – discussed in Chapter 16 – is an output of design transfer. Should the need be, PFMEA files are updated with any newly found information. The design transfer process is depicted in Figure 17.3.

Design Verification and Product Validation

Often team members refer to verification and validation activities as V&V. I have found that this abbreviation has indeed blurred the boundaries between these two very distinct areas. There seems to be a degree of ambiguity in regard to what sets of activities are within the boundaries of verification and what sets of activities are within the boundaries of validation. Many suggest that "paper" designs are *verified* prior to production; once on the manufacturing floor, then, processes to deliver the product are *validated* (Anleitner 2011). Similarly, some equate *verification* with inspection,[1] whereas *validation* is considered to be testing (MEDICept 2010).

Russell (2013) offers the following definitions based on what has been defined in ISO 9000:2005 and in ISO 9001:2008: "In general, verification is checking or testing, and validation is the actual performance of its intended use. . . . Verification is the authentication of truth or accuracy by such means as facts, statements, citations, measurements, and confirmation by evidence. An element of verification is that it is independent or separate from the normal operation of a process. . . . Validation is the demonstration of the ability of the system-processes under investigation to achieve planned results" (Russell 2013, pp. 281–282). Note that Russell's definition of verification includes both inspection and testing; as such, it is more extensive than that of Anleitner

In the context of the product development roadmap, validation focuses on the finished product and is a set of activities to ensure that customer needs – as reflected in the product

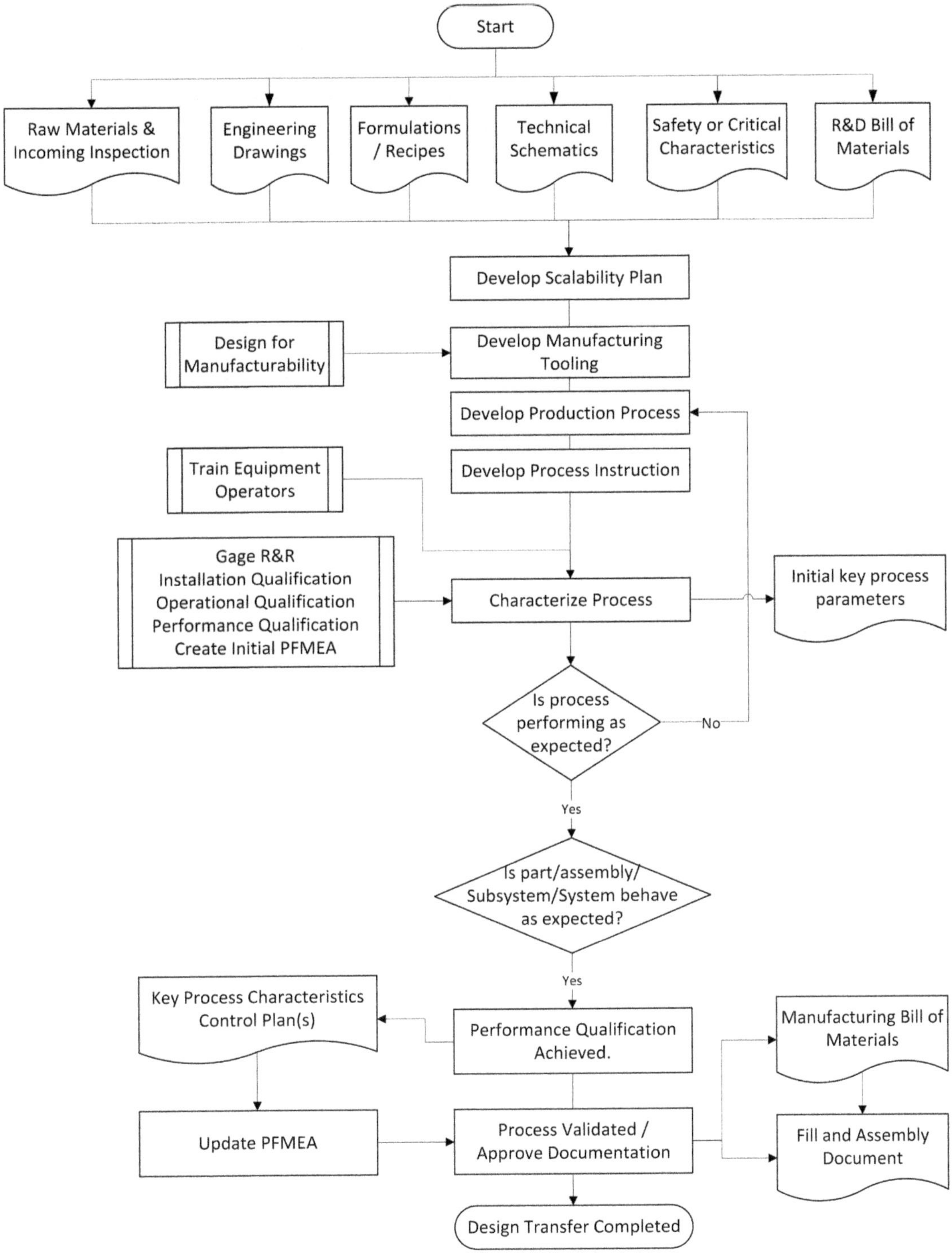

FIGURE 17.3
Design transfer process

requirements document – are met. Verification, however, focuses on requirements at system and subsystem levels and sets out to ensure that these requirements are met by analytical or empirical means.

In short, the purpose of *design verification* is to prove that "design outputs meet design requirements." This may be done by either inspection or testing of each design requirement on the entire system or by each subsystem or even assembly. The purpose of *design validation* is to ensure that the product design conforms to "defined user needs and intended uses." This may be done by testing each customer requirement as stated in the product requirements document.

> In the context of the product development roadmap, *validation* focuses on the finished product and is a set of activities to ensure that customer needs are met. *Verification*, however, focuses on requirements and sets to ensure that these requirements are met by analytical or empirical means.

There are possibly three major pitfalls in the design verification and validation activities that may be easily avoided. The first is that these tests should be performed on production-equivalent assemblies, subsystems, and/or products. The catchphrase is *production equivalent*. Many qualification tests may be completed using engineering samples. The results of these tests are often used either as the basis of design modifications or as the justification to freeze the design and start the design transfer phase. However, both V&V are formal activities that require that production tooling and procedures be in place. It is this requirement that creates a feedback loop[2] to design transfer. If issues are discovered during either verification or validation, design transfer activities may be updated or repeated. That is why it is often said that design transfer is complete when V&V is done.

In Chapter 22, I will elaborate more on the nature of testing and its development; however, as a reminder, I like to say that all V&V testing should be documented in test plans or protocols. Furthermore, they have to be performed using calibrated equipment;[3] should any fixtures be required for data collection, they should be validated prior to use. Finally, each V&V test should identify by serial numbers or other means the components or systems used. This traceability to test samples should remain intact in case of a future need to either audit or examine the test articles for further analysis. Figure 17.4 provides a pictorial means of connecting design input and requirements to various verification and validation activities.

Traceability to Requirements

One of the challenges with verification activities is documentation due to the sheer volume of the test cases that need to be executed. Often, during final design reviews and later during audits, it is difficult to provide an overview of all the activities that have taken place and to ensure that all requirements have indeed been verified and/or validated.

An effective tool to provide an overview of pictures and at the same time identify any potential gaps is called a traceability matrix. The basis of this matrix is to identify requirements in the first column and to provide objective evidence of the completed work in the last column. The columns in between are additional information that may be crucial or beneficial to specific teams and/or projects. Table 17.1 provides two examples of such a matrix. It goes without saying that these matrices cannot be effectively managed

> A traceability matrix is an effective tool to provide an overview of pictures and at the same time identify any potential gaps in the verification of requirements.

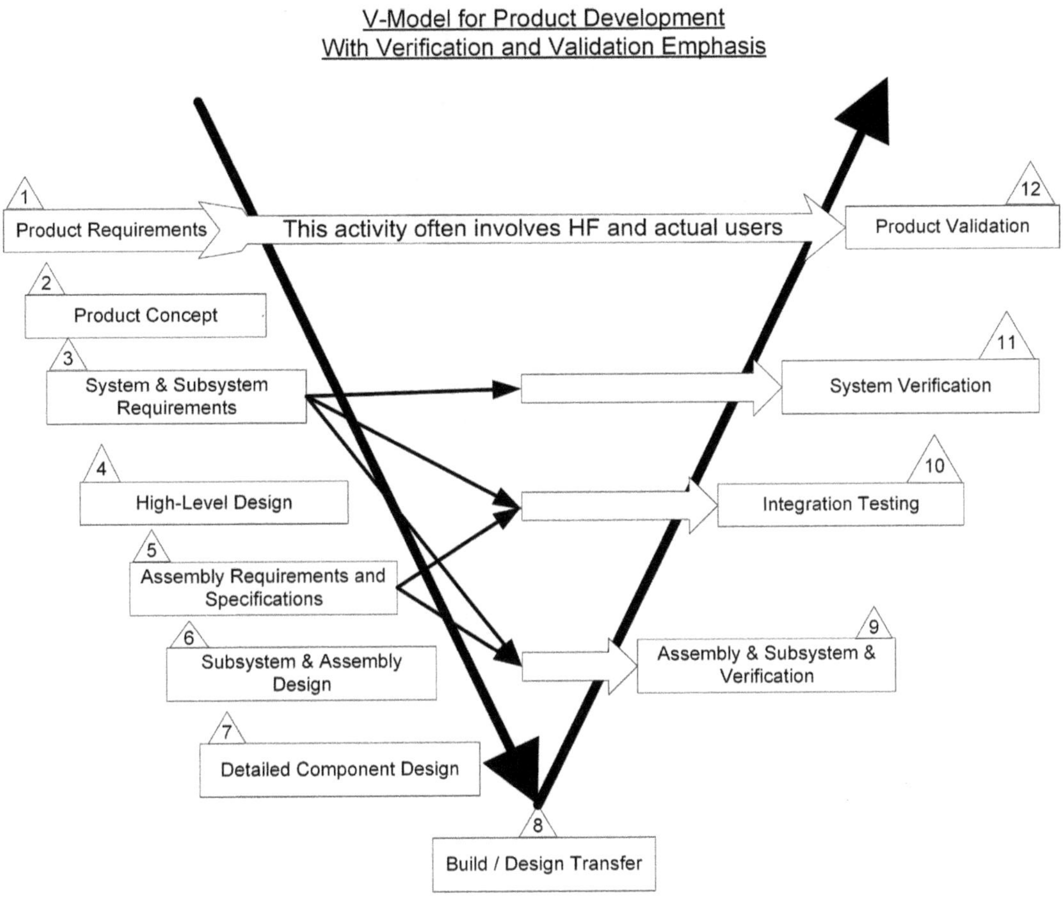

FIGURE 17.4
Verification and validation in the V-model
Note that HF stands for human factors

with general-purposed spreadsheets such as MS Excel for large requirement sets. Software tools such as DOORS by IBM or Quality Center by HP may be more appropriate.

Verification or Validation Process

The verification or validation process, as depicted in Figure 17.5, is rather intuitive. It starts by identifying the requirements associated with the unit under test (UUT). The unit under test may be an assembly, subsystem, or the system for verification testing or the product for validation testing. Input information includes safety and/or critical characteristics as well as the UUT's use cases.[4] With this information, test plans are developed. Inputs into the test plans are test cases, sample sizes, and finally acceptance criteria. Supporting documents such as process validation documents may be used to provide objective evidence that verification tests are appropriate. The next steps are to develop the traceability matrix and execute the test regiment. The "happy path" is the test sequence whereby nothing goes wrong, acceptance criteria are met, and the verification or validation is concluded

TABLE 17.1

Two Examples of Traceability Matrix

Requirement ID #	Description	Test Case/Method	Acceptance Criteria	Sample Size	Reference	Results	Comments
2.1.1	Assembly shall resist a pressure loading of 35 psi with deflections under 0.03in with 95% confidence and 90% reliability.	Pressure test per SOP 123456 to 35 psi	Resulting deflection should be less than 0.03in, with 95% confidence and 90% reliability.	15	Verification Study # 23456	Requirement was verified successfully.	None
–	–	–	–	–	–	–	–
–	–	–	–	–	–	–	–
–	–	–	–	–	–	–	–
–	–	–	–	–	–	–	–

Requirement ID #	Description	Design Output	Risk Analysis	Inspection	Process Validation	Design Verification	Comments
2.1.1	Assembly shall resist a pressure loading of 35 psi with deflections under 0.03in with 95% confidence and 90% reliability.	Drawing # 1231718 provides design details	DFMEA Study 2.44 PFMEA Study 3.67	See Engineering Analysis 5422	Protocol # 12345	Verification Study # 23456	None
–	–	–	–	–	–	–	–
–	–	–	–	–	–	–	–
–	–	–	–	–	–	–	–

(The second table is based on a traceability matrix provided by MEDICept 2010)

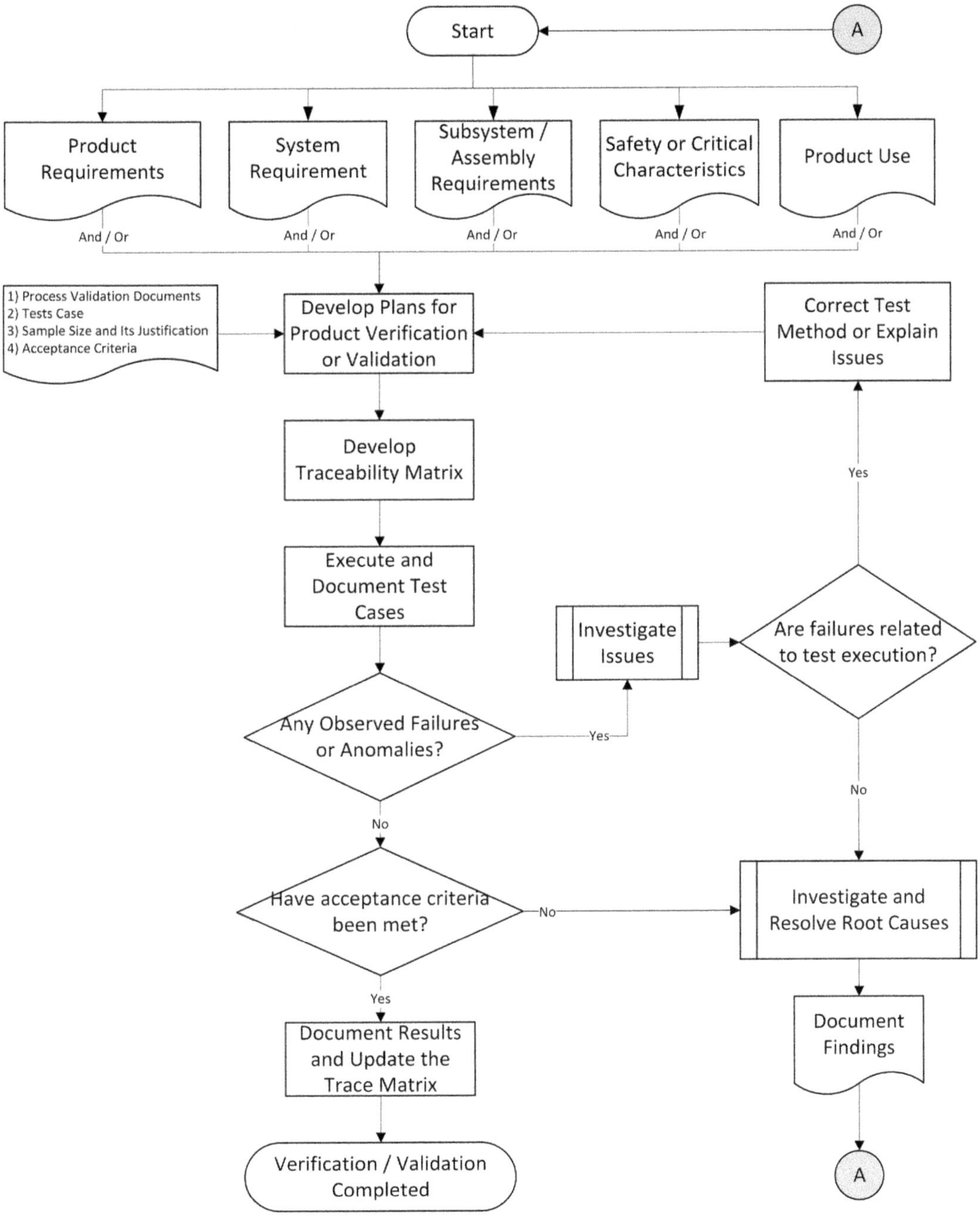

FIGURE 17.5
Verification and/or validation process

successfully. Happy path is not always the case, and quite frequently, tests fail. However, test failures are not always catastrophic and indicative of product failures. There are situations in which tests fail due to anomalies. These irregularities must be understood, and accounted for, and then the test must be repeated.

For instance, in measuring signal frequencies on some electronics boards, a superimposed 60 Hz signal on the desired signal may be observed, which is caused by fluorescent lighting frequency. Anomalies of this nature are often telltale signs that the test process has not been validated robustly. When causes of failures are understood and resolved, the test may be restarted.

There is also a possibility that the unit under test fails. This failure may be due to previously unknown sustained damage. Suppose that the unit under test was transported from a different location to the lab area and it sustained damage during transport. Again, as before, once the root cause of the failure is understood to be unrelated to the requirements and design specifications, the unit may be replaced with a known good unit, and the test is restarted.

Finally, it is also possible that the test proceeds as expected, but the acceptance criteria are not met. This is the most difficult situation to resolve. On the one hand, the results suggest that the design outputs do not meet the requirements. On the other hand, a design team would not enter into verification (or validation) without some degree of confidence that the tests would be successful. A verification failure (particularly at the system level) sheds doubt on all previous tests and their validity. It is important that an in-depth investigation be conducted and document the findings. The outcome of the investigation may be changes in test cases, acceptance criteria, or even design specification and redesign.

Summary

- First design transfer process ensures that all drawings, schematics, and specifications, as well as the bill of materials, are complete and error-free.
- Next design transfer process ensures that all critical design outputs have been identified.
- Finally, it ensures that all manufacturing and service procedures – including the use of fixtures – have been documented, and control plans are in place.
- Design verification provides evidence that "design outputs meet design requirements."
- Design validation provides evidence that the product design conforms to "defined user needs and intended uses."

Notes

1 There are three types of verification by inspection. These are visual inspection and confirmation, analysis, and functional demonstration. The sample size for inspection is one. I will expand on this in Chapter 22.
2 This is almost a chicken and egg scenario involving design transfer and verification. One needs to be in place before the other is complete.
3 Chapter 23 discusses metrology.
4 In other words, prior to developing the test protocol, the test personnel should know how the UUT is used within the context of the product.

Part VI

More on Agile

18

Market and Customer Changing Needs

Introduction

In the preceding chapters, we walked through various stages of product development, from requirements to concept to design to verification and validation and – though not mentioned explicitly – to product launch. The two chapters that follow will focus to some degree on approaches for sustaining marketed products for the duration of their design life. One element that needs attention now is a discussion (and a proposed method) on responding to changes in customer needs in a dynamic market.

V-Model and Agile

In Chapter 8, we reviewed how to rank, prioritize, and further cascade customer needs into product requirements that will be actualized in manufactured products. At its face value, this activity may be viewed as static that, once completed, can be logged and put aside. Today's realities, however, demand that product developers have their fingers on the pulse of the market. Today's reality is that advances in one market will impact customer expectations in another. For example, advances in wireless technology have led to the expectation that we should be able to connect wirelessly just about any electronic device whether a computer, a home appliance, or a medical device, and that we should be able to monitor them using our smart phones.

There is a growing population of business leaders who believe Agile methodologies will provide guidance on how we can navigate through this changing maze of customer needs and expectations. While Agile is widely incorporated into software development, its application and efficacy for hardware development are yet to be fully realized.

In Chapter 6, I introduced the V-model, and throughout the preceding chapters, I have demonstrated its application to hardware development. In a simplistic view, this model may be considered a glorified waterfall model. Those who hold this belief argue that all design steps must be taken before the development team can move to the next stage. For example, requirements must be completed prior to concept development, and concepts must be available before detailed design specifications are created. This argument is true and may even be made for the Agile methods as well – even in the software world – since to complete any task, its prerequisites need to be done first. The question is whether the V-model allows a return to a previous stage or

> In some waterfall models, once a gate is passed and approved, design may not go back to a previous stage. In a pure V-model, this is not the case, and in that sense, it is compatible with the Agile mindset, making changes to remain compatible with customer needs.

DOI: 10.1201/9781003564508-24

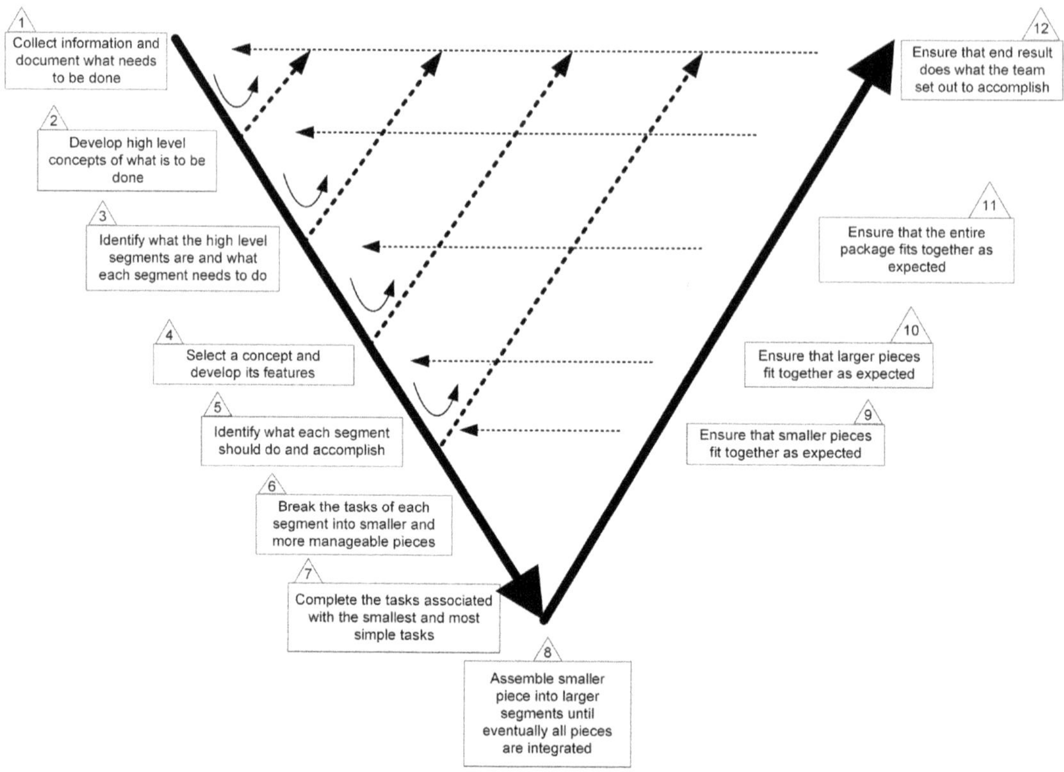

FIGURE 18.1
V-model's inherent flexibility, as at any stage of the "V" one may return to a previous stage in order to update a previous activity should flaws or inconsistencies be identified

completing some activities in parallel. As I have mentioned previously, the answer is a resounding affirmative.

Consider Figure 18.1, which depicts the V-model as shown previously in Figure 6.6. This model has two inherent flexibilities. First, user needs are cascaded into the product requirements and then into system and subsystem requirements. Subsystems and their modules may be developed in parallel without the need for sequencing them with other subsystems. Second, as each subsystem is being developed, deficiencies and/or ambiguities in the initial requirements and their flow-down may be identified. Their resolution necessitates a partial return to a previous stage only in the areas where these deficiencies or ambiguities require resolution; other modules or subsystems will not be affected. In fact, this may also happen at the time of subsystem or system integration, where technical issues may find their roots in the requirements.

A word of caution: this flexibility should not be misunderstood as a license for inadequate attention to details, a gross negligence of requirements, or bypassing design elements with the excuse of getting them done at a later date.

Managing Market Change

When we speak of responding to market changes and their associated needs, there are really two scenarios that we need to be mindful of. First, as a new product is being developed, new information surfaces that indicates the finished design may not satisfy the anticipated user needs. The second scenario is one of a recently launched marketed product that may no longer be needed (or desired) by a large portion of the market.

I have demonstrated that the V-model has the flexibility to incorporate changes in customer needs into the product development, though depending on the scope and time, this could be either seamless or rather costly. Hardware development requires prototypes during the design cycle and capital equipment during the manufacturing cycle; the later the change, the more costly it will be. Should this be the case and within the construct of an agile mindset, any changes in the requirement should be discussed in the next team retrospective (see Appendix A) and in the lessons learned session in order to understand the reason it was not identified earlier. A detailed impact assessment will identify impacted requirements, and by tracing its flow down into design specification, specific areas of change can be identified and planned in subsequent sprints.

Hardware development requires prototypes during the design cycle and capital equipment during the manufacturing cycle; the later the change, the more costly it will be. Should this be the case, and within the construct of an agile mindset, any changes in the requirement should be discussed and rationalized.

The element of responding to market changes for a recently launched product is considered under product upgrades below and its associated strategies.

Product Upgrades

It is quite likely that the business needs and strategies – possibly influenced by a deep understanding of personas – dictate that a product or a product line be launched into a market incrementally. It is in that sense that I like to provide some guidelines and indicate how an Agile mindset will enable an organization to achieve this goal.

Creating Product Feature Backlog

Typically, most organizations have a three- or five-year product development roadmap. Often, these roadmaps are expressed in general terms and lack specific details. While creating a feature roadmap is a similar exercise, it does require more in-depth definition of features to be included in each release cycle. Figure 18.2 suggests a list of activities in each cycle in order to prepare for the next. As suggested in this figure, not only should the features be discussed, but also resource needs, both in terms of employees and capital equipment needed for both testing and manufacturing should be included. Two elements that should not be forgotten are (a) the effort needed to complete relevant verification and validation and (b) the overall design life of the base products. Any platform will eventually become obsolete either due to component obsolescence or outdated base features.

In creating a product feature backlog, two elements should not be forgotten: (a) the effort needed to complete relevant verification and validation and (b) the overall design life of the base products.

Considering the significance of testing and capital equipment needed for this purpose, I have created a list of proposed activities as shown in Figure 18.3. This list is not exhaustive,

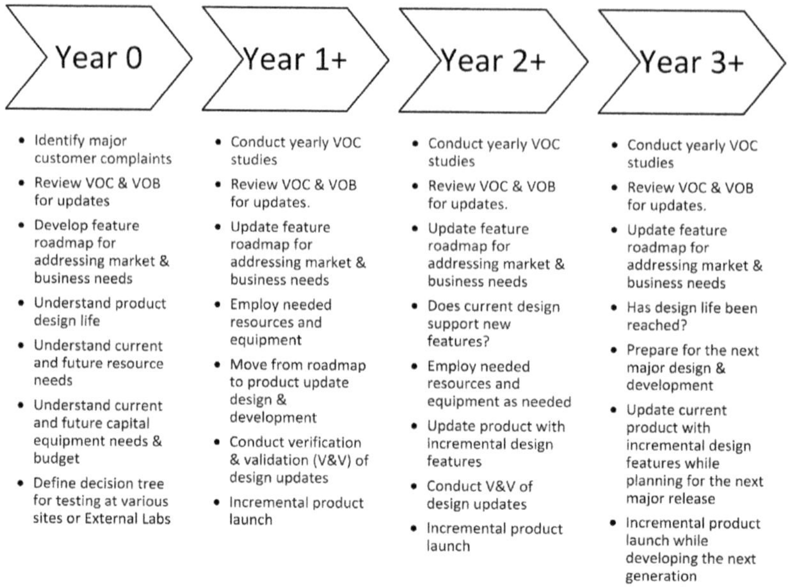

FIGURE 18.2
A product feature roadmap should anticipate not only what customers will be asking for but also resource and equipment needs

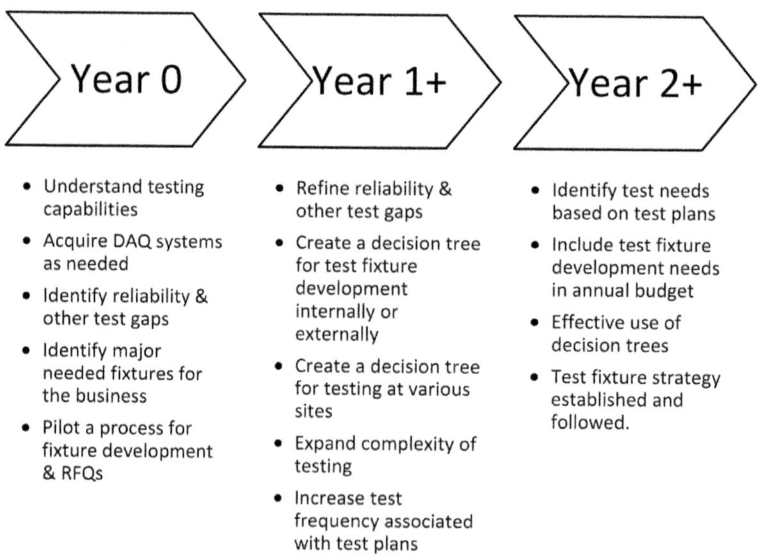

FIGURE 18.3
A test roadmap should accompany any feature roadmap to anticipate test needs and their impact on delivering product updates

but it should be considered as a starting point and include enterprise-specific elements. Test strategy should also include working with various external test houses and how they may provide additional or needed capital equipment on a short-term basis.

Feature Score Card

As mentioned, a detailed feature backlog and roadmap will include prioritized elements of customer needs. This, however, does not mean ignoring customer complaints. In other words, the feature backlog should include both new and sustaining elements. The question is how to create a balance between sustaining versus new features for a given set of resources and available funds. Tables 18.1 and 18.2 suggest an example of a scoring mechanism with which each feature – whether new or sustaining – may be given a semiobjective score.

For assessing sustaining activities associated with a product (Table 18.1), I have included robustness of available data (namely complaint or failure), complexity of change (both in terms of design and manufacturing), and how well the proposed change will in fact resolve the underlying issue and help reduce complaints or failures. The mindset behind this assessment is to identify changes that will have the lowest complexity with the highest return on issue resolution. For this reason, a change with the lowest complexity receives a high score, whereas low assessment of data or issue resolution is scored low.

A similar activity may be conducted for introducing product upgrades and new feature sets (Table 18.2). However, it is important to start with user needs and voice of customer prior to design decisions and develop traceability from these needs to specific design elements. A detailed impact assessment will identify additional design features that may also be impacted. Similarly, the mindset behind this assessment is to identify changes that will have the lowest complexity with the highest return on investment and customer satisfaction (or market significance). For this reason, a change with the lowest complexity receives a high score, whereas low assessment of return on investment or market significance is scored low.

Once this exercise is completed for both sustaining and new feature introductions, common themes and areas may be identified. For the example provided in Tables 18.1 and 18.2, these may be

- Enclosure
- Keypad
- AC Power Adaptor

Based on this input, resource and budget planning may be completed.

Agile Teams and Task Completion

An in-depth review of Agile is beyond the scope of this work; however, for the sake of brevity, a short review of this methodology is given in Appendix A, and a case study is presented in Appendix B. Having said this, once the feature sets for the new release are

TABLE 18.1

An Example of Scoring for Sustaining Activities Based on Customer Complaints

Module/Feature	Complaint Levels	Team Activity	Available Data	Complexity of Change	Issue Resolution	Score
Mechanical Assembly 1	7 field failures as well as observed failures in internal testing	Supplier had provided the wrong parts Retest is needed Set up process needs to be changed	3	1	5	15
Mechanical Assembly 2	2 field failures	Design modification of failed components Accelerated life test post redesign	3	3	3	27
PCBA	Internal testing indicates possible failures	Redesign of a few components Highly accelerated limit testing post redesign	1	1	2	2
Enclosure	Internal testing predicts the presence of cracks	Redesign of closure locking mechanism Aging studies post redesign	3	5	3	45
Sensor Assembly	Top 10 Complaints	PCBA redesign Accelerated life test post redesign	5	1	5	25
AC Power Adaptor	Top 10 Complaints	Connector redesign Highly accelerated limit testing post redesign	5	3	3	45
Mechanical Assembly 3	Top 20 Complaints	Resolve sticking issue by Modifying mold – process validation update	1	1	3	3
Keypad	Top 10 Complaints	Use 80% silver/20% carbon on connection pad Update manufacturing specifications process validation update	3	5	3	45
Mounting pin	Top 30 Complaints	Redesign pin Module verification testing	1	5	3	15

Score	Data	Complexity of Change	Issue Resolution
Low	1	5	1
Medium	3	3	3
High	5	1	5

TABLE 18.2

An Example of Scoring for New Feature Development Activities Based on Customer Needs

Customer Need	Impacted Module	Activity	Return on Investment	Complexity of Change	Significance to Market	Score
Easy-to-use interface	Mechanical Assembly 1	Redesign operator interface for ease of use	3	1	5	15
Reliable system	Enclosure	Design a more robust locking mechanism or establish a shorter warranty duration	3	5	5	75
	Sensor Assembly	Design a robust sensor against fluid ingress	1	1	3	3
	AC Power Adaptor	Design a robust solution to address current misuse issues	3	3	3	27
	Mechanical Assembly 3	Reduce manufacturing variability	1	1	1	1
	Keypad	Address sticking key failures	5	5	3	75
	Mounting pin	Reduce manufacturing variability	1	5	1	5

Score	Return on Investment	Complexity of Change	Significance to Market
Low	1	5	1
Medium	3	3	3
High	5	1	5

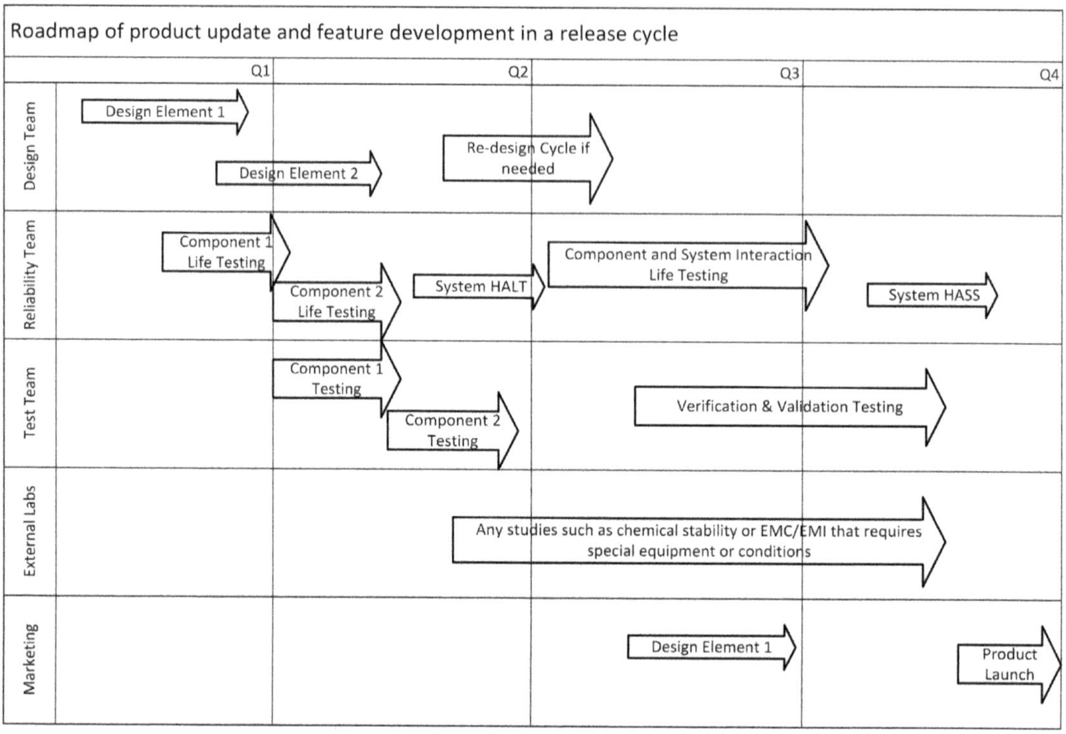

FIGURE 18.4
An example of planned activities (backlog) for each Agile team. In this example, the assumed release cycle is one year.

decided and the impact on specific subsystem requirements and design specifications are identified, backlogs for each agile team need to be created. Figure 18.4 provides an example of how each team plans to complete their tasks. Note that in this model, each team may complete their work at a different time prior to the end of the release cycle. This would mean that each team may then begin a new cycle of other product releases.

Summary

- A pure V-model is compatible with the Agile mindset, which is to make changes to requirements or design specifications to remain compatible with customer needs.
- Changes to hardware during development require efforts to verify and guard against unintended consequences and capital equipment during the manufacturing cycle; the later the change, the more costly it will be.
- In creating a product feature backlog to respond to market needs, there are two fundamental factors. The first is the overall design life of the base product, and the second is the effort needed to complete the feature.
- A feature scorecard will enable the development to prioritize what needs to be developed next.

Part VII

Sustaining a Marketed Product

19

Sustaining a Fielded Product

Introduction

In Chapter 17, I described the need for a formal process to transfer design of a product to manufacturing and at the same time V&V that all requirements and customer needs are met. Once design transfer and V&V are completed, there is a short period when the product is launched in a limited market. This limited launch is either a part of the ramp-up or a marketing study to gage the market response to the product. At the same time, it enables the design team to evaluate the behavior of the product in the field and correct design flaws if the need be.

Eventually, the product is fully launched in all its intended markets. Even though, there is much to celebrate, no one loses sight of the fact that the newly launched product needs to be sustained in the field.

To sustain a fielded product mean to keep a watchful eye on a number of different concerns. One the one hand, the sustaining team ensures that incoming components meet the required specifications and that there are no slip-ups in the manufacturing of the product. On the other hand, the same team looks at the behavior of the product in the field and the number of failed units that are returned for service and service effectiveness. All in all, the sustaining team – a cross-functional team with a similar make-up as a development team – tries to keep its finger on the pulse of the customer-base and to ensure that customers are satisfied with the product.

I like to divide the work of the sustaining team into two broad categories. The first category may include *complaint monitoring and trending, manufacturing* and *service concerns*. The second category – which in a way may be an outcome of the first – is called *change control* and *product recall.*

Complaint Monitoring and Trending

L.L. Bean boots are legendary for their long lives and the company's commitment to stand behind their product: "Our products are guaranteed to give 100% satisfaction in every way. Return anything purchased from us at any time if it proves otherwise. We do not want you to have anything from L.L. Bean that is not completely satisfactory" (downloaded on October 31, 2017, from https://www.llbean.com/llb/shop/513705).

> An organization cannot stand behind its products if it does not monitor factors that affect the quality of its products and study their trends.

An organization cannot stand behind its products if it does not monitor factors that

DOI: 10.1201/9781003564508-26

It should be noted that in general complaints are not *all* product failures but an indication of customer dissatisfaction.

affect the quality of its products and study their trends. Although these factors may vary from organization to organizations, the first and foremost concern for any organization is what its customers think of its product lines. They are concerned about the details of what customers complain about. Depending on the market served, customer-complaints may be as rudimentary as "I don't like it!" to as dangerous as "It caught fire!"

In general complaints are not necessarily indicative of product failures but an indication of customer dissatisfaction.[1] Often, the customer service department of an organization codifies customer complaints for ease of processing and data analytics. For example, "I don't like the product" may be coded as, say, CC-1234, and "the units heats up too much" may be coded as CC–1100. Figure 19.1 is an example of a Pareto chart of complaints generated for the Learning Station (as described in Chapter 9).

It is important to know that a Pareto chart – and indeed any graph of this nature – is time-bound. In other words, it is developed for specific time intervals. Complaints data reviews may be conducted on a daily, weekly, or monthly basis and their associated Pareto charts be generated.

From the chart shown in Figure 19.1, it is clear that three complaint codes (i.e. CC-1104, CC-1107, and CC-1103) are more prevalent than others. Successful organizations keep an eye on these first few items and require investigations to understand the underlying causes.

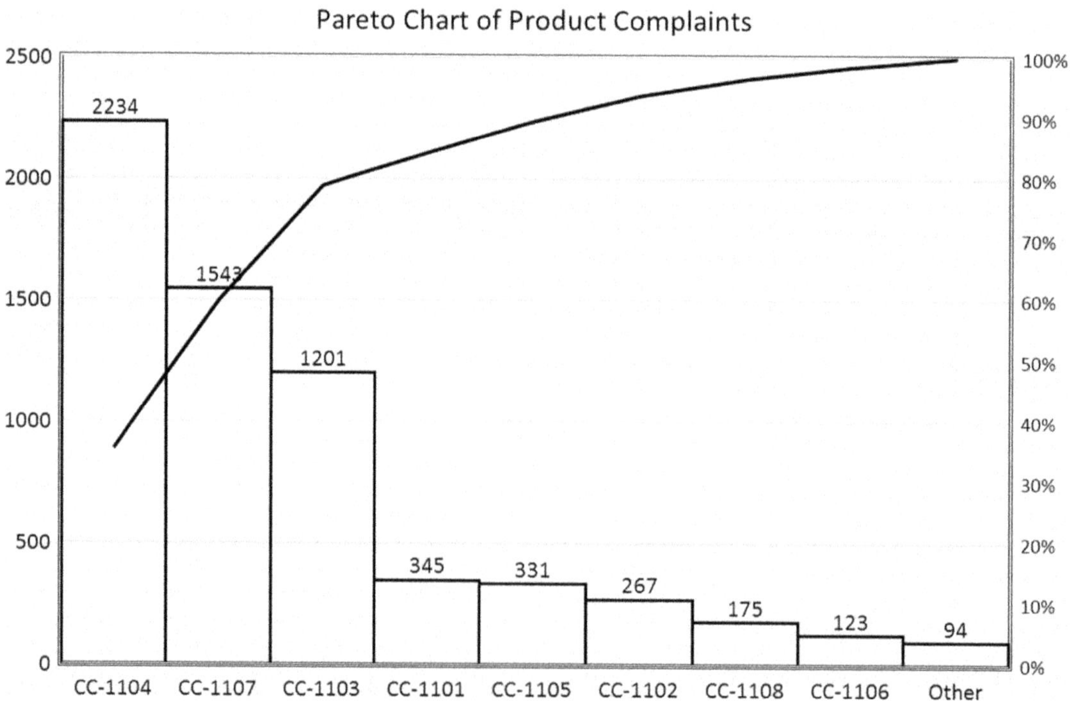

FIGURE 19.1
Example of a compliant Pareto chart

Product Risk

Earlier, I suggested that depending on the market served, customer-complaints may be as rudimentary as "I don't like it!" to as dangerous as "It caught fire!" It is important to explore the difference between these two complaints. The first speaks of a personal taste – if someone does not like a product, others may like or even love it. The second complaint refers to a product risk. Should this failure mode be actualized, someone's health or property may be in danger. There is a substantial difference between these two types and organizations should have mechanisms to deal with both.

An aspect of postlaunch trending is to identify complaints which may be related to product risk to ensure that the product is operating at safe levels. A review of news provides ample examples of how failure of a product has harmed its users. In the medical device industry, it is a requirements to evaluate product risk on a regular (often yearly) basis. Furthermore, the DFMEA and PFMEA files are updated with any newly identified failure modes.

Manufacturing Concerns

One of the major concerns of any manufacturing organization is manufacturing first-pass yield (FPY). In other words, what percentage of the attempted product builds are successful in the first pass and do not require rework? Factors affecting the FPY may be design-related (i.e., poor tolerance specifications) or manufacturing-process-related (e.g., line technicians are poorly trained) or a combination of the two (e.g., the design is too complex for the selected manufacturing process).

> A low FPY is indicative of waste and unnecessary costs.

Hence, understanding and monitoring FPY for any manufacturing organization is essential because low FPY is indicative of waste and unnecessary costs. Figure 19.2 is an example of a 12-month manufacturing first-pass yield. This graph shows a clear negative trend; however, by examining the graph closely, it becomes clear that in the first four months, the trend is positive. Starting with the fifth month (May), the negative trend begins; however, its impact on the annual trend is not felt until the eighth (August) or ninth (September) months. Even though there is an uptick in the yield in the tenth (October) month, the overall impact remains negative.

Manufacturing leaders try to understand the underlying causes of both positive and negative trends. This understanding will enable them to either sustain and grow the FPY or prevent and reverse its further decline. Similar to the Pareto of complaint charts, manufacturing studies the Pareto charts of all factors and mishaps that were encountered during the manufacturing periods. These factors may include component tolerance stack-ups, calibration, or in-process and end-of-line test failures.

Another two indicators that some manufacturing organizations review are the *out-of-box* and *early-life failures*. An out-of-box failure (OBF) occurs when a customer (or end user) encounters a non-operational product. An early-life failure takes place when a product fails shortly after being commissioned. These two metrics are indicative of the robustness of the manufacturing process. Excessive failures reported by these two indices are typically manifestations of either poor manufacturing techniques or lack of training.

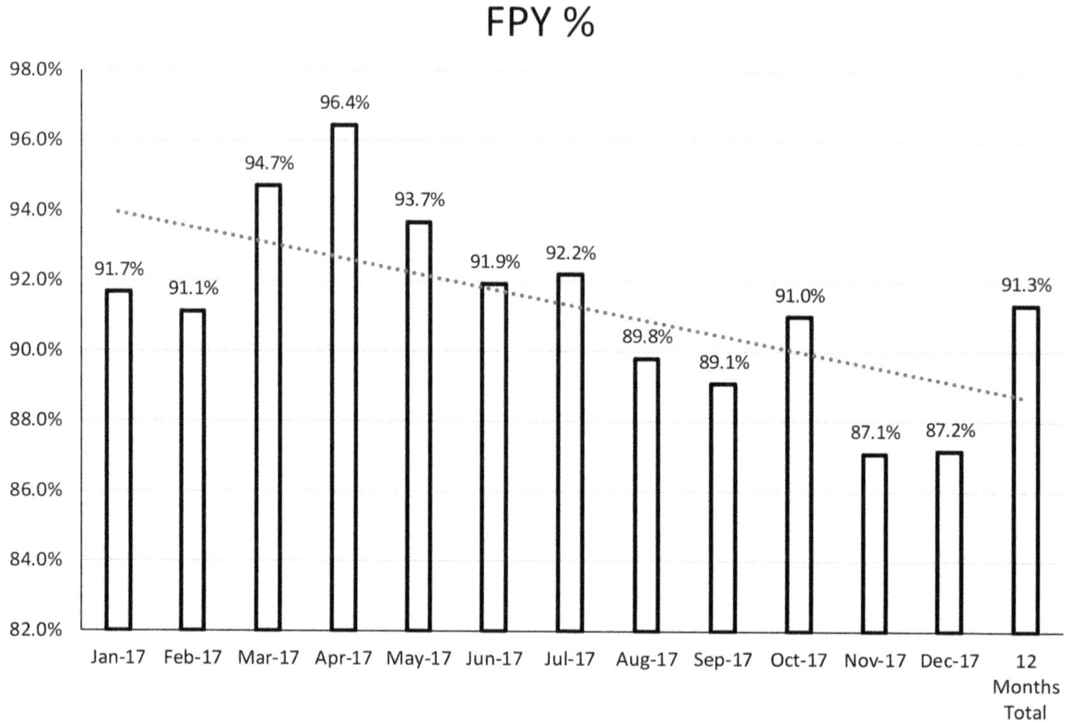

FIGURE 19.2
Example of a FPY chart

Supplier Issues

A contributing factor to both manufacturing yields both increased field failures (leading to additional service events) may be suppliers' poor quality. It is important to mention this here because suppliers (particularly of commodity items) always look for ways to lower their production cost and increase their profit margin. At times, these cost-saving efforts lead to increased part-to- part variations. This increase may have a detrimental negative impact on customers' first-pass yields.

Component Obsolescence and Shortage

Wilkins (2015) defined component obsolescence as "the inevitable consequence of the dependence of aerospace on a supply base whose major markets are outside of aerospace and whose technology life cycles are much shorter than those of other markets." Though he focused his definition on the aerospace industry, it holds for many other industries, such as medical and automotive, as well. Ansari and Sandborn (2014) indicated that 350,000 commonly used parts were obsoleted in 2013 alone.

> A factor that leads to part shortages is natural disasters.

In simple language, the problem is this: the product development team designs a device, specifies components, performs extensive verification studies, and launches the product into its intended markets. It is expected that this product would serve the market for a minimum of ten years.[2] A short while post launch, the sustaining team receives a note from one of its suppliers that a critical component will no longer be available. A simple replacement may not be readily available. A willy-nilly part replacement with another may have a severe impact on quality and unacceptable in regulated industries such as avionics and medical.

A factor that leads to part shortages is natural disasters. In 2011, a devastating flood in Thailand damaged or destroyed many factories. Many manufacturing suppliers managed to bring their production back online after a few weeks; however, their production quantities were limited due to the destruction of much of their production and test equipment (Chesla et al. 2013). I do recall that in my own organization, there was a heightened sensitivity to the quality of the parts received from Thailand during that period. As a result, more stringent incoming inspections were instituted to alleviate the risk of low-quality products.

Clearly, organizations that are at risk of component obsolescence should capture this factor in their project risk and have plans in place to meet its challenges. Depending on the product, its relative market significance, as well as its maturity in its life cycle, options may include incorporation of a replacement component,[3] or life-time or last-time buys.[4] In some rare cases, an organization may decide to stop production when the needed components are no longer available.

Service Concerns

To a large extent, the service organization has similar concerns as the manufacturing organization in terms of the efficiency of the service process and follow metrics on out-of-box and early-life failures of the units that have been touched by this organization. Additionally, the service organization provides data on the products' reliability in general and warranty concerns in particular. These two metrics are important to an organization because, of the two, the former impacts the customer's perception of the product, and the latter impacts the organization's budget and finances.

Failures and Warranty Issues

Customers have come to expect that the products will fail and look for data and information to enable them to select their desired product from a manufacturer with the least number of failures for the price range that they are willing to pay. Often the "Made in USA" label is a sign of quality. The reason is partly due to the general belief that products coming from low-cost countries such as China are cheap and hence fail more frequently.

To attract these customers and put their minds at ease, corporations and their distributors provide a variety of incentives, including product returns as well as limited and/or extended warranties.

Returned products require evaluations by service to either ensure their full functionality or identify their failures and be repaired. These products may be sold at a lower price than a "new" product because they carry the burden of being *refurbished*. The combination of the cost of repairs and lower prices constitutes a financial loss to the organization. Products repaired and returned to customers under warranty also pose a financial loss to the organization because often the cost of repair is greater than the price of the sold warranty.

For these reasons, the service organization monitors the components that are repaired or replaced. The sustaining team reviews this data, and should adverse trends be observed, corrective and preventive actions may be taken. Figure 19.3 is an example of such monitoring for the Learning Station (as described in Chapter 9).

In the example as shown in Figure 19.3, the power supply as well as the user interface card of the Learning Station are failing at higher rates than other components. The sustaining team should investigate this disparity and ensure that these failure rates are acceptable. I discuss the means and methods of failure investigation in Chapter 25.

> Excessive failures under warranty lead to excessive costs to the manufacturers. Excessive failures outside of warranty lead to customer dissatisfaction and a low opinion of the brand.

Change Requests and Control

In Chapter 3, I noted that one of the responsibilities of a project manager is to develop and maintain the change control plan. This document provides guidance on the operational mechanisms needed for processing and documenting change requests. These requests are the results of observations made in the manufacturing and service organizations. They are often made to respond to observed excessive failures, manufacturing cost reductions, or to mitigate part obsolescence. In Chapter 21, I will review the details of both configuration

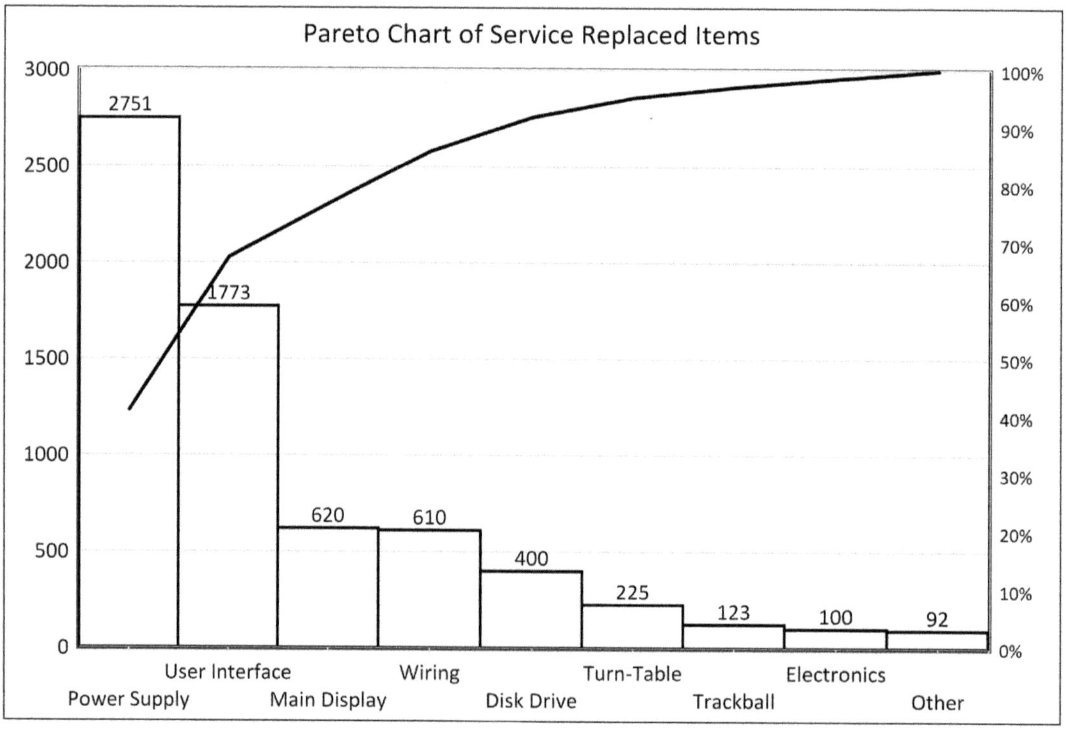

FIGURE 19.3
Example of a Pareto chart of parts replaced in service

management and change management. A major aspect of change control is to ensure that not only the proposed change is feasible. Another aspect of this control is to guard against any negative impacts to either product requirements or manufacturing and service processes.

Product Recalls

A product recall occurs when a corporation realizes that a customer or consumer may be endangered by the use of its product or that the corporate entity may be liable and subject to legal action. This is the most extreme action that may result from post-market monitoring and trending activities and is quite embarrassing for an organization.

> Product recalls are expensive, but more importantly, they may damage a corporation's reputation.

The requirements for product recalls are dictated by consumer protection laws in each country. In the United States of America, consumer protection is enforced at both federal and state levels through agencies such as the Federal FDA and the U.S. Department of Justice, as well as each state's Department of Consumer Affairs.

It should be noted that there are different classifications of recalls. The FDA defines these classifications as follows (https://www.fda.gov/Safety/Recalls/IndustryGuidance/ucm129337.htm).

- *Class I recall*: A situation in which there is a reasonable probability that the use of or exposure to a violative product will cause serious adverse health consequences or death.
- *Class II recall*: A situation in which use of or exposure to a violative product may cause temporary or medically reversible adverse health consequences or where the probability of serious adverse health consequences is remote.
- *Class III recall*: A situation in which use of or exposure to a violative product is not likely to cause adverse health consequences.
- *Market withdrawal*: Occurs when a product has a minor violation that would not be subject to FDA legal action. The firm removes the product from the market or corrects the violation. For example, a product removed from the market due to tampering, without evidence of manufacturing or distribution problems, would be a market withdrawal.
- *Medical device safety alert*: Issued in situations where a medical device may present an unreasonable risk of substantial harm. In some cases, these situations also are considered recalls.

Product recalls are expensive, and often the financial loss is experienced not only by the direct costs but also by a drop in share prices and a loss of market confidence. As Taguchi and Clausing (1990) have explained:

> "When a product fails, you must replace it or fix it. In either case, you must track it, transport it, and apologize for it. Losses will be much greater than the costs of manufacture, and none of this expense will necessarily recoup the loss to your reputation."

Summary

- Corporations monitor the performance of their products in the market through post-market activities such as complaint monitoring.
- Manufacturing and service metrics are needed to ensure that product quality is not reduced through deficiencies in either manufacturing or service processes.
- Part obsolescence is a major risk for organizations in regulated industries. Project risk plans should include mitigative activities.
- Product recalls are costly and will impact a corporation's image in the market place.

Exercises

1. Visit a local original equipment manufacturer site and learn more about their specific post-market monitoring and trending activities.
   a. What metrics do they collect for manufacturing and for their service organizations?
   b. How frequently do they meet to review post-market data?
2. Conduct an internet search of product recalls in a week
   a. Repeat the same search for three weeks. Do you find a trend, or are the recalls random in nature?

Notes

1 In many consumer product lines, a customer may return a product for no apparent reason.
2 Chesla et al. (2013) note that "Many products in the industrial, aerospace, automobile, military, and similar markets must be supported for 5 to 20 years or even longer."
3 This decision is costly and requires extensive verification activities.
4 Lifetime or last-time buys involve purchasing either enough components to last through the lifetime of a product or purchasing the existing available inventory.

20

Product Retirement and End of Life

Introduction

I have to admit that it is exciting and invigorating to be a part of a new product development team. Furthermore, being a part of a design team that supports and sustains a marketed product provides huge learning opportunities by providing exposure to all the design assumptions that were in fact problematic. I have had the pleasure of sustaining products that I have designed and developed; the experience has been humbling as well as rewarding.

Products are developed to serve a market need. As these demands evolve, existing market products may no longer be able to satisfy the customer base and need to be replaced with updates and newer versions. Eventually, a business may decide to retire a product or a product line. Although the decision to remove a product from the market is never taken lightly, the steps needed to cleanly obsolete the product internally may be confusing and ambiguous. This may be the reason why 90% of the sales of so many businesses are based on roughly 30% of items in their catalogs. In this chapter, I would like to propose an efficient means and process of retiring a product or its family.

End-of- Life Assessment and Planning

Among the many responsibilities of the sustaining product team is the monitoring and trending of various product quality metrics. Based on the information received, the team recommends various corrective actions either to the product design, manufacturing, service, or even the sales force. The sustaining team may also recommend obsoleting a product and establishing the time to stop sales and plan the execution of obsolescence activities. This recommendation may be based on a variety of factors. Primary among these factors is poor financial performance, including poor sales or price erosion and competitor pressures. Other reasons may be regulatory pressure, such as a change in rules. For instance, a product that is not restriction of hazardous substances (RoHS)[1] compliant may no longer be sold in Europe. Technology obsolescence and product replacements are among other reasons for stopping sales in anticipation of product obsolescence.

> Among the many responsibilities of the sustaining product team is the monitoring and trending of various product quality metrics. Based on the information received, the team recommends or enforces various corrective actions.

Prior to seeking management's approval to initiate product obsolescence, the team should conduct an impact assessment for removing a product from its market. This

DOI: 10.1201/9781003564508-27

assessment, and the plan that would follow, document the rationale for the initiation of product obsolescence and its alignment with business objectives. Once an initial approval to proceed is granted, the team begins to plan and communicate the end of sales and withdrawal activities to various stakeholders, including, if need be, customers and end users.

The final approval for the end-of-life (EOL) process is a confirmation that there are no external or internal commitments, for example, there are no outstanding unfilled orders, or there are no undue risks.[2] And that all ramp-down and phase-out activities have been completed.

Impact Assessment

A proper impact assessment plays a crucial role in a well-executed design change request and order. Consider this: product obsolescence is the most important design change request that may be executed for a product. Hence, a thorough impact assessment is required to identify how product obsolescence may impact the business. The areas of impact include branding and marketing, finance, distribution channels including partners, operations including manufacturing, service and supply chain, and finally design. The following are areas of concern that each team needs to focus on:

> Product obsolescence is the most important design change request that may be executed for a product. In that regard, the impact of obsolescence on the business must be understood prior to seeking management's approval to initiate this activity.

Business (Marketing, Branding, and Finance)

The impact on the business that should be evaluated is whether there is a replacement product offered by the enterprise and whether that replacement will be accepted and adopted by the market. Would a vacuum be generated that, if filled by the competitors, may have an adverse impact on other enterprise-marketed products? Other questions that may influence the assessment from a finance point of view are the cash flow concerns. Is the cost of internal support of the product at least equal to the revenues generated by the product? In my experience, this is not an easy question to answer because in many organizations internal costs are not monitored or tracked either carefully or accurately.

Logistics and Distribution Channels

The second area deserving of a thorough assessment is the distribution channels and distribution partners. Depending on the size and cost of a fielded product, it may not be practical for customers to simply throw the obsoleted products away. Indeed, should a customer be very happy with the product, they may just be resistant to any change. In this situation, an enterprise may just transfer the role of product support to a willing partner. In this case, any impact to the enterprise brand – due to the distribution partners' performance – should not be taken lightly. So much so that some organizations choose not to remove a product from the field but to replace it with an upgraded model and design. There may be other areas of logistics that would need assessments. Are there specialized services that need to be discontinued?

Operations and Design Team

The first two areas of impact assessment may be considered as market-related. Operations – in particular manufacturing, service, or supply chain – are product centric. A product removal from the market necessitates physical changes to both manufacturing and service floors. For instance, recall my previous example that I provided on metal injection molding of surgical instrument handles in Chapter 15. In that example, a number of pieces of equipment that were used in the older process were impacted because they needed to be removed from the floor to make room for other equipment. Manufacturing and service impact assessment studies whether a proposed change may have a ripple effect on other existing products and processes.

The role of supply chain impact assessment is to answer the following question. Are there components in the obsoleted product that are used elsewhere in other products? If so, will there be any cost impact to those other products?

Finally, it is the design team's responsibility to review all product-related documents, including design, manufacturing, and service, and to authorize the obsolescence of those documents that do not impact any other fielded product. This work should not be taken lightly because the number of existing documents for any existing electromechanical product could easily add up to hundreds, if not thousands.

End-of-Life Planning

Once a complete impact assessment is done, the EOL plan may be drafted that would reflect the work that needs to be done along with a designation of resource needs and timelines. An EOL plan may also include guidelines and elements of inventory management and material ramp-down (or sell-off), timing on last manufacturing builds, service inventory build-up,[3] and service closure date, and other related issues.

An EOL plan may also include guidelines and elements of inventory management and material ramp-down (or sell-off), timing on last manufacturing builds, service inventory build-up and service closure date, and other related issues.

Obsoleting a Product

In removing a product from a market to obsoleting it completely, there are both technical and business aspects. I would like to review them here briefly. Having said this, I like to indicate that the time frame for this activity may vary from weeks and months to years. In fact, in the case of medical devices, the removal process may be as long as a decade. The reason for this length of time is to ensure that the impact on the patients and the market is minimal.

Business Aspects

Obsoleting a product to a large extent is to communicate a business's and enterprise's goals and aspirations to its various stakeholders. It is an opportunity to provide a foundation of what is to come and how customers' new needs and expectations may be fulfilled by replacing an aging product with a newer one. This communication should alleviate customer concerns to the extent possible. Some businesses provide *buy-back* incentive programs, while others may give discounts for early adopters of new technologies and products. At some point in time and depending on the nature of the product, there are heightened activities to sell existing inventory, and beyond some point, the product is removed from catalogs, all sales come to a halt, and all existing inventory of finished products are discarded. As these activities are primarily within the domain of marketing, they are beyond the scope of this book.

Design and Technical Aspects

With the business team bringing issues to closure from their point of view, the technical team needs to review the EOL plan and put into place activities and actions needed to close the book on their end.

Design Team

The first implication of product obsolescence on the design team is freeing resources and their allocation so that they can work on other projects and programs. Also, depending on the timing of the ramp-down, various ongoing activities on the product (such as cost improvement activities) should be evaluated and prioritized. In some instances, *buy*-items associated with the product might have been obsoleted. Should this be the case, the supply chain team needs to arrange for last-time purchases of these items. Documents that are no longer used (as identified through the impact assessment) should be reviewed and obsoleted if appropriate. The pitfall in design documents is that there may be an impact on other products in terms of reuse of parts, components, instructions, and even tools and equipment. These impacts should be fully vetted and understood. The better and more complete design history files in place for the entire organization, the easier the impact on other products may be assessed.

Manufacturing

Similar to design, manufacturing should also assess the impact of product removal on product schedules, staff, assembly tools, and production equipment. Affected staff are assigned to other projects. Tools and equipment are typically decommissioned for other uses as well; otherwise, they are either retired or auctioned.

Other factors involve inventory of raw material on the incoming, work in progress, and the finished goods on hand. This information is typically communicated to the business group to develop marketing strategies to reduce these quantities. In addition, and in collaboration with both service and business teams, required quantities of field-replaceable units for service organization are forecasted and built. Finally, similar to design, manufacturing documents are retired and archived.

Supply Chain Management

Supply chain management is responsible for looking in two directions; first, at the supply side and incoming material, and second, at the distribution side and outgoing products. Their role is to evaluate the impact of product ramp-down on these two networks. Should the need arise, they negotiate piece and part pricing on the supply side and product portfolio positioning on the distribution side. Finally, all product-related supply chain management documents should be obsoleted.

Service

The actual ramp-down of service activities is probably the last set of activities for a product obsolescence. Typically, by this time, manufacturing has ramped down completely for some period of time, and so have almost all other facets of business and design, with possibly the exception of the risk management team.

Before closing the door on the service of a specific product line, this team should review the install base of the product and their needs over the service ramp-down period. This period may be used to train major customers[4] or other third-party service vendors.

This is also the time to close any unnecessary call centers and to revise any used call scripts for service, and conclude service contracts and warranties. Finally, any service-loaned equipment or rental units should be identified and returned to base for the closure activities.

Summary

- Once a product does not satisfy the needs of its customer base, it needs to be retired and pulled from the market.
- Among the reasons to retire a product, one is poor financial performance, including poor sales or price erosion and competitor pressures.
- Pulling a product from the market requires a plan and a well-staged series of activities. It begins with impact assessments in the areas of branding and marketing, finance, distribution channels including partners, operations (including manufacturing, service, and supply chain), and finally design.

Notes

1 RoHS is the acronym for Restriction of Hazardous Substances. RoHS is also known as Directive 2002/95/EC.
2 In the case of medical devices, obsoleting a product may constitute a risk to patients who rely on that product.
3 This may seem like an oxymoron. However, it is not uncommon for end of service to be several years behind end of manufacturing. In this period, products need to be serviced with components that were planned in the service inventory ramp-up period.
4 At times, institutional customers are so used to a particular product line that they are not willing to give it up, or the new product line may not be ready for deployment.

Part VIII

Skills Needed Across All Disciplines

21

Configuration and Change Management

Introduction

Before delving into details of configuration and change management, I would like to share a personal anecdote followed by an example.

A few years ago, I was part of a design team to develop a medical device. This particular equipment had utilized an electronic circuit board to drive an electric pulse at ultrasonic frequencies. The design team had completed the first and second engineering builds, and we had high confidence that the product would function as expected. On the basis of this gained knowledge and information, a decision was made to build enough units for subsystem verification testing. Lo and behold, during this activity, it became very evident that the subsystem did not behave as expected. Needless to say, this was not the sort of news that upper management considered lightly. As the entire team conducted an investigation to identify the root cause of this issue, it was discovered that one of the engineers had approved a "slight" deviation on an inductor without sharing the information with anyone else or giving it a second thought. I do not recall the exact deviation at the moment, but it had to do with the way the wire was wound on the inductor's core. Although the problem was rectified, it did cost both time and money – and a black eye for the team.

This problem could have been prevented if a process was in place to review each and every proposed change for its impact. Had the design engineer who approved the change (ironically a mechanical engineer approving changes of an electric component) requested a cross-functional review of the change, it might have been possible to reject the change and identify a different component or supplier.

> Configuration management is the *systematic approach to control* of design, manufacturing, and service documents and artifacts. It requires disciplined document maintenance and control from the start of product development.

Now for the example: Suppose that a product is being manufactured in several different manufacturing sites in different regions and being serviced by an independent service organization. Furthermore, assume that the design team had not maintained strict documentation on the design and component selection. As time has gone by, each manufacturing and service site has sourced some of the needed components locally without maintaining strict documentation. Now, an increasing number of complaints are recorded about the performance of the device, and there is a fear that some consumers might have been harmed. How would the design team identify the causes of the problem and develop a quick solution when no one can correctly identify the components of fielded products in any given region within a short period of time?

Here is another – though unrelated – scenario: presume that the primary product of one facility is a high-quality children's toy wagon. The wagon consists of front and back wheel assemblies, a steel-drawn body, and a handle. This manufacturer has a primary customer

DOI: 10.1201/9781003564508-29

whose requirement is for the body to be painted a certain color with its logo printed on its sides. The customer likes to expand their market by creating a similar wagon but with possibly a different color, logo, body, or even wheel style. Is it possible to create a document structure so that it would allow quick document replication and reuse of existing parts, should new product configurations be needed?

The term design history file (DHF) is often referred to as this collection of CM artifacts, which includes just about all the product-related documents.

The common thread between the anecdote and the examples may be summarized in four words: *systematic approach to change.* In a systems approach to product development, this is called *configuration management* (CM). It requires disciplined document maintenance and control from the start of product development. Any product-related document is considered to be a CM artifact. Although changes are expected to happen, they are instituted with a proper review of the impact of the change on the design, manufacturing, service, compliance, supply chain, etc. In other words, a proper review involves the cross-functional product-sustaining team.

Device master record (DMR) is a subset of the DHF that includes elements such as BOM, manufacturing instructions, and any tooling or fixtures that are needed.

The term DHF is often referred to as this collection of CM artifacts, which includes just about all the product-related documents. An example of a high level – product and system level – DHF tree is shown in Figure 21.1. A complete tree will include all the elements down to each component or part (not shown here).

Embedded within the DHF, there is a set of documents specific to manufacturing. This set is often referred to as DMR and includes elements such as BOM, manufacturing instructions, and any tooling that needed to be used.

Configuration Management

ANSI/EIA-649[1] defines CM as a "process for establishing and maintaining consistency of a product's performance, functional and physical attributes with its requirements, design and operational information throughout its life." The benefit of configuration management is that at any point in time a snapshot of the product's configuration is available. In other words, at any given point in time, the design team knows exactly what components have been used in which device by its serial numbers. Hence, there is a clear link between fielded products (by serial number) and the available documentation.

Now suppose that a change is proposed. Because a clear link between the product and its documentation exists, the impact of this change on the product, its documentation, manufacturing, and service processes may be assessed both precisely and in a short period of time.

The following elements are among the benefits of establishing configuration management as a part of design culture:

- CM requires that product requirements are properly defined and cascaded. This is in alignment with a systems approach to product development.
- There is a link and traceability from a part/assembly/subsystem/system (end items) to their associated documentation, such as specifications or drawings.

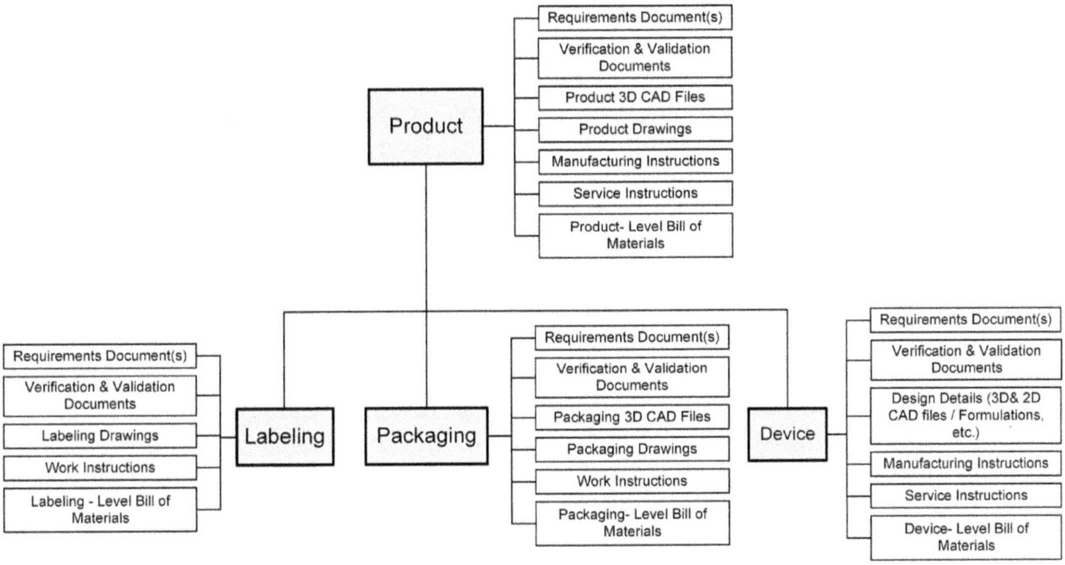

FIGURE 21.1
An example of elements of a design history file and configuration management at product and system level

- Once a product configuration baseline is established, any change request decision will be made on the basis of current and correct information. There is no longer either guesswork or long time lags to identify impacted areas.
- Efficient change processes may be established based on known product configurations.
- Based on a known product configuration, incorporated changes may be verified against requirements and then recorded throughout the product's life. This provides high confidence in the product information and its history.

Adversely lack of proper configuration management may lead to incorrect parts being installed during either manufacturing or service, affecting expected or known functionality or tolerances. Or, there would be increased service and manufacturing costs and time delays due to rework or inconsistencies and mismatches between instructions and parts or equipment. However, in my opinion, the greatest drawback of missing configuration management is the inability for quick and efficient impact assessments.

Configuration Management Process

There are two aspects to CM. The first aspect is the process of establishing the configuration of a product through its initial design and development activities – in a way, through the process that has been laid out throughout the chapters of this book. The second aspect of configuration management is the

> CM includes both the process of establishing the configuration of a product through its initial design and development activities and the systematic control of changes to the identified configuration for the purpose of maintaining product integrity and traceability throughout its life cycle.

systematic control of changes to the identified configuration for the purpose of maintaining product integrity and traceability throughout the product life cycle.

A formal configuration management process[2] consists of five steps: management and planning, configuration identification, configuration control, configuration status accounting, and finally, configuration verification and audit. These steps are described next.

Management and Planning

Management and planning represent the core CM activity for the enterprise and its relationships to the other activities. Inputs to management and planning consist of the authorization to initiate the CM program, communications with all of the other CM activities, and selected information and performance measurements received from the status accounting activity.

The CM activity is facilitated by the degree of management support provided, and the working relationships established with such other interfacing activities as program management, design and operations (including manufacturing and service). It is further facilitated by the resources assigned to the function, including such resources as automated tools, connectivity to a shared data environment, and other infrastructure elements.

The training and experience of the personnel and the guidance and resources they have at their disposal are also facilitators. The Management and Planning process may be constrained by a compressed time schedule for program execution, by a lack of needed people and tools, or by a lack of effective planning.

The outputs from this activity consist of CM management and planning information. Typically, this information includes the structure and content of the design history file and device master record structures, change control processes and procedures, and any other overview roles and responsibilities. It may also include interfaces to other programs and organizations.

Configuration Identification

This is the process of defining each baseline to be established during the hardware, software, or system life cycle. It describes the configuration items and their documentation that make up each baseline. This phase provides the foundation for all of the other functional activities by defining steps to specification identification, change control forms and templates, details of project baseline, and document control.

Specification Identification

Specification identification focuses on information such as document labeling and number schemes or the interrelationship between document versions and their releases. For instance, suppose that a part number is P1234. The specification identification may require that commercial specification documents associated with part numbers have a "_spec" suffix. Furthermore, if this part has been tested, the study protocol number should have a "_P" suffix and its report a "_R" suffix. Or that any numerical revisions (i.e., Rev 1, 2, etc.) are considered to be prerelease and any alphabetic revision (i.e., Rev A, B, etc.) are indicative of post release.

The main focus of specification identification, however, is the interrelationship between physical parts and their associated documents. Figure 21.2 provides a pictorial view of

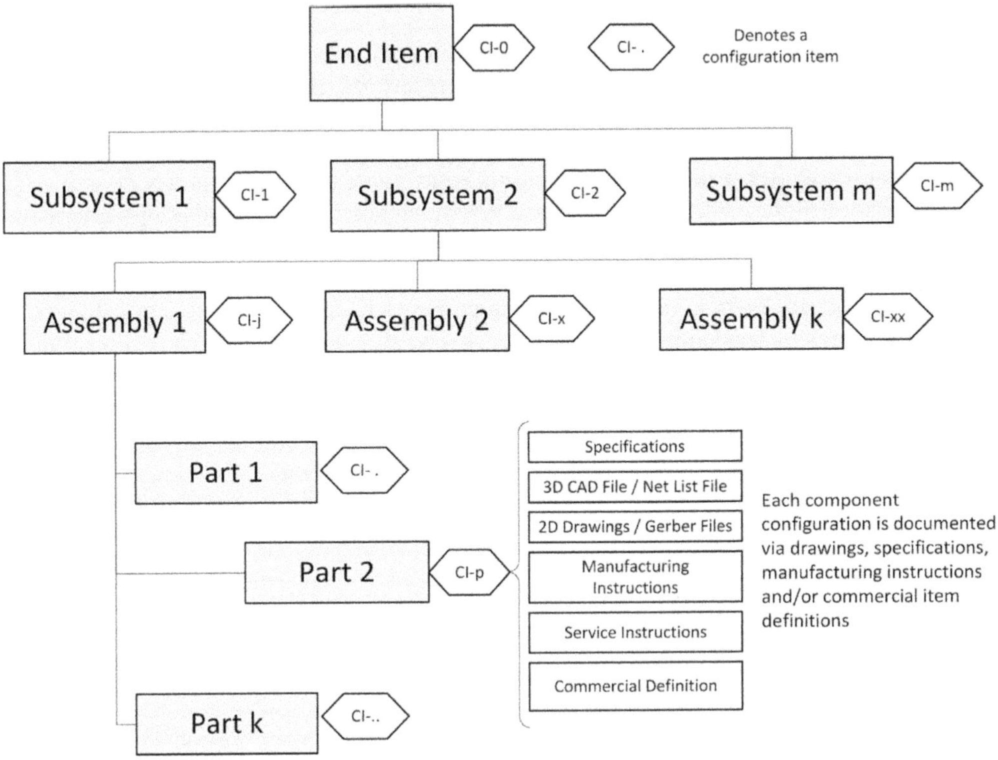

FIGURE 21.2
Interrelationship of physical part identification and its associated documentation

an example of a physical item hierarchy with the final product as the end item. In reality, end-item designations are made at the specification identification stage of configuration management and may be applied to a subassembly.

Change Control Forms

As mentioned earlier, an aspect of configuration management is change control. Elements of change control are impact assessment, change planning (including project schedule and resources), and design reviews. For consistency and ease of review, these documents should have a similar format between different projects. Change control forms are enterprise-level templates and forms that are utilized.

Details of Project Baseline

Project baseline identification is the identification of baseline stages and project gates that identify at what points in time (e.g., requirements allocation) the configuration is "frozen" and any subsequent changes would require formal change control. Furthermore, it dictates the purpose, mechanism, and content of the baseline. For example, a baseline needs to be developed so that a software program may move from a development site to a customer site. Its content should be the source file(s), and executable as well as the compiler. Finally, the roles and responsibilities, including the approval authority, need to be established. Figure 21.3 provides a pictorial view of configuration management within the V-model and possible baseline locations.

Document Control

The last aspects of configuration control are document access and control mechanisms, backup and recovery processes, and document retention policies. Proper configuration management requires that controls be put in place so that documents are stored properly and are maintained for an adequate length of time.

Another aspect of document control is identification of the difference between *released* and *effective* documents. The difference is that a released document is approved by all stakeholders, but its implementation may be held back for business reasons. An example of this situation is as follows. A component is redesigned to improve manufacturing cost structure. This component is functionally the same as the previous part but differs in fit and form. All design documents, along with verification activities, have been completed and documented. All related documents are released; however, they do not become effective till all the existing components are used. At that point in time the documentation for the new component becomes effective. Incidentally, if this is a major upgrade, a new baseline may also be required.

Configuration Control

Once the baselines are established in either product development or product sustaining cycles, changes require formal configuration assessments and control. This is the process

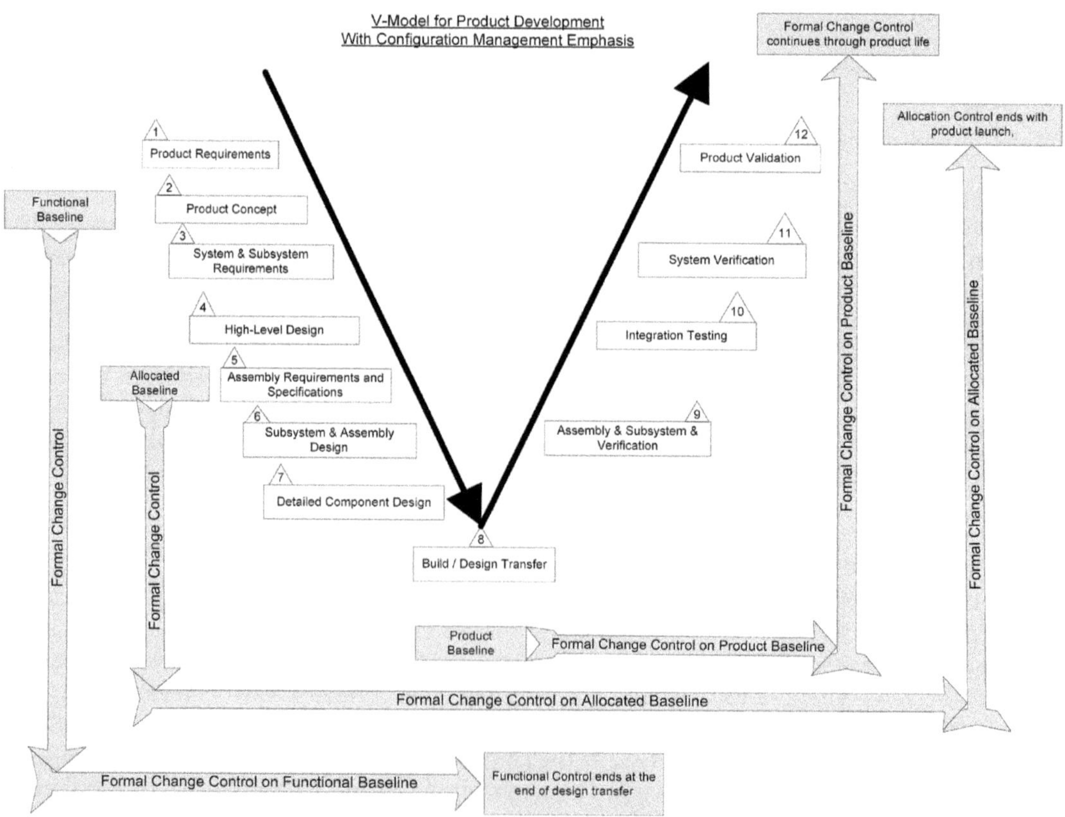

FIGURE 21.3
Configuration management baselines within the V-model

of evaluating, coordinating, and implementing proposed changes to the configuration items. Typically a configuration control board is in charge of these activities.

This board also governs the implementation of approved changes to baselined systems, hardware, software, and associated documentation. These changes may be requests to modify fielded items or manufacturing and/or service facilities. It receives and processes requests for design changes through technical change proposals or requests for deviations from various internal stakeholders such as technical, and operational, and external stakeholders such as contractors and suppliers.

This mechanism allows for a thorough review of requested changes or deviations and their approvals or disapprovals. Furthermore, it facilitates the necessary authorization and direction for change implementation.

Configuration Status Accounting

Configuration status accounting (CSA) is a database that maintains the status and history of proposed changes and the schedules for and status of configuration audits. It is also the vehicle to trace changes to the hardware, software, or system that may include such information as the as-designed, as-built, as-delivered, or as-modified configuration of any serial-numbered unit of the product as well as of any replaceable component within the product. This activity provides visibility into status and configuration information concerning the product and its documentation.

Performance measurement metrics on CM activities are generated from the information in the CSA database and may be provided to the management for monitoring the process and in developing continuous improvements.

Configuration Verification and Audit

Configuration verification is the process of verifying that deliverable items (e.g., hardware, software, or system baselines) match items that have been agreed upon and approved in the plan and that their requirements have been met.

Inputs to this audit include elements of the design history file. The output of this audit is the certification that

A) the product's performance requirements have been achieved by the product design, and
B) the product design has been accurately documented in the configuration documentation.

This process is also applicable to verification of incorporated technical changes. Successful completion of verification and audit activities results in a verified product and documentation set that may be confidently considered a product baseline, as well as a validated process that will maintain the continuing consistency of product to documentation.

Configuration Management P-Diagram

As with any process, there are input, output, control, and noise factors to consider for product-level CM, particularly once a baseline has been established. A P-diagram for the CM process is shown in Figure 21.4.

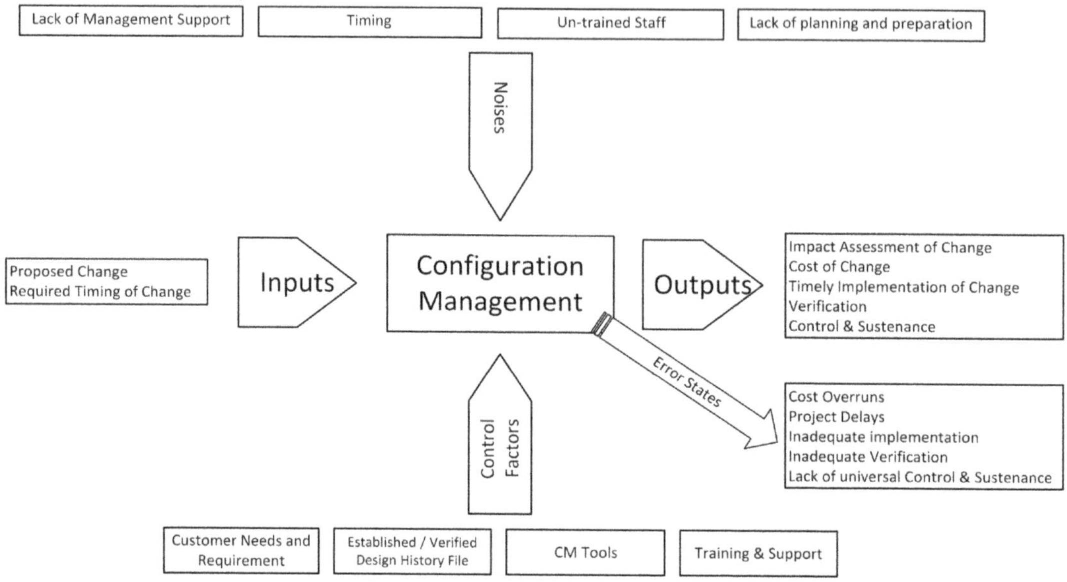

FIGURE 21.4
Configuration management process

- Inputs – Typically, for a baselined configuration, inputs are proposed changes, deviations, and the required time period for the requested change.
- Outputs – For a well-integrated configuration management process, outputs are an accurate assessment of change not only to the product design and its documentation, and to its manufacturing, service, and supply chain, but also to other products that may use the same impacted part, equipment, etc. In addition, the cost and timing of the requested change are determined before the change is implemented. Finally, the change is verified and, through control mechanisms, is sustained.
- Control Factors – To a large extent, control factors include established customer needs and product requirements, an established and verified design history file, proper CM tools, and staff training and support.
- Noises – Factors that inhibit or limit the configuration management process are lack of management support, lack of timing for the project, untrained staff, and a lack of planning and preparation.

CM is very efficient structure when it is running in an ideal state. Unfortunately, I have also witnessed how it has become a major hurdle in document control. As shown in Figure 21.4, one of the error states in CM is project delay. The reason that I have often observed is that one or two members of the cross-functional team need to be "sure" of the impact of the change and are not willing to make timely decisions.

> CM II the evolution of CM to go beyond the management of just the product but to bring the same level of efficiency to the enterprise as a whole.

Configuration Management and CMII

ISO 10007-2003, *Quality Management Systems – Guidelines for Configuration Management*; MIL-HDBK-61A, *Configuration Management Guidance*;

and MIL-STD-3046, *Interim Standard Practice, Configuration Management*, speak to configuration management of products. There are also a number of references that discuss configuration management of software; see, for instance, Aiello and Sachs 2011.

Guess (2006) defines the CM II process to be the evolution of the configuration management to go beyond the management of just the product but to bring the same level of efficiency to the enterprise as a whole. Stages of CM II differ from the traditional CM in that there are seven sub-processes as opposed to the five stages of CM mentioned earlier. These subprocesses are as follows:

1. Requirements management,
2. Change management,
3. Release management,
4. Data management,
5. Records management,
6. Document and library control, and
7. Enabling software tools.

As may be seen here, there is a great deal of similarity between the traditional CM and CM II; however, the fact that CM II starts with requirements management, its alignment with a systems approach to product development, and the V-model is easily observed (Figure 21.5).

The premise of the CM II model is that defects found during product verification are telltale signs of requirement deficiencies. Therefore, any changes that are initiated during the verification phase are corrective actions, expensive and time-consuming in nature. However, should changes be identified and initiated while the product is still in development, the development either aims to improve existing requirements or to extend the higher-level requirements. Changes during development are both less expensive to implement and take less time (Guess 2006). It is therefore advisable that more time be spent on requirements and their management than on rushing into physical item fabrication and assembly.[3]

Change Management and Control

> As change is inevitable, it is essential that a robust change management process be put in place to protect the known configuration of the product at any given time.

The nature of product development is often one of conflicting requirements and compromise; features and part selections are at times done to meet a particular regulatory concern or market need. Once a product is launched, it is hardly distributed in all potential markets all at once. For products with new designs or features, there are ramp-ups that are intentionally spread over many months, if not a couple of years. All these factors help reveal various needs for changes to product features, options, and/or design. Even for well-established marketed products, there are needs for making changes due to part and component obsolescence, as well as changes in manufacturing and/or techniques. As

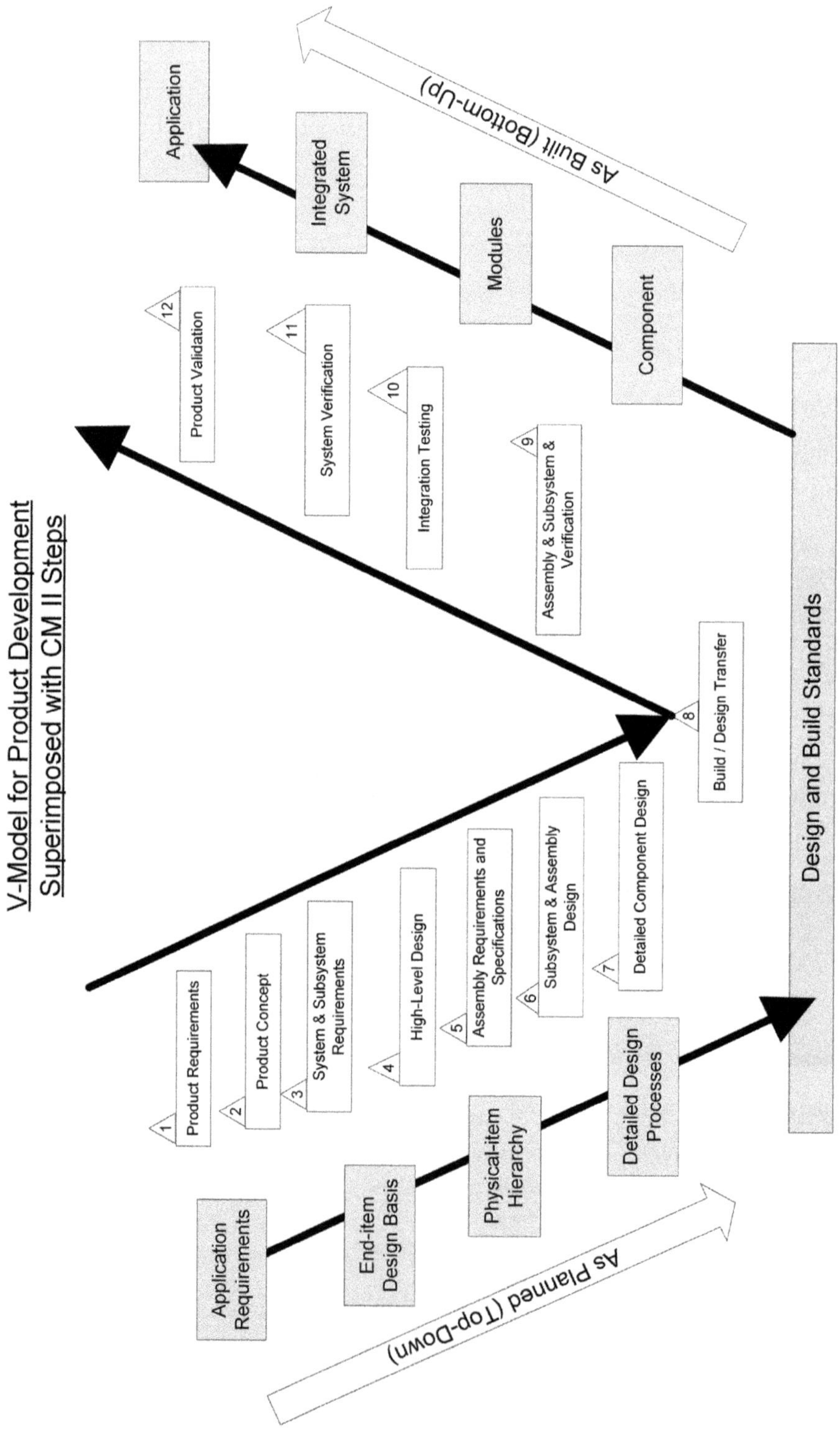

FIGURE 21.5
CM II process within the concept of the V-model

Elements of CM II are adopted from Guess (2006).

change is inevitable, it is essential that a robust change management process be put in place to protect the known configuration of the product at any given time.

In my view, there are three aspects of change control that should not be compromised under any circumstances. The first is documentation. Formatting of this information, or whether data is stored on paper or in an electronic database, is only secondary to actually capturing and recording information, events, and decisions made. There is an old adage that *if it is not documented, it did not happen*!

The second aspect of change control that should not be overlooked is impact assessment. Typically, impact assessment should be done by a cross functional team including members of the design team to assess the impact of a proposed change(s) to design specifications up to system and product requirements and representatives of manufacturing and service organizations to consider the impact on manufacturing/service procedures and even equipment used. Even the impact on risk, reliability, regulatory agencies (UL, FDA, TuV, etc.), and suppliers should be evaluated.

The final aspect of the change control is a final design review to ensure that all aspect of the intended change was in fact implemented. In this final review, typically, the agenda is to provide a laundry list of items that needed to be updated as a result of the impact assessment and then to provide objective evidence that the work was actually completed.

Change Control Process

The change control process begins with a written change request. This request should include the rationale for the requested change as well as the required time for change implementation.

A technical review board (TRB), comprised of a cross-functional team, reviews the proposed change request to answer the following questions:

- Is the proposed change technically feasible?
- What is the initial impact of the change?
- What would be an approximate duration to implement the change?
- What would be an approximate cost of the change to the business, including non-recurring design (or engineering) costs?

It is not uncommon that these questions are not answered in the first review. Often the TRB chair (called *Change Specialist 1* [CS1] in the language of CM II) interviews the requesting body or conducts research to develop the background information. Once TRB answers these questions, the CS1 records the decision of this board to either recommend this change or to reject it. If the change request is rejected, CS1 notifies the requesting body of the rejection and its reasons.

The responsibility to approve or disapprove a change request rests with the change control board (CCB). This board is a team of technical and business leaders who review the proposed changes that have been considered by the TRB. CCB reviews the change for its impact on the enterprise and business needs and concerns. If CCB approves a change, a priority is also assigned. Typically, the priority index enables the chair of the change implementation team (called *Change Specialist* 2 [CS2] in the language of CM II) to set the priority of the change to take place.

However, before the change is implemented, the change request goes back to TRB for a full impact assessment. Considering that the business has approved the change, a change

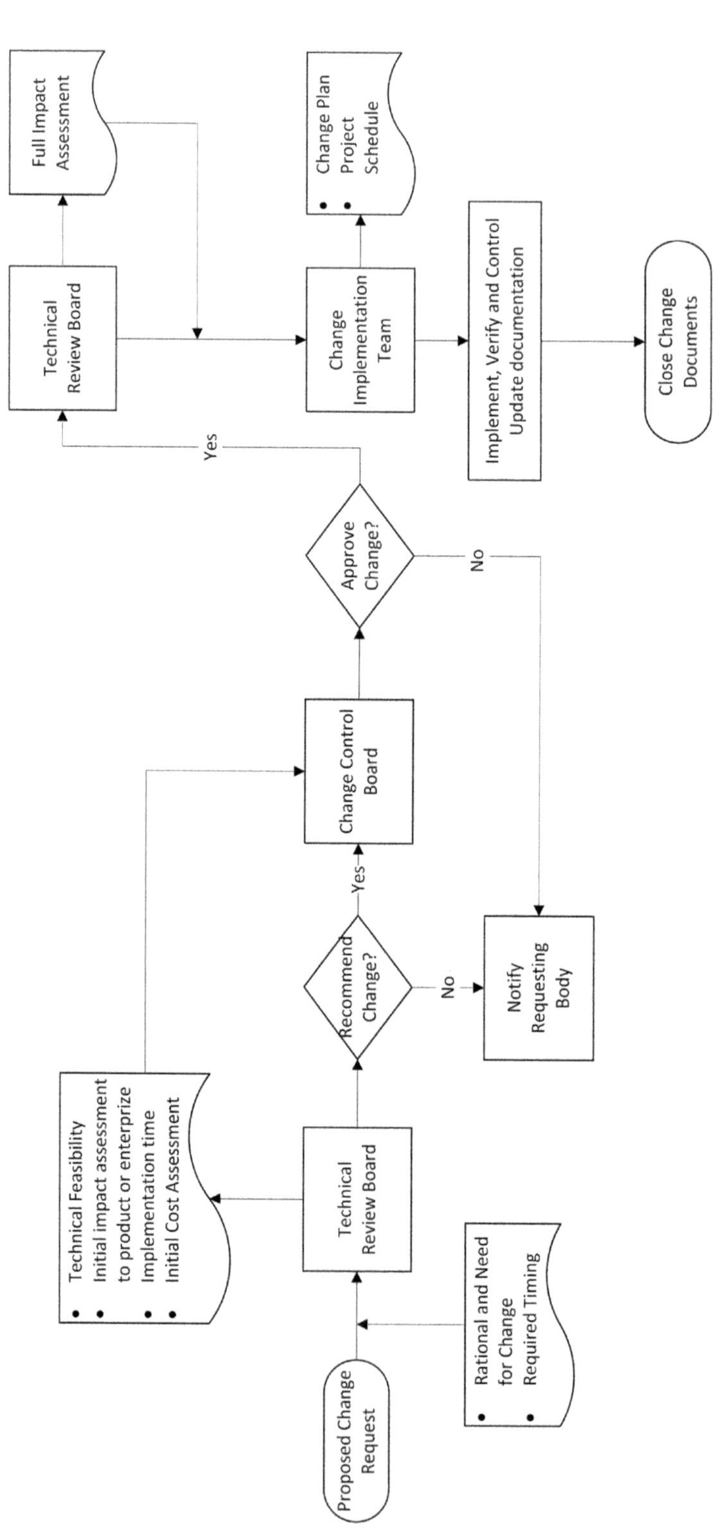

FIGURE 21.6
A change control process

owner is assigned who leads and captures the impact of the change to parts, documents, procedures, etc. Additionally, the change owner identifies the subtasks and their action owners along with an estimate of task durations.

The actual schedule and project management of the change take place as a part of the change implementation team (CIT) headed by the CS2. Change owners are members of the CIT. In this forum, details of the plan, its timing, and needed resources are planned and executed with the help of the TRB impact assessment document.

Once the change has been fully implemented, a design review is held to compare the elements of the change plan that had been written on the basis of the requested change with the assessed impact and the outcome of the implemented change and all its documentation. In this design review, an independent reviewer ensures that all work has been completed and no gaps exist. With this final review, the proposed change request is formally implemented, documented, and closed. Figure 21.6 provides a pictorial view of this process.

Summary

- In a systems approach to product development, a systematic approach to establishing and maintaining design and operational information of a product throughout its life is called Configuration Management (CM).
- The term design history file (DHF) often refers to the collection of CM artifacts.
- There are two aspects of CM; one is the process of establishing the configuration of a product, and the other is a systematic control of changes made to the established configuration.
- A process should be in place to review each and every proposed change for its impact not only on the product but also on its manufacturing and service procedures.

Notes

1 As quoted in MIL-HDBK-61A(SE) 2001.

2 Based on MIL-HDBK-61A.

3 This statement reminds me of Alice from *Alice in Wonderland*, who gave herself good advice but never followed it! The business and market pressures often override good engineering or science rules.

22

Data, Measurements, and Tests

Introduction

Earlier in the book, and particularly in Chapter 17, I had mentioned testing and, by association, conducting experiments. This implies that data is being collected and analyzed in order to make design or business decisions. The focus of this chapter is to take a closer look at data, and data collection and to provide guidelines on how to examine the impact of inputs on output, otherwise known as cause and effect, to ensure that all elements impacting the outcome of the test results are considered and accounted for so that, paraphrasing Edward Tufte (1997), it is an experiment and not merely an experience!

Let me share an anecdote. A few years ago, a colleague of mine was developing a mechanism for dispensing liquid in specific volumes. His liquid of interest was provided in relatively large plastic bags (about 2 or 3 gallons) with an attached (and sealed) tube. The fundamentals of his concept were simple. The tube was to be connected to a fixed-size orifice, which was controlled by an electronic valve. Gravity would provide the required driving force to the fluid. It was possible to calculate the flow rate based on fluid properties and the orifice size; hence, the dispensed volume was only a function of time. He had his transfer function between the volume dispensed and the valve open time. He had even taken into account the start-up and shut-down inertia. He built his engineering units and was running his initial verification studies.

He realized something was awry one morning when his calculations based on his transfer function indicated that he had dispensed more liquid than was contained in the source bag. When he had left his office the previous day, all seemed to be in order. He looked at the combined data for the two days and realized a great deal of variation, though the average was within acceptable limits. He shares his story with me.

I looked at his data and observed a bimodal distribution. It happened that if he ran his experiments in the morning, his results were substantially off-target. If he ran his experiments in the afternoon, his results were within acceptable limits. When the two datasets were segregated, it was obvious that an unaccounted noise factor was at work.

The liquid being dispensed was a nutrient with substantial levels of fat particles. The bags had to be refrigerated to prevent spoilage. Cold liquids were more viscous than the liquids at room temperature. This meant that when the bags were cold, the fluids would not flow as fast. Unaware of this fact, my colleague would bring a few bags from the storage area into his lab and set them aside. Often, he would take care of other office issues before beginning his test regimens. However, a number of times he started his tests before the liquid temperature had a chance to reach room temperature.

I have to admit when I first heard of this problem, I did think about temperature effects. However, I was not thinking about the liquid temperature. My hypothesis was that his electronics were not designed properly and that he was experiencing temperature drift due to the dissipated heat and increased internal temperatures. In other

DOI: 10.1201/9781003564508-30

words, as the electronic components heated up, their properties changed, leading to output variations. This is an example of noise factors that may be experienced during an experiment.

Noise Factors and Variations

Suppose that a circuit is made of three elements, R1, R2, and R3. For argument's sake, suppose that, in one setup, the voltage across resistor R2 needs to be measured, and, in the second configuration, the current should be measured. Now, suppose that R1, R2, and R3 are all functions of temperature. Clearly, some duration of time is needed before the temperature of the unit has stabilized. Should measurements be made prior to this thermal stabilization, wrong conclusions may be drawn. A more acute situation may be taking measurements under dynamic loads when heating or cooling of the components may take place depending on the loading of the device. A similar noise factor is that the measurement equipment may also be susceptible to either internal or external temperature variations.

Another complicating factor is the measurement equipment and its coupling to the unit under test. As shown in Figure 22.1, the impedance of the measurement unit does in fact couple with that of the unit under test. Hence, variation of the instrument's impedance due to temperature can adversely impact measured values.

A lack of awareness of a measurement instrument's impact on an experiment or other noise factors such as heat may lead to substantial errors. As engineers and scientists set up tests and experiments, they should be cognizant of noise factors, variations, and uncertainties.

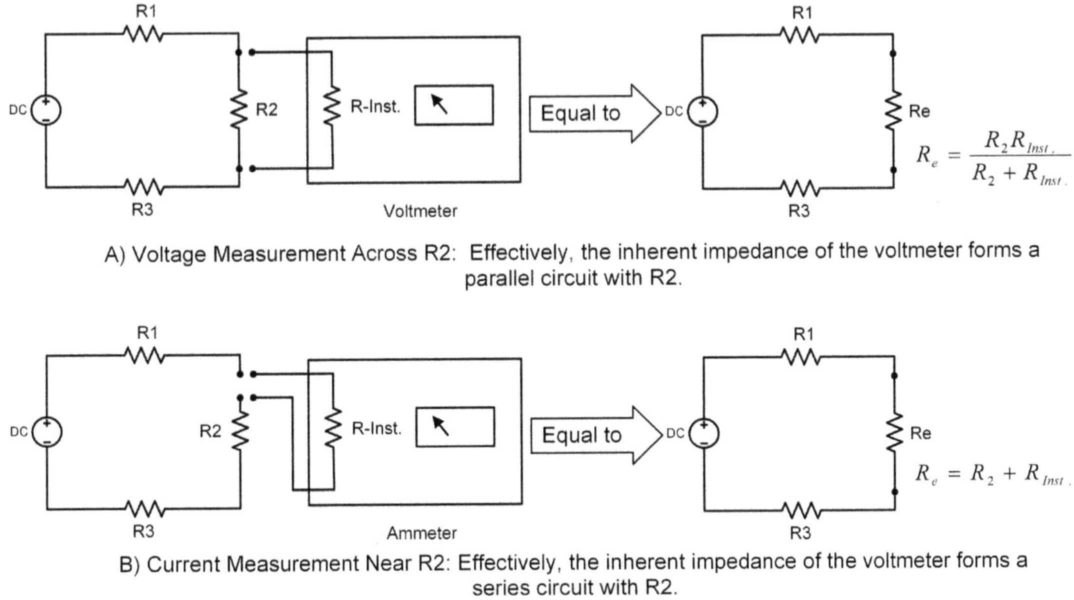

FIGURE 22.1
Impact of instrumentation on measured variable

Coupling of the instrument impedance to that of the unit under test presents another problem. Consider the equivalent resistor for the one shown in Figure 22.1. The voltage across R2 may be measured correctly if and only if the instrument resistance (R-Inst.) is infinite. In practice, the instrument resistance should be much larger than R2; however, the question remains: how large is large enough? The answer, of course, is that *it depends*! Figure 22.1B presents a different scenario. Here, to measure R2 correctly, R-Inst. should be zero. Again, we can only settle for a very small impedance, not zero.

A lack of awareness of a measurement instrument's impact on an experiment or other noise factors such as heat may lead to substantial errors. As engineers and scientists set up tests and experiments, they should be cognizant of noise factors, variations, and uncertainties.

Another source of uncertainty is associated with people. It so happens that if one operator makes the same measurement a number of times in a row, every single reading will be different. If different operators make the same measurement, each reading – again – will be different. Not only should test personnel and operators be aware of these factors, but to a large extent, they need to anticipate their impact and minimize it when possible.

For the rest of this chapter, I will focus on providing some background information on data types, levels of measurement and measurement systems, different types of product-related tests, and, finally, a review of elements of a robust test plan.

Level of Measurement and Basic Statistics

In the context of product development and/or product support and sustainment, engineers and scientists (among other interested parties such as marketing staff) collect information. The type of information that is collected depends to a large extent on the questions that are to be answered or the requirements that need to be satisfied.

Some questions and/or requirements need a "Yes/No" answer. These are questions that are frequently asked in marketing surveys: "Were you satisfied with our service?" Or, "will you recommend us to your friends?" Similarly, in a design-related inspection, the questions may be asked: "was the drawing (or formulation) approved/in-place/etc.?"

Other types of questions (or requirements) may not be simply answered by a binary response (i.e., "Yes/No" or even "0/1"). Questions such as one's preferred color, or the states where a product may be distributed require selection of an answer from a list of options. A very common multiple-answer questionnaire is this: "How satisfied were you with our services: select from 1 to 5; 1 indicating the least satisfied to 5 as the most satisfied." Another example of a multiple-answer question is selecting a favorite item from a list (such as colors of a product).

> Measurements have associated ranges, which must be reflected in our expectations. For instance, the force to open a cap from a soda bottle is 2.8 ± 0.1 lbs.

Finally, there are questions whose answers may not fit within a limited number of responses. For instance, "what is the required force to twist a cap from a soda bottle?" Or, "what is the accuracy of the company's insulin pump[1]?" The response to this type of data is typically numeric and may be selected from a theoretically infinite number of values over the range of a finite interval. In practice, one is limited by the resolution of the measurement system. For instance, if using a ruler to measure the length of a box, a practical limitation is the ruler's graduation. The actual length of the box falls on the space between two marks. Even the thickness of the graduation itself presents a finite length. Thus, measurements have associated ranges, which must be reflected in our expectations. Examples are:

- What is the company pump's accuracy? It is accurate within ± 5.5% of the user-specified flow-rate values.
- What is the required force to open a cap from a soda bottle? It is 2.8 ± 0.1 lbs.

In order to compare different data, it is important that uniform scales exist. Stanley Smith Stevens has been widely acknowledged as having defined nominal, ordinal, interval, and ratio scales, which are widely used among engineers and scientists (Velleman and Wilkinson 1993). *Nominal* measurements (such as blue, green, UK, China) cannot be put into a meaningful order from a low to high ranking. *Ordinal* measurements (such as a scale of strongly disagree to strongly agree) provide a measure of difference between successive values – albeit inexact. The values in this scale may be placed in a meaningful order. The next level of measurement topology is *interval*. As indicated by its name, the distances between measurements are meaningful, but they are relative to each other, not to a fixed reference point. Examples of these measurements may be time or temperature measured in Celsius or Fahrenheit. Similarly, the *ratio* topology requires that the distances between different measurements be well defined; however, additionally, this scale has a fixed reference value, i.e., a zero point.

Categorical Data

These various topologies have been placed into two broad measurement scales, although there is not a general agreement on what they should be called (Juran and De Feo 2010; Fowlkes and Creveling 1995). For argument's sake, let's call the first scale *categorical* and the second scale *quantitative*. Consistent with the scenarios that were just provided and Steven's definitions, categorical scale includes the nominal (and its special case binary) and ordinal scales. Quantitative scales include interval and ratio scales, which can be either discrete or continuous. From an informational point of view, binary data (such yes/no, high/yes, pass/fail) provide the least amount of information, whereas continuous quantitative data carry the most.

> There are two broad measurement scales. They are categorical and quantitative.

In situations where requirements dictate a binary output (i.e., the least amount of information about the outcome), it may be possible to set up the experiment or data collection in such a way as to obtain more information about the results. The following examples may provide some clarification.

Figure 22.2 depicts a series of markers that have been pad-printed on a plastic substrate. This particular tool may be used in some cosmetic surgeries to make anatomical measurements. Once used, the tool is washed and then ultrasonically cleaned. The ultrasonic cleaning causes degradation of the pad prints. In order to measure the life of these instruments as a function of the number of cleaning cycles, the pass/fail conditions of printed pads are converted to approximate 35%, 70%, and 100% damage readings (a 25%, 50%, and 75% scale is just as acceptable). Note that by shifting from a "pad is damaged" mindset to a "pad is x% damaged" mindset, the binary scale (pass/fail) has changed into a categorical scale with an increasing level of information about the outcome of the experiment. Now,

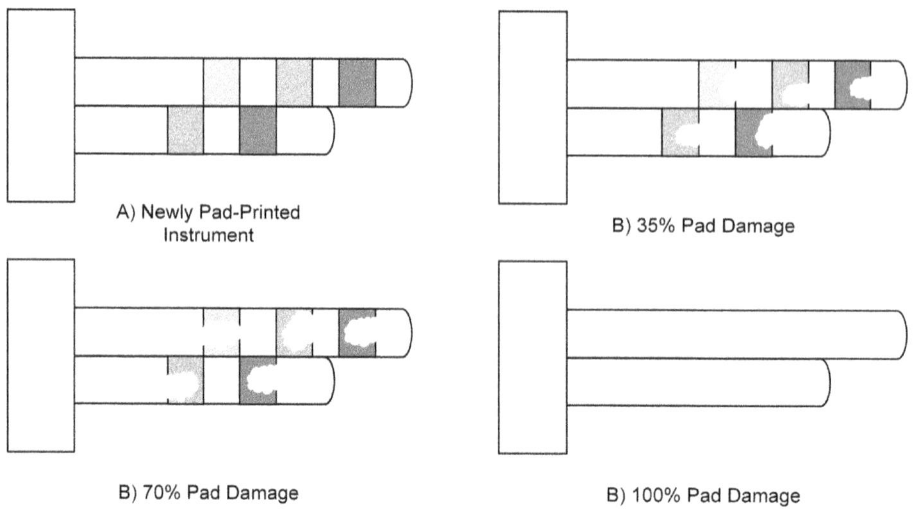

FIGURE 22.2
Conversion of binary to categorical data

imagine that if the actual pad-print area could be measured, a more accurate estimate for the life of this instrument could be made.

The second example is as follows. Another area where product-related data is collected is on the manufacturing floor, where parts are rejected on the basis of being defective. So, one statistic may show that the output of a particular process is 5% defective. Note that the only information being conveyed is that 5% of the outcome had to be either scrapped or reworked (depending on the product). If the requirements are written in such a way as to ask for the number of defects, a binary measurement level (i.e., defective or not) may be converted into a categorical level (defects at various locations on the part or rejections at various process stages).

For instance, suppose that the process for manufacturing the spatula shown in Figure 17.2 (Chapter 17) had the following steps as provided in Table 22.1. This table illustrates that although quantitative data may provide the highest degree of information – as I will shortly explain, it is possible to use categorical data to drive verification of requirements and/or identify potential root causes of manufacturing issues. For example, one may report that the process of making the spatula has a 3.25% scrap rate. However, as this overall data on defective parts is parsed to its details, it becomes obvious that the heat-treating step has the highest scraps, as indicated in Table 22.1. Further pictorial details indicate that almost all defects are observed in the middle of the part. Now, this is information that may be actionable. A failure mode for heat-treating flat parts is that they may warp if they are not staged correctly during the heat treatment process.

Quantitative Data

As is evident by its title, quantitative data may change seamlessly from one value to another, either discretely or continuously. Its power is that it enables engineers,

TABLE 22.1

A Second Example of Converting Binary Information to Categorical Data

Stage	% Defective	# of Defects per 1000 Units	Observed Areas for Defects
Stamping	0.10	3	
Trimming	0.25	10	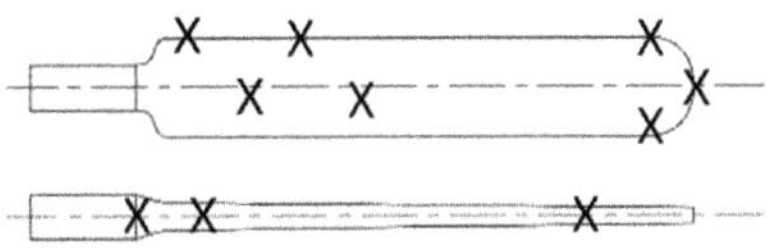
Heat Treating	2.00	20	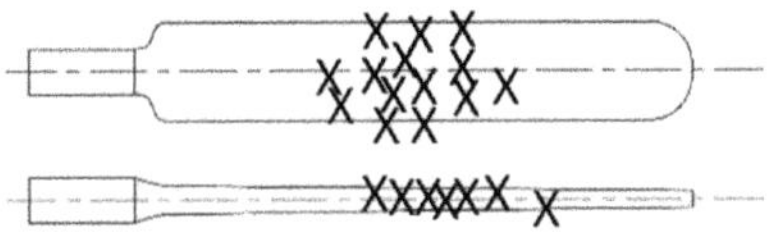
Polishing	0.90	27	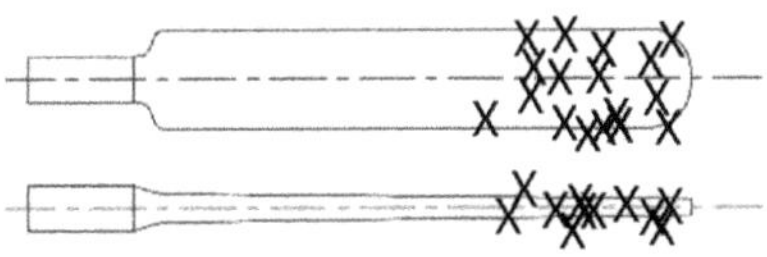
Total Scrap	3.25	32.5	–

scientists, and researchers to make logical comparisons and consequently make scientific decisions on the basis of their comparisons. To illustrate my point, consider a bag of peanuts. It may weigh 453.59 grams (~1.00 lb.), or it may weigh 1.00 lb. (~454.00 grams). The difference between the two is the use of the scale (grams vs. pounds) and the degree of precision used in measuring the bag (note: 453.5929 gr = 1.0000 lb.). Furthermore, converting and rounding the numbers from one scale to another may lead to precision differences.

Conversely, I can empty the contents of the peanut bag on a table and count the number of nuts in the bag – it may be 183 or even 184 (whole) peanuts. Say, I had 10 one-pound bags. Would they all weigh exactly 1.00 lb. or 453.59 grams? Would they all have 183 whole peanuts in them or 184? Furthermore, if I were to count the number of peanuts in every one of the one-pound bags, can I presume that every one-pound bag of peanuts would hold a number in the range that I have counted previously? Here is another question: if I were to randomly select 184 peanuts from a pile and place them in a bag, what should I expect the weight of that bag to be? How confident will I be in my estimate?

In a way, these questions are reflective of the need to identify how the data is distributed, its central value, and how the data is spread around this central value. These characteristics of quantitative data are defined by the following three terms: histogram, measures of central tendency, and measure of dispersion.

TABLE 22.2

Development of a Histogram Table

Bin	X	Population
2.25–2.75	2.5	0
2.75–3.25	3	1
3.25–3.75	3.5	2
3.75–4.25	4	2
4.25–4.75	4.5	3
4.75–5.25	5	6
5.25–5.75	5.5	3
5.75–6.25	6	2
6.25–6.75	6.5	1
6.75–7.25	7	0

Histogram

A histogram is a graphical representation of the cluster of data in a dataset. It is related to how the data is distributed around a central value. As an example, consider the following dataset:

$$\{3.2, 3.4, 3.7, 4, 4.1, 4.4, 4.4, 4.7, 4.8, 4.9, 4.9, 4.9, 5.2, 5.2, 5.3, 5.6, 5.7, 5.8, 6.2, 6.5\}$$

Now, eight bins may be formed and centered around X = 3.0 to 7.0 in intervals of 0.5. Next, the number of data points in each bin may be counted and recorded in each bin as shown in Table 22.2.[2] For instance, there are two data points in the 3.25–3.75 bin and six points in the 4.75 to 5.25 bin.

With this information, the histogram may be graphed as shown in Figure 22.3.

A histogram is a pictorial map of the data that provides a bird's-eye view. For instance, it is obvious that the data clusters primarily around 5.0. Furthermore, there are no reported data below 3.0 and above 6.5. In addition, one may note that data clusters population slowly grow toward the middle grouping and then slowly decrease away. This is a characteristic of a normal distribution. Later, I will provide some information on how one can take advantage of the properties of normal distribution to make product-related decisions. For now, there are characteristics of quantitative data that should be reviewed. Any quantitative dataset may be identified with two measures. One is called measure of central tendency, and the other is called measure of dispersion. In simple language, these are tools to evaluate whether there is a central cluster of data and if so, how far-spread the dataset is with respect to this central cluster.

A histogram is a pictorial map of the data that provides a bird's-eye view.

The three common measures of central tendency are mean, median, and mode.

Measures of Central Tendency

There are three common measures of central tendency. These are mean, median, and mode.

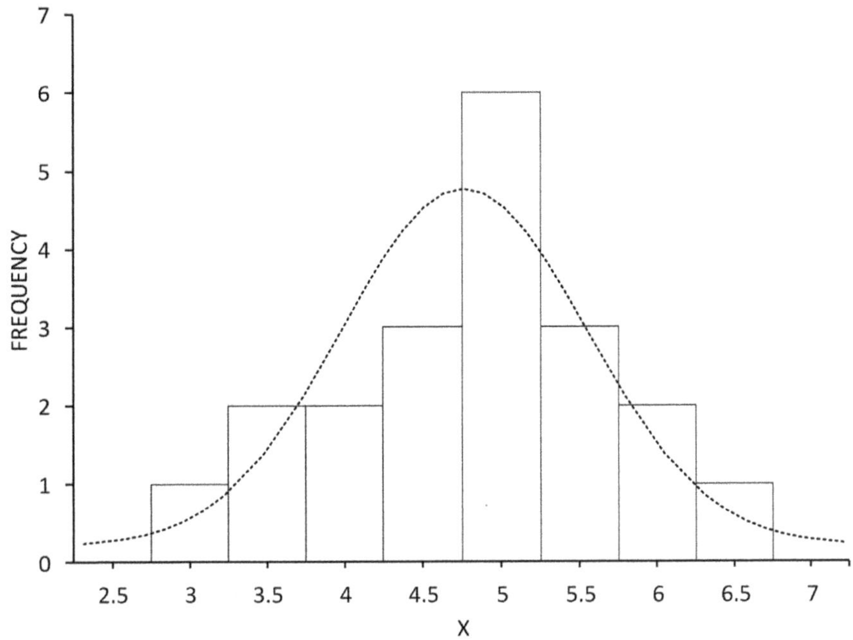

FIGURE 22.3
Histogram Superimposed by a Normal Distribution Curve

Mean is the sum of all data values (x_i) divided by the sample size[3] (N). It is also referred to as μ

$$\mu = \frac{\sum_{i=1}^{N} x_i}{N}$$

This is the most commonly used identifier of the central value of a dataset. As an example, consider the previous dataset once again:

$$\begin{Bmatrix} 3.2, & 3.4, & 3.7, & 4.0, & 4.1, & 4.4, & 4.4, & 4.7, & 4.8, & 4.9, & 4.9, & 4.9, \\ 5.2, & 5.2, & 5.3, & 5.6, & 5.7, & 5.8, & 6.2, & 6.5 \end{Bmatrix}$$

The mean for this dataset is 4.8 (rounded to one digit). In using this measure to represent data, one has to be careful because an extreme single value may distort any conclusions. For instance, if the last number in this set was 16.5 instead of 6.5, the mean would be calculated as 5.3. However, if we were to remove this large value from the dataset, the mean would still be 4.8.

Median is the middle value when the data is arranged from the lowest to highest value (or in reverse). If there are an even number of the data points, the median is the average of the two values in the middle. For the dataset above, the median is

$$\text{median} = \frac{4.9 + 4.9}{2} = 4.9$$

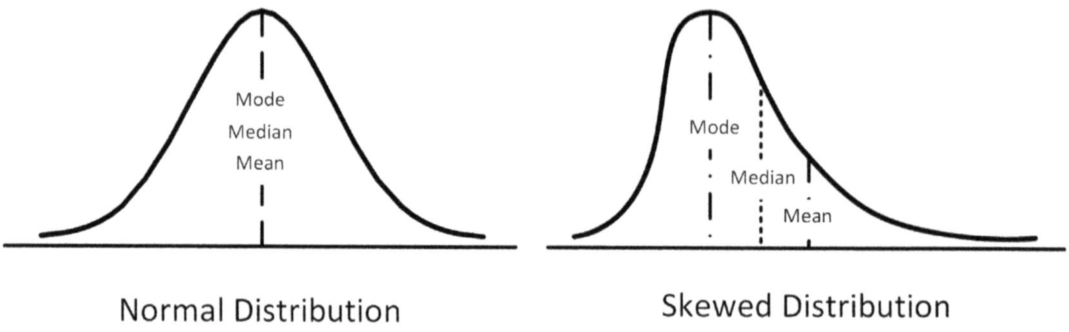

FIGURE 22.4
Normal versus skewed distributions

Note that the median is close to the mean for this dataset.

The last measure of central tendency discussed here is *mode*. The mode of a dataset is the most frequently occurring number in that set. In a histogram, it is the value associated with the tallest bar (or the bin with the highest number of units). In the example above, the mode of the set is

$$\text{mode} = 4.9$$

In a normal distribution, all three measures of central tendency are equal, whereas in a skewed distribution, these values are different from one another, as shown in Figure 22.4.

Measures of Dispersion

There are three common measures of dispersion. These are range, span, and standard deviation.

The range is the difference between the maximum and minimum value in a dataset. For the above example, the range is:

$$\text{Range} = 6.5 - 3.2 = 3.3$$

An alternative to range is called a *span*. Span is defined as the range associated with the 90% of the data in the center of the set. In other words, it is the range once the top 5% and bottom 5% of the data have been discarded. Span is not as sensitive to sample size as range is.

One of the shortcomings of the range is that it is highly dependent on sample size. A stronger measure is called *standard deviation* and is defined as follows:

$$\sigma = \sqrt{\sum_{i=1}^{N} \frac{(X_i - \mu)^2}{N}}$$

The strength of standard deviation, as may be observed in the equation above, is that it is based on using the entire dataset to provide a measure of variation about the mean.

Associated with standard deviation, there are two other measures that are used as well. They are variance (σ^2) and coefficient of variation (COV); and are defined as follows.

$$\sigma^2 = \sum_{i=1}^{N} \frac{(X_i - \mu)^2}{N} \text{ and COV} = \frac{\sigma}{\mu}$$

COV is typically reported as a percentage.

Population versus Sample

Statisticians often refer to two types of statistics, namely, descriptive and inferential statistics. In descriptive statistics, one is concerned only with the dataset at hand. Data is summarized in a meaningful way such that decisions and conclusions may be made without the expectation of expanding those conclusions and decisions beyond the existing dataset. This set may be small or large.

Inferential statistics is different in the sense that one's interest in studying the dataset is to extend, expand, and draw conclusions and decisions to larger groups and populations. In design-related experiments and testing, often inferential statistics is used because the goal is to predict the behavior of the total production units by testing a representative sample in a laboratory environment. To make distinctions between the two types of statistics, properties of populations such as mean or standard deviation are called *parameters*, whereas properties of samples are called *statistics*. These are shown in Table 22.3.

Recall the bag of peanuts example where I was looking at the average weight of a single peanut or the average number of peanuts in a pound. So long as I was only focused on the bags that I had collected, any data characteristics I would have calculated would be using descriptive statistics. In the last two questions, I attempted to describe the larger peanut population (the pile of peanuts) by examining the few sample bags that I had. This is in the realm of inferential statistics, and I need to ensure that the sample I use is representative of the entire population. The size and quality of the sample will provide a confidence level for my inference. A full treatment of inferential statistics, sampling, and its related issues is

TABLE 22.3

A Comparison of Population Parameters and Sample Statistics

Description	Population Parameters	Sample Statistics
Mean	$\mu = \frac{\sum_{i=1}^{N} x_i}{N}$	$\bar{x} = \frac{\sum_{i=1}^{n} x_i}{n}$
Variance	$\sigma^2 = \sum_{i=1}^{N} \frac{(X_i - \mu)^2}{N}$	$s^2 = \sum_{i=1}^{n} \frac{(X_i - \bar{x})^2}{n-1}$
Standard Deviation	$\sigma = \sqrt{\sum_{i=1}^{N} \frac{(X_i - \mu)^2}{N}}$	$s = \sqrt{\sum_{i=1}^{n} \frac{(X_i - \bar{x})^2}{n-1}}$
Coefficient of Variation	$\text{COV} = \frac{\sigma}{\mu}$	$\text{COV} = \frac{s}{\bar{x}}$

x_i is data point i, N, is the population size, n is the group sample size.

beyond the scope of this work; however, the major concern with sampling is that it should be a random selection. In new product development activities, getting truly random parts or products for testing may not always be practical.

Measurements

So far in this chapter, I have tried to convey that variation in any given process is to be expected. And that design team should develop an understanding of how variation may impact them both in the details of their design and in their experiments, as well as other aspects of product development and support. In Chapter 24, I will review how tolerances (expected variations in geometry or properties of components) may impact the performance of a product. Now in the context of this chapter, the question that needs to be entertained is this: "By testing a few samples in a laboratory environment, is it possible to draw the conclusion that all the launched products in that family will behave within the expected range?"

Variation in any given process is to be expected. The design team should develop an understanding of how variation may impact them both in the details of their design and in their experiments, as well as other aspects of product development and support.

Dealing with Variation in Data

Allow me to start this section with an example. Suppose that I have designed a pressure transducer that has an acceptable 1% error over its operating range of 5–15 psi (34.5–103.4 kPa). Suppose that to verify that the device meets its requirement, three tests are to be conducted with 10 devices measuring known pressures at 5, 10, and 15 psi. The results are shown in Table 22.4. This table also provides the mean for this dataset as well as a measure

TABLE 22.4

Pressure Transducer Example – Output of 10 Units

Nominal	Min	Max	Device 1	Device 2	Device 3	Device 4	Device 5	Device 6	Device 7	Device 8	Device 9	Device 10
5.000	4.950	5.050	4.961	4.961	4.985	5.026	5.035	4.964	4.977	5.034	4.979	5.013
10.000	9.900	10.100	9.910	9.983	9.995	9.984	9.921	9.923	9.902	9.989	9.991	9.997
15.000	14.850	15.150	14.859	14.942	14.914	14.933	14.851	14.948	14.935	14.905	14.886	14.889

Nominal	Min	Max	Mean	St-Dev
5.000	4.950	5.050	4.994	0.030
10.000	9.900	10.100	9.960	0.040
15.000	14.850	15.150	14.906	0.034

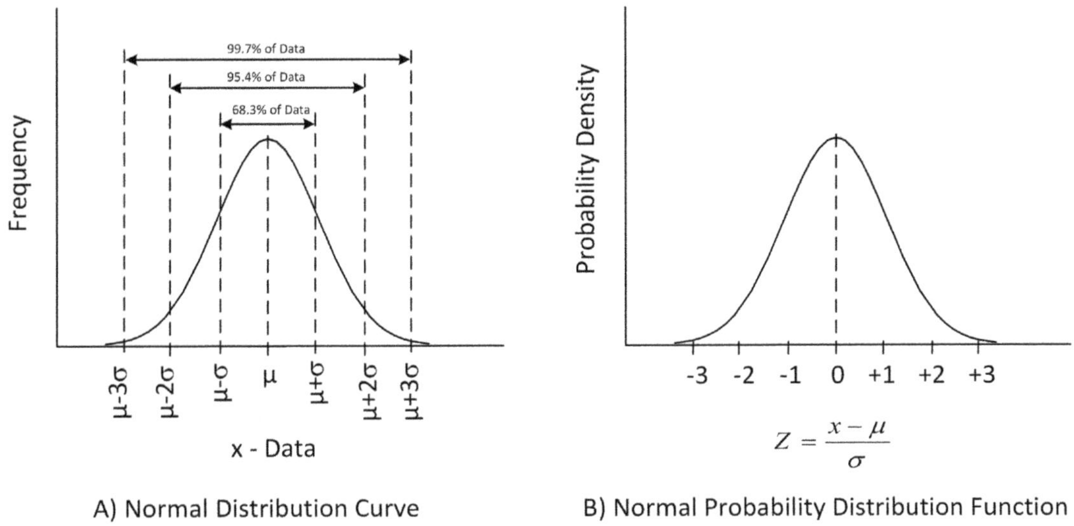

FIGURE 22.5
Characteristics of a normal distribution

of its spread; namely, the standard deviation. By comparing the mean value of the measured pressures for each known pressure with the acceptable minimum and maximum values (± 1%), one may easily conclude that the design meets requirements. However, that is not really the case. What has been demonstrated here is that the ten units tested are within specifications. At the moment, no conclusion may be drawn about the design yet.

Any quantitative data collected from experiments or manufacturing processes may be classified as a normal distribution (also known as a Gaussian or bell curve). As suggested earlier, there are two measures for characterizing a dataset; namely, central tendency and dispersion, or spread. For the example shown above, the mean value measured for a nominal 10 psi pressure is 9.960 psi with a standard deviation (measure of spread) of 0.040 psi. What does this mean?

Figure 22.5A provides an interpretation of the immediate conclusions that may be drawn from a normal curve: 68.3% of data is within one standard deviation (s or σ) of the mean ($\bar{x}$ or μ), 95.4% of data is within two standard deviations of the mean, and finally, 99.7% of the data is within three standard deviations of the mean. What would happen if this rule were applied to the pressure measurements of Table 22.4.

The results are shown in Table 22.5. At a 5 psi nominal pressure, both lower and upper specification limits (± 1% allowable) may be violated. However, at 10 and 15 psi nominal pressures, only the lower specification limits (–1% allowable) are exceeded. If the specification limits are exceeded, is it possible to estimate what percentage of the transducers will be faulty?

The answer to this question is, in a way, hidden in Figure 22.5B. It represents a normal probability density function (PDF) and is defined as

$$f(x) = \frac{1}{\sigma\sqrt{2\pi}} e^{-\frac{1}{2}\left(\frac{x-\mu}{\sigma}\right)^2}, \quad -\infty < x < \infty,$$

where μ is the population mean and σ is the standard deviation.[4]

TABLE 22.5

Pressure Transducer Example – Expected Variations

Nominal	Min	Max	Mean ($\bar{x}$)	St-Dev (s)	$\bar{x} - 3s$*	$\bar{x} + 3s$*
5.000	4.950	5.050	4.994	0.030	4.902	5.085
10.000	9.900	10.100	9.960	0.040	9.841	10.078
15.000	14.850	15.150	14.906	0.034	14.803	15.009

* Strictly speaking, this formula should use the tolerance interval because the values are calculated based on a sample and not the population. Tolerance interval is calculated based on sample size, confidence level, and population percentage. In this example, it is assumed to be three. A detailed discussion of this topic is beyond the scope of this book.

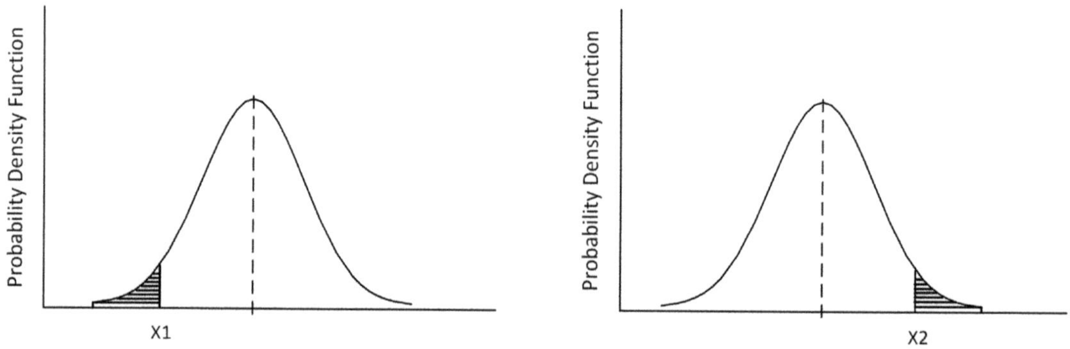

A) Shaded area under the curve depicts the probability of a sample measured with a value less than X1

A) Shaded area under the curve depicts the probability of a sample measured with a value greater than X2

FIGURE 22.6
Characteristics of a normal distribution

Figure 22.5B shows the standardized (or normalized) variable $Z = \frac{x - \mu}{\sigma}$ where x is any point in a dataset with a mean of μ and standard deviation σ. Z varies from $-\infty$ to ∞ but the probability density function (PDF) is always symmetric around $Z = 0$. The area under the PDF curve represents the probability of the number of units (often expressed as a percentage) that fall in that range. For instance, as shown in Figure 22.6A, the shaded area represents the probability of the number of units with a measured value equal to or less than X1. Alternatively, in Figure 22.6B, the shaded area represents the probability of the number of units with a measured value equal to or greater than X2.

Now, consider Figure 22.7. One of four conditions may take place:

1) Variations are within specified limits,
2) Variations exceed the lower specified limits,
3) Variations exceed the upper specified limits, and finally,
4) Variations exceed both the lower and upper specified limits.

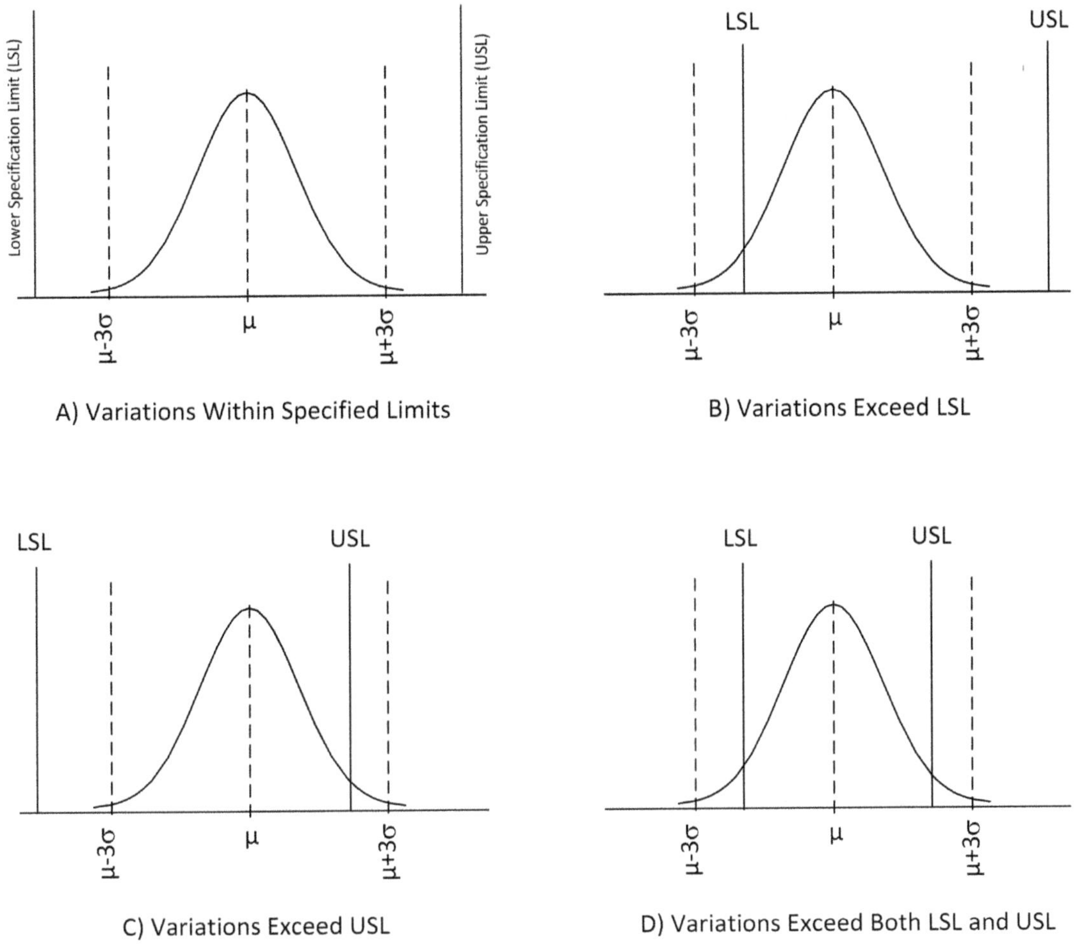

FIGURE 22.7
Data distributions and specifications limits

Now, the question that I raised earlier may be answered: "If the specification limits are exceeded, is it possible to estimate what percentage of the transducers will be faulty?" To make this calculation, it should be noted that one feature of the probability distribution function is that the area under the entire curve is unity $\left(\int_{-\infty}^{\infty} f(x)dx = 1\right)$. Furthermore, Z is a nondimensional value. Because of the properties of both PDF and Z, standard tables for PDF as a function of Z are available (see, for example, Dodson et al. 2009). For completeness sake, Table 22.6 provides a relatively small number of Z-PDF data curves. Note that although positive values of Z are shown, probability values are symmetric around $Z = 0$.

To calculate the probability of devices that may exceed specification, the area under the pdf curve to the right (or left) of the specification limit should be calculated as shown in Figure 22.8. To do so, first the Z value associated with the specification has to be calculated. Next, PDF associated with this Z is found in a Z-table. Should statistical software such as Minitab or SPSS be used, both the Z values and the probabilities are calculated automatically.

TABLE 22.6
Standard Normal (or Z-) Table

Z Value	Probability
0.000	50.00%
0.500	30.85%
1.000	15.87%
1.500	6.68%
2.000	2.28%
2.500	0.62%
3.000	0.135%

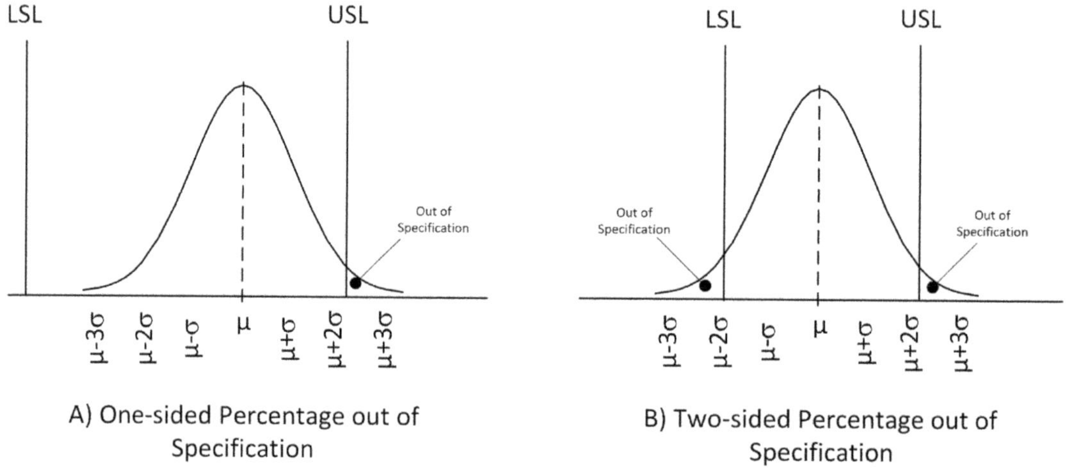

FIGURE 22.8
Calculations of out-of-specification values

Table 22.7 summarized the calculations for the pressure transducer and the values presented in Table 22.5. For this example, the majority of failures occur at the lower end of the specifications (7.14% at 5psi, 6.68% at 10psi, and 5.00% at 15 psi). Based on this information, design modifications may be required.

Confidence Level and Power of the Test

Earlier, I discussed populations and samples and that samples are taken from a population for ease of study. Then, what is learned from the sample is inferred on the larger population. For instance, in the pressure transducer example, the design team would use the conclusions made on ten samples to make decisions on the entire product and its production. It is only logical to ask whether the right decision was made.

> A Type I error (α) is the chance of rejecting a product when the product is good. Type I error (β) is the chance of accepting a product when the product has failed.

TABLE 22.7
Probability of Pressure Transducers That Fail Specification Limits

Nominal	Lower and Upper Specification Limits	Mean	St-Dev	Z Value	PDF (Percentage) Out of Specification
5	4.950	4.994	0.030	−1.467	7.14
	5.050	4.940	0.030	3.667	0.02
10	9.900	9.960	0.040	−1.500	6.68
	10.100	9.960	0.040	3.500	0.02
15	14.850	14.906	0.034	−1.647	5.00
	15.150	14.906	0.034	7.176	0.00

Two types of errors have been defined. *Type I* error – often denoted as α – is the chance of rejecting a product when the product is good. Type I error is also called producer's error. *Type II* error – denoted as β – is the chance of accepting a product when the product has failed (detecting a significant difference) and should be rejected (Mathews 2005).

Associated with Type I and II errors, two other terms are defined. The chances of making the correct decision based on the statistics are called *confidence level* (CL) and *power of the test* (PT). Confidence level is the chance of making the correct decision in accepting a good part, i.e., $CL = 1 - \alpha$. Similarly, the power of the test is rejecting a bad part when the part should be rejected, i.e., $PT = 1 - \beta$. A typical confidence level is 95%, and the power of the test is often set at 90%.

Confidence Intervals

Now that confidence levels and power of the test are defined, the question of how close the sample mean and standard deviations are to the population values may be answered. The relationship between the population and sample statistics for mean and standard deviation values is as follows (Dodson et al. 2009):

For large sample sizes (n > 30)

$$\mu = \overline{x} \pm Z_{\alpha/2} \frac{s}{\sqrt{n}}$$

For small sample sizes (n < 30)

$$\mu = \overline{x} \pm t_{\alpha/2} \frac{s}{\sqrt{n}}$$

For a 95% confidence level $\alpha = 0.05$, $Z_{0.05/2} = 1.96$. Values of $t_{0.05/2}$ are dependent on the sample size. For instance, for a sample size of 25, $t_{0.05/2} = 2.064$, and for a sample size of 10, $t_{0.05/2} = 2.262$.

Table 22.8 provides the lower and upper limits for the mean of the pressure transducer population example previously studied. This table shows that while the samples tested say at 5 psi nominal has a mean of 4.994 psi, the population may have a mean as low as 4.972 psi and as high as 5.015 psi at a 95% confidence.

Pressure Transducer Example (n = 10)

TABLE 22.8

Confidence Intervals for the Calculated Mean

Nominal	Mean	St-Dev	μ Lower Limit	μ Upper Limit
5.000	4.994	0.030	4.972	5.015
10.000	9.960	0.040	9.931	9.988
15.000	14.906	0.034	14.881	14.931

For brevity's sake, the tables associated with Z and t are not duplicated here because I anticipate the calculations will be done using software such as MS-Excel or other packages such as Minitab. However, anyone interested may use a reference on statistics or conduct an internet search for these tables.

The confidence interval on the standard deviation uses the χ^2 distribution as follows.

$$\left(\sqrt{\frac{n-1}{\chi^2_{\left(\frac{\alpha}{2},n-1\right)}}}\right)s < \sigma < \left(\sqrt{\frac{n-1}{\chi^2_{\left(1-\frac{\alpha}{2},n-1\right)}}}\right)s$$

For a reference on χ^2 distribution or this topic, see Dodson et al. 2009; Bethea and Rhinehart 1991 or Sheskin 2007.

Sample Size

Closely associated with confidence levels and power of the test is sample size. Needless to say, the more samples used, the higher the level of confidence in the results, provided that the samples are randomly selected and are representative of the entire process. However, in the product development process, samples may be expensive and, at times, time-consuming to produce. Hence, it is prudent to consider the relationship between sample size, confidence level, and variation. Or, similarly, the triad between sample size, power of test, and significant differences may be summarized here.

Type I error:

For a fixed sample size n: as variations increase, confidence decreases.

For a fixed confidence $(1\text{-}\alpha)$: as variations increase, sample size must increase.

For a fixed variation (σ): as sample size decreases, the confidence level drops.

Type II error:

For a fixed sample size n: as the significant differences increase, the power of the test decreases.

For a fixed power of test $(1-\beta)$: as sample size decreases, the ability to identify a significant difference declines.

For a fixed significant difference (Δ): as sample size decreases, the power of the test also decreases.

Sample Size Calculations

How many samples should be selected for testing? And here is the sad response: *It depends!*[5] It depends what type of study is being conducted. Since this discussion has focused on comparison studies, Mathews (2005) offers the following equation for sample size calculations:

$$n \geq \left[Z_{\alpha/2} \frac{s}{\delta} \right]^2$$

where "n" is the required sample size to meet a confidence level of $(1\text{-}\alpha)$ and an acceptable error interval of δ with a previously known standard deviation (s). Should "n" be calculated to be less than 30 samples, the calculation may be verified using $t_{\alpha/2}$ values instead of $Z_{\alpha/2}$.

Similarly, general statistical software packages may be used to calculate sample sizes needed for making comparison studies. In the case of Minitab, the required inputs are the significant level (α), the estimated (known) standard deviation (s), significant differences (Δ), required power of test $(1\text{-}\beta)$ The output is the desired sample size (n).

Comparing Two Datasets

In the foregoing discussion, the distribution of a dataset was compared to a single limit (in here, either upper or lower specification). There are other situations where two datasets have to be compared to one another. For instance, as in the previous pressure transducer, suppose that two different teams have made similar measurements. Understandably, the mean and standard deviation of the two datasets are different due to operator or measurement variations. It is also possible that design modifications have been made, and now the previous dataset needs to be compared to the current one. Figure 22.9 suggests the three different situations that may exist: (a) overlapping distributions, (b) distributions distinctly separate, and finally (c) distributions that have different spreads.

Comparison of the two datasets requires understanding of statistical methods such as analysis of variance (ANOVA) or Levene's test. Unfortunately, a full review of this subject is beyond the scope of this book.

In the language of gage repeatability and reproducibility (gage R&R), the measurement system should be both capable and stable.

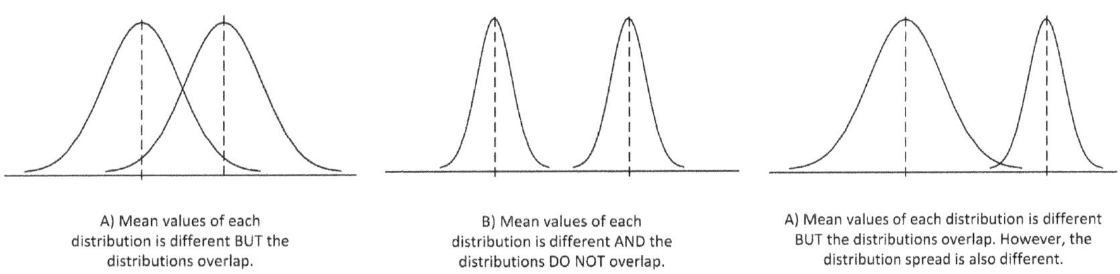

FIGURE 22.9
Comparison of two datasets

Measurement System Analysis (MSA)

Up to now in this chapter, the focus has been primarily placed on variations in data and developing means of drawing the right conclusions from them. It is intuitively clear that there are two sources of variation in product measurements. One source of variation comes from the product itself: part-to-part variation, lot-to-lot material variations, assembly process, operator-to-operator variations, etc. The second source of variation is the measurement process itself. The focus of this segment is on the measurement process and its variations.

> Bias is a difference between the measurement mean and a theoretically known target value. Precision is a measure of the spread of measured values. Accuracy refers to a measurement system with low or zero bias.

In the language of gage repeatability and reproducibility (gage R&R), the measurement system should be both *capable* and *stable*. By stable, I mean that repeated measurements do not change over time. Furthermore, changes in operators or environment do not have an adverse effect on the measured values either. Capable means that measurements are both accurate and precise. The terms "capable" and "stable" are not universal terms. Authors such as Wheeler (2006) use different terminology: consistency, precision, and bias. Consistency is defined as the predictability of the measurement process – a term similar to stable. Precision refers to the variations of a consistent measurement process, and finally, bias is the mean of a consistent measurement process. Another term that is often implied is gage linearity, which means that the gage bias is either constant or proportional to the measured value over the working range of the gage.

Accuracy, Precision, and Bias

Figure 22.10 depicts a pictorial view of the terms accuracy, precision, and bias. Of these three, bias is the most elusive to quantify. Bias is a difference between the measurement mean and a theoretically known target value. In Figures 22.10A and 22.10B, the measurement mean and target (the dashed line) coincide; hence, in these two cases, bias is zero.

Precision is a measure of the spread of measured values. As shown in Figures 22.10A and 22.10C, higher spreads are considered imprecise. Finally, accuracy refers to a measurement system with low or zero bias.

Gage Repeatability and Reproducibility

Gage R&R is a systematic approach to understanding and quantifying variations associated with a measurement system. A gage R&R study will enable the following (Burdick et al. 2005, p. 2):

"(i) Determine the amount of variability in the collected data that is due to the measurement system.
(ii) isolate the source of variability in the measurement system, and
(iii) assess whether the measurement system us suitable for use in the broader project or application."

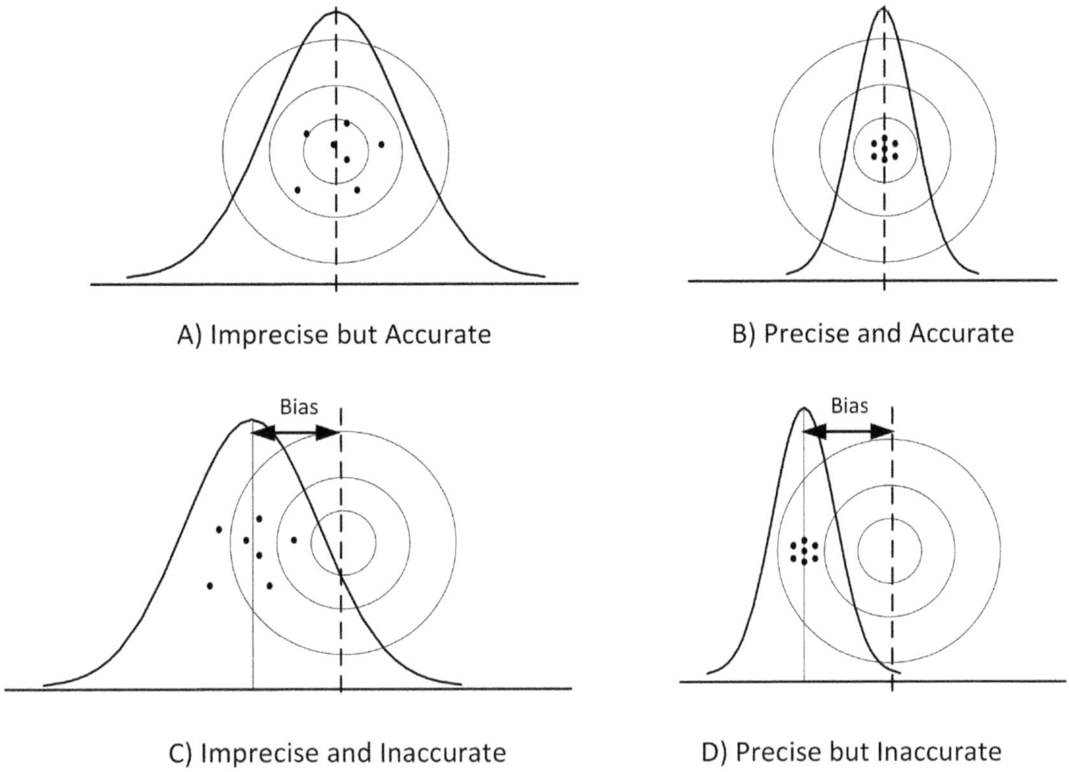

FIGURE 22.10
Precision, accuracy, and bias – using a "gun-shot in a bull's eye" example

Repeatability in the context of gage R&R means that variations of successive measurements under the same conditions are low and within acceptable limits. This implies that the same equipment and operator are used to measure the same value several times.

Reproducibility means that variations in the average of measurements taken by different operators using the same instruments are within acceptable limits. It is implied here that different operators use the same instructions and techniques and are measuring the same aspect of the part.

The general approach to a gage R&R is to select a set of samples as well as multiple operators and a series of measurements. The best set of samples is one that includes both acceptable and unacceptable values. The question to be answered is how much variation exists when one operator measures the same part over and over and, how the results compare to variations obtained when different operators measures the same part. Although different strategies may be used in a gage R&R study, minimally, it requires two different operators, five different parts, and two measurements per part. Unless measurement is destructive, it is further recommended that operators use the same parts for measurements.

As in any test, a protocol is developed instructing each operator on what to measure and how to record the results. For instance, two operators are instructed to measure a specific feature of a part twice. The study dictates that five parts be measured. Once the study is completed, twenty data points are collected and recorded. Analysis of this data is done using ANOVA – a description of which is beyond the scope of this book. Typically Minitab (Mathews 2005) or other statistical software packages are used for this analysis. The output of a gage R&R analysis is a list of contributors to the variations. These contributors include part-to-part variations as well as the total gage R&R variations, which are a sum of their individual parts: repeatability, reproducibility, and finally operator variability. Another outcome of the analysis is the *number of distinct categories*. This number suggests whether or not the gage under study can accurately quantify characteristic values of individual parts. A strong gage R&R has the following characteristics:

1) gage total study variation less than 20%,
2) gage total tolerance less than 30%, and
3) number of distinct categories equal to or greater than 5.

Should the outcome of a gage R&R fall outside of these characteristics, the measurement system may not be either capable or stable.

A Few More Words on Variations

In this chapter, there have been a number of references to variations. For completeness sake, it should be remembered that there are two types of variations. The *common-cause* variation is the overall outcome of a given process. For instance, a wave-solder operation produces a given solder profile consistently across all the leads that come into contact with the wave. Although individual solder profiles are not exactly the same, they resemble each other. Now, suppose that the solder pool temperature is at a lower level than required (for reasons that are unknown at the moment). As the solder attaches to the leads, a profile forms that is different from the typical expected shape. This variation is called *special cause* because there were circumstances that were different from typical.

In the context of both process validation and product testing, anomalies are typically associated with special cause variations, whereas overall process outcome and product performance belong to the common cause variation. Although variation is to be expected, excessive variation of any kind leads to customer dissatisfaction and ultimately product acceptance failure.

Elements of a Robust Test Plan

Prior to closing this chapter, I would like to discuss elements of a robust test briefly. The element that differentiates a test from a measurement is that a test – as implied by its name – either passes or fails. Should this need to happen, clear acceptance criteria must be set. In my opinion, by contemplating the acceptance criteria, test personnel have a chance to set the scope and purpose of the test. Is the test set up to merely collect a series of measurements, or is the test set up to verify a requirement or validate a user need? In a way, a test plan begins by providing a clear and concise description of the purpose of the activities and the scope of the measurements. The purpose may also be considered as a "problem statement."

The element that differentiates a test from a measurement is that a test – as implied by its name – either passes or fails. A test plan begins by providing a clear and concise description of the purpose of the activities and the scope of the measurements.

If requirements are poorly written, it would not be surprising if the test results are less than expected. I do highly recommend that the test personnel spend some time to review the scope and purpose statement of a test plan to ensure that (a) good requirements do exist and (b) test activities are aligned with meeting test requirements. Here is a simple example to illustrate my point. Suppose that a test should be set up to verify that a product is sterile. Note that a definition of sterility is not provided, nor the means of creating the sterility. How can a test plan be created?

Another element of a robust test design is defining a precise set of activities that accomplishes the required tasks. This requires questioning whether the proposed methods can in fact achieve what they are intended to do and whether there is sufficient accuracy in the proposed methods. Suppose that the purpose of a test is to measure a design feature that is 1.000±0.002in wide. Which would be a more appropriate tool, a graduated ruler or a caliper accurate to four decimal places? It is obvious that it would be inappropriate to use a graduated ruler to make this measurement because of the degree of precision and accuracy that is needed.

Admittedly, the inaccuracies and/or inabilities of the measurement systems may not always be as evident as this example; however, it does make sense to think about and ensure the validity of the test methods. If there are any doubts, a gage R&R should be conducted.

Related to test method validation is calibration of the equipment used for the measurement. If a piece of equipment is not properly calibrated, there may be a chance that the bias is excessive, which would then render the results useless.

It is important to include both operators and the environment in the test plan as well. The experience of the operators will dictate whether measurement variations would be excessive or not. Should the need be, test dry runs are conducted to answer questions or clarify steps in the test procedures. Changes in the test environment may impact the sensitivity of the equipment or the units under the test. Typically, environmental conditions are monitored and recorded. Should any anomalies be observed, the role of the environment may be ruled out (or in) by comparing the time stamp of the anomaly with the lab environment conditions.

For many tests, techniques and methods have already been developed. For instance, the American Society for Testing and Materials (ASTM) has many standard test methods (such as, say, ASTM D5276 – Drop Test of Loaded Containers by Free Fall). Uses of these standard test methods (called compendia) provide justifications that acceptable practices of scientific and/or engineering communities are being followed.

In short, a robust test plan has the following elements

1) Scope and purpose of the study tying the need to conduct the test to a requirement or a problem statement.
2) Test method and a justification of why the selected method is appropriate for obtaining the required results.
3) Number of needed test samples and the rationale for the sample size.
4) Number of test operators and the test procedures for their training.
5) A list of calibrated equipment and their calibration certificate.
6) Data analysis techniques; for instance, would it be sufficient to just get an average value of, or should there be a detailed statistical analysis.
7) Acceptance criteria.

There are two other factors that I do not choose to include in this test. First, there is always a possibility that a failure may be observed. Any and all failures should be investigated and reported. We learn more about our products from their failures than we learn from their smooth operations. The second element that I chose not to place on the list is test deviations. Many times (and despite dry runs), anomalies and issues arise that require changes and modifications to the test plan. These deviations should be properly documented and explained so that the next individual running a similar test may take advantage of the lessons learned from previous experience.

Summary

- A lack of awareness of a measurement instrument's impact on an experiment or other noise factors such as heat may lead to substantial errors.
- Another source of uncertainty in measurements is associated with people. If different operators make the same measurement, each reading will be different.
- Scientists and engineers collect information and data; however, the type of information collected depends to a large extent on the questions that are to be answered.
- In order to compare different data, it is important that uniform scales exist.
- Categorical scale includes the nominal (and its special case binary) and ordinal scales.
- Quantitative scales include interval and ratio scales, which can be either discrete or continuous.
- Quantitative data may change seamlessly from one value to another, either discretely or continuously.
- A histogram is a graphical representation of the cluster of data in a dataset. It is a pictorial map of the data that provides a bird's-eye view.
- In a normal distribution, all three measures of central tendency (i.e., mode, median, and mean) are equal; whereas, in a skewed distribution, these values are different from one another.
- Statisticians often refer to two types of statistics; namely, descriptive and inferential statistics. In descriptive statistics, one is concerned only with the dataset at hand.
- Inferential statistics is different in the sense that one's interest in studying the dataset is to extend, expand, and draw conclusions and decisions to larger groups and populations.
- Type I error (also called producer's error) is the chance of rejecting a product when the product is good.
- Type II error (also known as consumer's error) is the chance of accepting a product when the product has failed and should be rejected.
- Gage R&R is a systematic approach to understand and quantify variations associated with a measurement system.

Notes

1 Note that some have the tendency of putting this question in a binary format: "Is the company's insulin pump accurate?" The sarcastic answer is, "Well, how accurate do you want it to be?!" Realistically, accuracy may not be measured in a binary form unless some type of criteria has been defined.
2 To develop a more descriptive graph, I have added an empty bin at each end of the table.
3 In statistics, there is a difference between what is called a *population* and what is called a *sample*. I will explain the difference shortly. For now, I am presenting the formula for population statistics.
4 These equations hold for samples as well.
5 It depends on what the goals of the study are and what is in scope. For instance, for a comparison study, sample size calculations are different from if a reliability study is conducted.

23

Metrology and Conformity Assessment

Contributed by Dr Mohsen Torabi*

Introduction

In Chapter 22, we considered collected data from experiments and/or tests. Furthermore, in analyzing this data, we examined key characteristics such as the mean, standard deviation, and the overall shape of the distribution. We also discussed probability specifications pertinent to each sample and introduced MSA, including gage R&R, highlighting its importance in evaluating the quality of a measurement system.

In this chapter, we delve into metrology – the science of measurement – focusing on its fundamental requirements for achieving reliable, high-quality data any time a measurement is made in a product's life cycle, be it in R&D or production. For many engineers and scientists, this area remains largely unexplored, yet it is central to products that meet specifications, regulatory standards, and customer expectations.

What Is Metrology?

Metrology is the science of measurement and its application across the product life cycle. From a theoretical perspective, it establishes a common understanding of fundamental measurement units (e.g., SI units), their realization, and the traceability of base and derived units. From a practical perspective, it informs technical and industrial applications, from manufacturing and quality control to regulatory compliance.

The most critical component of any measurement is its uncertainty, which quantifies the confidence we can place in a reported value. Simply put, the lower the measurement uncertainty, the closer we can expect the reported value to be to the true value of the measurand. Uncertainty is the cornerstone of *conformity assessment*, determining whether a product falls within specified limits (tolerance). It is noteworthy to consider that when we say "a product," by which "a" we mean a single unit, we should consider specific risk. If we say "a product," and by "a" we mean a product line or category, we should examine global risk. To have a better understanding of what we mean by uncertainty of measurement and traceability and how it impacts our measurement and measurand, consider the following simple example.

DOI: 10.1201/9781003564508-31

Example

Consider calibrating a digital balance at 1 kg. Suppose that we use a calibrated weight of 1 kg with a 1 g uncertainty (usually with 95% confidence) and make a measurement on the balance. This implies that the actual weight may vary from 0.999 kg to 1.001 kg. Furthermore, suppose that we place the weight on the balance and it reads 1.001 kg. Can we actually say how much of this error (i.e., 0.001 kg) is due to the uncertainty of the calibrated weight or the inherent error of the digital balance? If not, how do we perform conformity assessment on pass or fail for the balance?

Now presume that our calibrated weight has an uncertainty of 1 mg. In other words, its actual weight may vary from 0.999999 kg to 1.000001 kg. Now, if we were to place this weight on the digital balance and get a 1.001 kg result, how would we calculate the uncertainty of the measurement? How do we calculate the error of the measurement? And how do we perform conformity assessment on pass or fail for the balance?

Now, we can deduce that as further measurements are made, the uncertainty begins to propagate to all future measurements. This propagation is known as *traceability*,[1] a central concept in metrology.

Metrology investigates how accumulated uncertainties affect downstream measurements in R&D labs, manufacturing, and product evaluation. Without understanding these uncertainties, there is no quantitative basis for assessing measurement quality nor for determining product conformity. Traceability and uncertainty analysis form the foundation of metrology and, by extension, product development.

Key Concepts in Metrology

There are some key concepts that we need to be familiar with when discussing metrology and conformity assessment. Herein, I would like to introduce tolerance, accuracy, measurement uncertainty, and error (or deviation).

Tolerance and Accuracy

In Chapter 22, we discussed the concepts of accuracy and tolerance. We recall that tolerance defines the permissible deviation of a value from its nominal specification. It may be bilateral (± limit) or unilateral (single-sided limit). Accuracy refers to the expected closeness of a measured value to the true value. While true accuracy is often unknowable, for practical purposes, accuracy and tolerance are often treated interchangeably in measurement contexts.

Measurement Uncertainty

Generally, measurement uncertainty arises from three primary sources:

1. Measurement Standard – any calibration standard carries inherent uncertainty, whether physical, electrical, or chemical. Expanded uncertainty is commonly expressed with a coverage factor for a given confidence level. It is typical to use a coverage factor of two ($k = 2$) to provide a 95.4% confidence level.

2. Unit Under Test (UUT) – the characteristics of the UUT itself contribute to uncertainty. These characteristics may include repeatability, resolution, and environmental sensitivities.
3. Measurement Process – the methodology, operator, and instrumentation can introduce additional uncertainties.

In practice, uncertainties are categorized as Type A or Type B. Type A category is evaluated using statistical analysis of repeated measurements, such as, say, repeatability. Type B category is obtained from reference sources such as calibration certificates, such as resolution or standard uncertainty.

Metrologists typically summarize these sources in an uncertainty budget table, using a root-sum-square approach to calculate combined and expanded uncertainty for the measurement system.[2]

Uncertainty versus Error

Uncertainty arises from unknown or variable factors that cannot be fully corrected, whereas error (or deviation) represents a known, fixed discrepancy between the measured value and the standard value.

To explain this further by way of an example, consider testing a platinum resistance thermometer (PRT) against a standard platinum resistance thermometer (SPRT).[3] Briefly, the calibration process is to place the thermometer in a bath of some sort. This bath can be water or molten salt, depending on the temperature ranges needed for calibration. Then the temperature values measured by the two thermometers are compared. Some discrepancies may be due to environmental factors such as temperature gradients within the bath, or tip alignment introduces *uncertainty*, whereas a consistent offset between the PRT and SPRT is considered an *error*.

The point is that errors should be corrected or accounted for, while uncertainty must be quantified and managed. Both of these processes – correcting the measurement values or recalculating uncertainties for downstream measurements – are resource intensive. The decision to implement these corrections or adjustments should be guided by the potential impact on the measurement's accuracy and the overall quality of the process.

Test Accuracy Ratio and Test Uncertainty Ratio

The early work on measurement decision risk was performed in the 1950s by Alan Eagle (1954), Frank Grubbs and Helen Coon (1954), and Jerry Hayes (1955) at the U.S. Naval Ordnance Laboratory. The outcome of their work was a complex mathematical model that was not readily usable for contractors and manufacturers. Hayes (1955) simplified this approach by correlating the ratio of the accuracy of the UUT to a comparable standard to the probability of true or false acceptance. For him, a 3:1 ratio will lead to a 1% probability of false acceptance. Later calculations included reliability of tolerances and extended this ratio to 4:1. Today, this ratio is known as the test accuracy ratio (TAR). TAR is the ratio of the accuracy of a tool, or unit under test (UUT), and the reference standard used to calibrate the UUT. In simple terms, a 4:1 TAR, means that the standard is four times more accurate than the tool being calibrated.

In the 1990s, with advances in computational power, TAR evolved into test uncertainty ratio (TUR) where uncertainty of measurement is incorporated into the calculations. ILAC-G8:09/2019 (2019) provides the following definition of TUR:

"Test Uncertainty Ratio (TUR) is the ratio of the tolerance, TL, of a measurement quantity, divided by the 95% expanded measurement uncertainty of the measurement process. Using mathematical notation, this parameter is defined as $TUR = TL/U_{95\%}$."

Example

Assume that we have a gage block with a nominal value of 1 inch and a tolerance of ±20 μinch. We use a micrometer with an accuracy of ±5 μinch as our measurement standard at the 1-inch test point. Also, assume that our measurement system, which comprises measurement standard, measurement process, and UUT, has an expanded uncertainty of ±3 μinch.[4] Keep in mind that accuracy is different from uncertainty of measurement system, and the two are independent of each other. Now, suppose that we measured the gage block using the micrometer, and the result is 1.000013 inch. Thus, the deviation of the gage block is 0.000013 inch or 13 μinch. The question is whether we can accept the gage block as calibrated or out of calibration. To do so, we need to calculate the TAR and TUR values as follows.

$$\text{TAR} = \frac{\text{gage block tolerance limit(TL)}}{\text{Accuracy of micrometer}} = \frac{20}{5} = \frac{4}{1} = 4$$

$$\text{TUR} = \frac{\text{gage block tolerance limit(TL)}}{U_{95\%} \text{ of measurement system}} = \frac{20}{3} = \frac{6.67}{1} = 6.67$$

Clearly, as expected, TAR and TUR result in two different values. However, TAR intuitively speaks to the accuracy of measurement and TUR speaks to the levels of uncertainty and probability of false acceptance. To answer the question of whether the gage is in calibration, we need to assess its conformity and probability of the error in this decision.

Conformity Assessment and Decision Risk

Conformity assessment evaluates whether a product meets its specifications. Due to measurement uncertainty, every decision carries a risk, either of probability of false acceptance (PFA) or probability of false rejection (PFR). There are two approaches to calculating these probabilities: either specific risk or global risk.

In metrology, specific risk is the associated probability of making a correct/incorrect conformity assessment when performing only one measurement for one test point. Global risk is the associated probability of making a correct/incorrect conformity assessment when performing measurements on (theoretically) an infinite number of units for one test point.

In the following sections, I will show how uncertainty of the measurement system will impact each of these risk approaches and how the measurement can result in false acceptance or false rejection because of the uncertainty of measurement.

Specific Risk

Specific risk considers one measurement at one test point. I like to discuss this topic by means of an example. Imagine we have a gage block with a nominal value of 1 inch and a tolerance of ± 20 µinch. Our measurement system is a micrometer, and the measurement is conducted in a controlled environment. Let's assume that the expanded uncertainty of the measurement is ± 6 µinch (unlike the previous example), while the accuracy of the micrometer is ± 5 µinch. The micrometer reading for the gage block in this example is 1.000016 inch. Does the gage block conform to its specifications? Does it pass the calibration or manufacturer quality control?

At first glance, the answer might seem straightforward. One might simply compare the measured deviation of + 16 µinch against the ± 20 µinch tolerance and conclude that the gage block is within specifications and therefore acceptable. However, to provide a definitive answer, we need to delve a bit deeper. First, let us calculate TAR:

$$\text{TAR} = \frac{\text{gage block tolerance limit (TL)}}{\text{Accuracy of micrometer}} = \frac{20}{5} = \frac{4}{1} = 4$$

Although TAR is 4, to determine whether the gage block conforms to its specifications and passes calibration, we should construct histograms of the standard values (i.e., measurement system values), its expanded uncertainty (*TUR*) and the measurand value and specification.

Figure 23.1 shows this representation. Here the standard value is 1.000016 inches, the measurand value is 1.000000 inch with ± 20 µinch tolerance, and the solid line represents the ± 6 µinch expanded uncertainty of measurement. If we accept this gage block as a good unit, as it is shown from this graph and the curve of expanded uncertainty, it is possible that the "actual" value of the measurand was outside of the specification limit. The calculation of the probability of a (true or) false accept or a (true or) false reject can be done by integrating the area under the curve. For this example, if we accept this gage block as a good unit, PFA is roughly 9%, which is the shaded area under the curve.

Specific risk is mainly only applied in calibration labs, where calibrated instruments are used for downstream measurements, since decreasing the PFA means increasing the

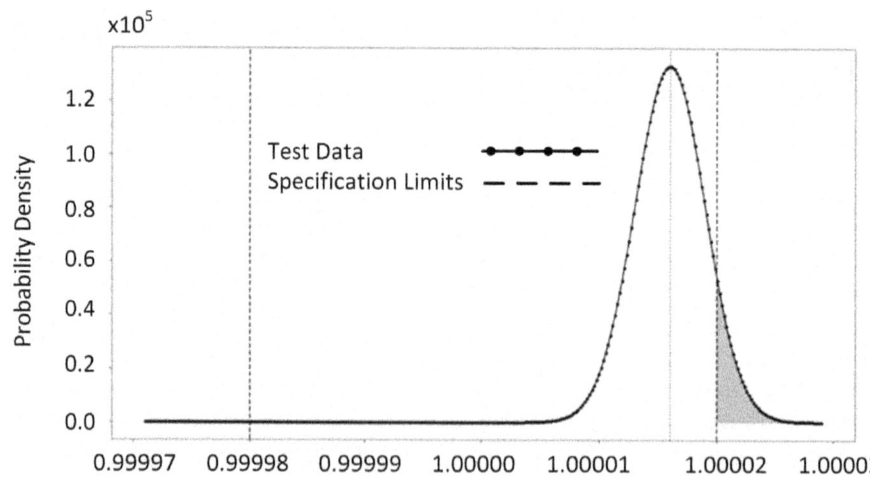

FIGURE 23.1
Uncertainty of measurement system and measurement value for the gage block. This figure shows the test distribution uncertainty, and probability of true and false acceptance using specific risk model

PFR, and this is costly for the organizations. Also note that TAR or *TUR* does not appear in conformity assessment for specific risk, and only measurement uncertainty is the main tool to perform conformity assessment.

Global Risk

Global risk considers measurements for a production line for a specific measurement. Global risk calculations include uncertainty of measurement, measurement value, and nominal value of measurand and its specifications, i.e., upper and lower tolerances. In addition the historical in-tolerance probability (ITP) of UUT is also a contributing factor. The mathematical construction of global risk is more complicated than specific risk and involves what is called joint probability.

Figures 23.2a and 23.2b show the PFA and PFR versus ITP for different *TURs* using the global risk approach. These two figures are classical plots in measurement science when depicting global risk. Note that these figures have been plotted assuming the following:

- Normal distribution for process
- Normal distribution for measurement
- Homogeneous sample characteristics
- Sufficient number of samples (theoretically infinite number)
- Available history of measurement or calibration
- No measurement system bias
- Symmetric tolerance limits

Guardbanding

The discussion of specific and global risks demonstrated that any pass or fail decision has a probability of being wrong due to uncertainties of the measurement system. Since elimination of uncertainties is a physical impossibility, we need to mitigate the risks. Metrologists have adopted guardbanding as a means of reducing the risks. For specific risks, guardbanding is the action of creating limits on tolerance intervals to decrease the PFA below acceptable levels. For global risk, performing guardbanding is more complicated due to the inherent mathematical complexity of the joint probability equations that govern global risk formulation. Unfortunately, any further discussion of this topic is beyond the scope of this chapter. The interested reader may refer to available literature on this topic (Crowder et al. 2020).

Effect of Standard Error

As discussed in the "Error vs Uncertainty" section, the measurement system should be either adjusted for its error (also called bias or deviation) or the measurand values should be corrected based on the error of the measurement system. The reason for making this adjustment or correction is that even small systematic errors in measurement shift distributions, amplifying PFA and PFR. For example, in the specific risk our gage block example,

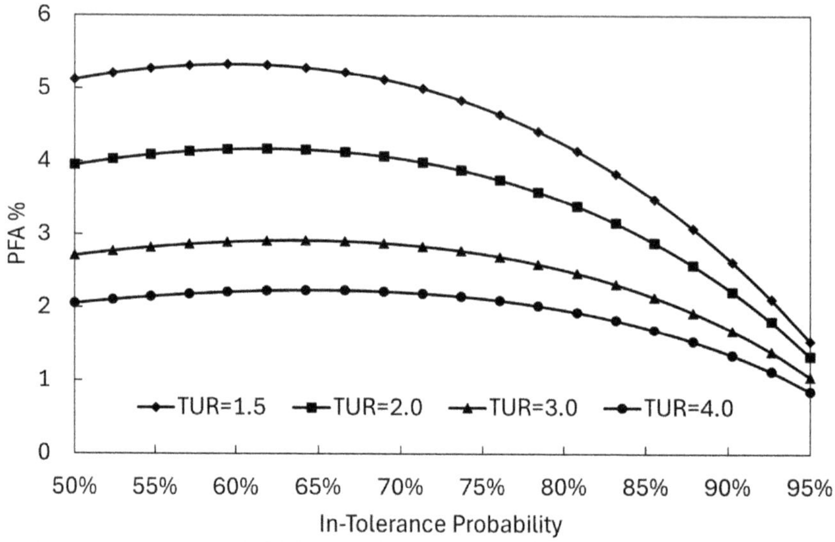

A) Probability of False Acceptance (PFA)

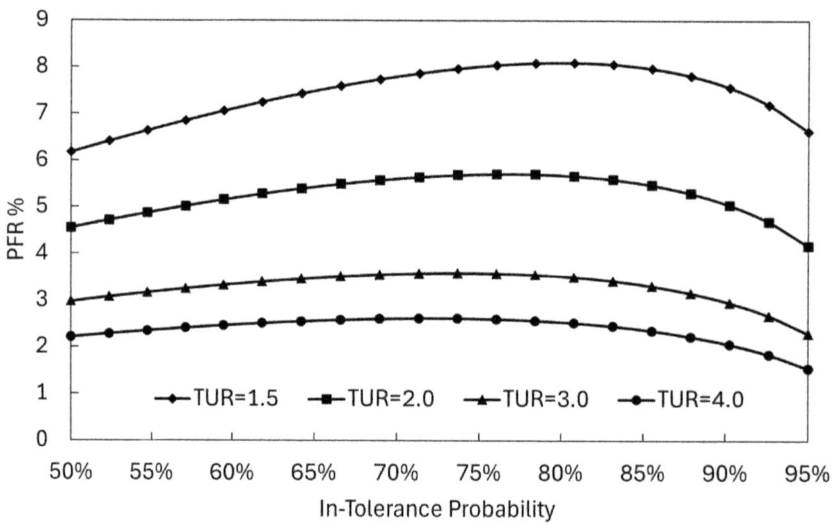

B) Probability of False Rejection (PFR)

FIGURE 23.2
PFA and PFR versus ITP with different *TUR* values. This figure shows the PFA and PFA using the global risk model for various *TUR* and ITP scenarios. This shows that the higher the *TUR*, the lower would be the value of PFA or PFR

a -3 µinch bias in the measurement system can increase PFA by 7-fold, and for the global risk, the same bias can double PFR and modestly increase PFA in production processes. Again, any further discussion of this topic is beyond the scope of this chapter. The interested reader may refer to JCGM 100:2008 (2008).

Summary

In this chapter, I highlighted the central role of metrology in measurement science and conformity assessment. Metrology provides the structure for realizing SI units,[5] ensuring traceability, and establishing consistency across technical, industrial, and regulatory domains. The discussion emphasized measurement uncertainty as the cornerstone of this discipline. Every measurement carries uncertainty, which defines the level of confidence in reported results and ultimately influences product quality and regulatory compliance. Without accounting for uncertainty, measurements cannot be meaningfully applied to conformity decisions. I also distinguished between error (deviation) – a known and often correctable – and uncertainty, the inherent variability in the process. Practical tools such as uncertainty budgets expanded uncertainty, and *TUR* were shown as essential in relating measurement performance to tolerance and in enabling structured conformity assessments.

Equally important is the connection between uncertainty and decision risk. Because all measurements contain uncertainty, there is always the possibility of false acceptance or false rejection. These risks can be evaluated at both specific and global levels, with techniques such as guardbanding offering ways to reduce PFA, though at the expense of higher rejection rates. The role of measurement bias was also highlighted, since even small biases can distort risk calculations and undermine decision-making.

In summary, metrology is not simply the practice of measurement but the science of managing uncertainty, risk, and conformity assessment. By applying its principles and tools, organizations make better conformity assessments, reduce decision risk, and safeguard the quality and reliability required in modern manufacturing and regulation. Metrology is more than measurement; it is managing uncertainty, bias, and risk to enable confident conformity assessment, and relying on quality measurements for root cause analysis.

Notes

* Dr. Mohsen Torabi is a Senior Engineering Leader at Baxter International, where he leads global metrology, calibration, and equipment qualification programs across North America and Europe. He holds a Ph.D. in Mechanical Engineering and an MBA in Strategy and Leadership. Dr. Torabi has authored more than 100 scholarly publications, with research that bridges academic rigor and real-world engineering applications across metrology, thermal-fluid systems, and regulated product development. His expertise spans measurement science, quality systems, risk management, design of experiment, thermofluidic analysis and testing. He is passionate about advancing engineering excellence through innovation, leadership, and continuous improvement.

1 Not to be confused with traceability in requirements down to design specifications and verification.

2 There is also another approach, which is the GUM approach and more fundamental than the uncertainty budget but less practical. That's why not so many people in the industry use that approach, and it is mainly for research purposes and derived uncertainty calculations.

3 Temperature is a fundamental parameter, which impacts physical characteristics of nearly every material. Hence, devices measuring temperature require accurate calibration. To do so, calibration laboratories rely heavily on high-precision thermometers such as platinum resistance thermometer (PRT) and standard platinum resistance thermometer (SPRT). SPRTs provide a wider range of temperatures with greater precision than PRTs.

4 This value is typically calculated from the chain of measurements and standards used and comes with the calibration certificate. Details are beyond the scope of this work.

5 Imperial units will be calculated and converted based on SI units after realization.

24

Tolerance Specification and Analysis

Introduction

I mentioned various techniques for tolerance analysis in Chapter 12 with an indication that more details would be found here in this chapter. Admittedly, most technical people think of mechanical tolerances when this topic comes up. The fact remains that any material used for any physical product will have variations in its specifications. These variations – if not checked or controlled – will have an impact on the quality of the final product. This chapter will focus on mechanical aspects of tolerance, but any member of a design team will benefit from the general principles set in this chapter.

In the early days of modern engineering, there were two groups of people responsible for developing the detailed engineering drawings – the so-called blueprints. The first group was the engineers responsible for the design of the product from the concept down to each component. Once the design was on paper, it was handed over to the second group, often referred to as draftsmen. This second group was responsible for finishing the design by providing specifications for the required hardware (such as screws, nuts, bolts, etc.), specifying tolerances on nominal dimensions provided by engineers, and creating the physical two-dimensional drawings. It was not uncommon for an engineer-in-training to spend time working as a draftsman and learning from the old-timers.

> It is well understood that component variation cannot be avoided. However, pioneers such as Deming and Taguchi taught that reducing variation would result in an increase not only in manufacturing yields but also in the end customers' perception of quality.

This did not happen in the early days of my career. Even though, my engineering schooling included the old-fashioned training in engineering graphics, by the time I began my work as an engineer, CAD software was used commonly, and shortly after, just about all engineers I knew were experts in using 3D CAD software to develop concepts. And, by the touch of a button, 2D drawings could be created from the 3D files. The task of placing nominal dimensions on the prints was simplified tremendously. Tolerances were identified based on the design engineer's (or his supervisor's) knowledge – typically, focused on the interaction of neighboring components. Gross tolerance issues or problems were often identified in the first and second builds and corrected. However, no one would predict manufacturing yield failures due to tolerance mismatch. In a way, relatively low manufacturing yields (and rework) were accepted as a fact of life. If yields were exceptionally low due to tolerance mismatch, an accepted solution was (and still is in some areas) one of the following: either to tighten the tolerance band and incur additional manufacturing cost or to develop different bins for components with slightly different dimensions that could be matched for successful assemblies. In other words, one bin may contain parts on the low end of the tolerance range, another for the mid-range of tolerance, and finally, the third for parts on the high

DOI: 10.1201/9781003564508-32

end of the tolerance. Through experience, manufacturing staff develop the knowledge of which combination of part tolerances work together. This method often involves a high degree of rework as well as more in-line and final inspections. In short, it is not efficient.

The goal of this chapter is to identify factors in design that contribute to manufacturing variability. By controlling these factors in the design, it is possible to control manufacturing variability and to maintain high yield production rates. Hochmuth et al. (2015) offer an overview of tolerance in mechanical engineering that speaks to this variability and approach to optimization.

Tolerance Design in the V-Model

It is well understood that component variation cannot be avoided. However, pioneers such as Deming and Taguchi taught that reducing variation would result in an increase not only in manufacturing yields but also in the end customers' perception of quality. In other words, either variations should be so small that they are not perceived, or they should be within a customer-acceptable range. Once customers' tolerance for variation is properly understood, then it may be translated into an acceptable product performance variation range.

It is, therefore, no surprise that proper specification of tolerances would need to be traced back to customer needs (or acceptance). Typically, the threshold for this tolerance is dependent on the financial impact to the end user. The more expensive a product, the higher the expectation for its performance.

In Chapters 11 and 12, I mentioned the role of developing equations for various transfer functions associated with a product. At the beginning of the design cycle, these equations enable the design team to develop an understanding of the design boundaries associated with various input factors. However, they may also be used to check the impact of part-to-part variation on the overall performance of the assemblies. At this stage of design, sensitivity calculations may be conducted to pinpoint specific dimensions whose variations may have a detrimental impact on the assembly performance.

Once, the design team decides on the appropriate tolerance values for critical dimensions, the manufacturing processes should be evaluated to ensure that these tolerances can be met. From this point forward, the focus would shift from the paper design to manufacturing process capabilities, first at component and assembly levels, and eventually at system and product levels. Ultimately, the financial impact (i.e., product cost) on the customer would be determined. This approach to tolerance design and its relationship to the V-model is shown in Figure 24.1.

Similar to other V-models, this is an iterative approach. It is not practical to start a product development activity without some understanding of the customers cost point. In turn, this may dictate the choice of one manufacturing approach over another. Various manufacturing processes have inherent limitations that can only provide a given tolerance "band" around a specific nominal dimension (or value). This *band* represents the inherent capability of that manufacturing process.

Various manufacturing processes have inherent limitations that can only provide a given tolerance "band" around a specific nominal dimension (or value). This band represents the inherent capability of that manufacturing process.

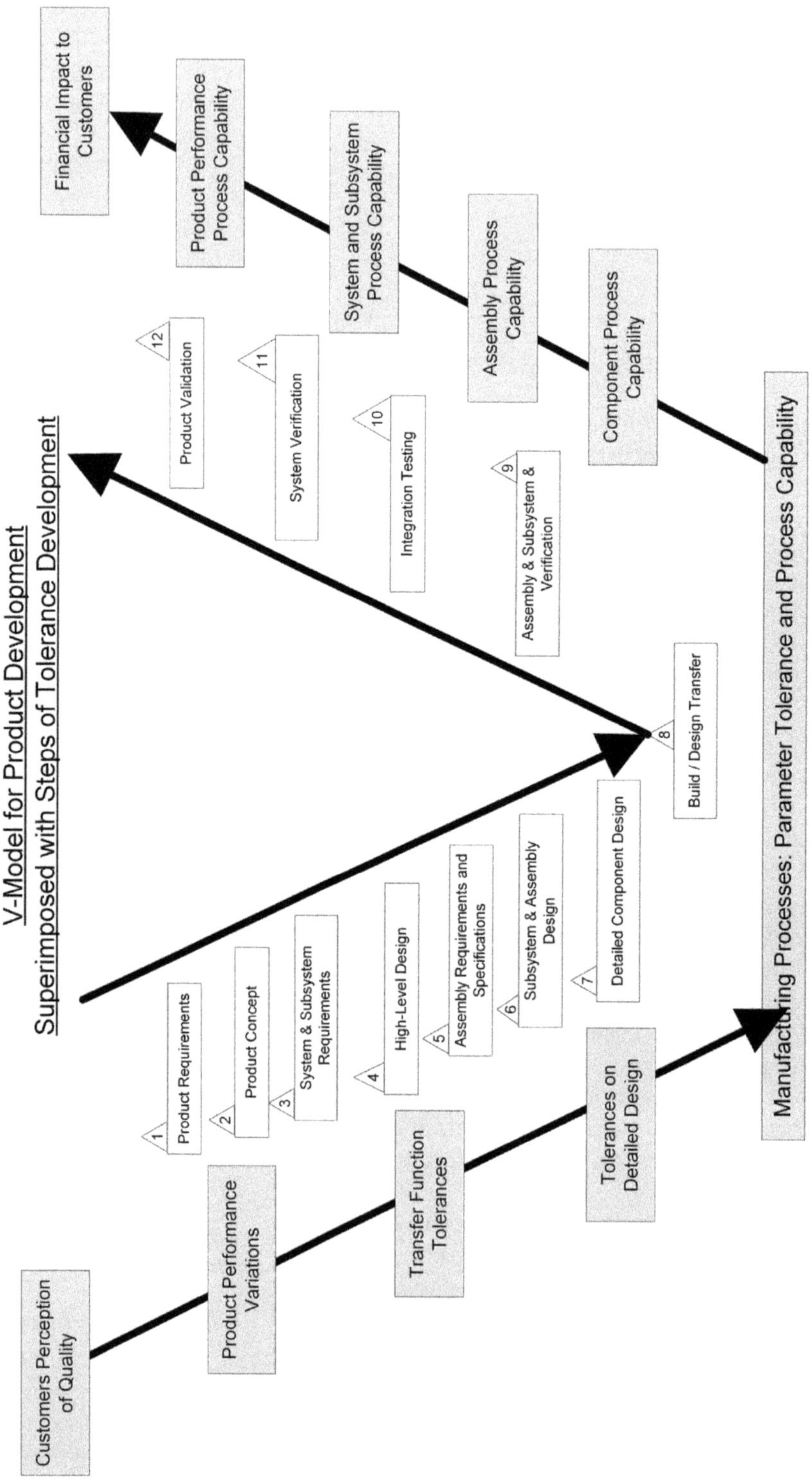

FIGURE 24.1
Tolerance specification with the V-model

Adopted from Creveling 1997.

Should the impact of this tolerance band (i.e., product variation) cause excessive variations in functions associated with the component, then the problem should be resolved by changes in the manufacturing process, or through design innovation, or possibly other compromises that need to be made.

From a hands-on design point of view, there are seven steps to tolerance specification and analysis as follows.

1) Understanding customers mindset for acceptable quality
2) Translating the customers' mindset into product requirements
3) Conducting sensitivity studies using various assembly transfer functions
4) Specifying tolerance in detailed design. This includes the following three considerations:
   a. Tolerances specified for "make" items
   b. Holes and Shafts Tolerances
   c. Tolerance and Cost
5) Analyzing specified tolerances. There are two categories of dimensional tolerances: linear and nonlinear – sometimes referred to as static and dynamic. There are three methods of calculating each, as shown below.
   a. Worst Case
   b. Root Sum Squared
   c. Statistical approaches such as the Monte Carlo technique
6) Although many engineers may stop at step 5, there are two other steps yet to be explored. The first step after tolerance analysis is a study of the manufacturing process capabilities as applied to
   a. Part fabrications
   b. Assembly and subsystem fabrication, and
   c. Delivering what was designed.
7) Finally, Taguchi's quality loss function may be deployed to understand the relationship between part-to-part variation and the ultimate cost to the product.

Steps 1 through 3 have been discussed to some degree in other sections of this book. The rest of this chapter will focus on steps 4 through 7.

Tolerance Specifications

I recall one of my dilemmas as a junior engineer was the specified tolerances on an engineering drawings – particularly for injection-molded parts. Once, a manufacturing engineer many years my senior gave me this advice: "if you are designing glass-filled plastics, say 30% glass, you should be able to maintain a tolerance of ± 0.005in (±0.127mm). If the plastic is un-filled, then use two or three times as much (i.e., tolerance of ± 0.010in to ± 0.015in)." At the time that I received this advice, I was developing various sensors for automotive applications. The approximate size of our products hardly exceeded the size of one's palm and hand. This advice worked very well for me for many years. A corollary of this *rule* was that I had to use secondary operations if the tolerance band that could be maintained by the process was too wide for my specific application. In other words, I would design the feature oversized, and then I would use secondary machining operations to bring it to the size and tolerance that was needed. This should not be surprising. Many design engineers who work with cast metals such as aluminum or zinc count on secondary operations without so much as a second thought. We just do not consider it as frequently for plastics.

Later, I learned that similar rules of thumb exist for various manufacturing processes. Each manufacturing process has certain inherent capabilities to deliver a tolerance band. Although this band may be extended, it may not be tightened. To obtain a tighter tolerance band, often, a different and typically more expensive process should be used.[1]

In fact, many manufacturing processes have been developed to such a degree of sophistication that their tolerance grades and limit deviations have been captured in ISO 286–1 and ISO 286–2. Known as *international tolerance grades*, and designated as IT01, ITO, IT1, through IT18, they define various tolerance bands for known manufacturing processes in a given size limit. The smaller the *IT* number, the tighter the tolerance band; and conversely, the larger the *IT* number, the wider the band. For instance, if a feature of size has a nominal dimension between 3 mm and 6 mm, a process that is capable of an IT01 grade will produce features sizes with a tolerance of ±0.0004mm. Conversely, a different process that is capable of an IT16 grade will produce the same feature of size to a tolerance band of 0.760 mm. Based on this argument, gages and instruments require processes that are capable up to IT5 or IT6. Most industrial processes are capable of producing grades of IT5 to IT12. Clearly, as more precision is required, IT values closer to 5 are used. Most as-used castings (including die-casting) are within the IT11 to IT16 bracket.

Kota (2011) suggested the following relationship between the tolerance band and the IT grade.

$$T = 10^{0.2(ITG-1)}\left(0.45\sqrt[3]{D} + 0.001D\right)\times 10^{-3} \tag{24.1}$$

where, T is the tolerance band in millimeters, D is the nominal dimension in millimeters, and ITG is the IT grade for the intended manufacturing process.

The takeaway of this discussion is this: the design engineer can calculate the tolerance band that a given process is capable of producing based on its ITG number. Next, this number may be compared to the specified tolerance value for the features of interest. Should the calculated value be less than the specified tolerance band, the selected process is capable of meeting the design need; otherwise, if the calculated tolerance value is larger than the specified band, the proposed process is not capable, and a more precise process should be considered.

It should be noted that there are costs associated with either of the two solutions. A more precise manufacturing process is costly both in terms of the capital cost of the equipment used as well as the higher piece price due to the required precision. However, precision components may be used in an automated assembly; hence, the cost of labor may be reduced. Conversely, less precise processes would require lower equipment capital costs and would have lower piece prices, though they may lead to higher scrap rates or selective manufacturing where by less automation may be used. Hence, there may be increased labor costs. In most real-life cost evaluation situations, these two opposing factors must be weighed against each other.

TABLE 24.1

Calculated Tolerance Bands Based on ITG values

ITG Value	Calculated Process Tolerance Band	
	mm	in
5	0.014	0.0005
6	0.022	0.0009
7	0.035	0.0014
8	0.055	0.002
9	0.087	0.003
10	0.138	0.005
11	0.219	0.009
12	0.347	0.014
13	0.550	0.022
14	0.871	0.034
15	1.381	0.054
16	2.189	0.086
17	3.469	0.137
18	5.498	0.216

Another topic that has been treated extensively in various ISO standards (including ISO 286–2) is holes and shafts due to their importance in machinery and machine designs. These tolerances typically define various degrees of clearance (e.g. slip) to interference fits.

Example

Let's consider that a critical dimension of a component needs to be maintained at 3.94 in ± 0.02in (100.00mm ± 0.51mm). The design engineer needs to determine whether a die-casting or an injection molding process would be appropriate.

Based on Equation 24.1, we can calculate the tolerance band for a variety of processes:

$$T = 10^{0.2(ITG-1)}\left(0.45\sqrt[3]{100} + 0.001 \times 100\right) \times 10^{-3}$$

For either die-casting or injection molding processes, the ITG value between 11 and 13 may be used. For ITG = 13, the calculated tolerance band is $T = 0.55$ mm (0.022 in). This tolerance value is slightly above the line. For ITG = 12, T = 0.347 mm (0.014 in). Thus, the selected manufacturer should be able to produce to ITG levels of 12 or better. Table 24.1 provides calculated tolerance bands for other IT grades from 5 to 18 for a nominal value of 100 mm.

(Nominal Dimension = 100 mm, Required Tolerance ± 0.51 mm)

Process Capability

Equation 24.1 is a tool for the design engineer to narrow the selection of potential manufacturing processes. In the previous example, the design engineer makes the inherent assumption that his (her) chosen supplier can – in fact – deliver to the selected international grade number (ITG). In

> Frequently, a process begins with producing conforming product dimensions; however, as time goes by, and for a variety of reasons, part-to-part variations increase.

practice, however, manufacturing equipment, their age, and the conditions under which these equipment are operated influence part-to-part variations greatly. Frequently, a process begins with producing conforming product dimensions; however, as time goes by, and for a variety of reasons, part-to-part variations increase. A missmatch between required tolerances and product variations translates into additional costs and lower overall quality because nonconforming parts need to be either reworked or scraped altogether. To catch these parts, typically additional part inspections are needed, leading to additional costs.

There is a term and two associated metrics that speak to this phenomenon. It is called process capability and is denoted by either C_p (capability ratio) or C_{pk} (centered capability ratio). Capability ratio metric (C_p) compares the *specified* variation of a feature of size (i.e., the difference between the largest acceptable dimension and the lowest acceptable dimension) and compares it to the standard deviation of a sample of the manufactured parts. The centered capability ratio metric (C_{pk}) determines if the produced feature dimension is centered between the upper and lower limits or whether the feature dimension is closer to either the upper or lower specified limits (Wheeler 2000a, 2000b).

Mathematically, these metrics are calculated as follows:

$$C_p = \frac{USL - LSL}{6\sigma} \tag{24.2}$$

$$C_{pk} = \text{Min}\left\{\frac{USL - \mu}{3\sigma}, \frac{\mu - LSL}{3\sigma}\right\} \tag{24.3}$$

Where USL is the upper tolerance limit, LSL is the lower tolerance limit, and σ is the standard deviation of the measurements of the feature under consideration. μ is the measurement average value. Associated with the process capability metrics, there are two other variables that are defined as:

$$UCL = \mu + 3\sigma$$

$$LCL = \mu - 3\sigma$$

UCL stands for upper control limit and LCL stands for lower control limit. These two metrics provide an indication of the highest and lowest values that may be expected from any given process. These are very powerful metrics to consider. If UCL is less than USL and if LCL is greater than LSL, then all the components that are produced are within the specified tolerance values and – at least in theory – all parts should be assembled easily.

Figure 24.2 provides a pictorial overview of a variety of conditions that may take place. Figure 24.2 (A) speaks to an ideal situation where good design margins exist. The process is capable of producing parts that are well within specified design tolerance limits (i.e., $LSL < LCL$ and $USL > UCL$). The larger the difference between these limits, the higher the value of C_p and C_{pk} will be. Ideally, C_p should be equal to or larger than 1.6 and C_{pk} should be equal to or larger than 1.3. At the limit, $LSL = LCL$ and $USL = UCL$. In this case, $C_p = C_{pk} = 1$ as calculated below.

Note that in the case of e or f, although C_{pk} is below 1, C_p may have any value.

$$C_p = \frac{USL - LSL}{6\sigma} = \frac{UCL - LCL}{6\sigma} = \frac{(\mu + 3\sigma) - (\mu - 3\sigma)}{6\sigma} = 1$$

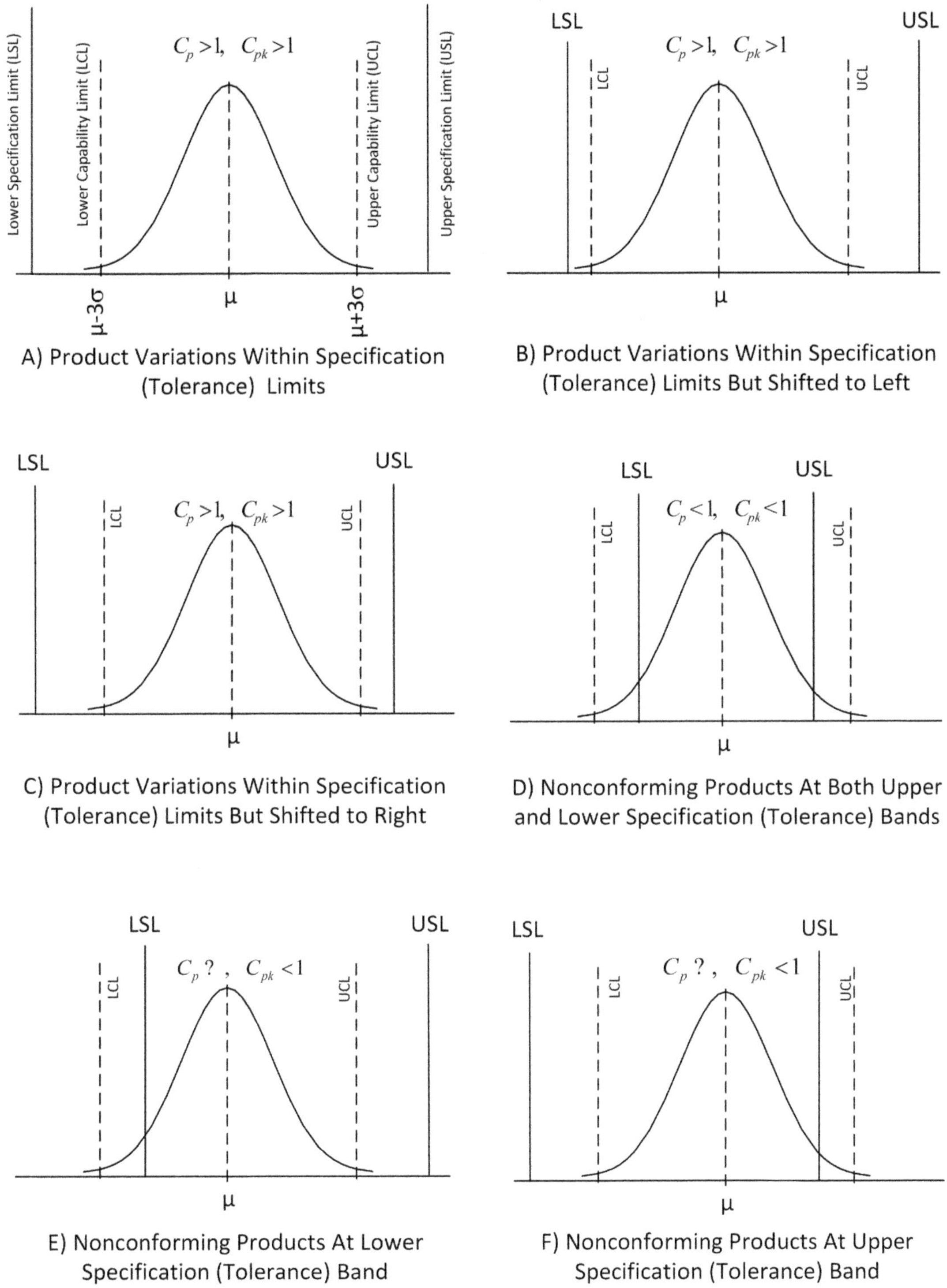

FIGURE 24.2
Product variations, process capability metrics, and tolerance limits

Similarly,

$$C_{pk} = \text{Min}\left\{\frac{USL-\mu}{3\sigma}, \frac{\mu-LSL}{3\sigma}\right\} = \text{Min}\left\{\frac{UCL-\mu}{3\sigma}, \frac{\mu-LCL}{3\sigma}\right\}$$

$$C_{pk} = \text{Min}\left\{\frac{(\mu+3\sigma)-\mu}{3\sigma}, \frac{\mu-(\mu-3\sigma)}{3\sigma}\right\} = 1$$

Figures 24.2B through 24.2F depict a variety of parts produced and their relationship to the specified tolerances. There is a condition that may exist but is not shown here: the situation where the nominal dimension of the produced feature is not even within the specified tolerance. In this scenario, the process is said *not to be in control*. It describes a situation that is beyond the scope of this work.

CASE STUDY

To illustrate the concept of process capability metrics, consider the sheet metal bracket as shown in Figure 24.3. There are three critical features of size that need to controlled in this component. These are the dimensions of the base (5.00 in ± 0.02 in), the angle of the left side (45° ± 2°), and the perpendicularity of the right hand wall (± 0.04 in).

Suppose that a sample of thirty pieces was fabricated and measured.[2] The average value as well as the standard variation of each measurement is as shown in Table 24.2.

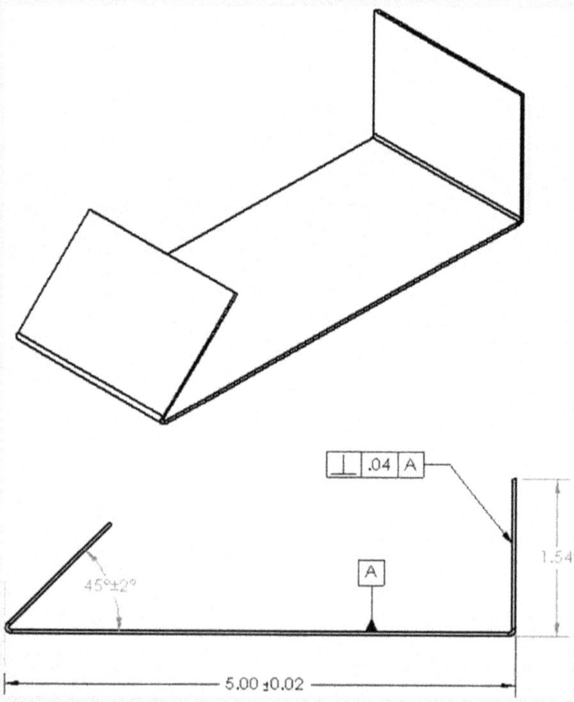

FIGURE 24.3
Sheet metal example

Nominal Values: Base = 5.00 in, Left Angle = 45, Right Perpendicularity = 0 (i.e., angle = 90).

TABLE 24.2
Process Capability Calculations for the Sheet Metal Bracket Shown in

	Base	Left Angle	Right Perpendicularity
Average	4.993	46	0
Standard Deviation	0.0038	0.6650	0.0133
USL	5.02	47	0.04
LSL	4.98	43	−0.04
UCL	5.004	47.995	0.040
LCL	4.982	44.005	−0.040
C_p	**1.75**	**1.00**	**1.00**
$\frac{USL-\mu}{3\sigma}$	2.37	0.50	1.00
$\frac{\mu-LSL}{3\sigma}$	1.14	1.50	1.00
C_{pk}	**1.14**	**0.50**	**1.00**

The results shown in Table 24.2, indicate that the selected sheet-metal fabrication process can easily meet the dimensions specified for the bracket base with virtually no probability of producing nonconforming parts $(C_p = 1.75, C_{pk} = 1.14)$. However, the same process meets the specified perpendicularity requirements with a 0.3% failure rate ($C_p = C_{pk} = 1.$ means that 99.7% of produced parts are within 3 standard deviations [3σ] of the nominal value). Finally, this process requires adjustment for the 45° left-hand wall fabrication in order to produce acceptable parts $(C_p = 1., C_{pk} = 0.5)$. A $C_p = 1$ indicates that the process is capable of producing a 3σ variation, but because $C_{pk} = 0.5$, the variation is skewed to one side (in this case leaning on the high side). It may be possible to make needed adjustments to improve the output; otherwise, scrap or rework rates may be quite high.

Tolerances should not be defined in either a silo or in a vacuum. Specified tolerances should not only reflect user needs and/or requirements but also be realistic for the selected manufacturing process.

In a Nutshell

Tolerances should not be defined in either a silo or in a vacuum. On the one hand, specified tolerances should reflect user needs and/or functional requirements. On the other hand, selected processes should be capable of providing desired tolerances without overburdening the cost structure and finances.

There are three common tolerance analysis methods. These are *worst case analysis* (WCA), *root sum squared* (RSS) method, and *Monte Carlo analysis*.

Before leaving the subject of process capabilities, I would like to add that even though the example used here was specific to geometric measurements of parts and their relationship to their specified values, the same mindset and arguments apply equally to *observed* outcomes of a process versus *expected* outcomes. Finally, it should be mentioned that the field of process capability is quite rich, a topic whose full treatment is well beyond the scope of this book.

Tolerance Analysis

The revolution in manufacturing that people such as Henry Ford had helped to advance was brought about by the concept of replaceable components. The basis of an assembly line approach to manufacturing rests on this concept. I had first discussed this concept in Chapters 12 and 13. In these two chapters, among others, I suggested that part-to-part variations are unavoidable. One way that design engineers control part-to-part variations is to specify tolerances on the engineering drawing. The foregoing argument focused on the relationship between the specified tolerances and the capability of the manufacturing process to deliver the required tolerances.

The next step in design is to analyze whether the specified components would form their intended assembly for all ranges of their specified tolerances. Typically, there are three common techniques that are used to make the assembly assessment. These are *worst case analysis* (WCA), *root sum squared* (RSS) method and *Monte Carlo analysis*. I will be discussing these methods briefly.

Worst Case Analysis

Once nominal values of various design elements (or features) have been determined and their expected variations (or tolerances) specified, it is customary to conduct some type of a tolerance analysis. The simplest form of this analysis is called *Worst Case Analysis* (WCA). WCA may be applied to mechanical assemblies to ensure that all components may be assembled correctly. It may also be applied to electrical circuits to ensure that the circuit can function as expected.

This approach is based purely on the influence of the maximum and/or minimum value of various design elements and their impact on the ultimate feature to be calculated. For instance, in a mechanical assembly, there are typically two questions that the design team is concerned:

- Will the parts assemble as expected, or is there excessive interference between some or all components?
- Once assembled, will there be excessive play (i.e., gaps) between some or all components?

  To illustrate this point, consider the assembly as shown in Figure 24.4. Three blocks, namely blocks A, B, and C, are assembled within the U-channel L. The success or failure of this assembly not only depends on the nominal sizes of L, A, B, and C, but also on their tolerances as well.

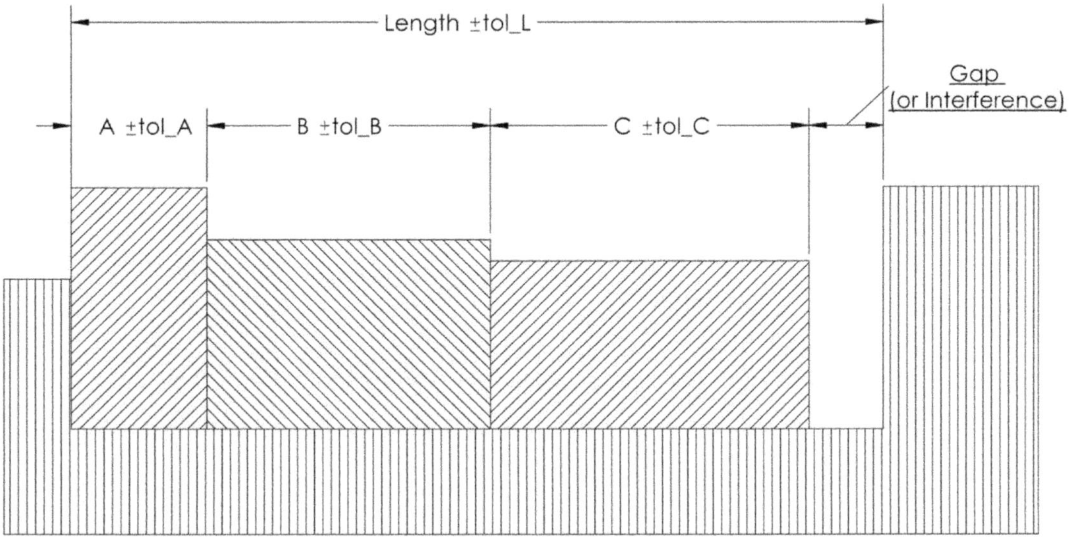

FIGURE 24.4
An assembly block for tolerance calculation demonstration

Mathematically, the maximum and minimum gaps may be calculated by using the largest opening (i.e., the width of the U-channel) and the smallest blocks. This is expressed as:

$$\text{Gap}_{\text{maximum}} = (L + \text{tol\_L}) - [(A - \text{tol\_A}) + (B - \text{tol\_B}) + (C - \text{tol\_C})]$$

Rearrange the terms to obtain

$$\text{Gap}_{\text{maximum}} = [L - (A + B + C)] + [\text{tol\_L} + (\text{tol\_A} + \text{tol\_B} + \text{tol\_C})]$$

Alternatively, the minimum gap is calculated by using the narrowest opening (i.e., the smallest width of the U-channel) and the largest blocks, as shown below:

$$\text{Gap}_{\text{minimum}} = (L - \text{tol\_L}) - [(A + \text{tol\_A}) + (B + \text{tol\_B}) + (C + \text{tol\_C})]$$

Rearrange the terms to obtain

$$\text{Gap}_{\text{minimum}} = [\text{L} - (\text{A+B+C})] - [\text{tol\_L} + (\text{tol\_A} + \text{tol\_B} + \text{tol\_C})]$$

In this scenario, one could summarize the nominal and tolerances on the Gap as follows:

$$\text{Gap}_{\text{nominal}} = [L - (A + B + C)]$$

$$\text{Gap}_{\text{tolerance}} = \pm[\text{tol\_L} + (\text{tol\_A} + \text{tol\_B} + \text{tol\_C})]$$

Note that a gap with a negative value suggests that an interference between components exists.

Electrical Circuits

To illustrate the applicability of this technique to electronics, consider the RCL circuit as shown in Figure 24.5. Equation (24.4) holds for this circuit:

$$Z = \sqrt{R^2 + \left(\omega L - \frac{1}{\omega C}\right)^2} \quad (24.4)$$

Z is the total impedance of the circuit, ωL and $\frac{1}{\omega C}$ are the inductive and capacitive reactance, respectively. The maximum current is calculated using Equation (24.5).

$$I = \frac{V}{Z} \quad (24.5)$$

The same approach may be applied to calculating the worst-case values of Z and I_{max} using various tolerances:

$$Z_{max} = \sqrt{(R + \text{tol\_}R)^2 + \left[(\omega + \text{tol\_}\omega)(L + \text{tol\_}L) - \frac{1}{(\omega + \text{tol\_}\omega)(C + \text{tol\_}C)}\right]^2}$$

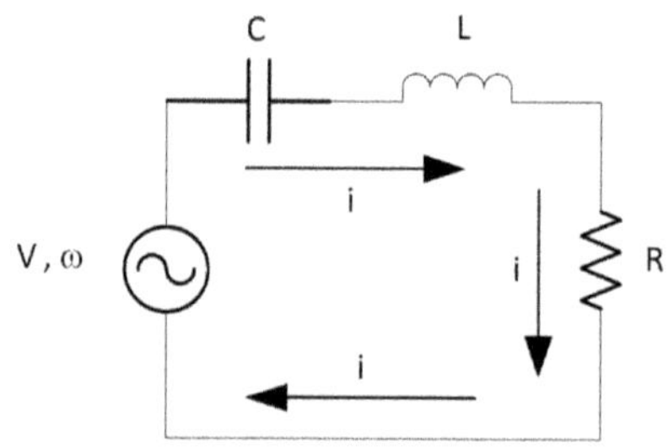

FIGURE 24.5
An RCL circuit for tolerance calculation demonstration

$$Z_{min} = \sqrt{(R - \text{tol\_}R)^2 + \left[(\omega - \text{tol\_}\omega)(L - \text{tol\_}L) - \frac{1}{(\omega - \text{tol\_}\omega)(C - \text{tol\_}C)}\right]^2}$$

$$I_{max} = \frac{V + \text{tol\_}V}{Z_{min}}$$

$$I_{\text{min}} = \frac{V - \text{tol\_}V}{Z_{\text{max}}}$$

It is clear that the nominal values of Z and I as well as their tolerance bands are calculated as follows.

$$Z_{\text{nominal}} = \frac{Z_{\text{max}} + Z_{\text{min}}}{2}$$

$$Z_{\text{tolerance}} = \pm\left(\frac{Z_{\text{max}} - Z_{\text{min}}}{2}\right)$$

Similarly,

$$I_{\text{nominal}} = \frac{I_{\text{max}} + I_{\text{min}}}{2}$$

$$I_{\text{tolerance}} = \pm\left(\frac{I_{\text{max}} - I_{\text{min}}}{2}\right)$$

Nonlinear Worst Case Analysis

The RCL circuit is an example of a nonlinear WCA. The addition of nonlinearity poses additional challenges in determining the variability of an assembly variation (or expected function) due to component variability and tolerances. Suppose that the relationship between the assembly variable (say, the maximum current in the circuit) and the independent component variables (i.e., resistors, capacitors, etc.) may be expressed as

$$y = f(x_1, x_2, \ldots, x_n)$$

Then, Greenwood and Chase (as quoted in Creveling 1997) define the tolerance on y as

$$y_{\text{tol}} = \left|\frac{\partial f}{\partial x_1}\right| \text{tol}_1 + \left|\frac{\partial f}{\partial x_2}\right| \text{tol}_2 + \ldots + \left|\frac{\partial f}{\partial x_n}\right| \text{tol}_n$$

The nominal values are defined as

$$y_{\text{Nominal}} = \left|\frac{\partial f}{\partial x_1}\right| x_1 + \left|\frac{\partial f}{\partial x_2}\right| x_2 + \ldots + \left|\frac{\partial f}{\partial x_n}\right| x_n$$

A full treatment of nonlinear WCA is beyond the scope of this book. For more details on nonlinear WCA, see Creveling 1997.

Root Sum Squared Method

There are two drawbacks of WCA. The first is that although the worst conditions are analyzed, the distribution of the assembly variable is not determined. Should the worst-case conditions become unacceptable, there are no metrics providing evidence of what percentage may be rejected. The second drawback is that all dimensions occur at their maximum (or minimum) condition at the same time. This is clearly a rare or improbable case. More realistically, assembled components are picked randomly, and their sizes could be at any location on the tolerance band.

A drawback of WCA is that all dimensions occur at their maximum (or minimum) condition at the same time. This rarely happens. RSS overcomes this issue by providing a way to calculate a tolerance band. The advantage of a Monte Carlo analysis is that a distribution of the expected value may be calculated.

In the *Root Sum of Squares* (RSS) method, the nominal value of the assembly variable is calculated based on nominal values of each contributing component; however, the tolerance band of the assembly variable is calculated using the square of the standard deviation (also known as variance).

$$s = \sqrt{\sum_i (s_i)^2}$$

Since the tolerance band is directly related to the standard deviation (often the tolerance band is equal to three times the standard variation), the following relationship also holds.

$$y_{\text{tolerance}} = \sqrt{\sum_i (x_{tol,i})^2}$$

Where $y_{\text{tolerance}}$ is the expected tolerance of the assembly variable due to the specified tolerances of the individual components. $x_{tol,i}$ is the specified tolerance of the i-th component. It should be noted that the nominal value of y is calculated based on nominal values of x_i.

Sensitivity Analysis

The process of evaluating the impact of component dimensions on the assembly variables is called sensitivity analysis and is often presented as a percentage contribution to $y_{\text{tolerance}}$. In general it is expressed as

$$P_i = \frac{(x_{tol,i})^2}{(y_{\text{tolerance}})^2}$$

Where P_i is the contribution of the i-th component to the tolerance of the assembly variable.

Sensitivity analysis provides information on which variable contributes the most to the assembly variable. In turn, this knowledge may be used to redesign the component so that the overall variability may be reduced. For more information, see Creveling 1997; Ullman 2010.

CASE STUDY

Recall the assembly as shown in Figure 24.4. Now, suppose that each block has a specified dimension and tolerance band as shown in Figure 24.6.

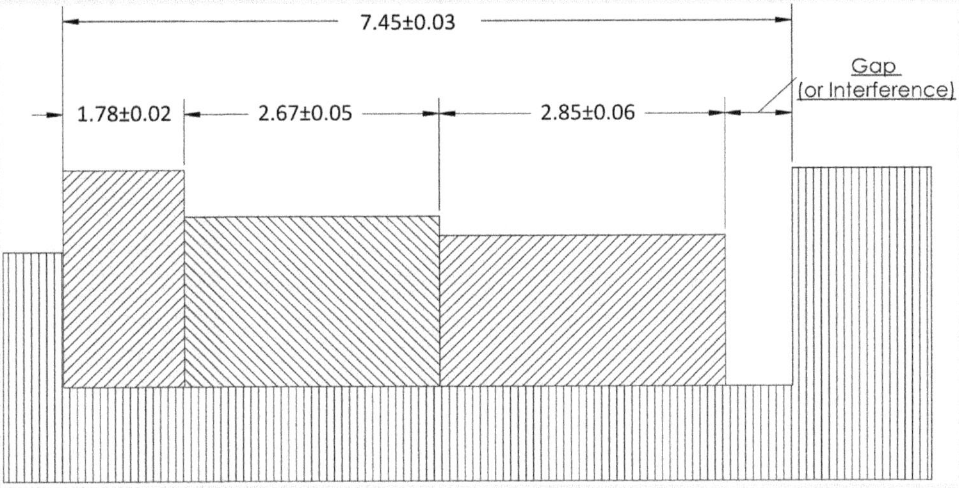

FIGURE 24.6
Assembly block example for tolerance calculations

Tables 24.3 and 24.4 provide the nominal values of the gap as well as its tolerance band using WCA and RSS. Although the nominal value for the gap is identical between the WCA and RSS methods, it is clear that the WCA approach predicts a much wider tolerance band (even an interference) than the RSS method. The RSS method is accurate within a three-sigma variation, whereas WCA assumes that all components are at their respective maximum or minimum values.

In addition to the nominal and tolerance values of the gap, Table 24.4 provides a sensitivity analysis. It shows that component C is the largest contributor to the gap variation, followed by component B. Should a redesign be warranted, these two components are likely candidates for redesign.

TABLE 24.3

Calculation of Nominal Values of the Gap in Figure 24.6 Using WCA

	Nominal	tolerance	min	max		Nominal	tolerance	max	min
A	1.78	0.02	1.76	1.8	Gap	0.15	0.16	0.31	−0.01
B	2.67	0.05	2.62	2.72					
C	2.85	0.06	2.79	2.91					
L	7.45	0.03	7.42	7.48					

TABLE 24.4

Calculation of Nominal Values of the Gap in Figure 24.6 Using RSS

	Nominal	tolerance	tol²	P		Nominal	tolerance	max	min
A	1.78	0.02	0.0004	5.4%	Gap	0.15	0.09	0.24	0.06
B	2.67	0.05	0.0025	33.8%	Sample Calculations				
C	2.85	0.06	0.0036	48.6%					
L	7.45	0.03	0.0009	12.2%					
		sum	0.0074	100.0%					

Gap Nominal = 7.45 − (1.78 + 2.67 + 285) = 0.15

Gap tol. = $(0.0004+0.0025+0.0036+0.0009)^{1/2}$ = 0.086

$$P_A = \frac{0.0004}{0.0074} = 5.4\%$$

Note that calculations have been done using computer accuracy; however, results have been rounded post computation (e.g., 0.086 is rounded to 0.09).

It should be noted that in RSS, it is assumed that component variations and the resulting assembly variable follow normal distributions. Hence, it is possible to calculate the proportion of assembly variables (e.g., gap or interference) that falls within a given range or falls above (or below) a certain threshold.

Monte Carlo Analysis

The Monte Carlo analysis technique was first discussed in Chapter 12. It is a relatively simple technique that was initially developed to study (or model) gambling outcomes (Bethea and Rhinehart 1991). This approach can be used effectively to model the outcome variations based on input changes and variations. In the case of tolerance analysis, input values are the nominal values of each design feature along with its expected tolerance band. Additionally, component variation distributions, if known, may be used as inputs.[3]

Once nominal values for each variable along with specified distribution curves for each variable are specified, then random values for each variable are selected and the outcome is calculated. Clearly, to develop a proper distribution of the assembly variable, a large number of calculations (often in the thousands or tens of thousands) should be carried out.

Table 24.5 provides a set of sample calculations for the example shown in Figure 24.6. It contains 15 trials, where random sizes of parts A, B, C, and L were selected and the gap calculated. Based on these 15 trials, the average (or nominal) gap size is 0.16, and its standard deviation is 0.05.[4]

Should the process capability be 3σ (i.e., my tolerance band is three times the standard deviation of the components produced by the process), then maximum Gap is 0.32 and the minimum Gap is 0.0. As the number of trials increases, the average (nominal) Gap and its standard deviation converge. For 50 trials, the nominal value is 0.14 and the standard deviation is 0.05; for 100 as well as 600 trials, the nominal value and the standard deviation remain at 0.15 and 0.05, respectively, leading to a maximum Gap of 0.30 and a minimum of 0.00.

In engineering design, the so-called law of unintended consequences is always at play because a design is often required to satisfy conflicting requirements.

TABLE 24.5
Sample Calculations in a Monte Carlo Analysis

Trial	A	B	C	L	Gap
1	1.77	2.64	2.80	7.43	0.22
2	1.80	2.68	2.84	7.43	0.12
3	1.77	2.72	2.87	7.44	0.08
4	1.77	2.65	2.79	7.47	0.26
5	1.80	2.70	2.81	7.44	0.13
6	1.79	2.66	2.89	7.45	0.11
7	1.79	2.72	2.82	7.44	0.11
8	1.78	2.64	2.86	7.47	0.20
9	1.79	2.65	2.84	7.44	0.17
10	1.80	2.63	2.84	7.44	0.18
11	1.79	2.68	2.83	7.42	0.13
12	1.77	2.65	2.90	7.46	0.15
13	1.76	2.66	2.90	7.43	0.11
14	1.78	2.63	2.86	7.44	0.18
15	1.76	2.66	2.80	7.46	0.24
				Average	0.16
				St. Dev.	0.05
				Max	0.32
				Min	0.00

Values of A, B, C, and L are generated using a random number generator.

Maximum and minimum values are based on 3x standard deviation calculation range.

Unintended Consequences of Tolerance Specifications

In engineering design, the so-called law of *unintended consequences* is always at play because a design is often required to satisfy conflicting requirements. An example of such conflicting requirements may be found in Figures 24.7 through 24.9. Figure 24.7 depicts a design of several linkage arms that is covered by a housing. A bolt passes through holes in the linkages as well as the housing. The bolt not only provides an axis of rotation, it also provides a way to assemble the housing.

Due to the possible wide tolerances on the parts used in the assembly, the target dimension may vary from loose to interference (interference occurs when the target dimension is negative, indicating that two parts are trying to occupy the same space). Based on the dimensions shown in Figure 24.7, a Monte Carlo analysis was conducted as shown in Figure 24.8. It is realized that for a portion of population of assemblies, there will be

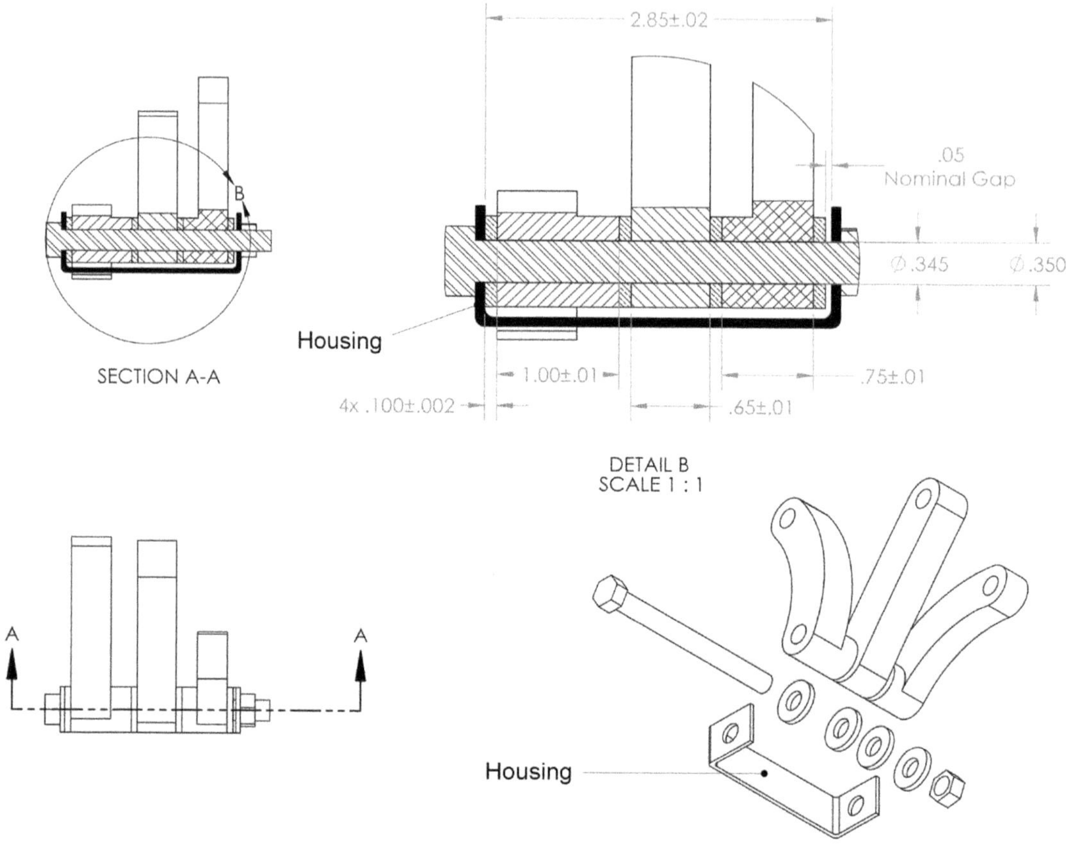

FIGURE 24.7
An example to illustrate various tolerance analysis techniques

an interference between the housing and the rest of the assembly, leading to scraped assemblies.

The team decided to modify the component sizes so that the probability of scraps is reduced. The unintended consequence of this change, however, is shown in Figure 24.9. This figure shows that as the assembly is made and the bolt is tightened, the housing begins to bow out due to excessive gap values.

Another unintended consequence is the impact of environmental conditions such as temperature, pressure, humidity, etc. It is important that the impact of temperature and material behavior be evaluated when tolerance analysis is conducted. The impact of humidity or other factors – though just as important – may be more difficult to evaluate.

One last thought: whenever an analysis is done, it is helpful to adjust the tolerances and process capability values (C_p) on various parts to see how the stack up is affected.

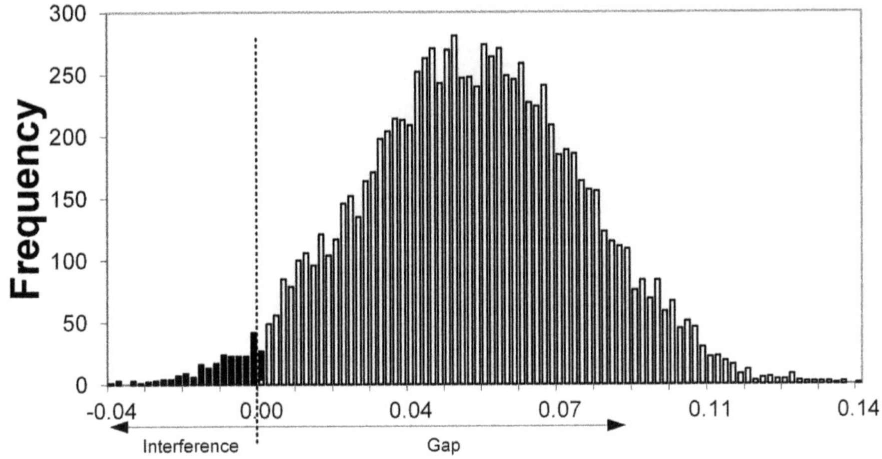

FIGURE 24.8
A Monte Carlo analysis results for the gap size in Figure 24.7

This analysis indicates an interference of 0.04 to a maximum gap of 0.14 may exist.

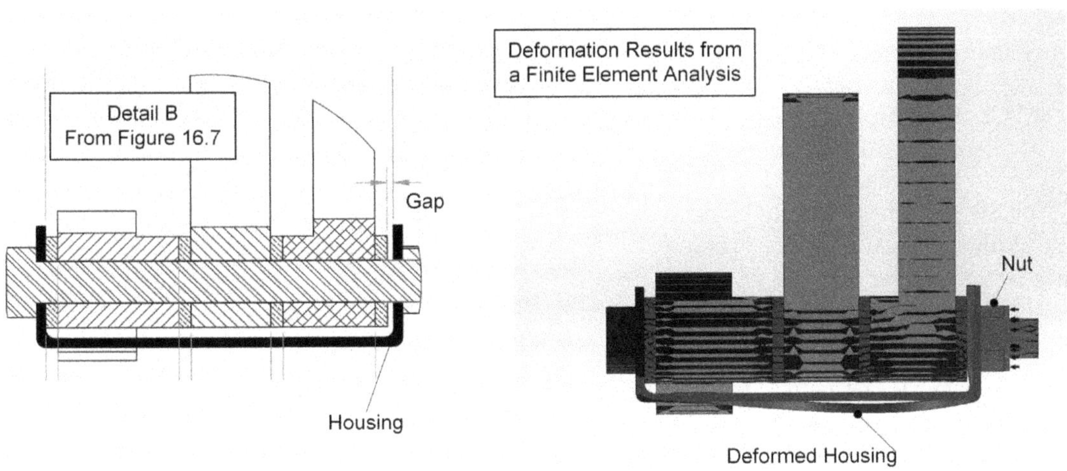

FIGURE 24.9
An often overlooked issue: deformation of housing due to tightening of the nut on the right-hand side

Cost-Based Tolerance Analysis

Issues such as the deformed housing as shown in Figure 24.9 are but an example of many types of failures that may occur as a result of improper tolerance specifications or analysis. The impact of failures may be experienced as low manufacturing (or service) first-pass yield (FPY) or as field failures. In *Practical Guide to Reliable Packaging of Electronics* (Jamnia 2025),

I have suggested that engineers do not intentionally design faulty products. Failures are often due to factors that are over-looked by the design team. That being the case, failures have financial impacts. One area that is often overlooked by design engineers is tolerance stack-ups and their consequences. Low FPY and the associated scrap or rework due to mismatched parts impact both the labor and material expenses and ultimately the overall manufacturing costs. I like to make my point using a simple example below.

Consider the assembly in Figure 24.6. Suppose that for this design, it has been realized that the Gap variation of −0.01 to 0.31 (with a nominal 0.15) may lead to excessive housing deformations. Furthermore, suppose that the user requirements document has been modified to reflect that housing deformation should be maintained below a certain threshold. Engineering has determined that to meet the user requirements, the gap should not exceed 0.10. However, an interference of 0.01 is acceptable (i.e., $-0.01 \leq \text{Gap} \leq 0.10$, and $\text{Gap}_{\text{Nominal}} = 0.045$)..

This requirement suggests that the components on the higher side of the tolerance should be used. How should the design engineer approach this situation? Let's look at a few different case studies.

CASE STUDY 1

The first approach may be to change the nominal values of each component so that the nominal value of the Gap is 0.045. Table 24.6 provides the results of this action (using WCA for simplicity). In this example, nominal dimensions of A and B increased, L decreased, and C remained the same. The result is that although the nominal value of the gap was reduced to 0.05 (not quite 0.045), the maximum and minimum values of the gap are radically beyond acceptable limits. In fact, it is clear that part-to-part variation is causing the problem.

TABLE 24.6

Calculation of Nominal Values of the Gap in Figure 24.6 Using WCA.

	Nominal	tolerance	min	max		Nominal	tolerance	max	min
A	1.80	0.02	1.78	1.82	Gap	0.05	0.16	0.21	−0.11
B	2.70	0.05	2.65	2.75					
C	2.85	0.06	2.79	2.91					
L	7.40	0.03	7.37	7.43					

Note: Components Are Redesigned to Meet Gap Nominal Requirements

One proposed solution may be to sort components based on size and place them in different color-coded bins or buckets. Then, the parts belonging to a given color bucket are assembled together. It is understood that there may be components that are too large or too small to belong to any given bucket. Large parts may be reworked and evaluated again for size; smaller parts may need to be scrapped. This approach may work, but manufacturing yields are low. Furthermore, there are labor costs associated with sorting, binning, and ultimately reworking or scrapping parts. Often, these costs remain hidden and are considered as part of manufacturing over-head.

Although manufacturing yield may increase, part interchangeability is reduced significantly. Service organizations experience the impact of this decision because the product may no longer be easily repaired. As a result, the cost of service increases.

CASE STUDY 2

Recognizing that part-to-part variation should be reduced, processes that are used to fabricate parts B, C, and L are optimized to be equivalent to the same process used for part A (i.e., tolerances are the same for part A). Table 24.7 provides the results. This is a move in the right direction in that the design is closer to target values. However, there are still non-conforming parts that produce gaps that are too large or interferences greater than intended. It is possible that the binning process, proposed in the previous scenario, may be more effective here with fewer scraps or reworked parts. The manufacturing yield may increase, leading to an overall cost reduction.

Just as before, service organizations may feel the impact as well, although not as severely as in the first case study. As parts interchangeability is compromised, service repair yields (SRY) may be impacted because either parts are not assembled together or, in the case of electromechanical units, they may not be calibrated – again suffering either rework or assembly scrap. In this scenario, while manufacturing would enjoy cost reductions, service organizations will be faced with an increase.

I would like to make a point here. A reduction of tolerances, in general, is not just a paper exercise. If manufacturing machinery is not capable of meeting the specified tolerance, secondary operations may be needed. Even higher precision equipment may need to be purchased. Alternatively, the work may need to be out-sourced.

TABLE 24.7

Calculation of Nominal Values of the Gap in Figure 24.6 Using WCA.

	Nominal	tolerance	min	max		Nominal	tolerance	Max	Min
A	1.80	0.02	1.78	1.82	Gap	0.05	0.08	0.13	−0.03
B	2.70	0.02	2.68	2.72					
C	2.85	0.02	2.83	2.87					
L	7.40	0.02	7.38	7.42					

Note: Processes Are Modified to Meet Gap Requirements

CASE STUDY 3

Table 24.8 shows the results of a more drastic move. The change in this approach is to increase the dimension of part C from 2.85 to 2.855 and to reduce part tolerance from 0.02 to 0.010 and 0.015. This change is not merely going from two digits to three digits on the tolerance values. It is a change in manufacturing process capabilities and precision. It is a more expensive process; however, the outcome is that 100% of the assemblies are conforming products, and there are no scraps or rework – either in manufacturing or service.

In this scenario, there may be initial and up-front costs in terms of capital equipment and development of processes to provide the required precision. Once manufacturing and service begin, so long as they are stable and in control, the overall costs of manufacturing and service are minimized and near optimum ranges.

TABLE 24.8
Calculation of Nominal Values of the Gap in Figure 24.6 Using WCA.

	Nominal	tolerance	min	max		Nominal	tolerance	Max	Min
A	1.800	0.010	1.790	1.810	Gap	0.045	0.050	0.095	−0.005
B	2.700	0.010	2.690	2.710					
C	2.855	0.015	2.840	2.870					
L	7.400	0.015	7.385	7.415					

Note: Processes Are Made More Precise to Meet Gap Requirements

Quality Loss Function

Even though there may be a dispute on the definition of quality, that is, conformance to requirements or customer's perception (Guess 2006), the element of cost is always associated with quality. It is within the product's price structure that its cost point is set. And, based on cost, other requirements or specifications are developed.

There is a prevailing assumption that so long as components are within their specifications, they are acceptable and may be used to produce acceptable parts. However, as I demonstrated in the previous section with the deformed housing example, a customer's perception of quality may be greatly affected if parts dimensions deviate greatly from the target values. Even though the example here focuses on dimensional tolerances, similar conclusions may be drawn for any component, assembly, and eventually system functions and their accepted variations.

Taguchi, Fowlkes, Creveling, Wheeler and others (Fowlkes and Creveling 1995; Creveling 1997; Wheeler 2000a, 2000b) proposed mathematical relationships between manufacturing variations and excess costs. At a high level, the equation reads as follows:

$$Excess\ Costs = \int excess\ cost\ \ function\ x\ probability\ model$$

This high-level equation may be reduced to a simple quadratic equation for the quality loss function as output variables (i.e., y's) deviate away from the target value.

$$L(y) = k(y - m)^2$$

Where m is the target value, y is the output variable, and k is defined as

$$k = \frac{\text{Cost of a defective product}}{(\text{tolerance})^2} = \frac{A}{\Delta^2}$$

Pictorially, this is depicted in Figure 24.10.

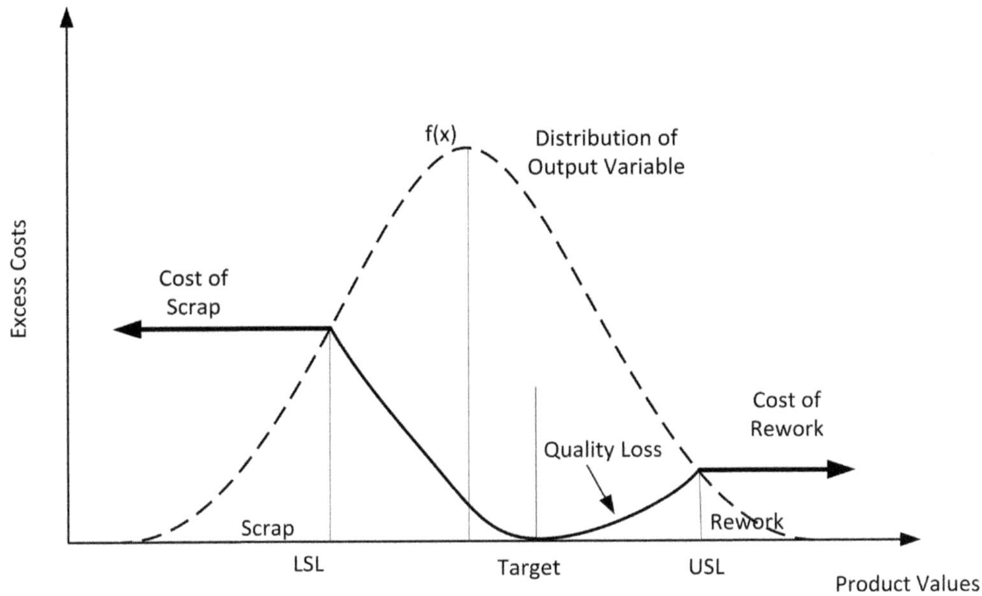

FIGURE 24.10
Excess costs of production combined with a normal probability model

Adapted from Wheeler 2000a, 2000b

Example

Consider the conditions as expressed in Case Study 3. Suppose that the business has a warranty period during which a dissatisfied customer may return the product for any reason. This return requires the business to clean and inspect the product first and then place it for resale at a price discount. The average cost of cleaning, inspection, and price discount is \$150. What is the quality loss for a product with a gap of 0.055 in?

$$\text{Cost of a defective product} = A = \$150$$

$$\text{tolerance} = \Delta = 0.05$$

$$k = \frac{\text{Cost of } a \text{ defective product}}{(\text{tolerance})^2} = \frac{A}{\Delta^2} = \frac{150}{0.05^2} = 60{,}000$$

$$L(y) = k(y - m)^2$$

$$L(y) = 60000(y - 0.045)^2$$

For $y = 0.055$

$$L(0.055) = 60000(0.055 - 0.045)^2 = 60000(0.01)^2$$

$$L(0.055) = \$6.0$$

I like to emphasize a point here. The gap used in this example is 0.055 in which is well within the accepted gap tolerance band (–0.005 in to 0.095 in), and yet, there is a monetary loss associated with this small deviation from the target value – not too many people consider this loss. Clearly, this loss needs to be balanced with the cost of manufacture at higher precisions and lower variations.

Any further in-depth review of this topic is beyond the scope of this book; however, interested readers are encouraged to read Fowlkes and Creveling 1995; Creveling 1997; Wheeler 2000a, 2000b.

Summary

- Product and part-to-part variation cannot be avoided. Hence, either variations should be so small that they are not perceived, or they should be within a customer-acceptable range.
- ISO 286–1 and ISO 286–2 standards capture many manufacturing processes tolerance grades and limit deviations known as international tolerance grade.
- Process capability is a measure of the inherent variations of a manufacturing process within a required part specification.
- Part specifications (or tolerances) should not be defined in either a silo or in a vacuum. On the one hand, specified tolerances should reflect user needs and/or functional requirements. On the other hand, selected processes should be capable of providing desired tolerances without overburdening the cost structure and finances.
- The three common tolerance analysis techniques are worst case analysis (WCA), root sum squared (RSS) method, and Monte Carlo analysis.
- Low first-pass yield and the associated scrap or rework due to mismatched parts impact both the labor and material expenses and ultimately the overall manufacturing costs.
- There is a prevailing assumption that so long as components are within their specifications, they are acceptable and may be used to produce acceptable parts. However, a customer's perception of quality may be greatly affected if parts dimensions are deviated greatly from the target values.

Notes

1 It should be clear that when I talk about specifying (or even changing) tolerances, I am specifically talking about *make* items under the control of the design team.
2 To assess product variation, samples should be selected randomly, though this may not be practical in some cases.
3 Otherwise, a normal distribution is assumed.
4 For this example, the actual units (i.e., mm or in) are irrelevant.

25

Failure Analysis and Product Improvements

Introduction

Successful product verification, validation, and ultimately launch – the goal of all design efforts – yields little information on which to base design improvements. Failures, on the other hand, contribute a wealth of data on "what to improve" or "what to design against" in subsequent efforts. The feedback of information obtained from the analysis of failures is one of the principal stepping stones of progress. The nature and underlying causes of failures, once identified and corrected, lead to product improvements. Various organizations develop their own unique methods of tracking failure metrics (and their rates), which are often monitored through quality reviews and corrective action preventive action (CAPA) events.

In general, failure reporting is a subset of reliability data review, which not only includes reports of the manner in which a product fails but also an understanding of the successful duration of the fleet's operation. Hence, analysis of failure becomes more meaningful when it is done under the umbrella of a reliability analysis where the entire failed and operational population is considered. The analysis of reliability data has three main purposes:

- To verify that the device is meeting its performance requirements or goals.
- To discover deficiencies in the device to provide the basis for corrective action.
- To establish failure histories for comparison and for use in future product development activities.

> Failures contribute a wealth of data on "what to improve" or "what to design against" in subsequent efforts.

As mentioned, reliability metrics represent an aggregate of part failures in a product population rather than a single failure or a single product. Considering this accumulation, it may be difficult to scope any failure investigation[1] and the area to focus on. The goal of this chapter is to shed light on the challenges faced during product improvement investigations and to provide guidance on working through these challenges.

This guideline emphasizes prioritization of investigation and analysis of failures based on a variety of factors, such as apparent frequency or impact on customers. There may also be other influences, such as classification of failures according to categories of design/part procurement, manufacture, or assembly and inspection. Regardless, the investigation must provide essential information on the following:

1. What failed,
2. How it failed,

DOI: 10.1201/9781003564508-33

3. Why it failed, and
4. How future failures can be eliminated.

Investigation Process

In Chapter 6, I introduced the DMAIC toolkit and at the same time, I mentioned that this tool is best suited for systematically driving to root cause analysis of failures and improvements. Again, as mentioned earlier, DMAIC has five steps, which I repeat and then expand on.

> DMAIC is a Six Sigma toolkit that is best suited for systematically driving to root cause analysis of failures and improvements.

1. Define – In this step, the problem statement is defined clearly and concisely. This is crucial to the success of the DMIAC activities. In fact, the team should dwell in this stage as long as possible and come back to it in order to ensure that the problem statement is complete. It should focus on what has failed (or needs to be improved), who has observed the issue, and its extent. The define stage corresponds to blocks 1 (Failure Observation), and 2 (Failure Documentation) in Figure 25.1. Completion of these three blocks enables the investigator to develop a strategy to collect data and/or an approach to the manner in which inquiries are to be conducted.
2. Measure – As is determined by its name, in this stage, relevant data is collected. This step corresponds to blocks 3 (Failure Verification), 4 (Failure Isolation) and 5 (Suspect Item Replacement) of Figure 25.1. If additional facts are identified that affect the problem statement, Define is revisited and the problem statement is updated. The more comprehensive the problem statement, the richer the collected data.
3. Analyze – Once sufficient levels of data are collected, their analysis begins. In Figure 25.1, blocks 6 (Suspect Item Verification), 7 (Failure Analysis), 8 (Data Search), and 9 (Establish Root Cause) correspond to the Analyze stage. The focus should be placed on identifying factors that have a significant impact on functions that are critical to quality. By studying these factors and their impact, a root cause of failure is generally identified.
4. Improve – On the basis of what has been learned, a design change may be proposed. Prototypes are created and evaluated for expected outcomes. Should the need be, designs of experiments may be run to evaluate and document the impact of proposed changes on the essential transfer functions. Blocks 10 (Determine Corrective Action), 11 (Incorporate Corrective Action), and 12 (Operational Performance Test) belong to these steps of DMAIC.
5. Control – The DMAIC process identifies root cause(s) of the failure. However, if the problem has happened once, it is likely to happen again. To close the loop, control mechanisms must be identified and put in place to prevent a relapse and a return to the state under which failure took place. Block 13 in Figure 25.1 asks this question and ensures that the identified problems are properly addressed.

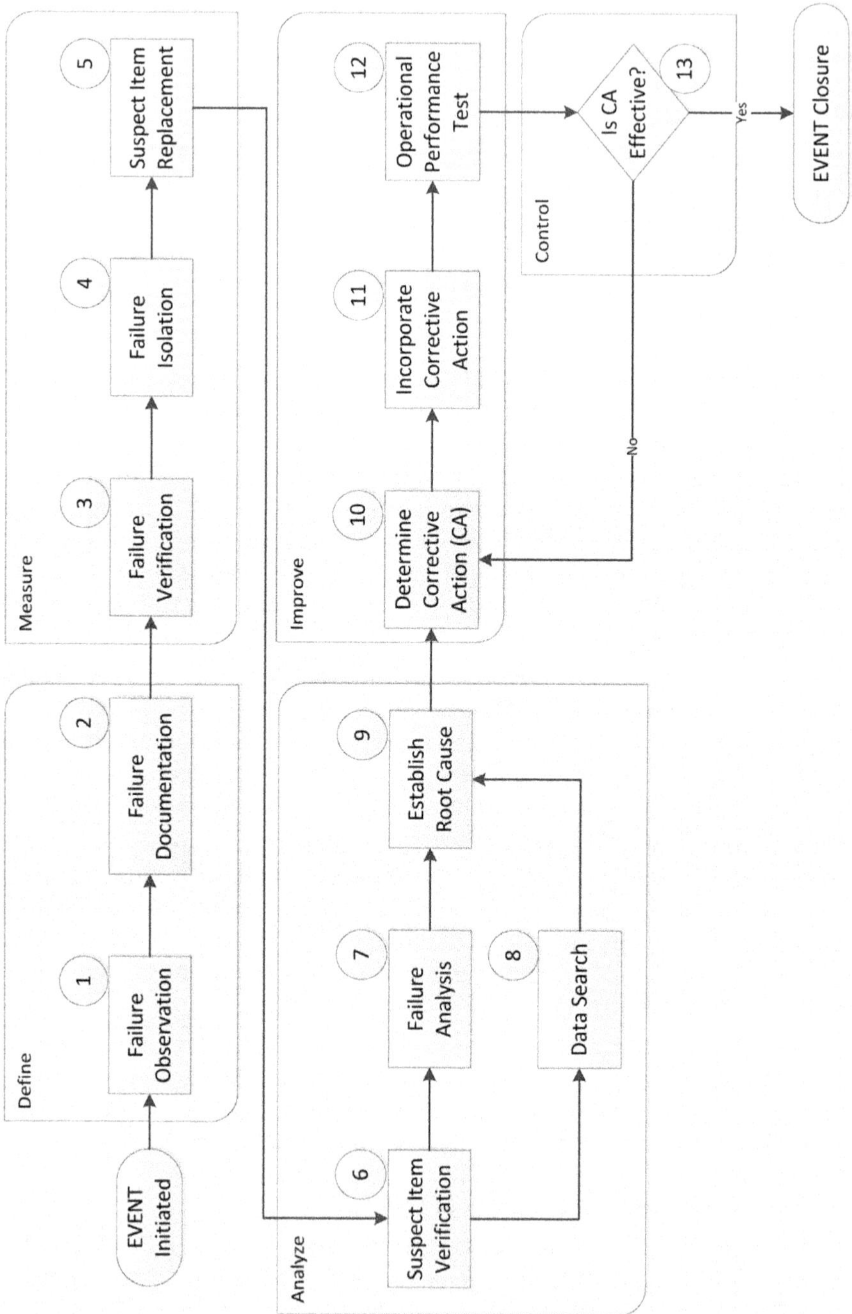

FIGURE 25.1
Reliability-based investigation process flowchart

The following is a fuller description of the steps and stages of DMAIC as depicted in Figure 25.1.

Failure Observation and Determination

The event (e.g., reliability metric trigger) or the decision to investigate should be documented and include the following information:

a. Date of field data review, when it was determined that an investigation was required,
b. Product line, and all applicable product codes, applicable to the reliability data under investigation
c. The subsystem associated with the trigger item.
d. Reliability metric value (i.e., mean time between failure [MTBF], failure rate) during the triggering month
e. Reliability metric threshold(s) and the value that was exceeded

This information may be summarized in a problem statement.

Example Problem Statement

On 17 May 2017, during a review of the product performance data, it was observed that the Learning Station User Interface Subsystem MTBF value for January 2017 was 225 months. Furthermore, this subsystem was exhibiting a negative trend for five consecutive months.

Failure Isolation and Scope Determination

> Failure isolation can prioritize the top failed component(s) of the subsystem, rather than investigating all subsystem components/parts.

Considering that the primary goal of this type of investigation is to improve product performance, and because a subsystem may consist of several components or parts, it is important to determine which subsystem components will be included in the scope of the investigation. Therefore, the failure isolation can prioritize the top failed component(s) of the subsystem rather than investigating all subsystem components/parts. This decision should be driven by the reliability metrics, with the goal of making a design change that will improve reliability metrics. The scope determination must be documented with rationale as to why certain components/parts were deemed out of scope. Often, the use of a Pareto chart of parts that failed or replaced helps the decision-making process.

After the scope is determined, the problem statement should be updated to include the components that will be investigated.

Example Problem Statement

On 17 May 2017, during a review of the Product performance data, it was observed that the Learning Station User Interface Subsystem Mean Time Between Failure (MTBF) value for Jan 2017 was 225 months. Furthermore, this subsystem was exhibiting a negative trend for five consecutive months. The top most failed component in this subsystem is the Controller Boards, which causes 40% of Enclosure failures.

Failure Verification

Because product performance metrics (such as reliability or failures) are calculated from service and field data, it is important to verify failures for an initiated event. The first step of this process is to obtain failed samples from the servicing center(s). If possible, obtain the parts from the time periods when the negative trends were observed. Failed samples obtained from the service center(s) require testing to ensure that they indeed did fail, and the product performance metrics accurately reflect the fielded product.

Once the failure has been confirmed, a failure analysis of the defective item(s) should be performed to establish the responsible failure mechanism.

Failure Analysis and Root Cause Determination

Once the failure has been confirmed, a failure analysis of the defective item(s) should be performed to establish the responsible failure mechanism. This can be done in a variety of ways; however, much of the technical staff tends to focus on what they or their close teammates can (or cannot) do; many opt to look for a solution in their own labs. This approach is practical when the same business has also manufactured the failed component or assembly. Otherwise, a very knowledgeable ally is the manufacturer because they have a more intimate knowledge of potential causes of failure modes. At times, performing a search of existing data to uncover similar failure occurrences in this or related items may provide a historical perspective of the observed failure mode/failure mechanism. Have these failure modes been evaluated and/or addressed before?

Additionally, design, manufacturing, or service changes – either in process or tools – may have caused an increase in failure rates of the component and/or subsystem. To identify any sources, the following should be reviewed:

1. Service process changes
2. Manufacturing process changes (of both the subsystem and the component)
3. Supplier changes
4. Design changes to the system and subsystem
5. System and subsystem reliability trending in the past 12–24 months
6. Geographic location of products in the field
7. Component age

Root Causes

Using the collected data, it may be determined whether the observed failure(s) were due to a product or process change or whether they were due to narrow design margins. Additionally, the root cause of the top failure mode(s) for the top failed component(s) may be identified.

Example

The root cause for the Learning Station User Interface board failures is a poor design of the older versions of PN 260315, the interconnect assembly. This interconnect ties the user interface board to the touch screen. Rev D and earlier versions of the power cable use a terminal that makes contact with the connector on the one side, which may be exposed to fluid ingress. This design can lead to corrosion or buildup of minerals over time.

Corrective Actions and Verification

Determine the necessary action(s) to prevent future failure recurrence (i.e., design change, process change, procedure change). The decision regarding the appropriate corrective action should be made by a cross-functional design team.

Once a corrective action has been identified, and prior to its implementation on fielded products, its effectiveness at eliminating the issues at hand should be verified. It is important to authenticate that the financial impact of the proposed corrective actions will justify the resources and time dedicated to implementing them. Another factor is to verify that when the actions are implemented, no new issues are introduced. The following steps will verify the recommended actions:

Part of the investigation process is to create effectiveness criteria for each recommended action. This criterion is used to evaluate the actions once implemented and will determine if they effectively eliminated the issue(s) that caused the event initiation.

1. Incorporate the recommended corrective action into the original test system/equipment. *If the action is to make a design change, put the new design into a unit.*
2. Retest of the system/equipment with the proposed corrective action modification incorporated. *For a design change, expose the product to the same field conditions that resulted in the issue under investigation.*
3. Determine if the proposed corrective action is effective at solving the problem that caused decreased performance of the system or subsystem. If needed, perform accelerated life testing to ensure that the new design is more reliable than the existing design. If the issue recurs, it can be concluded that the design change is not effective. However, if the issue does not occur, the design change can be deemed a success and is ready for implementation on a larger scale.

Effectiveness and Control

Part of the investigation process is to create effectiveness criteria for each recommended action. This criterion is used to evaluate the actions once implemented and will determine if they effectively eliminated the issue(s) that caused the event initiation. Testing the effectiveness of a performance-related design change may require the use of accelerated life testing on the old design and the new design to compare failure modes and times to failure. Additionally, once the changes are applied to the fielded products, they need to be monitored.

Investigation Tools and Tips

Keys to a Successful Investigation

There are several "keys" that make the failure reporting and investigation process effective. These include the following:

1. The discipline of the report writing itself must be maintained so that an accurate description of failure occurrence and proper identification of the failed items are ensured.

2. The proper assignment of priority and the decision for failure analysis must be made with the aid of the design team and systems engineers.
3. The status of all failure analyses must be known. It is of prime importance that failure analyses be expedited as priority demands and that corrective action be implemented as soon as possible.
4. The root cause of every failure must be understood. Without this understanding, no logically derived corrective actions can follow.
5. There must be a means of tabulating failure information for determining failure trends and the mean times between failures of system elements. There should also be a means for management visibility into the status of failure report dispositions and corrective actions.
6. The system must provide for high-level technical management concurrence in the results of failure analysis, the soundness of corrective action, and the completion of formal actions in the correction and recurrence prevention loop.

Maintaining accurate and up-to-date records through the implementation of the data reporting, analysis, and corrective action system provides a dynamic, expanding experience base for others to use in future designs.

Data Collection and Retention

Maintaining accurate and up-to-date records through the implementation of the data reporting, analysis, and corrective action system provides a dynamic, expanding experience base. This experience base, consisting of test failures and corrective actions, is not only useful in tracking current programs but can also be applied to the development of subsequent hardware development programs. Furthermore, the experience data can be used to:

1. Assess and track reliability
2. Perform comparative analysis and assessments
3. Determine the effectiveness of quality and reliability activities
4. Identify critical components and problem areas
5. Compute historical part failure rates for new design reliability prediction (in lieu of using generic failure rates found in MIL-HDBK-217, for example)

CAPA Core Team

In all systematic failure investigations, a core team is established to oversee the effective functioning of the investigation and proper execution of corrective action (see Figure 25.2). It also provides increased management visibility and control of the investigation. The team typically consists of a group of cross-functional representatives with a sufficient level of responsibility to ensure that failure causes are identified with enough detail to generate and implement effective corrective actions, which are intended to prevent failure recurrence and to simplify or reduce the maintenance tasks. The main responsibilities of a core team consist of setting priorities, establishing schedules, assigning

In all systematic failure investigations, a core team is established to oversee the effective functioning of the investigation and proper execution of corrective action.

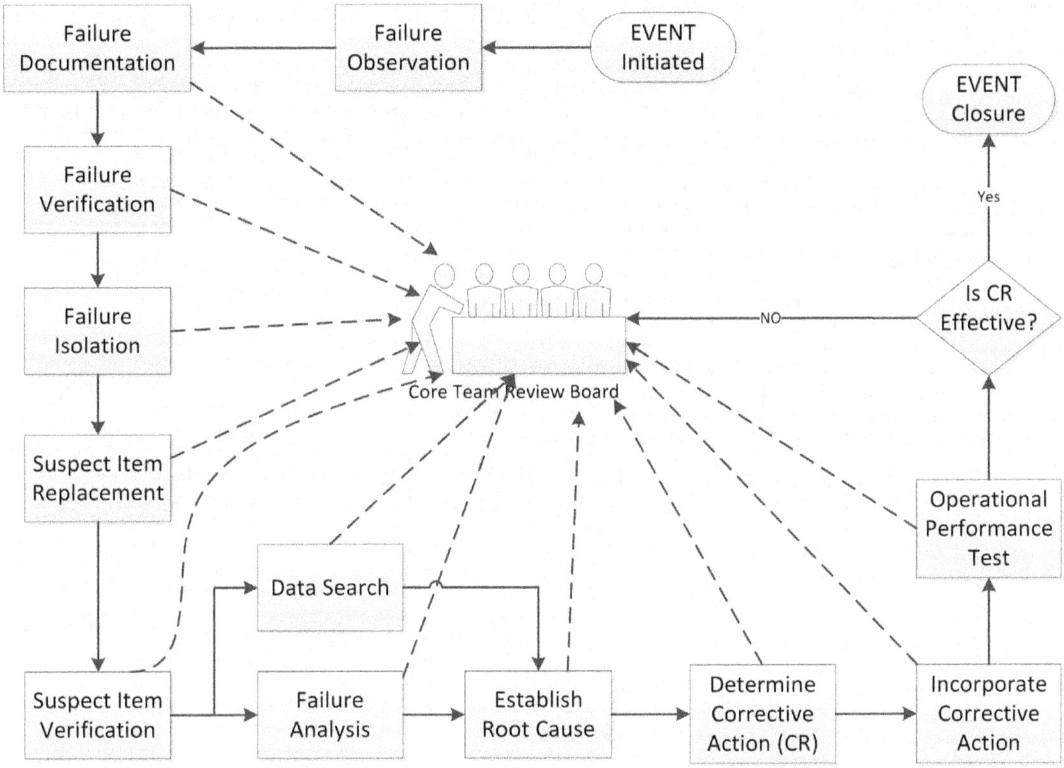

FIGURE 25.2
Failure reporting and the role of the core team

specific responsibility, and authorizing adequate funding to ensure the implementation of any necessary changes when dealing with complex and difficult problems.

In a performance-based investigation, the core team is especially important, and helps the investigator(s) to improve the reliability and maintainability of hardware and associated software by the timely and disciplined utilization of failure and maintenance data. A successful core team will include (but is not limited to) higher authority representatives from the following functions: reliability engineering, systems engineering, and quality. It should also contain a product expert for the product under investigation.

The core team should be involved in all major steps of the investigation, with the purpose of providing guidance and expertise to enable the investigator to make concrete decisions based on the available data. The core team is to be consulted during the stages outlined in dashes in Figure 25.2 and listed below:

- Determine scope
- Root Cause of Trigger
- Root Cause of Top Failure Modes
- Determination of Recommended Actions
- Verification of Actions
- Effectiveness of Actions

When consulting the core team at the various investigation steps, it is important to provide them with adequate yet concise information. Be sure to present only the pertinent data, in an unbiased manner, so that they can make decisions and give advice for how to move forward with the investigation. Tools like a four-blocker can be helpful for communicating the issues that require insight and collaborative decisions from the team.

Summary

- Product failures contribute a wealth of data on "what to improve" or "what to design against" in subsequent design iterations.
- Reliability metrics represent an aggregate of part failures in a product population, rather than a single failure or a single product.
- DMAIC toolkit has five steps. These are Define, Measure, Analyze, Improve, and Control.
- In all systematic failure investigations, a core team is established to oversee the effective functioning of the investigation and proper execution of corrective action.

Note

1 I am very aware that at times, there may be excessive failures of a component that sets a panic mode in the entire team if not the enterprise. These situations are beyond the scope of product improvement activities that I am discussing here.

Appendix A. What Is Agile?

Introduction

At the time of writing this book, Agile has become quite a buzzword in the new product development community. What is Agile? Or, to put it better, what is Agile methodology? It turns out there is not a clear-cut answer to this question. The *Merriam-Webster* dictionary defines agile to mean "marked by ready ability to move with quick easy grace" or "having a quick resourceful and adaptable character."

If we were to keep these definitions of agile in mind, we would realize that – in the context of product development – any activity or mindset that leads to a more efficient and timely delivery of a product or product family to market may be considered to be "agile." In fact, there is truly no single accepted approach that one may call an Agile methodology.

A Brief History

In my view, the history of Agile methodology finds its roots in attempts to manage large-scale projects of 1940s and 1950s and control cost overruns when the initial versions of a gated, or phased, approach to funding projects were established. By this time, project managers had recognized that the root cause of many delays and unwarranted expenses could be found in a lack of sequential planning of activities. Often team members were unclear of project task dependencies as teams worked in their own functional groups or silos. A delay in one task could have had a domino effect on other task deliverables. Worse yet, a work packet might have needed to be revised because of an oversight leading to further delays and expenses.

To avoid these issues, large-scale projects were divided into phases with well-defined tasks and deliverables. Entry into the next phase of a project required detailed review and approval of deliverables by management. Needless to say, project funds for each phase would be released once approvals were obtained. As mentioned in Chapter 6, Cooper introduced his Stage-Gate® process in the early 2000s after having developed a standard five-gate process for product development.

As projects became more complicated, close collaboration between various technical fields was required to meet the requirements of each phase of the project. This cross-functional cooperation led to what is called concurrent engineering today. Concurrent engineering required project team members to share information and communicate closely to achieve common goals.

Strictly speaking, phase-gate methods are business processes. They do not dictate details of product development tasks and activities. This is left to project management to establish what activities come first and what needs to be completed next – just like a *waterfall* stream of activities. The first step always begins with scoping the product to be developed, followed by planning activities. For the technical teams, the first step is to establish and "lock-down" product requirements. Once a project is ongoing, any changes in requirements may require revising any completed tasks, leading to project delays. Shortly, we will learn that Agile methodologies for software development challenge and revoke this tenet of traditional waterfall approaches.

In the late 1990s, an adaptation of waterfall methodologies called the V-model gained currency within the software development communities. This book demonstrates the application of the V-Model to product development – not just software. The advantage of the V-model over the waterfall model is to establish clear delineation of design activities versus manufacturing, as described earlier in this work. For software development, it provides a clear understanding of periods of code development and its associated testing.

A major criticism of either the waterfall or the V-model is the extensive effort needed at the beginning of the project to plan and create requirements and the perceived lack of flexibility to make any changes once these documents have been approved. Agile methodologies seek to provide an alternative, more flexible approach.

If we were to call the waterfall method a heavy-handed approach or heavyweight, efforts to develop products with a lightweight mindset, iteratively and/or incrementally, go back to 1930s with an approach called "plan-do-study-act" (PDSA) cycles (Larman and Basili 2003). This method was then adopted by Deming and later by Glib and Zultner for software development (Larman and Basili 2003).

The efforts to advance PDSA or lightweight methods continued into 1970s, 1980s, and 1990s. By mid-9190s, a number of "Agile" methods for software development, such as rapid application development (RAD), feature-driven development (FDD), extreme programming (XP), or Scrum, had gained currency. Larman and Basili (2003) provide an interesting account of the history of various methods throughout these decades.

It is, then, no surprise that in early 2001, a group of software development leaders[1] representing different methodologies such as XP, SCRUM, DSDM, adaptive software development, crystal, FDD, pragmatic programming, and other sympathetic individuals met to identify an alternative to the documentation-driven, heavyweight software development waterfall processes. The outcome of this meeting was the *Manifesto for Agile Software Development* (http://agilemanifesto.org/) as well as the *Agile Alliance* (https://www.agilealliance.org/) to support people and organizations interested in advancing this approach.

Agile Manifesto: Values and Principles

The Agile manifesto is as follows:

We are uncovering better ways of developing software by doing it and helping others do it. Through this work we have come to value:

Individuals and interactions	*over*	*processes and tools*
Working software	*over*	*comprehensive documentation*
Customer collaboration	*over*	*contract negotiation*
Responding to change	*over*	*following a plan*

That is, while there is value in the items on the right, we value the items on the left more.

It is important to consider the principles behind this manifesto, particularly as we seek to apply this methodology to hardware development and other aspects of product life cycle management (http://agilemanifesto.org/principles.html):

1. Our highest priority is to satisfy the customer through early and continuous delivery of valuable software.
2. Welcome changing requirements, even late in development. Agile processes harness change for the customer's competitive advantage.
3. Deliver working software frequently, from a couple of weeks to a couple of months, with a preference for the shorter timescale.
4. Business people and developers must work together daily throughout the project.
5. Build projects around motivated individuals. Give them the environment and support they need, and trust them to get the job done.
6. The most efficient and effective method of conveying information to and within a development team is face-to-face conversation.
7. Working software is the primary measure of progress.
8. Agile processes promote sustainable development. Sponsors, developers, and users should be able to maintain a constant pace indefinitely.
9. Continuous attention to technical excellence and good design enhances agility.
10. Simplicity – the art of maximizing the amount of work not done – is essential.
11. The best architectures, requirements, and designs emerge from self-organizing teams.
12. At regular intervals, the team reflects on how to become more effective, then tunes and adjusts its behavior accordingly.

As I examine these principles throughout the body of this work, I will demonstrate that some of these principles may need to be adopted for hardware development or other aspects of the product life cycle and should not be applied blindly. For instance, consider the principle that *working software is the primary measure of progress.* In hardware development, it may be months before the first prototypes are fabricated.

For now, let's review what the Agile process teaches in the software world.

Agile Process: Plan-Do-Review

The Agile process is based on a "Plan-Do-Review" mindset. I use the term *mindset* rather intentionally. If the right mindset is not in place, then chances of success may be rather limited. I will discuss this mindset later in this appendix, but for now, let it suffice to say that, unlike the traditional waterfall methods, in Agile, project planning is done to the extent that it can get started. This is probably the most difficult aspect of Agile for organizations to grasp, adapt to, and adopt. Let me explain further.

As shown in Figure A1, a project starts with defining a product profile. In developing this profile, the team meets to discuss the persona's (i.e., customers) overarching needs, followed by the creation of a profile of a product that would respond to these needs. This

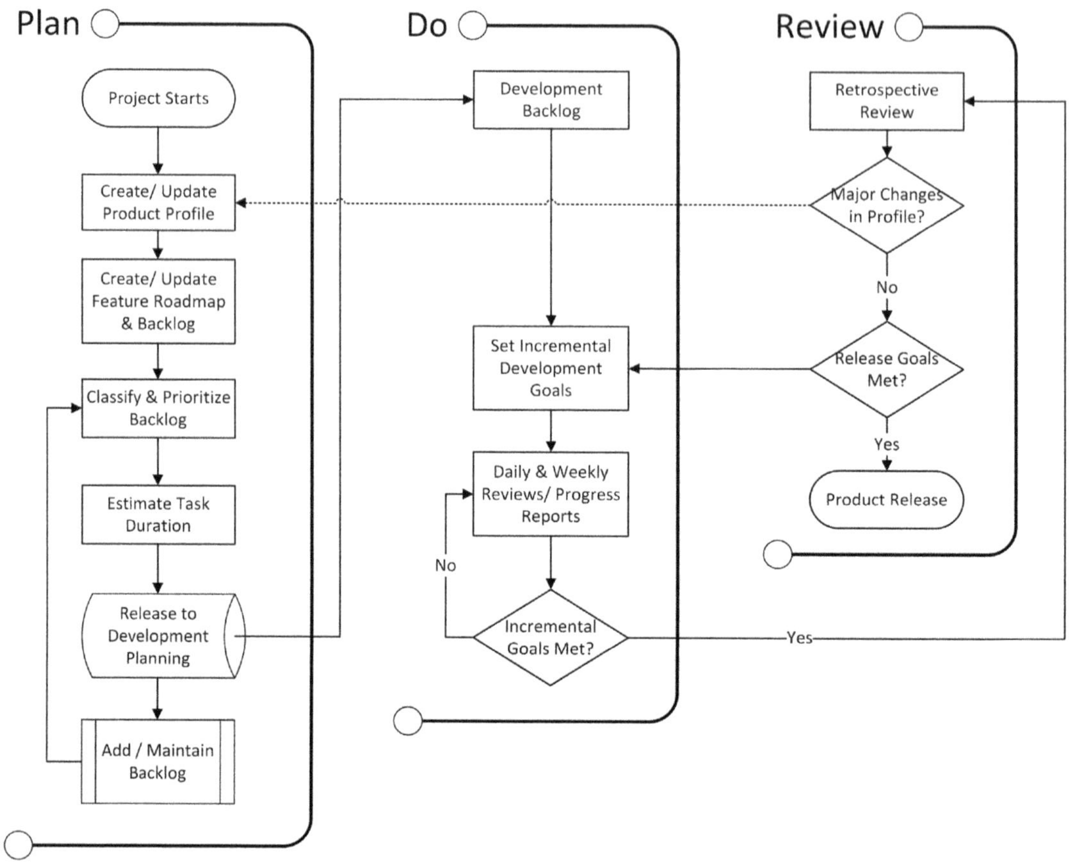

FIGURE A1
The Agile process

profile serves as a *north star*, ensuring that as future features are added to the backlog, a decision tree guides what is implemented and what is rejected.

Once a product profile is understood, the next step is to create a feature roadmap and a backlog of activities or tasks to be completed. I will discuss the roadmap in the following section in more detail. Once the first set of tasks has been identified, they need to be classified and prioritized. Classifying a task simply means to identify which element of the roadmap it belongs to. Prioritizing it means determining whether it needs to be assigned immediately or whether it can wait for a later date. Once tasks are prioritized, the duration to completion of each task is estimated. At the onset of a project, a workload/task packet is given to the development team.

Once a packet is provided to the development team, planning continues by reviewing the backlog and by adding additional features or removing them as needed. Planning continues by providing additional work packets to the development until such time that there are no more significant elements remaining in the planning backlog.

The *Do* aspect of Agile begins at the time when the development team receives the first work packet. This team will place the assigned tasks in a backlog and set incremental goals to deliver what is required. One significant element of an Agile approach is establishing daily and weekly touchpoints so that the team can discuss and review their progress and escalate any roadblocks that they may have. To facilitate a close collaboration between team members, Agile principles recommend that team members be co-located within the same office space. While it may be possible for small corporations to co-locate the team in the same area, for many large and multinational businesses whose development teams are spread across the world, co-location may not be practical. I am not sure if this presents a problem per se. If it does, a practical solution may take time to develop.

Finally comes the *Review* segment of the Agile process. Once an incremental work packet is completed, the team meets for a retrospective review. In this meeting, the team would discuss any lessons learned and any improvements to the process that may be needed. Additionally, the team entertains whether any changes to the product profile are needed. Have the customer's requirements/needs changed? This one element has been hailed as the true mark of Agile: the ability to make change while the project is ongoing. If there are major changes, the profile is updated and the planning section is resumed; otherwise, the development team continues with completing the workload in their backlog until all release goals are met and the product is released.

Roadmap and Backlog

One of the first steps in any project planning (be it Agile, V-model, or waterfall), is that a product (or feature) roadmap needs to be developed. In project management, the term *work breakdown structure* (WBS) is often used to denote this map. WBS is used to identify milestones, major deliverables, and their relationships in a project. In a waterfall (and even V-model) project management, WBS is further developed to identify additional details and is ultimately used to create project schedules and their associated Gantt charts.

In contrast, in Agile, as soon as the WBS identifies high-level tasks along with their dependencies, and it can be used to chart a pathway to the destination (i.e., it has become a *Roadmap*) work can begin by developing a backlog of activities and prioritizing what needs to be done. A detailed WBS, project schedule, and Gantt chart are not needed.

For instance, Figure A2 depicts a simplistic (but not necessarily accurate) roadmap for embedded software operating an infusion pump. At the top of the map lie the following four elements: user interface, drug library, control and feedback and safety and cyber security. This map may be followed to identify high level tasks and deliverables to the desired outcome – which is to display infusion data and pump condition, presumably on a screen.

A backlog may simply be developed by placing the elements of a roadmap on a list – as shown in Figure A3. Figure A3 also depicts the next element in the *Plan* section of Agile, which is to estimate the timeframe needed to complete a given task. Before communicating this backlog to the development team, it needs to be prioritized.

Note that planning activities do not stop here. For software development, other elements and features may be added to the roadmap or the backlog. In the above-mentioned example, at some point, voice activation or operation may become desirable. This new feature may be added to the backlog and prioritized at any point. Additionally, it may be incorporated either in the existing release or in future releases, depending on priorities and release dates.

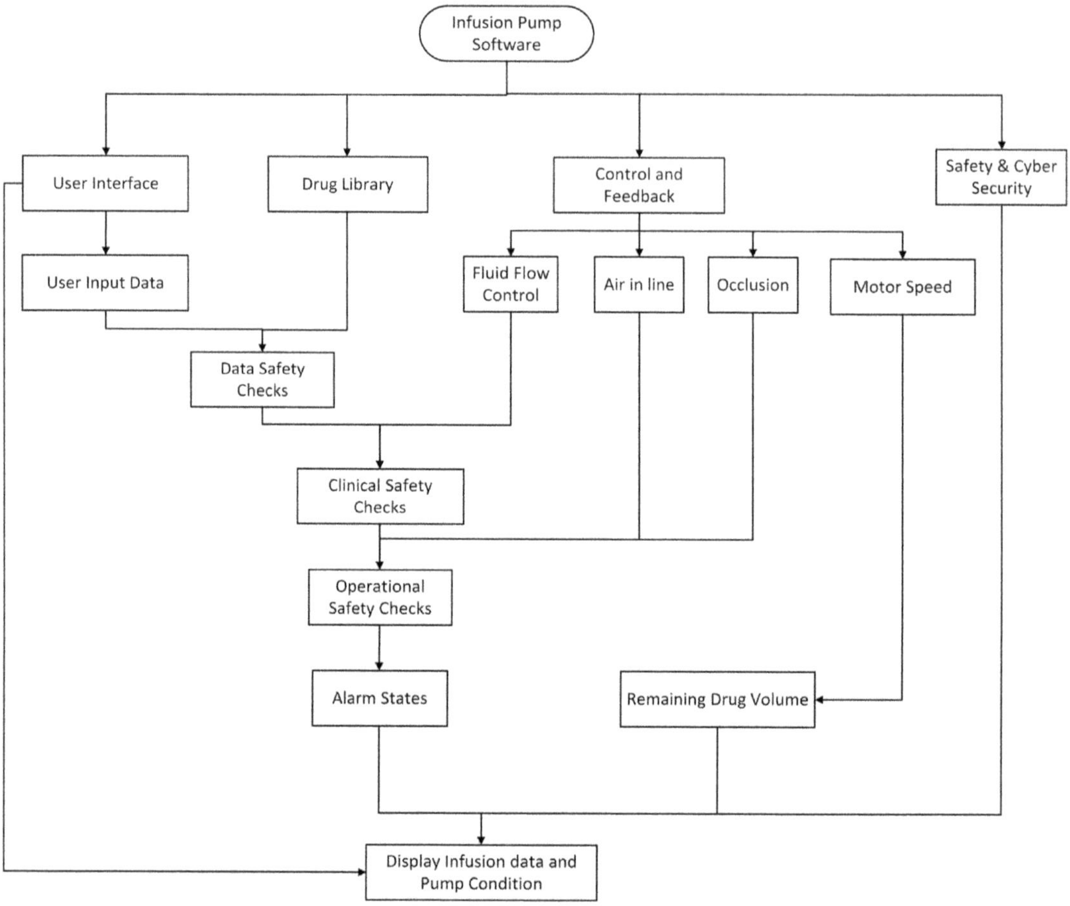

FIGURE A2
An example of a roadmap

Agile Roles and Responsibilities

Agile roles and responsibilities are very similar – if not the same – as any other product development models:

- Product Owner: This person understands the market and reflects the voice of customers and business. A primary responsibility for this role is to set development priorities.
- Project Manager: The responsibilities of this role include overall project ownership, creation of reports and metrics, and being a liaison between the development team and stakeholders to escalate roadblocks. Project managers often facilitate various Agile meetings as well. Within the context of Scrum, this role is called Scrum master or project facilitator.
- Development Team: This is the group of individuals who get the work done.

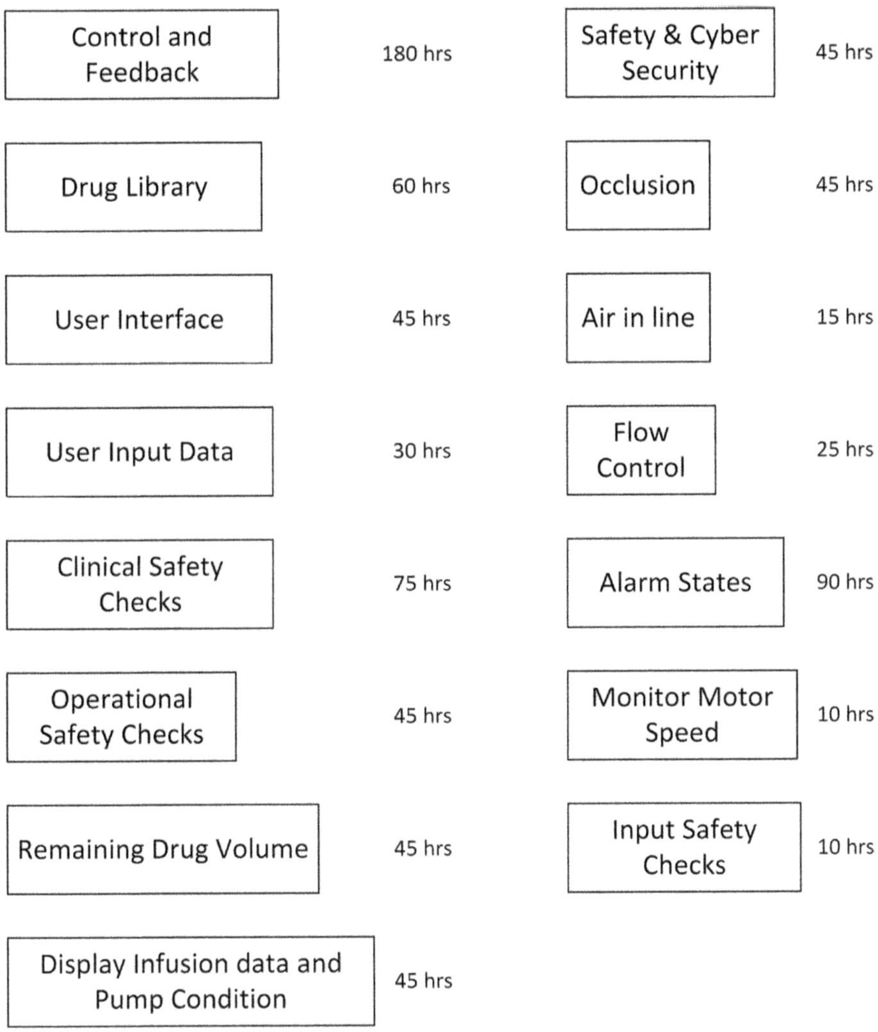

FIGURE A3
An example of a backlog

In addition to these roles, some include stakeholders and Agile mentors. Similarly, stakeholders are often members of executive management with the authority and ability to resolve roadblocks for the team and/or make high-level decisions. An Agile mentor is someone with experience who can provide guidance for the team.

Scrum and Kanban

In a way, the Agile mindset dictates that development activities begin as soon as possible. This would mean entering the *Do* phase. While there is no universally accepted process for completing tasks, most people think of Scrum (to a large extent) and Kanban (to a lesser extent) when they think of Agile. A full review of Scrum or Kanban is beyond the scope of this work. Interested readers may consider Drumond (2022), who has provided a succinct overview. Another substantial reference has been provided by Rubin (2013). For Kanban, see Brechner (2015).

Scrum

Scrum is a framework that values continuous learning. Its foundation in Agile is to take small steps and at each step communicate with stakeholders and customers whether their needs have been addressed.

Figure A4 provides a depiction of the Scrum process. Scrum not only depends on a strong *development team* but also on a *product owner* and a project manager (often called a *Scrum master*). The responsibility of the Scrum master – first and foremost – is to provide training and mentoring for the entire organization and to ensure that the team understands the Agile mindset and optimizes the process for the work at hand. While the Scrum master's focus is on the team deliverables and the so-called burndown rate, the product owner focuses on the overall product backlog and reflecting both business and customer needs by constantly updating and optimizing this list. In a way, this role is to be a liaison between customers, business, and the development team. It is also the product owner who decides when to release the product to the customer and initiate the next development cycle.

The development team is the group of individuals who get the work done. They start by creating a sprint backlog from the product roadmap and product backlog. Then, they plan each sprint for what can be realistically delivered in a given sprint duration. In software development, sprint duration is typically taken to be two weeks but it can vary from a week to several weeks. In the example shown in Figure A5, sprint duration is set at four weeks.

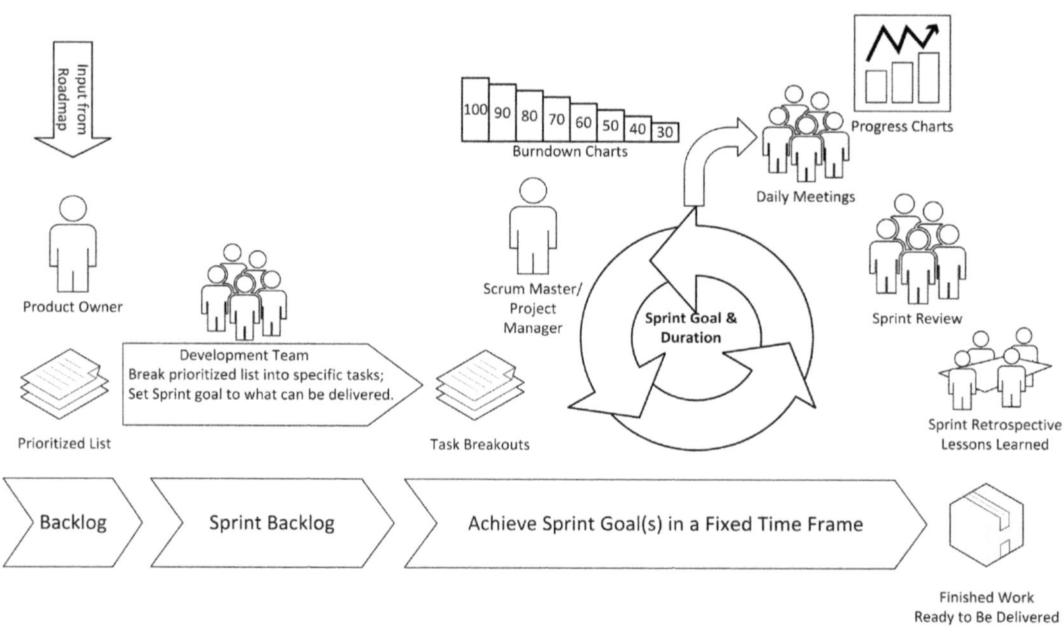

FIGURE A4
An overview of Scrum

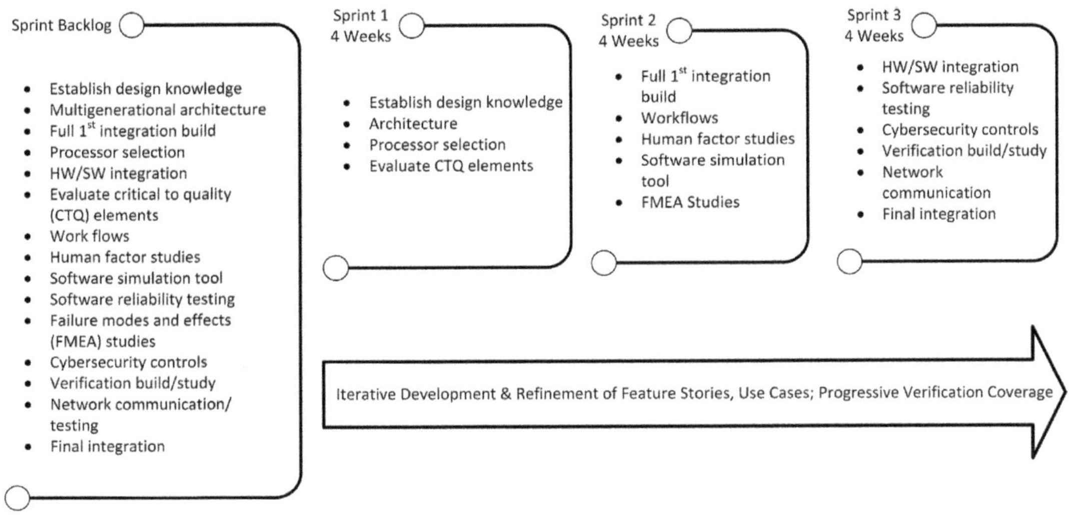

FIGURE A5
An example of sprint backlog and planning

I mentioned that the development team plans the sprints based on the product backlog. Having said this, I like to note that this plan is directional in nature. Since the product backlog may change, so does the sprint backlog, potentially impacting all upcoming sprints (but not the current one).

It has been recommended (Drumond 2022) that team sizes remain small (between 5 and 9 members) and co-located to develop a certain level of camaraderie. Also, team members have differing backgrounds but are cross-trained. With this team makeup, anyone may step up and help another team member to ensure on-time delivery. While in software development this may be feasible, this would be a tall order to fill for hardware development (including combination products) with a group of truly cross-functional teams from a variety of fields and regions.

Kanban

Kanban was implemented at Toyota in the 1940s as a tool for just-in-time production. It is based on a *pull* production mindset. It begins with the customer placing an order. The order would pull finished goods from the inventory. This in turn signals production to manufacture units to replace the sold items in inventory. To assemble units, subassemblies are pulled, which in turn will need modules and eventually components that need to be brought in from the warehouse. Once warehouse levels fall below a certain limit, a signal is sent to suppliers and more raw material is ordered. It is almost as if a ripple moves through the manufacturing facility from inventory back to the warehouse. In its early days at Toyota, the need for more parts was communicated by passing a card from one team to another. In Japanese, this *card* (that) *everyone can see* is called Kanban. Some people have translated Kanban as *a signboard* or *visual signal.*

Regardless of how to translate this word, in today's world, it is taken to be a method to visualize work packets and an optimized workflow. In fact, Kanban works if workflow process is standardized. A project begins by planning and determining what tasks need to be completed. These tasks are written on separate cards and placed on a Kanban board in the *Do* column. Then a team member selects a workpacket and places it in the *In-Progress* column. Once completed, the member who completed the task will move the card from In-Progress to the *Done* column. An example (based on the backlog in Figure A3) is given in Figure A6.

I like to make two points here. First, the *Do, In-Progress, Done* columns are only a suggestion. The number and the names of columns can be modified to reflect the needs of the team. An alternative is shown in Figure A7. In my own work, I have used as many as seven columns to reflect the work in progress.

The second point that I would like make is to draw the contrast between Kanban and Scrum. In Scrum, focus is placed on delivering work packets within a given time period called a sprint. Hence, the project team selects only tasks that may be delivered within that time period. Larger tasks are divided into suitable subtasks. In Kanban, focus is placed on work in progress and cycle time as opposed to finishing work in an artificial preset duration. Through an inspection of a properly set board, bottlenecks and roadblocks are readily identified and resolved. Therefore, full transparency and open and full communication is paramount to the success of project completion.

Roles and responsibilities in Kanban are similar to Scrum. The product owner sets the priorities of the "To Do" list, the development team does the work, and the project manager tracks progress and reports the results to the rest of the enterprise. For more information on Kanban, see Anderson (2010).

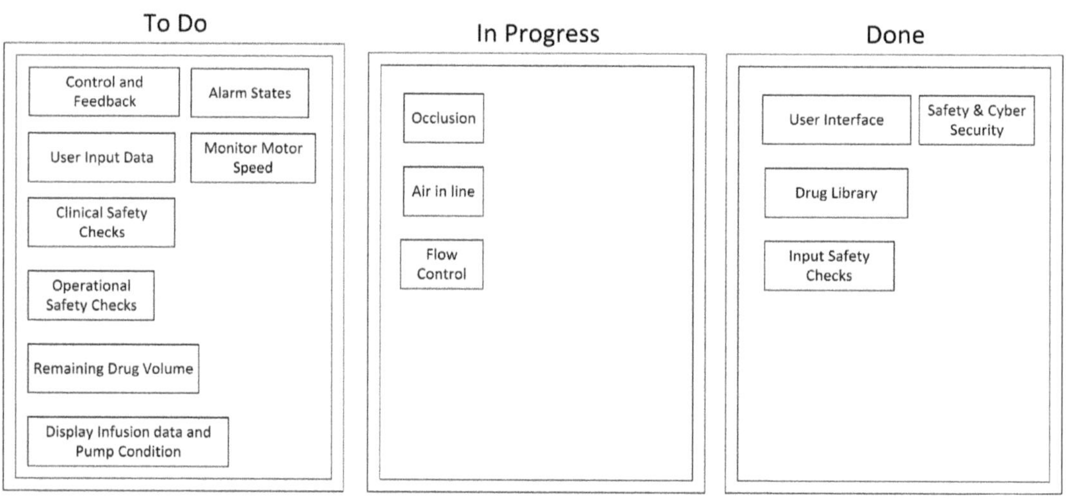

FIGURE A6
An example of a Kanban board

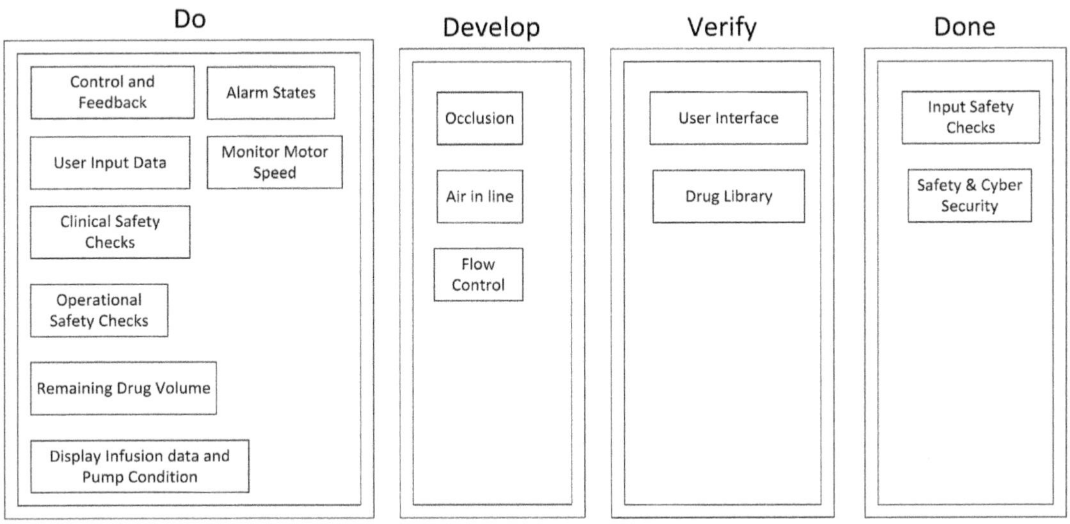

FIGURE A7
An example of an alternative Kanban board

Agile Project Management

Regardless of which Agile approach is selected, an Agile project manager has to manage seven phases of the project. These phases are depicted in Figure A8 within the context of the V-model:

1) Vision: The project manager leads activities to define a new product to be developed in line with market needs and the enterprise strategy.
2) Product Roadmap: The project manager works with the product owner and management to define feature sets to be released.
3) Release Planning: Based on market and customer needs, the enterprise needs to deliver products to remain viable. Traditionally, this meant a major release with many new or updated features. Agile mindset suggests more frequent releases but in shorter intervals. Again, it is the role of the project manager to plan these frequent releases with the input of the product owner.
4) Development Planning: The next phase is to plan features that will make up the next release. The project manager has oversight, but this is an activity owned by the product owner with the input of the development team.
5) Daily Meetings: The purpose of these meetings is to review roadblocks. The project manager escalates issues that cannot be solved by the team to stakeholders and works to resolve them.
6) Team Reviews: Some may refer to these meetings as design reviews. The main purpose of these meetings is to take a critical look at the product (software within the context of this chapter) to ensure that design risks associated with the current development cycle have been either retired or mitigated. The project manager facilitates this meeting.

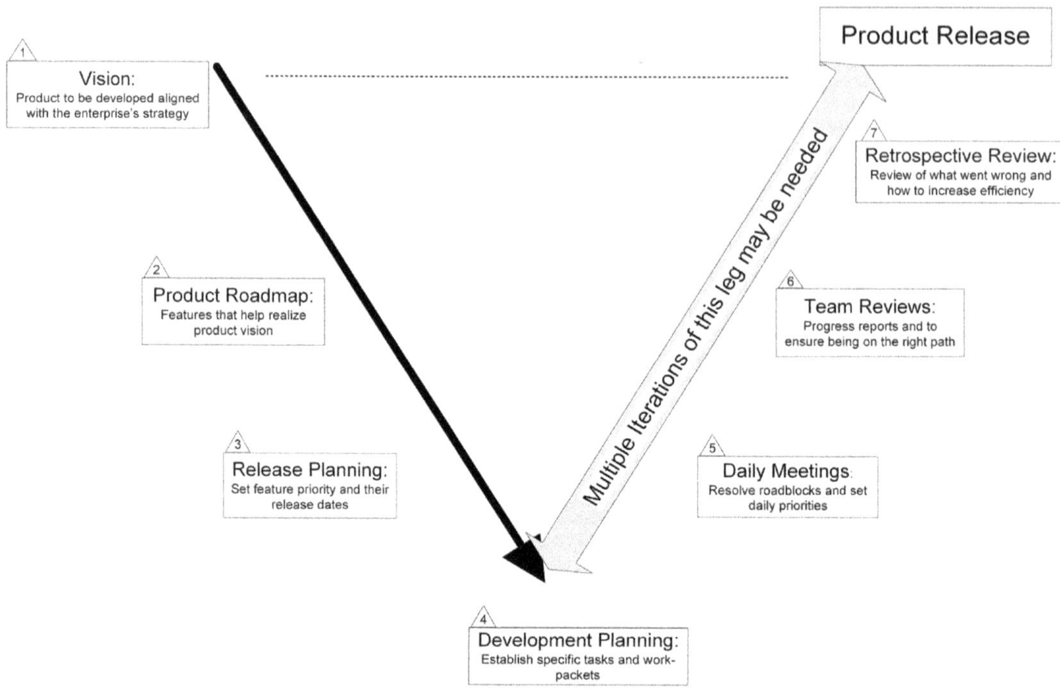

FIGURE A8
Agile project management phases depicted on a V-model

(Adapted from Lyton, M.C., and Ostermiller, S.J., *Agile Project Management*, 2nd Ed., John Wiley and Sons, NJ, 2017)

7) Retrospective Reviews: Once a development cycle (e.g., sprint within Scrum) is completed, the project manager facilitates a review meeting to examine the mechanics of what was done and how improvements may be made. Once this review is done, the next development cycle may begin.

Reported Metrics

In addition to typical metrics such as cost and progress reported by an Agile project manager, there are metrics that are unique to Agile. The following are the most common:

1) Burndown or burnup charts. Recall that at the onset of a project, a roadmap and backlog are created. Whether a product is developed using Scrum or Kanban, it may not be released into the market unless identified priority features have been completed. By tracking completion of these work packets against time, a burndown or burnup graph may be developed. This metric provides a visual and possibly early indicator of whether a project may be completed in time or not.
2) Team Velocity: This metric measures how quickly the team completes their tasks within a development cycle. It is generally used within Scrum because the sprint duration is fixed. Within Kanban, the metrics to report are work in progress and takes time, that is, the time it takes to finish one work packet and starting the next one.

Agile Leadership

The eleventh Agile principle suggests that *the best architectures, requirements, and designs emerge from self-organizing teams*. So, who is the Agile leader if the team is self-organizing? In theory, every team member is a leader. However, in the right environment, someone within the team naturally begins to act as the leader, coordinating activities with the rest.

It is important to make a distinction between leading and managing. Often, the two terms are intermixed, but these include two different sets of activities. While a full review of this topic is beyond the scope of this work, a brief overview may be beneficial. Kotter (2001) defines the traits of a leader as someone who aligns, motivates, and inspires people and sets and gives them direction for action. As a minimum, a leader owns his work packet and acts to deliver. They do not wait for or need someone else to tell them what to do. A manager, as defined by Kotter (2001) plans, budgets, organizes and staffs, provides solutions to problems, and exerts a degree of control.

In that sense, there needs to be a synergy between team leaders and enterprise management to foster project execution that exceeds any expectations. In that light, McKinsey & Company has published an article that speaks of five Trademarks of Agile Organizations (Aghina et al. 2001). These marks are:

1) A strategy of shared purpose and vision across the organization. This enables seizing opportunities by having flexible resource allocations.
2) A network and a structure of empowered teams who have clear and accountable roles with hands-on governance.
3) A process for rapid and action-oriented decision-making, standardized work, information transparency, and continuous learning.
4) Dynamic people with role mobility who have a servant leadership mindset.
5) A mindset to evolve technology architecture, systems, and tools to develop the next generation of technology.

In short, Agile leadership should exist not only at the team level but also throughout an organization for this methodology to take root and succeed.

Corporate Culture and Agile

Often, Agile is thought of as the approach that will shorten product development cycle time. Yet in practice, many organizations are puzzled when, in practice, Agile does nothing to reduce timelines. The root of this failure may be found in an incomplete understanding of the Agile mindset as well as the corporate culture needed for it to flourish.

Sahota (2012) points to this issue: "As word gets out about Agile, it follows a common pattern observed with many technological adoptions where there is hype and disillusionment" (Sahota 2012, p. 4). He continues by suggesting that the number one barrier to adopting Agile is, in fact, a needed culture change. I will come back to this point shortly, but for now let me elaborate a bit.

Consider, for instance, the Agile value that *individuals and interactions* have precedence over *processes and tools.* For many organizations today with their standard operating procedures, as well as decision-making hierarchies, it is extremely difficult – if not impossible – to give a certain degree of control to a small team – or a group of small teams for larger projects – for fear of a myriad of "what if they don't / do . . ." situations and scenarios.

Or, the Agile value of *customer collaboration* having priority over *contract negotiation.* A typical response from sales or marketing teams maybe, "do you mean that design and technical teams meet directly with customers?. . . are you mad?"

I agree with Sahota (2012) that to be Agile, there needs to be a fundamental shift in corporate thinking and culture. Before discussing this shift, let's first define what we mean by culture. Broadly speaking, a company's culture may be defined as the sum total of activities, approaches, and ways of conducting day-to-day business in an attempt to succeed in the market. A company's culture reveals to us what is important in how people interact with each other as well as how they get their work done.

Figure A9 provides a culture model proposed by Schneider (1999). This model places corporate cultures against two axes: a vertical axis defining the spectrum from heeding conceptual possibilities on the one end and focusing on reality (what is actual) on the other end; and a horizontal axis from focusing on people on the one end to focusing on the corporation and enterprise on the other end. This creates four quadrants that define (or place cultures) depending on their relative location on these horizontal and vertical axes:

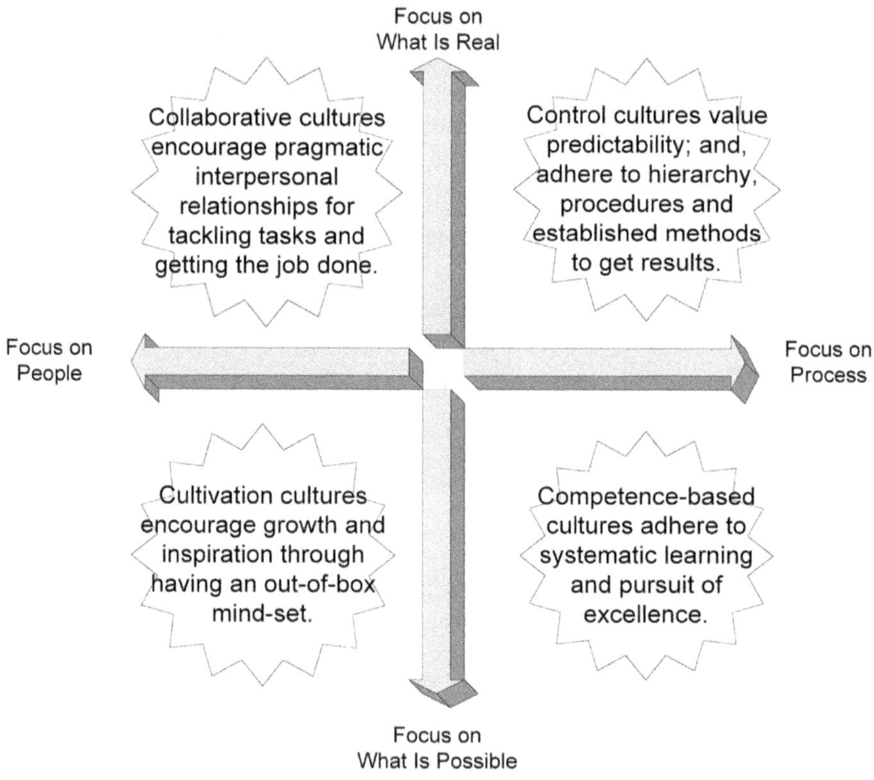

FIGURE A9
A culture model as proposed by and adopted from Schneider (1999).

Schneider, W.E., *The Reengineering Alternative: A Plan for Making Your Current Culture Work*, McGraw Hill, NY, 1994.

1) An organization is considered as having a *control* culture (first quadrant) if their focus is on "reality" and "enterprise." These organizations have typically strict work instructions and standard operating procedures. Additionally, they are hierarchical, and decisions need to be made at higher levels.
2) The second quadrant is defined by an organization's focus on its people and their well-being as well as the realities of the marketplace. This organization gets its goals achieved via a *collaborative* culture.
3) Should an organization be driven by a mission of encouraging its members/employees to grow as the organization achieves its goals, then its culture is defined as *Cultivation* – third quadrant.
4) Finally, the fourth quadrant belongs to an organization that strives to be the most competent in their field by allowing its associates to be out-of-the-box thinkers yet strive to achieve organizational goals. They are said to have a *competence* culture.

Needless to say, most companies exhibit a dominant culture as indicated above but do not deny the presence of the other three culture quadrants or subcultures as long as they serve the dominant culture. In this model, Schneider suggests that cultures that have one axis in common may be compatible but incompatible if no common axis exists. For instance, collaboration and competence culture are opposites, whereas collaboration and control are likeminded.

Sahota (2012) has come to conclude that Agile culture is about collaboration and cultivation. In other words, the focus should be on people and team members. A thoughtful review of Agile values and principles affirms this conclusion. It is important to consider this point as an organization moves to transform their mindset into Agile. It may need to happen in stages, possibly by redefining the agile values and principles as needed or adjusted to their own organization.

Note

1 A group of 17 in total. Go to http://agilemanifesto.org/ for a list of their names.

Appendix B. A Case Study for Employing Agile Methodology

Introduction

The primary focus of this book has been to provide an organizational and a systems structure needed to drive a common and consistent set of outputs via best practices. Additionally, I have introduced the element of Agile methodology and its application to electromechanical and/or combination products and demonstrated how it can be integrated into a systems approach to product development. In this section, I will provide a case study where a team of design for reliability engineers applied Agile to first grow the reliability of a medical device and then to demonstrate and verify system reliability requirements.

Nuances of Hardware Development

Recall the agile manifesto provided in Appendix A. It began by suggesting that the goal of agile is to uncover "better ways of developing software by doing it and helping others do it." Then, the manifesto continued by providing what is known as Agile Values. The second and third values are that *working software* and *customer collaboration* are preferred over *comprehensive documentation* and *contract negotiation*, respectively. Associated with these values, there is a set of principles. The first three of the principles are closely related to these two values:

1. Our highest priority is to satisfy the customer through early and continuous delivery of valuable software.
2. Welcome changing requirements, even late in development. Agile processes harness change for the customer's competitive advantage.
3. Deliver working software frequently, from a couple of weeks to a couple of months, with a preference for the shorter timescale.

Once we contemplate the intricacies of these mentioned values and principles, we will conclude that these may not be directly transferable to hardware development. It required adaptation.

Hardware versus Software Development

At times, I have found it amusing that in response to my question, *what does R&D produce*, my junior colleagues often respond with a *product of course*. While an original equipment manufacturer (OEM) delivers a product to its customers, unlike software, the output of an OEM's R&D activities is comprehensive documentation and specifications ready for the transfer of the paper design into a physical entity by the OEM's operations and manufacturing teams. At its face value, what has been delivered is not a working unit, nor has it been an external entity – as suggested or expected by Agile's second and third values.

Let's consider a second scenario. Suppose that a contract design house is developing a turn-key solution for a single client. In this scenario, as expressed by Agile, we can develop a greater level of collaboration and communication between the customer and the design house. But the product to be delivered to the customer may not be a working unit – rather design deliverables such as drawings and specification.

Another software agile element that is significantly different from hardware is the issue of *changing requirements even late in development*. While, in theory, this may make sense, in practice, each change requires significant impact assessment. For instance, while it may be rather simple to change the color of a housing of a device, it can be significantly more difficult to reduce its weight or add an additional sensor simply because a customer requirement has changed.

With these examples in mind, I am of the opinion that what was developed in the early 2000s as Agile values and principles for software development may not be readily applicable to hardware development. Each organization needs to develop its own set of agile values and principles by identifying who their customers are – either internal or external; what constitutes deliverables; and to what degree changes in requirements may be allowed or incorporated. All with their ways of getting work-done, that is, culture in mind.

Team Makeup, Scrum, and Integration

Another element impacting successful deployment of Agile is team formation and dynamic. In Appendix A, I mentioned that team sizes should remain between five and nine members and be co-located to develop close working relationships. I had also added that (in the context of software development) team members should have different backgrounds but be cross-trained so that anyone may step up and help another team member to ensure on-time delivery. This is neither feasible nor practical in hardware development (including combination products) when the team is composed of truly cross-functional individuals from various fields and regions.

A practical approach may be to form teams around various design elements. For instance, a team to focus on sensors, another to focus on enclosure design, a third on power train, and so on. This would mean that a number of teams and meta-teams need to function simultaneously – each with its own set of backlogs and product owners. Depending on the complexity of the product under development, there needs to be an overarching mechanism to ensure that the various subsystems are being integrated at the right junctions in time and that system development is on track.

Figure B1 provides a generic roadmap for hardware development. Depending on the product and the team makeup, this figure proposes four sprints to complete the work. While this roadmap is generic to a large extent, it will help define the overall backlog and scope the work that needs to be accomplished. Figure B2 provides a view of a plan on how the backlog may be distributed within various iterations and integration phases. This is a critical part of planning and a potential pitfall for those who are less familiar with the agile mindset. At this time, there is a tendency for experienced "water fall" project managers to create Gantt charts for the entire project and beat the drum to get work done. And agile is all but forgotten in name. Should this pitfall be avoided, a meta-team of development team leads may be formed as shown in Figure B3. As I mentioned, in this overall integration Scrum, the development team members are, in reality, the team leads from the smaller and more focused Scrum teams. The purpose of this team is to ensure clear communication between various sub-teams.

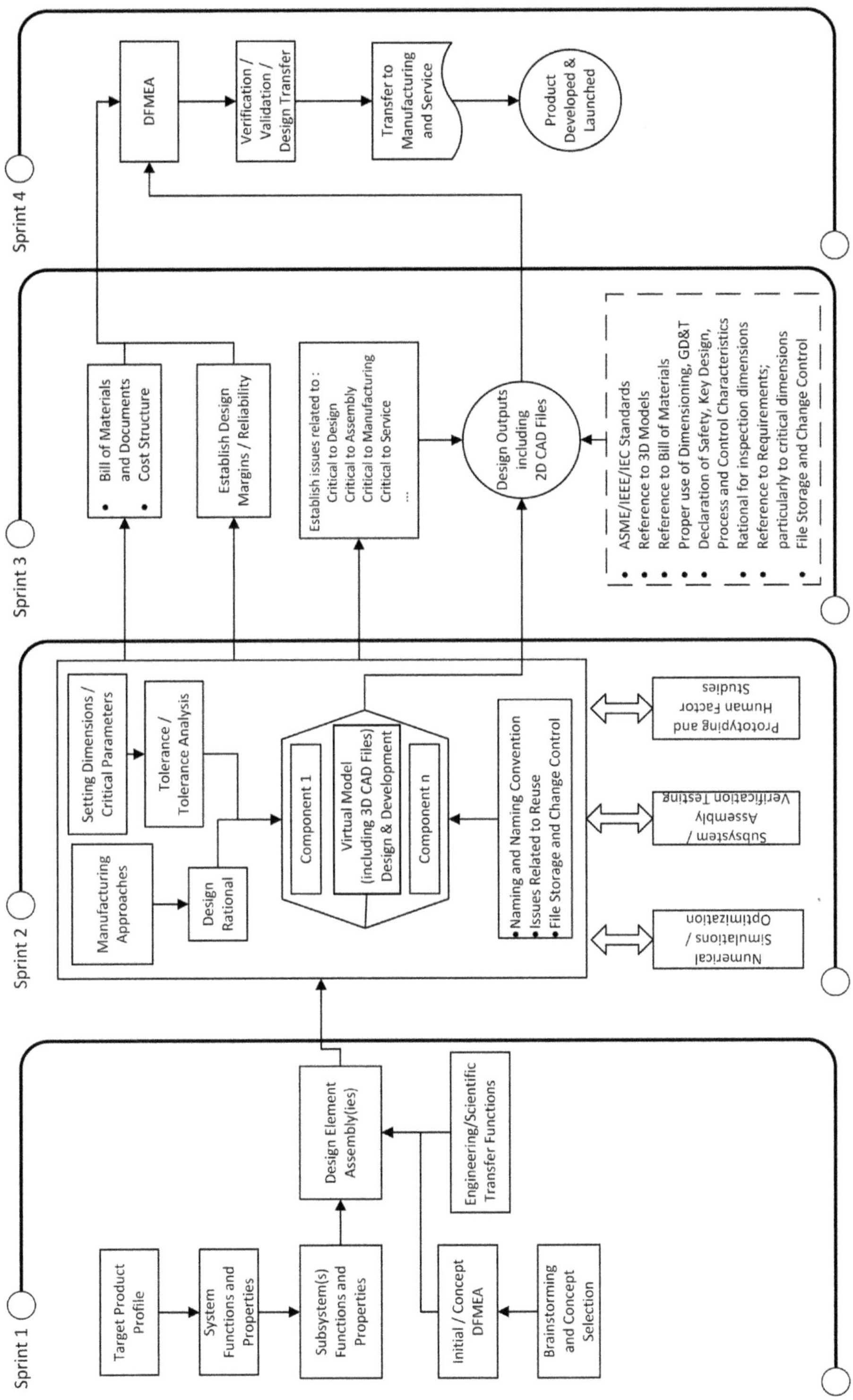

FIGURE B1
Generic hardware roadmap. This is the roadmap that has been discussed extensively in the book

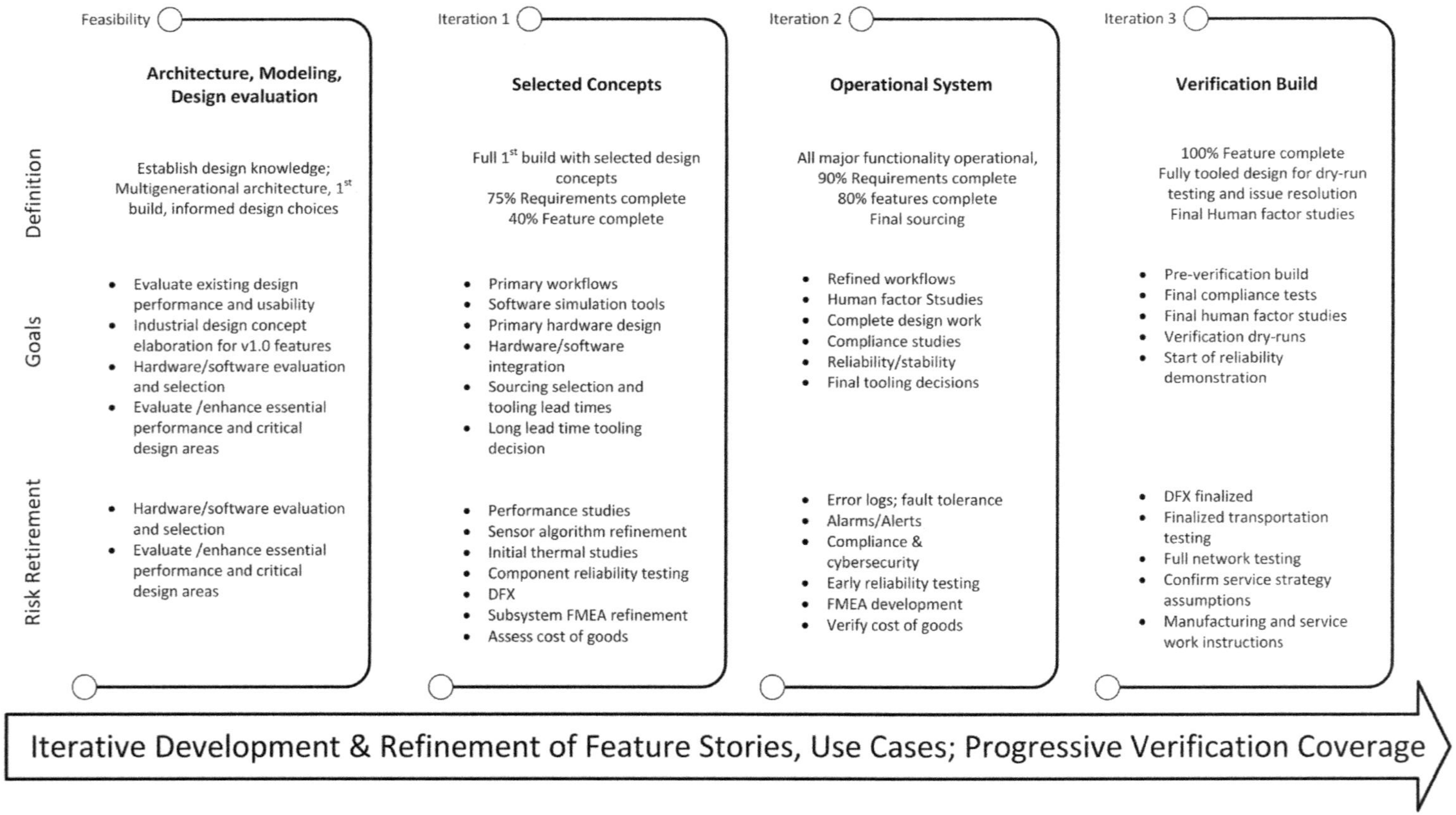

FIGURE B2
Integrated system Scrum. It is important to spend time to develop an understanding of the details of the backlog and how various subsystems and module developments need to merge

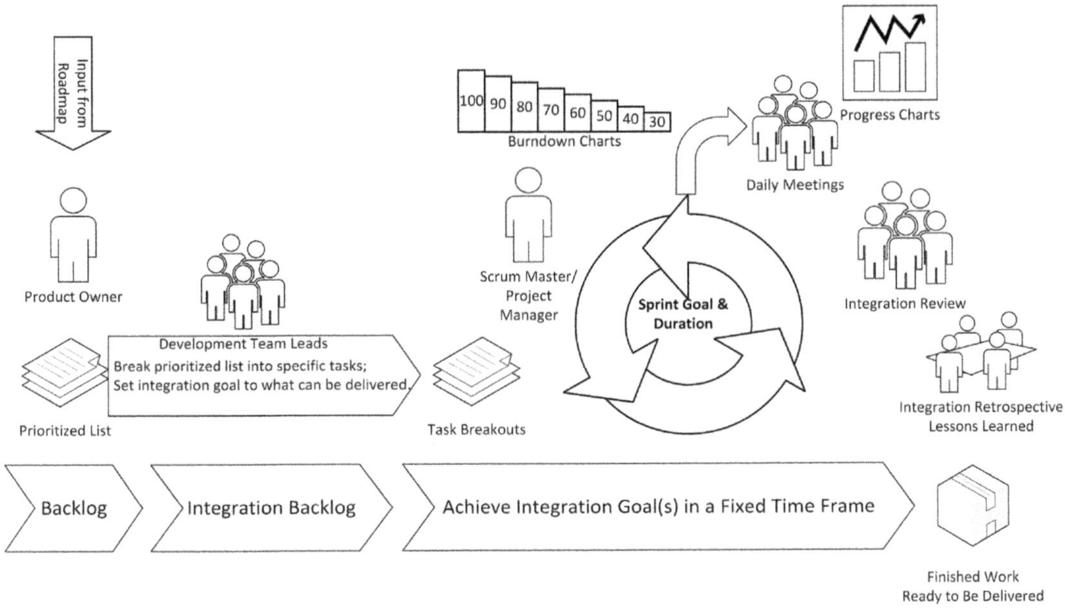

FIGURE B3
A generic overall integration Scrum – here subsystem and module leads participate to maintain the conversation at a relatively high level

Reliability Testing Using Agile Mindset

On the basis of the strategy suggested earlier, the overall integration team was formed with its sub-teams categorized, including the reliability. The reliability team consisted of eight members: a reliability architect, a reliability lead, a reliability analyst, two test engineers, and three test technicians. As shown in Figure B4, the architect had the responsibility of creating the backlog as well as setting the overall direction for the team. The reliability lead had the responsibility of setting day-to-day activities and expectations and holding daily meetings. On a weekly basis, both the architect and the lead met with the greater integration team to share issues or roadblocks as well as get feedback impacting reliability.

Once the team had formed and responsibilities were identified, the architect and the lead set up a series of technical workshops to ensure that a certain degree of cross-training was in place in case a team member became unavailable either short- or long-term.

Figure B5 depicts a design for reliability (DFR) backlog of activities. This backlog list should be based on the overall design and development for the product, which in turn is based on the business plan and business case for this undertaking.

It is important to clarify to the extent possible needed inputs, deliverables, resources, and finally estimated tasks for each deliverable. It is only at this time that a sprint may be properly planned, scoped, and its duration set.

A noteworthy element is that reliability has strong dependencies on available samples for testing. While DFR activities may have their own set of sprints, the need for samples

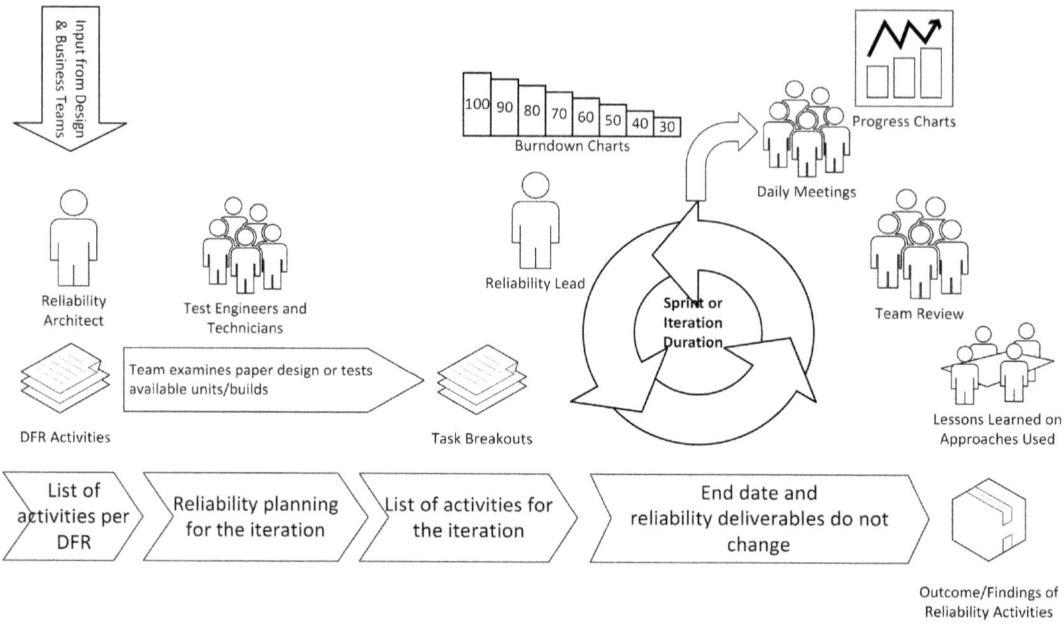

FIGURE B4
A design for reliability Scrum

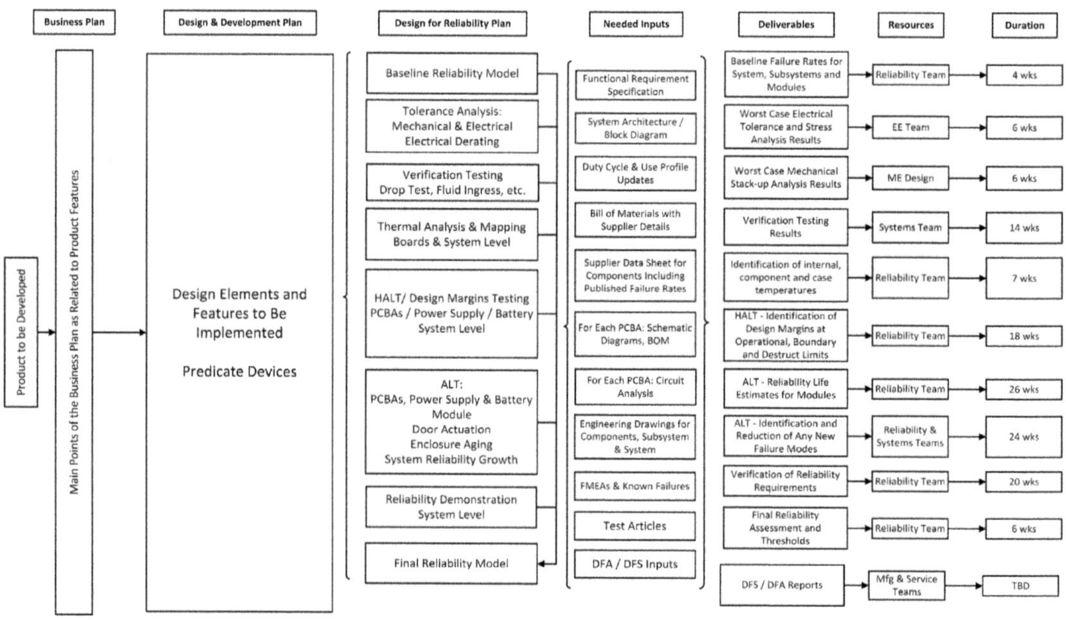

FIGURE B5
A DFR backlog – note that this backlog includes completion of activities by external resources. It is important that this need is clearly communicated and planned for in sprint activities of other teams

and their number must be communicated early for the overall product planning. Within this context, an interdependence of reliability sprint planning with other design teams exists. This will require careful consideration and clear communication. Figure B6 provides this sample request and planning for the first three iterations of this project.

Hand-in-hand with planning and requests for samples comes developing a vision for what tests will be conducted later in the program in order to verify requirements. Figure B7 provides a summary of when ALT, HALT, reliability demonstration testing (RDT) may be conducted. As the product development nears its final months and weeks, the reliability team will plan for highly accelerated stress testing as well as highly accelerated stress screening in order to identify and resolve any preproduction and production quality issues. For a more in-depth discussion of reliability, see Jamnia and Atua (2020).

Figure B8 provides an overview for the duration of some of the reliability tests. This is a reality check, which I will discuss in the lessons learned section below. To believe that hardware development activities may be fitted into two to four week cycles – as in software – is, in my opinion – foolhardy and potentially leads to disappointments and possibly letting go of the agile approach all together.

Iteration 1

Test \ Test Article	Sub Assy	
Life Testing	3	Sample Size

Iteration 2

Test \ Test Article	Clutch Assy	Drive Train	Loading Mech	Door Mech	PCBA 1	PCBA 2	Sensor Assy	Power Adaptor	
Life Testing	5	5	5	5	5	5	5		Sample Size
HALT / Thermal Cycling								3	

Iteration 3

Test \ Test Article	Mechanism 1	Mechanism 2	PCBA 3	PCBA 4	PCBA 5	Sensors & PCBA 1	Sensors & PCBA 2	Clutch Assy	Drive Train	Loading Mech	Door Mech	Motor and Drive Train	Power Adaptor	Display Testing	
Life Testing	5	5						5	5	5	5	5	5	5	Sample Size
HALT / Thermal Cycling			6	6	6	6	6								

FIGURE B6
A design for a reliability integration plan – requested samples for life testing as well as for HALT and thermal testing – is shown

Integration 4			Integration 5			Integration 6		
Test	UUT	Output	Test	UUT	Output	Test	UUT	Output
ALT	System	Failure Mode Identification Early Life Expectancy Estimated	Mini RDT	Sysetm	Verification Readiness	RDT	System	Requirements Verification Failure Rate and Life Expectancy
			ALT	PCBAs	Failure Rate	ALT	PCBAs	Failure Rate
	Power Adapters	Failure Mode Identification Early Life Expectancy Estimated		Power Adapters	Failure Rate Life Expectancy		Power Adapters	Requirements Verification
	Door Actuations	Failure Mode Identification Early Life Expectancy Estimated		Door Actuations	Life Expectancy		Door Actuations	Requirements Verification
HALT	System	Failure Modes Operational and destruct limits	HAST	System	Pre-Production issues	HASS	System	Production Quality Issues
	PCBAs	Failure Modes Operational and destruct limits		PCBAs	Pre-Production issues		PCBAs	Production Quality Issues

FIGURE B7
A design for reliability test breakdown

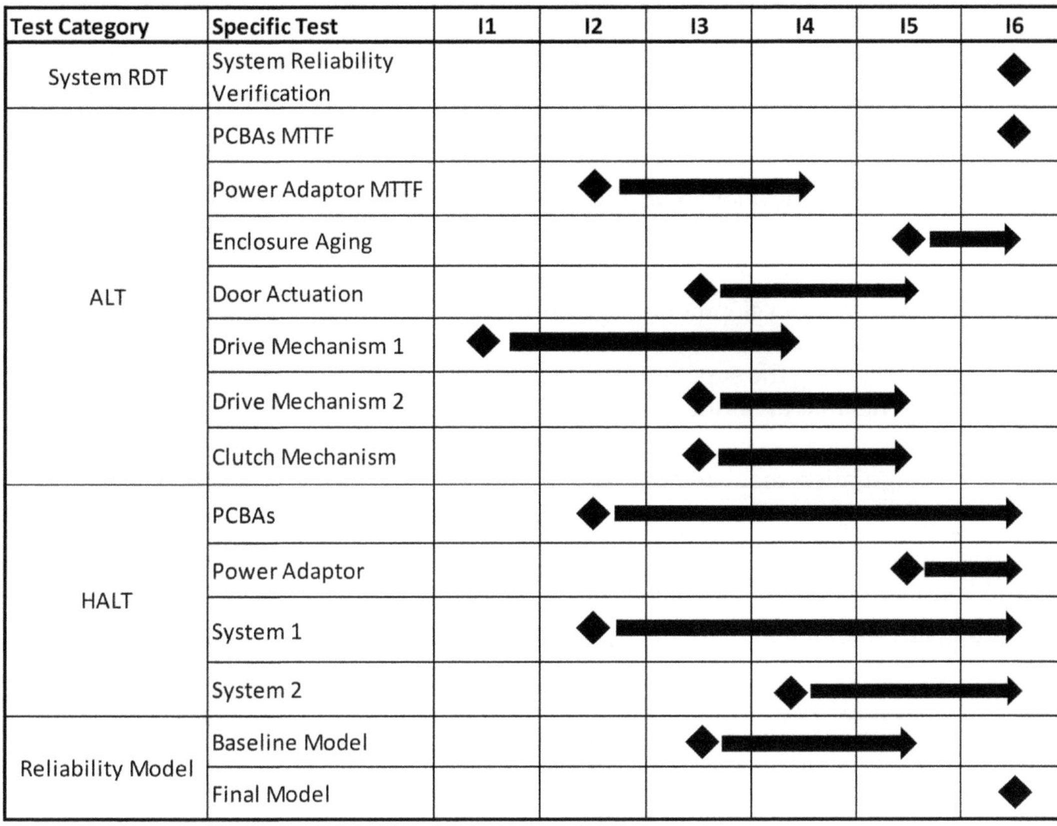

FIGURE B8
A design for reliability test categories across integration activities

Lessons Learned

The reliability team, in collaboration with the rest of the design teams, completed their tasks successfully within 20 months – a duration that under traditional circumstances might have taken as long as 36 to 40 months. There were some lessons learned that are worth mentioning here.

Now let's talk about the Agile values. What does this mean from a hardware point of view:

1. In software development, the emphasis is on having the shortest sprint duration possible.
   a. In hardware development, this should apply to paper design alone. The supplier and other lead times must be taken into account when the overall schedules are developed.
   b. This is also true of any in-house or outsourced reliability testing.

2. In software development, the finished work is the software.
   a. In hardware development, the finished work does not need to be an engineering build to begin with. In the first iteration, the finished work should be the paper design.
3. In software development, test-fix-redesign is emphasized.
   a. In hardware development, the initial testing of the paper design should be done through numerical simulations and statistical tolerance analysis.
4. In software development, documentation is de-emphasized.
   a. In hardware development, engineering drawings and technical specifications should be well defined before the first engineering build to ensure that everyone, including the suppliers, has a clarity of vision on what is being fabricated. Elements that are critical to quality need to be clearly identified prior to each build.
   b. Configuration management is paramount to the integrity of units tested and the correlation of results to the actual build.

In summary, I may add the following:

- Agile is a methodology that is capable of speeding product development time to market.
- A major pitfall is to misunderstand the purpose and the mindset of Agile methodology, which is to allow the teams to drive the deliverables and not the schedule.
  - It is important to develop your own set of Agile values and principles based on the company culture and what may or may not be acceptable practices.
- At times, the new product development team needs to slow down before they can take advantage of speeding up.
- It is essential to understand the overall product design backlog and how it would fit within various iterations.

Appendix C. Reagent Requirement Decomposition Example

In Chapter 12, I provided the requirements for the decomposition process; however, the example that was provided had an electromechanical bias. For completeness sake, I would like to share how to decompose the requirements for Schiff reagent – a chemical reagent that was illustrated in Chapter 6. Schiff reagent is used in the field of histology to color biological samples used in pathology.

At the product level, there are two requirements that impact the Schiff reagent design. The first – and the obvious – is the requirement that the product should provide a high-contrast coloring of biological tissues. The second – and more subtle – requirement is that the product has a size limitation; and a third, which is not mentioned here, is related to the material of its container and its opacity to filter specific frequencies of light in order to prolong its shelf life. In this example, I will decompose the performance and chemical component requirements in Tables C.1.

TABLE C.1

An Example of a Reagent Requirements Decomposition

Requirement	Description		
Product	The Schiff reagent shall give a high contrast to biological tissue samples and be packaged in 1-liter bottles.		
System	The reagent shall produce a strong, high-contrast magenta reaction product with aldehyde groups in fixed tissue with minimal nonspecific background staining.		
Performance Requirement	The reagent shall give a high contrast with the following parameters		
	Contrast	5:1	Optical Density Ratio
	Specificity	≤ 20% of positive OD	
	Achieve full reaction	≤ 10 minutes at 25 °C	
	Penetration and homogeneity	5–8 µm of formalin-fixed paraffin-embedded tissues	
Reagent Components	Components must meet the minimum parameters in the table below		
	Pararosaniline		
		Purity	≥ 98%
		Form	solid dye
		organic impurities	≤ 1% UV fingerprint
	Sulfite source		
		Source	Sodium metabisulfite $(Na_2S_2O_5)$
		Purity	≥ 98%
		Stoichiometry	4 × sulfite to dye
		Water	reagent-grade
	Hydrochloric acid (HCl)		
		Molarity	diluted to 0.75 M
		Purity	analytical grade
	Diluent Water		
		Resistivity	≥ 18 MΩ cm
	Preservative		
		Sodium benzoate	0.02%

References

Ackoff, R.L., From Data to Wisdom, *Journal of Applied Systems Analysis*, 16, pp. 3–9, 1989.

Adner, R., *The Wide Lens; A New Strategy for Innovation*, Portfolio/Penguin, NY, 2012.

Aghina, W., Ahlbäck, K., De Smet, A., Fahrbach, C., Handscomb, C., Lackey, G., Lurie, M., Murarka, M., Salo, O., Seem, E., and Woxholth, J., *The 5 Trademarks of Agile Organizations*, December 2001, Downloaded on 7/30/2022 from https://www.mckinsey.com/business-functions/people-and-organizational-performance/our-insights/the-five-trademarks-of-agile-organizations.

AIAG, *Potential Failure Mode and Effects Analysis (FMEA), Reference Manual*, Chrysler Corporation, Ford Motor Company, General Motor Corporation, Dearborn, MI, 1995.

Aiello, B., and Sachs, L., *Configuration Management Best Practices: Practical Methods that Work in the Real World*, Addison-Wesley, Upper Saddle River, NJ, 2011.

Alexander, M., *Agile Project Management: A Beginner's Guide*, 2017, Downloaded on 8/28/2017 from https://www.cio.com/article/3156998/agile-development/agile-project-management-a-beginners-guide.html.

Altshuller, G., *40 Principles Extended Edition, TRIZ Keys to Technical Innovation*, Technical Innovation Center, Worcester, MA, 2005.

Anderson, D.J., *Kanban: Successful Evolutionary Change for Your Technology Business*, Blue Hole Press, Washington, 2010.

Anleitner, M.A., *The Power of Deduction, Failure Modes and Effects Analysis for Design*, ASQ Quality Press, Milwaukee, WI, 2011.

Ansari, R., and Sandborn, P., *Managing Electronic Component Obsolescence*, Webinar Sponsored by SiliconExpert Technologies and CALCE, January 2014, Downloaded on 11/4/2017 from https://www.siliconexpert.com/sites/default/files/webinar_download/Obsolescence_012314_Webinar.pdf.

Anthony, S.D., *The Little Black Book of Innovation; How It Works; How to Do It*, Harvard Business Review Press, MA, 2012.

Apogee, *SDI Tools*, Statistical Design Institute, 2011, Downloaded from http://www.stat-design.com/Software/SoftwareOverview.html.

AS ISO 10007-2003, *Quality Management Systems – Guidelines for Configuration Management*, Australian Standard, Sydney, Australia, 2003.

ASME National Standard, *Dimensioning and Tolerancing, ASME Y14.5-1994*, ASME, New York, 1995.

Badiru, A.B., *Project Management – Systems, Principles, and Applications*, Taylor & Francis, 2012, Downloaded from http://app.knovel.com/hotlink/toc/id:kpPMSPA00B/project-management-systems/project-management-systems.

Baer, D., Why the Myers-Briggs Personality Test is Misleading, Inaccurate, and Unscientific, *Business Insider*, June 18, 2014, Downloaded on 7/31/2017 from http://www.businessinsider.com/myers-briggs-personality-test-is-misleading-2014-6.

Baniasad, S., *Design for Reliability, Private Communications*, Baxter Healthcare Corp., Round Lake, IL, 2015.

Beauregard, M.R., Mikulak, R.J., and McDermott, R.E., *The Basics of Mistake-Proofing*, Quality Resources–A Division of the Kraus Organization Limited, New York, 1997.

Bethea, R.M., and Rhinehart, R.R., *Applied Engineering Statistics*, Marcel Dekker, New York, 1991.

Bland, D.J., *What is an Empathy Map*? December 21, 2020, Downloaded on 12/27/2022 from https://www.accenture.com/us-en/blogs/software-engineering-blog/what-is-an-empathy-map.

Bralla, J.G., *Design for Excellence*, Knovel, 1996, Downloaded from http://www.knovel.com/hotlink/toc/id:kpDE000012/design-excellence/design-excellence (accessed 2015).

Brechner, E., *Agile Project Management with Kanban*, Microsoft Press, Washington, 2015.

Brown, J.L., *Empathy Mapping: A Guide to Getting Inside a User's Head*, June 27, 2018, Downloaded on 12/27/2022 from https://www.uxbooth.com/articles/empathy-mapping-a-guide-to-getting-inside-a-users-head/.

Burdick, R.K., Borror, C.M., and Montgomery, D.C., *Design and Analysis of Gauge R&R Studies, Making Decisions with Confidence Intervals in Random and Mixed ANOVA Models*, Society for Industrial and Applied Mathematics, Philadelphia, PA, 2005.

Cabane, O.F., *The Charisma Myth: How Anyone Can Master the Art and Science of Personal Magnetism*, Portfolio/Penguin, NY, 2012.

Chesla, B., Gibbs, N., and Cairns, L., Component Obsolescence: Is Your Process Up to the Challenge?, *Electronic Design Network (EDN)*, May 10, 2013, Downloaded on 11/4/2017 from https://www.edn.com/design/components-and-packaging/4414186/Component-obsolescence-Is-your-process-up-to-the-challenge-.

Clapperton, S., Are Personality Tests Worth the Investment? Here's 5 Reasons They are, *LinkedIn*, April 1, 2015, Downloaded on 7/31/2017 from https://www.linkedin.com/pulse/personality-tests-worth-investment-heres-5-reasons-sarah-clapperton.

Cohen, A., *Apparatus for Simulating a High-Five*, United States Patent Number: 5,356,330, Issued on October 18, 1994.

Collis, T., and Clohessey, J., *Design for Manufacturing, Private Communications*, Baxter Healthcare Corp., Round Lake, IL, 2015.

Condra, L.W., *Reliability Improvement with Design of Experiments*, 2nd Ed., Revised and Expanded, Marcel Dekker, New York, 2001.

Cooper, A., *The Inmates are Running the Asylum: Why High-Tech Products Drive Us Crazy and How to Restore the Sanity*, Sams–Pearson Education, IN, 2004.

Cooper, R.G., New Products – What Separates The Winners From The Losers and What Drives Success, in K.B. Kahn (Ed.), and G. Catellion, and A. Griffin (Assoc. Eds.), *The PDMA Handbook of New Product Development*, 2nd Ed., pp. 3–28, John Wiley and Sons, NJ, 2005.

Cooper, R.G., *Winning at New Products, Accelerating the Process from Idea to Launch*, 3rd Ed., Basic Books, NY, 2001.

Cooper, R.G., Edgett, S.J., and Kleinschmiddt, E.J., *Portfolio Management for New Products*, 2nd Ed., Basic Books, NY, 2001.

Crawford, C.M., and Di Benedetto, A., *New Products Management*, 7th Ed., McGraw Hill, MA, 2003.

Creveling, C.M., *Tolerance Design: A Handbook for Developing Optimal Specifications*, Addison-Wesley, Reading, MA, 1997.

Creveling, C.M., Slutsky, J.L., and Antis, D., Jr., *Design for Six Sigma in Technology and Product Development*, Prentice Hall PTR, Upper Saddle River, NJ, 2003.

Crowder, S., Delker, C., Forrest, E., and Martin, N., *Introduction to Statistics in Metrology*, Springer, NY, 2020.

Daniel, A., and Kimmelman, E., *The FDA and Worldwide Quality System Requirements Guidebook for Medical Devices*, 2nd Ed., American Society for Quality (ASQ), WI, 2008.

De Feo, J.A., Quality Planning: Designing Innovative Products and Services, in J.M. Juran, and J.A. De Feo (Eds.), *Juran's Quality Handbook: The Complete Guide to Performance Excellence*, 6th Ed., pp. 83–136, Mc Graw Hill, New York, 2010.

Del Vecchio, R.J., *Understanding Design of Experiments*, Hanser/Gardner Publishers, New York, 1997.

Dodson, B., Dovich, R.A., Edenborough, N.B., Gee, G., Walton, K.K., Marr, R., McLinn, J.A., O'Leary, D., Richardson, W.R., Wies, G., and Wortman, B., *The Reliability Engineer Primer*, 4th Ed., Quality Council of Indiana, IN, 2009.

Dodson, B., and Schwab, H., *Accelerated Testing: A Practitioner's Guide to Accelerated and Reliability Testing*, SAE International, Warrendale, PA, 2006.

Dornberger, U., and Suvelza, G.J.A., *Managing the Fuzzy Front-End of Innovation*, Intelligence 4 Innovation, International SEPT Program, the Leipzig University, Leipzig, Germany, 2012.

Drumond, C., *Scrum, Learn to Scrum with the Best of 'em*, Atlassian Agile Coach, 2022, Downloaded on 7/16/2022 from https://www.atlassian.com/agile/scrum.

Eagle, A.R., *A Method of Handling Errors in Testing and Measuring*, Industrial Quality Control, March 1954.

The Economist Editorial, Personality Testing at Work, Emotional Breakdown, Can Leaders be Identified by Psychometrics? *The Economist*, April 6, 2013, Downloaded on 7/31/2017 from https://www.economist.com/news/business/21575817-can-leaders-be-identified-psychometrics-emotional-breakdown.

Edmondson, A.C., *The Fearless Organization: Creating Psychological Safety in the Workplace for Learning, Innovation, and Growth*, Wiley, New York, NY, 2018.

Emmons, R.A., Is Spirituality an Intelligence? Motivation, Cognition and the Psychology of Ultimate Concern, *The International Journal for the Psychology of Religion*, 10(1), pp. 3–26, 2000.

Engineering Materials and Standards, *Worldwide Failure Modes and Effects Analysis, System–Design–Process Handbook, Technical Affairs, Automotive Safety and Engineering Standards*, Ford Motor Company, Dearborn, MI, 1992.

Evans, G., *Ideation to Project Creation: Taming the "Fuzzy Front End" of New Product Development*, Project Management Institute, Central Virginia Chapter, July 16, 2003.

FDA 21 CFR 820.30, *Design Control Guidance for Medical Device Manufacturers*, FDA Center for Device and Radiological Health, March 11, 1997, Downloaded from http://www.fda.gov/medicaldevices/deviceregulationandguidance/guidancedocuments/ucm070627.htm.

Feller, R.L., *Accelerated Aging; Photochemical and Thermal Aspects, Research in Conservation*, No. 4, The Getty Conservation Institute, 1994.

Forsberg, K., and Mooz, H., *Systems Engineering for Faster, Cheaper, Better*, A Publication of Center for Systems Management, Inc., Reprinted by SF Bay Area Chapter of INCOSE, 1998.

Foster, M.K., and Dadez, J., *Advanced GD&T Workshop Notes: ASME Y14.5-2009, Rev. Adv_v2_v2*, Applied Geometrics, Inc., Harwood Heights, IL, 2011a.

Foster, M.K., and Dadez, J., *Fundamental GD&T Workshop Notes: ASME Y14.5-2009, Rev. 2009_v3_v4*, Applied Geometrics, Inc., Harwood Heights, IL, 2011b.

Fowlkes, W.Y., and Creveling, C.M., *Engineering Methods for Robust Product Design: Using Taguchi Methods in Technology and Product Development*, Addison-Wesley, Reading, MA, 1995.

Furr, N., and Dyer, J., *The Innovator's Method; Bringing the Lean Start-Up Into Your Organization*, Harvard Business Review Press, MA, 2014.

Gardner, H., A Case Against Spiritual Intelligence, *The International Journal for the Psychology of Religion*, 10(1), pp. 27–34, 2000.

Gardner, H., Reflections on Multiple Intelligences: Myths and Messages, *Phi Delta Kappan*, 77, pp. 200–209, 1995, Downloaded on 7/15/2017 from https://learnweb.harvard.edu/WIDE/courses/files/Reflections.pdf.

Gawron, V.J., *Human Performance, Workload, and Situational Awareness Measures Handbook*, 2nd Ed., Taylor & Francis, 2008, Downloaded from http://app.knovel.com/hotlink/toc/id:kpHPWSAMHI/human-performance-workload/human-performance-workload (accessed 2015).

Genium Publishing Corp., *Modern Drafting Practices and Standards Manual*, Genium Publishing Corporation, Amsterdam, New York, 2009.

GHTF/SG3/N99-10:2004, *Quality Management Systems – Process Validation Guidance*, 2nd Ed., The Global Harmonization Task Force, January 2004, Downloaded on 5/15/2016 from http://www.imdrf.org/docs/ghtf/final/sg3/technical-docs/ghtf-sg3-n99-10-2004-qms-process-guidance-04010.pdf.

Goleman, D., *Emotional Intelligence, Why It Can Matter More Than IQ*, The 10th Anniversary Ed., Bantam Dell, NY, 2005.

Grenny, J., Patterson, K., McMillan, R., Switzler, A., and Gregory, E., *Crucial Conversations*, 3rd Ed., McGraw Hill, New York, NY, 2022.

Grewal, D., and Salovey, P., Feeling Smart: The Science of Emotional Intelligence, *American Scientist*, 93, pp. 330–339, 2005.

Griffin, A., Obtaining Customer Needs for Product Development, in K.B. Kahn (Ed.), and G. Catellion, and A. Griffin (Assoc. Eds.), *The PDMA Handbook of New Product Development*, 2nd Ed., pp. 211–227, John Wiley and Sons, Hoboken, NJ, 2005.

Grubbs, F.E., and Coon, H.J., *On Setting Test Limits Relative to Specification Limits*, Industrial Quality Control, March 1954.

Guess, V.C., *CMII for Business Process Infrastructure*, 2nd Ed., CMII Research Institute, Scottsdale, AZ, 2006.

Hall, C.S., Lindzey, G., Loehlin, J.C., and Manosevitz, M., *Introduction to Theories of Personality*, John Wiley and Sons, NJ, 1985.

Harley, S., *How to Say Anything to Anyone*, Greenleaf Book Group, TX, 2013.

Harvard Business Review Editors, *HBR Guide to Delivering Effective Feedback*, Harvard Business School, MA, 2016.

Hatley, D., Hruschka, P., and Pirbhai, I., *Process for System Architecture and Requirements Engineering*, Dorset House Publishing, New York, 2000.

Hayes, J.L., *Factors Affecting Measurement Reliability. Technical Memorandum No. 63–106*, U.S. Naval Ordnance Laboratory, October 24, 1955.

HBR Staff, *Five Critical Roles in Project Management, Project Management Article*, Reprint Harvard Business Review, H038ZB, Published on HBR.ORG, November 3, 2016.

Hemmerich, K.J., General Aging Theory and Simplified Protocol for Accelerated Aging of Medical Devices, *Medical Device and Diagnosis Industry Online Journal*, July 1, 1998, Downloaded on 12/28/2022 from https://www.mddionline.com/design-engineering/general-aging-theory-and-simplified-protocol-accelerated-aging-medical-devices.

Henzold, G., *Geometrical Dimensioning and Tolerancing for Design, Manufacturing and Inspection*, 2nd Ed., Butterworth-Heinemann, MA, 2009.

Henzold, G., *Geometrical Dimensioning and Tolerancing for Design, Manufacturing and Inspection – A Handbook for Geometrical Product Specification Using ISO and ASME Standards*, 2nd Ed., Elsevier, 2006, Downloaded from http://app.knovel.com/hotlink/toc/id:kpGDTDMIA5/geometrical-dimensioning/geometrical-dimensioning.

Hicks, C.R., *Fundamental Concepts in the Design of Experiments*, 2nd Ed., Holt, Rinehart and Winston, NY, 1973.

Hnatek, E.R., *Practical Reliability of Electronic Equipment and Products*, Marcel Dekker, New York, 2003.

Hochmuth, R., Meerkamm, H., and Schweiger, W., *An Approach to a General View on Tolerances in Mechanical Engineering*, Downloaded on 12/31/2015 from http://adcats.et.byu.edu/GuestPapers/beitrag_ipd_byu2.pdf.

Hooks, I.F., and Farry, K.A., *Customer Centered Products; Creating Successful Products through Smart Requirements Management*, AMACOM, New York, 2001.

IEC 60812:2006, *Analysis Techniques for System Reliability – Procedure for Failure Mode and Effects Analysis (FMEA)*, National Standards Authority of Ireland, Dublin, Ireland, 2006.

IEEE Std. 1624, *IEEE Standard for Organizational Reliability Capability*, IEEE Reliability Society, New York, 2008.

ILAC-G8:09/2019, *Guidelines on Decision Rules and Statements of Conformity*, International Laboratory Accreditation Cooperation (ILAC), 2019.

Jamnia, M.A., *Practical Guide to the Reliable Packaging of Electronics; Thermal and Mechanical Design and Analysis*, 4th Ed., CRC Press, Boca Raton, FL, 2025.

Jamnia, M.A., and Atua, K., *Executing Design for Reliability and Robustness*, CRC Press, Boca Raton, FL, 2020.

Jamnia, M.A., and Chen, J., *Variable Position Sensor Employing Magnetic Flux and Housing Therefore*, United States Patent Number: 7,165,450 B2, January 23, 2007.

Jamnia, M.A., Parker, P.H., Bollig, W.L., Berkun, C., and Matthis, M.J., *Torque Limiting Wrench for Ultrasonic Scaler Insertion*, United States Patent Number: 7,159,494, 2007.

JCGM 100:2008, *Evaluation of Measurement Data – Guide to the Expression of Uncertainty in Measurement (GUM 1995 with Minor Corrections)*, Joint Committee for Guides in Metrology (JCGM), 2008.

Jones, P., *Budgeting, Costing and Estimating for the Injection Moulding Industry*, Smithers Rapra Technology, 2009, Downloaded from http://www.knovel.com/hotlink/toc/id:kpBCEIMI01/budgeting-costing-estimating/budgeting-costing-estimating (accessed 2015).

Juran, J.M., and De Feo, J.A. (Eds.), *Juran's Quality Handbook: The Complete Guide to Performance Excellence*, 6th Ed., McGraw Hill, New York, 2010.

Kahn, K.B. (Ed.), and Catellion, G., and Griffin, A. (Assoc. Eds.), *The PDMA Handbook of New Product Development*, 2nd Ed., p. 591, John Wiley and Sons, NJ, 2005.
Klinger, D., Nakada, Y., and Menendez, M., *AT&T Reliability Manual*, Van Nostrand Reinhold Company, New York, 1990.
Koen, P.A., The Fuzzy Front End for Incremental, Platform and Breakthrough Products, in K.B. Kahn (Ed.), and G. Catellion, and A. Griffin (Assoc. Eds.), *The PDMA Handbook of New Product Development*, 2nd Ed., pp. 3–28, John Wiley and Sons, NJ, 2005.
Kota, S., *ME 452 – Design for Manufacturability, Course Notes, University of Michigan*, 2011, as Quoted in Wikipedia, Downloaded on February 2, 2016 from https://en.wikipedia.org/wiki/IT_Grade#cite_note-1 (last modified on April 22, 2016).
Kotter, J.P., What Leaders Really Do, *Harvard Business Review*, December 2001, Downloaded on 8/21/2022 from https://hbr.org/2001/12/what-leaders-really-do.
Krulikowski, A., *Fundamentals of Geometric Dimensioning and Tolerancing*, 2nd Ed., Delmar Cengage Learning, Albany, NY, 2007.
Kubiak, T.M., and Benbow, D.W., *Certified Six Sigma Black Belt Handbook*, 2nd Ed., American Society for Quality (ASQ), 2009, Downloaded from http://app.knovel.com/hotlink/toc/id:kpCSSBBHE6/certified-six-sigma-black/certified-six-sigma-black (accessed 2015).
Lamb, T., *Ship Design and Construction*, Vols. 1–2, Society of Naval Architects and Marine Engineers (SNAME), 2003–2004, Downloaded from http://app.knovel.com/hotlink/toc/id:kpSDCV0001/ship-design-construction/ship-design-construction (accessed 2015).
Larman, C., and Basili, V.R., Iterative and Incremental Development: A Brief History, *Computer*, 36(6), pp. 47–56, June 2003, Downloaded on 7/31/2022 from https://www.craiglarman.com/wiki/downloads/misc/history-of-iterative-larman-and-basili-ieee-computer.pdf.
Lencioni, P., *The Five Dysfunctions of a Team, A Leadership Fable*, Jossey-Bass Publishers, NJ, 2002.
Lencioni, P., *The Ideal Team Player: How to Recognize and Cultivate the Three Essential Virtues, A Leadership Fable*, Jossey-Bass Publishers, NJ, 2016.
Lencioni, P., *Overcoming the Five Dysfunctions of a Team, A Field Guide for Leaders, Managers, and Facilitators*, Jossey-Bass Publishers, NJ, 2005.
Lillie, R.D., Simplification of the Manufacture of Schiff Reagent for use in Histochemical Procedures, *Stain Technology*, 26(3), pp. 163–165, 1951.
Linstone, H.A., and Turoff, M. (Eds.), *The Delphi Method: Techniques and Applications*, Addison-Wesley, Reading, MA, 1975.
Littauer, F., and Sweet, R., *Personality Plus at Work: How to Work Successfully with Anyone*, Revell, MI, 2011.
Lotian, A., *Insightful Strategies*, Marketing Material Insights Corporation, 2000.
Lyton, M.C., and Ostermiller, S.J., *Agile Project Management*, 2nd Ed., John Wiley and Sons, NJ, 2017.
Mannan, S., *Lees' Loss Prevention in the Process Industries, Vols. 1–3 – Hazard Identification, Assessment and Control*, 4th Ed., Elsevier, 2012, Downloaded from http://app.knovel.com/hotlink/toc/id:kpLLPPIVH2/lees-loss-prevention/lees-loss-prevention (accessed 2015).
Martin, J.N., *System Engineering Guidebook: A Process for Developing Systems and Products*, CRC Press, Boca Raton, FL, 1997.
Mathews, P.G., *Design of Experiments with MINITAB*, ASQ Quality Press, Milwaukee, WI, 2005.
Matthews, D.D., *The A3 Workbook: Unlock Your Problem-Solving Mind*, CRC Press, NY, 2010.
Maxwell, J.C., *The 17 Essential Qualities of A Team Player*, Thomas Nelson Publishers, 2002, a Book Review by Business Summaries October 13, 2006, Copyright EBSCO Publishing Inc., 2010.
McQueen, M., *MindStuck: Mastering the Art of Changing Minds*, Amplify Publishing, Herndon, VA, 2023.
MEDICept, *Design Verification – The Case for Verification, Not Validation*, MEDICept, Inc., Ashland, MA, 2010, Downloaded on 11/24/2015 from http://www.medicept.com/blog/wp-content/uploads/2010/10/DesignVerificationWhitePaper.pdf.
MIL-HDBK-61A(SE), *Configuration Management Guidance*, Department of Defense, Washington, DC, February 7, 2001.

MIL-HDBK-338B, *Electronic Reliability Design Handbook Equipment*, Department of Defense, Washington, DC, 1988.

MIL-STD-882E, *Department of Defense Standard Practice, System Safety*, Department of Defense Standard Practice, Washington, DC, 2012.

MIL-STD-3046(ARMY), *Configuration Management, Interim Standard Practice*, Department of Defense, Washington, DC, March 6, 2013.

MIL-STD-31000A, *Technical Data Packages*, Department of Defense Standard Practice, Washington, DC, February 26, 2013.

Monte, C.F., and Sollod, R.N., *Beneath the Mask: An Introduction to Theories of Personality*, 7th Ed., John Wiley and Sons, NJ, 2003.

Morris, G., *DFSS for Thermal Management, Part IX – DFSS for Thermal Management: Regression – Part 1*, Downloaded on 7/27/2004 from http://www.coolingzone.com/Guest/News/NL_JUN_2004/Garron/Garon_June_04.html.

Munro, R.A., Maio, M.J., Nawaz, M.B., Ramu, G., and Zrymiak, D.J., *Certified Six Sigma Green Belt Handbook*, American Society for Quality (ASQ), 2008, Downloaded from http://app.knovel.com/hotlink/pdf/id:kt00ATZ7XK/certified-six-sigma-green/idov (accessed 2015).

Myers, R.H., and Montgomery, D.C., *Response Surface Methodology: Process and Product Optimization Using Designed Experiments*, 2nd Ed., John Wiley and Sons, New York, 2001.

Nagle, T.T., and Holden, R.K., *The Strategy and Tactics of Pricing: A Guide to Profitable Decision Making*, 3rd Ed., Prentice Hall, Upper Saddle River, New Jersey, 2002.

NASA, *Systems Engineering Handbook, NASA/SP-2007-6105 Rev 1*, 2007, Downloaded on 9/22/2016 from http://ntrs.nasa.gov/archive/nasa/casi.ntrs.nasa.gov/20080008301.pdf.

Nelson, W.B., *Accelerated Testing, Statistical Models, Test Plans, and Data Analysis*, John Wiley & Sons, Chichester, 2004.

O'Connor, P.D.T., and Kleyner, A., *Practical Reliability Engineering*, 5th Ed., John Wiley & Sons, Chichester, 2012.

Ogrodnik, P.J., *Medical Device Design, Innovation from Concept to Market*, Elsevier, Oxford, UK, 2013.

Pahl, G., Beitz, W., Feldhusen, J., and Grote, K.H., *Engineering Design: A Systematic Approach*, Springer, Berlin, Germany, 2007.

Pancake, M.H., Human Factor Engineering Considerations in New Product Development, in K.B. Kahn (Ed.), and G. Catellion, and A. Griffin (Assoc. Eds.), *The PDMA Handbook of New Product Development*, 2nd Ed., pp. 406–416, John Wiley and Sons, Hoboken, NJ, 2005.

Pillay, A., and Wang, J., *Technology and Safety of Marine Systems*, Elsevier, 2003, Downloaded from http://app.knovel.com/hotlink/toc/id:kpTSMS0003/technology-safety-marine/technology-safety-marine (accessed 2015).

PMI, *A Guide to the Project Management Body of Knowledge*, 6th Ed., Project Management Institute, PA, 2017.

Poli, C., *Design for Manufacturing: A Structured Approach*, Butterworth-Heinemann, Oxford, 2001.

Pugh, S., *Total Design: Integrated Methods for Successful Product Engineering*, Addison-Wesley, Wokingham, England, 1991

Quinn, J.B., and Zien, K.A., *Innovation Explosion: Using Intellect and Software to Revolutionize Growth Strategies*, The Free Press, NY, 1997.

Rath, T., *Strengths Finder 2.0*, Gallup Press, NY, 2007.

ReVelle, J.B., *Quality Essentials: A Reference Guide from A to Z*, ASQ Quality Press, WI, 2004.

Ritchey, T., Analysis and Synthesis – On Scientific Method based on a Study by Bernhard Riemann, *Systems Research*, 8(4), pp. 21–41, 1991, Downloaded on 4/4/2015 from http://www.swemorph.com/ma.html.

Ritchey, T., *General Morphological Analysis: A General Method for Non-Quantified Modeling, Adapted from the Paper: Ritchey, T. Fritz Zwicky, Morphologie and Policy Analysis*, Presented at the 16th EURO Conference on Operational Analysis, Brussels, Belgium, 1998, © Swedish Morphological Society, 2002 (Revised 2013), Downloaded on 4/4/2015 from http://www.swemorph.com/ma.html.

Ritchey, T., *Wicked Problems: Structuring Social Messes with Morphological Analysis*, 2005, Adapted from a Lecture Given at the Royal Institute of Technology in Stockholm, 2004, Downloaded on 4/4/2015 from http://www.swemorph.com/ma.html.

Rivin, E., and Fey, V., *Innovation of Demand, New Product Development Using TRIZ*, Cambridge University Press, Cambridge, UK, 2005.

Rosenau, M.D., Jr., and Githens, G.D., *Successful Project Management: A Step by Step Approach with Practical Examples*, 4th Ed., John Wiley & Sons, NJ, 2005.

Rowley, J., The Wisdom Hierarchy: Representations of the DIKW Hierarchy, *Journal of Information Science*, 33(2), pp. 163–180, 2007.

Rubin, K.S., *Essential Scrum, A Practical Guide to the Most Popular Agile Process*, Addison-Wesley Publishers, NJ, 2013.

Russell, J.P., *ASQ Auditing Handbook – Principles, Implementation, and Use*, 4th Ed., American Society for Quality (ASQ), 2013, Downloaded from http://www.knovel.com/hotlink/toc/id:kpASQAHP02/asq-auditing-handbook/asq-auditing-handbook (accessed 2015).

Saaty, T.L., *The Analytical Hierarchy Process: Planning, Priority Setting, Resource Allocation*, McGraw-Hill International, New York, 1980.

Sahota, M.K., *An Agile Adoption and Transformation Survival Guide: Working with Organizational Culture*, Lulu.com Publishers, 2012.

Salovey, P., and Grewal, D., The Science of Emotional Intelligence, *Current Directions in Psychology Science*, 14(6), pp. 281–285, 2005.

Salovey, P., and Mayer, J.D., Emotional Intelligence, *Imagination, Cognition, and Personality*, 9, pp. 185–211, 1990.

Schneider, W.E., *The Reengineering Alternative: A Plan for Making Your Current Culture Work*, McGraw Hill, NY, 1999.

Sen, P., and Yang, J.-B., *Multiple Criteria Decision Supporting Engineering Design*, Springer-Verlag, London, 1998.

Sheskin, D.J., *Handbook of Parametric and Nonparametric Statistical Procedures*, 5th Ed., CRC Press, MA, 2007.

Shorten, D., Pfitzmann, M., and Kaushal, A., *Make Versus Buy: A Decision Framework*, Booz Allen Hamilton, 2006, Downloaded on 7/3/2015 from http://www.strategyand.pwc.com/media/uploads/MakeVersusBuy.pdf.

Shulyak, L., Introduction to TRIZ, in G. Altshuller (Ed.), *40 Principles Extended Edition, TRIZ Keys to Technical Innovation*, pp. 15–20, Technical Innovation Center, Worcester, MA, 2005.

Simmons, C.H., Maguire, D.E., and Phelps, N., *Manual of Engineering Drawing – Technical Product Specification and Documentation to British and International Standards*, 3rd Ed., Elsevier, 2009, Downloaded from http://app.knovel.com/hotlink/toc/id:kpMEDTPSD4/manual-engineering-drawing/manual-engineering-drawing (accessed 2015).

Sobek, D.K., II, and Smalley, A., *Understanding A3 Thinking: A Critical Component of Toyota's PDCA Management System*, CRC Press, NY, 2008.

Stabile, S.J., The Use of Personality Tests as a Hiring Tool: Is the Benefit Worth the Cost?, *U.PA. Journal of Labor and Employment Law*, 4(2), p. 279, 2002.

Stamatis, D.H., *Failure Mode Effect Analysis: FMEA from Theory to Execution*, 2nd Ed., ASQ Quality Press, WI, 2003.

Stockhoff, B.A., Research and Development: More Innovation, Scarce Resources, in J.M. Juran, and J.A. De Feo (Eds.), *Juran's Quality Handbook: The Complete Guide to Performance Excellence*, 6th Ed., pp. 891–950, McGraw Hill, New York, 2010.

Stone, D., and Heen, S., *Thanks for the Feedback: The Science and Art of Receiving Feedback Well*, Penguin, NY, 2015.

Taguchi, G., and Clausing, D., Robust Quality, *Harvard Business Review*, No. 90114, January–February 1990, p. 65.

Tague, N.R., *The Quality Toolbox*, 2nd Ed., ASQ Quality Press, Milwaukee, WI, 2004.

Teixeira, M.B., *Design Controls for the Medical Device Industry*, 2nd Ed., Taylor & Francis, 2014, Downloaded from http://www.knovel.com/hotlink/toc/id:kpDCMDIE01/design-controls-medical/design-controls-medical (accessed 2015).

Thor, P., Jolly, M., Montgomery, J., Owings, S., Heimer, S., Kelley, S., Bartlett, T., and Magnuson, M., Humidity as a Use Condition for Accelerated Aging of Polymers, *Medical Device and Diagnosis Industry Online Journal*, February 19, 2021, Downloaded on 4/18/2022 from https://www.mddionline.com/testing/humidity-use-condition-accelerated-aging-polymers.

Triptych, *SDI Tools, Statistical Design Institute*, Downloaded from http://www.stat-design.com/Software/SoftwareOverview.html.

Tufte, E.R., *Visual and Statistical Thinking: Displays of Evidence for Making Decisions, A Reprint of Chapter 2 of Visual Explanations: Image and Quantities, Evidence and Narrative by the Same Author*, Graphics Press, Cheshire, CT, 1997.

Ullman, D.G., *The Mechanical Design Process*, 4th Ed., International Edition, McGraw-Hill, NY, 2010.

Ullman, D.G., *Scrum for Hardware Design, Supporting Material for Mechanical Design Process*, Published by David G. Ullman, Oregon, 2019.

Velleman, P.F., and Wilkinson, L., Nominal, Ordinal, Interval, and Ratio Typologies are Misleading, *The American Statistician*, 47(1), pp. 65–72, 1993.

Von Seggern, E.A., *Method and Apparatus for Sonic Dental Drilling*, United States Patent Number: 3,133,351, Issued on May 19, 1964.

Weiss, S.I., *Product and System Development: A Value Approach*, John Wiley and Sons, Hoboken, NJ, 2013.

Wheeler, D.J., *Beyond Capability Confusion*, 2nd Ed., SPC Press, TN, 2000a.

Wheeler, D.J., *EMP III: Evaluating the Measurement Process and Using Imperfect Data*, SPC Press, Knoxville, TN, 2006.

Wheeler, D.J., *The Process Evaluation Handbook*, SPC Press, Knoxville, TN, 2000b.

Wheelwright, S.C., and Clark, K.B., *Revolutionizing Product Developments: Quantum Leaps in Speed, Efficiency, and Quality*, The Free Press, NY, 1992.

Wilkins, C., *Obsolescence and Life Cycle Management for Avionics, Final Report, DOT/FAA/TC-15/33*, Federal Aviation Administration, William J. Hughes Technical Center, Aviation Research Division, Atlantic City International Airport, NJ, November 2015, Downloaded on 11/4/2017 from http://www.tc.faa.gov/its/worldpac/techrpt/tc15-33.pdf.

Wilson, B.A., *Design Dimensioning and Tolerancing*, Goodheart-Wilcox, Tinley Park, IL, 2001.

Xiao, K., *Analytical Scientists in Pharmaceutical Product Development: Task Management and Practical Knowledge*, John Wiley & Sons, NJ, 2021.

Index

K

L

M

O

P

Q

R

S

T

U

V

W

Z

For Product Safety Concerns and Information please contact our EU
representative GPSR@taylorandfrancis.com
Taylor & Francis Verlag GmbH, Kaufingerstraße 24, 80331 München, Germany

www.ingramcontent.com/pod-product-compliance
Ingram Content Group UK Ltd.
Pitfield, Milton Keynes, MK11 3LW, UK
UKHW052217190926
475702UK00006B/1063

* 9 7 8 1 0 3 2 8 9 5 2 1 5 *